Molecular Modeling and Dynamics of Bioinorganic Systems

NATO ASI Series

Advanced Science Institutes Series

A Series presenting the results of activities sponsored by the NATO Science Committee, which aims at the dissemination of advanced scientific and technological knowledge, with a view to strengthening links between scientific communities.

The Series is published by an international board of publishers in conjunction with the NATO Scientific Affairs Division

A Life Sciences **B Physics**	Plenum Publishing Corporation London and New York
C Mathematical and Physical Sciences **D Behavioural and Social Sciences** **E Applied Sciences**	Kluwer Academic Publishers Dordrecht, Boston and London
F Computer and Systems Sciences **G Ecological Sciences** **H Cell Biology** **I Global Environmental Change**	Springer-Verlag Berlin, Heidelberg, New York, London, Paris and Tokyo

PARTNERSHIP SUB-SERIES

1. Disarmament Technologies	Kluwer Academic Publishers
2. Environment	Springer-Verlag / Kluwer Academic Publishers
3. High Technology	Kluwer Academic Publishers
4. Science and Technology Policy	Kluwer Academic Publishers
5. Computer Networking	Kluwer Academic Publishers

The Partnership Sub-Series incorporates activities undertaken in collaboration with NATO's Cooperation Partners, the countries of the CIS and Central and Eastern Europe, in Priority Areas of concern to those countries.

NATO-PCO-DATA BASE

The electronic index to the NATO ASI Series provides full bibliographical references (with keywords and/or abstracts) to more than 50000 contributions from international scientists published in all sections of the NATO ASI Series.
Access to the NATO-PCO-DATA BASE is possible in two ways:

– via online FILE 128 (NATO-PCO-DATA BASE) hosted by ESRIN, Via Galileo Galilei, I-00044 Frascati, Italy.

– via CD-ROM "NATO-PCO-DATA BASE" with user-friendly retrieval software in English, French and German (© WTV GmbH and DATAWARE Technologies Inc. 1989).

The CD-ROM can be ordered through any member of the Board of Publishers or through NATO-PCO, Overijse, Belgium.

3. High Technology – Vol. 41

Molecular Modeling and Dynamics of Bioinorganic Systems

edited by

Lucia Banci
University of Florence,
Department of Chemistry,
Florence, Italy

and

Peter Comba
Universität Heidelberg,
Anorganisch-Chemisches Institut,
Heidelberg, Germany

Kluwer Academic Publishers

Dordrecht / Boston / London

Published in cooperation with NATO Scientific Affairs Division

Proceedings of the NATO Advanced Research Workshop on
Molecular Modeling and Dynamics of Biological Molecules Containing Metal Ions
San Miniato (Pisa) Italy
March 15–21, 1997

A C.I.P. Catalogue record for this book is available from the Library of Congress

ISBN 0-7923-4824-9

Published by Kluwer Academic Publishers,
P.O. Box 17, 3300 AA Dordrecht, The Netherlands.

Sold and distributed in the U.S.A. and Canada
by Kluwer Academic Publishers,
101 Philip Drive, Norwell, MA 02061, U.S.A.

In all other countries, sold and distributed
by Kluwer Academic Publishers,
P.O. Box 322, 3300 AH Dordrecht, The Netherlands.

Printed on acid-free paper

Table of Contents

List of Contributors

Emil Alexov
Department of Physics, City College of New York, 138th Str. and Covent Avenue, New York 0031, USA

Frederic Allain
Laboratoire de Chimie Théorique, Université Pierre et Marie Curie, 4, place Jussieu, 75005 Paris, France

Roger Attias
Laboratoire de Chimie et Biochimie, URA 400 CNRS, 45 Rue des Sants-Peres, 75270 Paris 06, France

B. Asbóth
Institute for Biochemistry and Protein Research, Agricultural Research Center, P.O. Box 170, H-2101 Gödöllő, Hungary

Pascale Auge-Barrere-Mazouat
Laboratoire de Chimie Théorique, Université Pierre et Marie Curie, 4, place Jussieu, 75005 Paris, France

Lucia Banci
Department of Chemistry, University of Florence, via Gino Capponi, 7, 50121 Florence, Italy; e-mail: lucia@riscl.lrm.fi.cnr.it

Jacqueline Berges
Laboratoire de Chimie Théorique, Université Pierre et Marie Curie, 4, place Jussieu, 75005 Paris, France

P. Beroza
Department of Molecular Biology, The Scripps Research Institute, 10666 N. Torrey Pines Road, La Jolla, CA 92037, USA

Ivano Bertini
Department of Chemistry, University of Florence, via Gino Capponi, 7, 50121 Florence, Italy; e-mail: bertini@riscl.lrm.fi.cnr.it

Z. Böcskei
Department of Theoretical Chemistry, Eotvos Lorand University, P.O. Box 32, 1518 Budapest, Hungary

David Case
Department of Molecular Biology, The Scripps Research Institute, 10666 N. Torrey Pines Road, La Jolla, CA 92037, USA; e-mail: case@scripps.edu

Jean-Claude Chottard
Laboratoire de Chimie et Biochimie, URA 400 CNRS, 45 Rue des Sants-Peres, 75270 Paris 06, France

Timothy Clark
Computer Chemie Centrum des Instituts für Organische Chemie, Universität Erlangen-Nürnberg, Nägelsbachstrasse 25, 91052 Erlangen, Germany ;e-mail: clark@organik.uni-erlangen.de

Thomas Cleveland
Department of Chemistry, University of Wisconsin, 1101 W. University Ave., Madison, WI 53706, USA

Peter Comba
Anorganisch-Chemisches Institut, Universität Heidelberg, Im Neuenheimer Feld 270, 69120 Heidelberg, Germany; e-mail: comba@akcomba.oci.uni-heidelberg.de

Csilla Csiki
Chemistry Department, Conneticut College, Box 5624, New London, CT 06320, USA

Robert J. Deeth
Department of Chemistry, Inorganic Computational Chemistry Group, Universtiy of Warwick, Coventry CV4 7AL, UK; e-mail: r.j.deeth@warwick.ac.uk

Miguel-Angel Elizondo-Riojas
Centro Universitario Contra el Cancer, Hospital Universitario "Dr. José Eleuterio Gonzales", Universidad Autónoma de Nuevo León, Monterrey, N. L., Mexico

Timothy K. Firman
Department of Chemistry, University of Wisconsin, 1101 W. University Ave., Madison, WI 53706, USA

M. Fuxreiter
Department of Theoretical Chemistry, Eotvos Lorand University, P.O. Box 32, 1518 Budapest, Hungary

Peter Gedeck
Computer Chemie Centrum des Instituts für Organische Chemie, Universität Erlangen-Nürnberg, Nägelsbachstrasse 25, 91052 Erlangen, Germany

Florence Gonnet
Laboratoire de Chimie Théorique, Université Pierre et Marie Curie, 4, place Jussieu, 75005 Paris, France

Giovanni Gori Savellini
Department of Chemistry, University of Florence, via Gino Capponi, 7, 50121 Florence, Italy

Marilyn R. Gunner
Department of Physics, City College of New York, 138th Str. and Covent Avenue, New York 0031, USA; e-mail: gunner@gunner.sci.ccny.cuny.edu

Michael B. Hall
Department of Chemistry, Texas A&M University, College Station, Texas 77843-3244, USA

Ernst-Walter Knapp
Institut für Kristallographie, Freie Universität Berlin, Takustrasse 6, 14195 Berlin, Germany; e-mail: knapp@chemie.fu-berlin.de

Peter Kollman
Department of Pharmaceutical Chemistry, University of California at San Francisco, 513 Parnassus Ave., San Francisco, Ca 94143, USA; e-mail: pak@cgl.ucsf.edu

Jiri Kozelka
Laboratoire de Chimie et Biochimie, URA 400 CNRS, 45 Rue des Sants-Peres, 75270 Paris 06, France; e-mail: kozelka@citi2.fr

Clark R. Landis
Department of Chemistry, University of Wisconsin, 1101 W. University Ave., Madison, WI 53706, USA; e-mail: landis@chem.wisc.edu

Harald Lanig
Computer Chemie Centrum des Instituts für Organische Chemie, Universität Erlangen-Nürnberg, Nägelsbachstrasse 25, 91052 Erlangen, Germany

J. Li
Department of Molecular Biology, The Scripps Research Institute, 10666 N. Torrey Pines Road, La Jolla, CA 92037, USA

Bernd Melchers
Institut für Kristallographie, Freie Universität Berlin, Takustrasse 6, 14195 Berlin, Germany

Ian J. Munslow
Department of Chemistry, Inorganic Computational Chemistry Group, Universtiy of Warwick, Coventry CV4 7AL, UK

Gabor Náray-Szabó
Department of Theoretical Chemistry, Eotvos Lorand University, P.O. Box 32, 1518 Budapest, Hungary; e-mail: naray@para.chem.elte.hu

L. Noodleman
Department of Molecular Biology, The Scripps Research Institute, 10666 N. Torrey Pines Road, La Jolla, CA 92037, USA

Karen M. Norenberg
Chemistry Department, Conneticut College, Box 5624, New London, CT 06320, USA

Wieslaw Nowak
Molecular Biophysics Group - Institute of Physics, N. Copernicus University, ul. Grudziadzka, 5, 87100 Torun, Poland; e-mail: wiesiek@phys.uni.torun.pl

Veronica J. Paget
Department of Chemistry, Inorganic Computational Chemistry Group, Universtiy of Warwick, Coventry CV4 7AL, UK

Fritz Parak
Fakultät für Physik E17, Technische Universität München, James-Frank Straβe, 85747 Garching, Germany

Daniel M. Root
Department of Chemistry, University of Wisconsin, 1101 W. University Ave., Madison, WI 53706, USA

Antonio Rosato
Department of Chemistry, University of Florence, via Gino Capponi, 7, 50121 Florence, Italy

Gudrun Schürer
Computer Chemie Centrum des Instituts für Organische Chemie, Universität Erlangen-Nürnberg, Nägelsbachstrasse 25, 91052 Erlangen, Germany

Christina M. Shoemaker
Chemistry Department, Conneticut College, Box 5624, New London, CT 06320, USA

Per Siegbahn
Department of Physics, University of Stockholm, Box 6730, 11385 Stockholm, Sweden; e-mail: ps@physto.se

H. Süli-Vargha
Research Group of Peptide Chemistry, Hungarian Academy of Sciences, 1518 Budapest 112, P.O. Box 32, Hungary

Andrei L. Tchougréeff
Department of Quantum Chemistry, Karpov Institute of Physical Chemistry, Vorontsovo pole, 10, 103064 Moskow, Russia; e-mail: andrei@cc.nifhi.ac.ru

Katalin Várnagy
Department of Inorganic and Analytical Chemistry, Lajos Kossuth University, 4010 Debreen, P.O. Box 21, Hungary; e-mail: varnagy@tigris.ktle.hu

Claus-Wilhelm von der Lieth
DKFZ, Abteilung Zentrale Spektroskopie, Im Neuenheimer Feld 280, 69120 Heidelberg, Germany; e-mail: w.vonderlieth@dkfz-heidelberg.de

Arieh Warshel
Department of Chemistry, University of Southern California, Los Angeles, CA 90089-1062, USA; e-mail: warshel@invitro.usc.edu

Snezana Zaric
Department of Chemistry, University of Beograd, Studentski trg. 16, 11001 Beograd, Serbia, Yugoslavia; e-mail: zaric@chemvx.tamu.edu

Marc Zimmer
Chemistry Department, Conneticut College, Box 5624, New London, CT 06320, USA; e-mail: mzim@conncoll.edu

Preface

The computation of structures and properties of bioinorganic compounds has experienced enormous progress in recent years. Novel approaches to model metal centers, new techniques to compute charges and solvation, and efficient algorithms to screen energy surfaces of large and complex structures have lead to a situation, in which it has now become possible and useful to discuss structures, electronic properties and reaction mechanisms of metalloproteins on the basis of computational studies. The good agreement between experimentally observed properties and computed parameters has increased the credibility of computational methods to a degree where experimentalists and computational chemists agree that the most demanding problems in bioinorganic chemistry may only be solved in a concerted way. Typical examples are the determination of solution structures of metalloproteins, the insight to the function of metalloenzymes and the design of mutated metalloproteins and transition metal-based drugs.

The chapters in this book are based on lectures presented at an Advanced NATO Research Workshop on Molecular Modeling and Dynamics of Biological Molecules Containing Metal Ions (San Miniato, Pisa, Italy, March 15 - 21 1997), organized by Lucia Banci, Katalin Várnagy, Peter Comba and Gabor Naray-Szabo. The chapters include a review of new force field-based approaches and their application to the computation of structures, electronic properties and dynamics of bioinorganic compounds, the discussion of quantum chemical and integrated QM/MM methods for understanding metalloenzyme functions, the presentation of methods used to compare electrostatic interactions and the evaluation of their importance for enzyme reaction mechanisms and experimental studies. The wide variety of topics and the state-of-the-art approaches presented have stimulated the participants of the NATO workshop, and it is hoped that the same will be true for the readers of this book.

The NATO Advanced Research Workshop was sponsored by the High Technology Division of NATO. We are grateful for this support and additional sponsorship by the University of Florence, the Cassa di Risparmio di San Miniato, IBM Semea and TRIPOS. The help of Dr. Giovanni Gori Savellini and Mrs. Karin Stelzer in the production of this book and the proof-reading by Ms. Anne Ramlow are gratefully acknowledged.

Lucia Banci
Florence

Peter Comba
Heidelberg

June 1997

SOLUTION STRUCTURES OF PROTEINS CONTAINING PARAMAGNETIC METAL IONS

IVANO BERTINI and ANTONIO ROSATO
Department of Chemistry, University of Florence
via Gino Capponi 7, 50121 Florence, Italy

1. Introduction

Metal ions containing unpaired electrons (i.e. paramagnetic) are rather common in biological systems. In particular, electron transfer metalloproteins must have at least one redox state with unpaired electrons. The quantum-mechanical treatment of these systems is cumbersome and the results are uncertain [1,2], although the density functional approach has provided some interesting data [3].

The solution structure of proteins may normally be obtained by NMR spectroscopy [4,5]. The presence of metal ions containing unpaired electrons causes a broadening of NMR signals [6-9] and therefore makes it difficult to observe NOE's, from which upper distance limits are obtained. In macromolecular systems the line broadening due to paramagnetism depends on the electron relaxation time (τ_S), when the dominating mechanism is the electron-nucleus dipolar coupling [10,11], and on the magnetic field and the rotational correlation time of the molecule (τ_r), when the dominating mechanism is the dipolar coupling between the nucleus and the average magnetic moment induced by the external magnetic field on the metal ion [12,13]. The equation describing

L. Banci and P. Comba (eds.), Molecular Modeling and Dynamics of Bioinorganic Systems, 1–19.

the paramagnetic contribution to nuclear transversal relaxation due to the electron-nucleus dipolar coupling is [10,11]:

$$T_{2M}^{-1} = \frac{1}{15}\left(\frac{\mu_0}{4\pi}\right)^2 \frac{\gamma_I^2 g_e^2 \mu_B^2 S(S+1)}{r^6}\left[4\tau_s + \frac{13\tau_s}{1+\omega_S^2\tau_S^2} + \frac{3\tau_s}{1+\omega_I^2\tau_S^2}\right] \tag{1}$$

where S is the spin moment of the nucleus, γ_I is the gyromagnetic ratio of the nucleus, r is the metal to nucleus distance, τ_S is the correlation time for the electron relaxation, ω_I and ω_S are the Larmor frequencies of the nucleus and of the electron, respectively,and all other symbols have their usual meaning. In Table 1 the values of τ_S for various metal ions at room temperature are reported [7,9]. We have set 10^{-11} s as the upper limit of τ_S in order to obtain signal linewidths suitable for high resolution NMR. The other origin of line broadening is Curie relaxation, which is described by eqn. (2) [12,13]:

$$T_{2M}^{-1} = \frac{1}{5}\left(\frac{\mu_0}{4\pi}\right)^2 \frac{\omega_I^2 g_e^4 \mu_B^4 S^2(S+1)^2}{(3kT)^2 r^6}\left(4\tau_r + \frac{3\tau_r}{1+\omega_I^2\tau_r^2}\right) \tag{2}$$

where S is the spin moment of the nucleus, r is the metal to nucleus distance, ω_I is the Larmor frequency of the nucleus, τ_r is the correlation time for the reorientation of the molecule, and all other symbols have their usual meaning. Due to this contribution, when $S > 1/2$ and the magnetic field higher than 500 MHz, the linewidths of the signals of protons close to the metal ion may be broadened beyond detection (depending on the value of τ_r). Indeed, it may happen that the signal of a given proton is detected at 90 MHz more easily than at 600 MHz [14]. Thus, the use of different magnetic fields may be required, depending on the nuclear shell around the metal that we want to focus on.

TABLE 1. Electronic relaxation rates for various paramagnetic metal ions. The line broadening is calculated for a proton at a 5 Å distance from the metal ion, at 500 MHz ^1H frequency, assuming the only contribution is that due to the electron-nucleus dipolar coupling (eqn. (1)). Metal ions for which high resolution NMR is feasible are in bold.

Metal ion	τ_S (s^{-1})	Line broadening (Hz)
Ti^{3+}	**10^{-10}-10^{-11}**	**20-200**
VO^{2+}	10^{-8}	10000
V^{3+}	**10^{-11}**	**50**
V^{2+}	10^{-9}	5000
Cr^{3+}	$5\cdot10^{-9}$-$5\cdot10^{-10}$	3000-25000
Cr^{2+}	**10^{-11}-10^{-12}**	**20-150**
Mn^{3+}	**10^{-10}-10^{-11}**	**150-1500**
Mn^{2+}	10^{-8}	100000
Fe^{3+} (H.S.)	**10^{-9}-10^{-11}**	**200-12000**
Fe^{3+} (L.S.)	**10^{-11}-10^{-13}**	**0.5-20**
Fe^{2+} (H.S.)	**10^{-11}-10^{-13}**	**5-150**
Co^{2+} (H.S.)	**10^{-11}-10^{-13}**	**2-100**
Co^{2+} (L.S.)	**10^{-9}-10^{-10}**	**200-1000**
Ni^{2+}	**10^{-10}-10^{-12}**	**5-500**
Cu^{2+}	10^{-9}	1000-5000
Ru^{3+}	**10^{-11}-10^{-12}**	**2-20**
Re^{3+}	**10^{-12}-10^{-13}**	**5-20**
Gd^{3+}	10^{-8}-10^{-9}	20000-200000
Ln^{3+}	**10^{-12}-10^{-13}**	**1-100**

By using the density matrix formalism [15-17], it is possible to simulate most of the NMR experiments and to optimize either the pulse sequences or the various delays in a given pulse sequence, in order to maximize the signal intensities. With this procedure, the detection of distance constraints from NOE's could be also pursued in unfavorable conditions. At this point the NMR experiment provides the necessary constraints to obtain the solution structure of the protein part [18-20]. We are now going to discuss the use of new additional constraints based on the electron-nucleus hyperfine coupling. These provide proton-metal ion distance information and allow for improving the resolution of the protein part and obtaining direct information on the metal coordinates within the protein frame [20]. A class of these constraints depends on the reciprocal sixth power of the metal to proton distance and another class on the reciprocal third power of the same distance. The different dependences on the distance may provide precious information related to protein mobility, as the two classes of constraints provide different average positions as a function of time.

These approaches allow researchers to obtain accurate solution structures, which may be the starting point for molecular dynamics simulations with the aim of (*i*) further refining the structure and (*ii*) providing models for mobility.

2. The proton-metal ion dipolar coupling

The presence of a paramagnetic center in a protein affects longitudinal nuclear relaxation times essentially through the electron-nucleus dipolar coupling [21], which is described by Solomon's law [10,11]:

$$T_{1M}^{-1} = \frac{2}{15}\left(\frac{\mu_0}{4\pi}\right)^2 \frac{\gamma_I^2 g_e^2 \mu_B^2 S(S+1)}{r^6}\left[\frac{7\tau_s}{1+\omega_S^2\tau_S^2} + \frac{3\tau_s}{1+\omega_I^2\tau_S^2}\right] \tag{3}$$

Eqns. (1) and (2) provide a tool for obtaining metal to proton constraints. It has been discussed [22-26] whether the recovery of magnetization in non-selective inversion-recovery experiments on proteins is exponential, and thus allow us (*i*) to estimate the diamagnetic contribution to nuclear T_1 values, and (*ii*) to relate the paramagnetic contribution to a distance. Cross-relaxation effects which allow a proton to relax by exchanging energy with another proton, introduce a small error which can be accounted for during the calibration process [24,26]. Indeed, in analogy with what is done with structural constraints derived from NOE's, one should use the paramagnetic enhancement of the nuclear relaxation rate to extract an upper limit for the proton-metal distance, instead of trying to obtain an exact distance. In this way, possible problems arising either from the inaccuracy in the estimate of the diamagnetic contribution to the nuclear relaxation rate, or from cross-relaxation effects may be overcome. We proposed [24,26] to use a calibration procedure which is basically identical to that already commonly used [4,5]. The simultaneous calibration of NOE- and T_1-derived constraints allows one to make the two sets of constraints fully consistent. Two examples of the results obtained from the application of the above procedure are shown in Figs. 1 and 2 [24,26].

Caution needs to be used in the presence of strong electron spin density delocalization. In such cases the simple metal centered point dipole-point dipole approximation, which is described by eqns. (1-3), may not be valid anymore. For instance, in the case of iron-sulfur proteins, we discarded the β-CH_2 protons of the ligating cysteines [24,26], as they are probably affected by ligand-centered contributions [27]. In general, the use of T_1s lead to a substantial refinement of the structure. Moreover, such constraints lead to a significant improvement of the definition of the metal centers in distance geometry calculations [24,26].

Cross-relaxation effects do not affect the T_2 values of nucleear resonances. This suggests that T_2's may conveniently be used for the determination of solution structures. As it can be seen from eqns. (1) and (2), the dependence of the T_2 values on the nucleus-metal ion distance is the same as that of T_1 values (eqn. (3)). The effect of ligand-centered contributions to T_2 values is comparable to that to T_1 values. In contrast to the case of T_1s, T_2 values are also affected by Curie relaxation as well. Indeed, depending on experimental conditions and on the system under investigation, the T_2 of NMR signals may be dominated either by Curie or Solomon relaxation. However, the distance dependence of the Curie contribution to nuclear relaxation is the same as that of

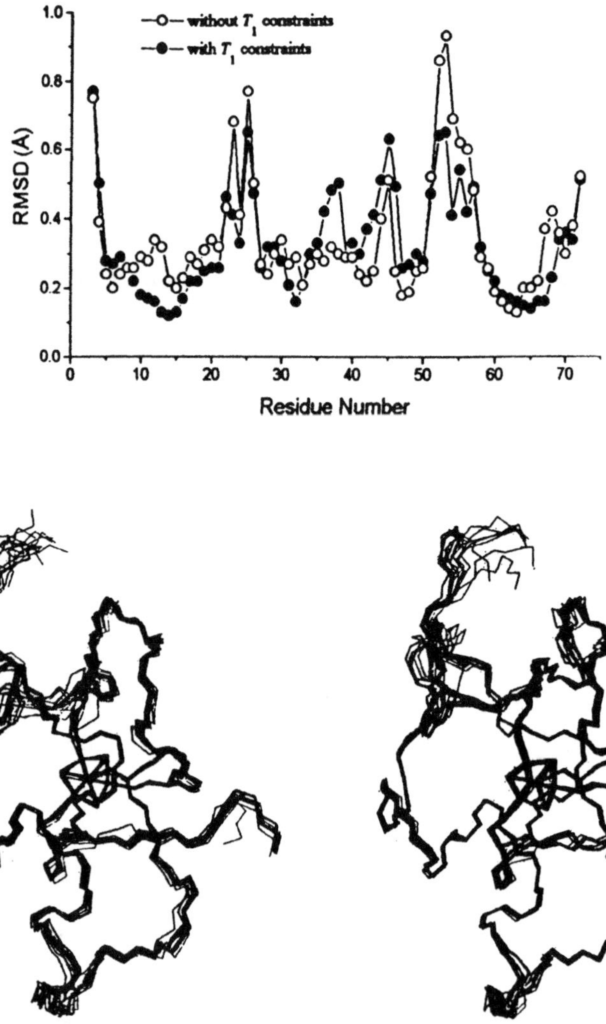

Fig. 1. Top: the RMSD of the backbone atoms obtained for the family of structures of HiPIP I from *E. halophila* with and without the use of T_1 constraints. Bottom: Plot of the backbone of the families obtained with (right) and without (left) T_1 constraints [24].

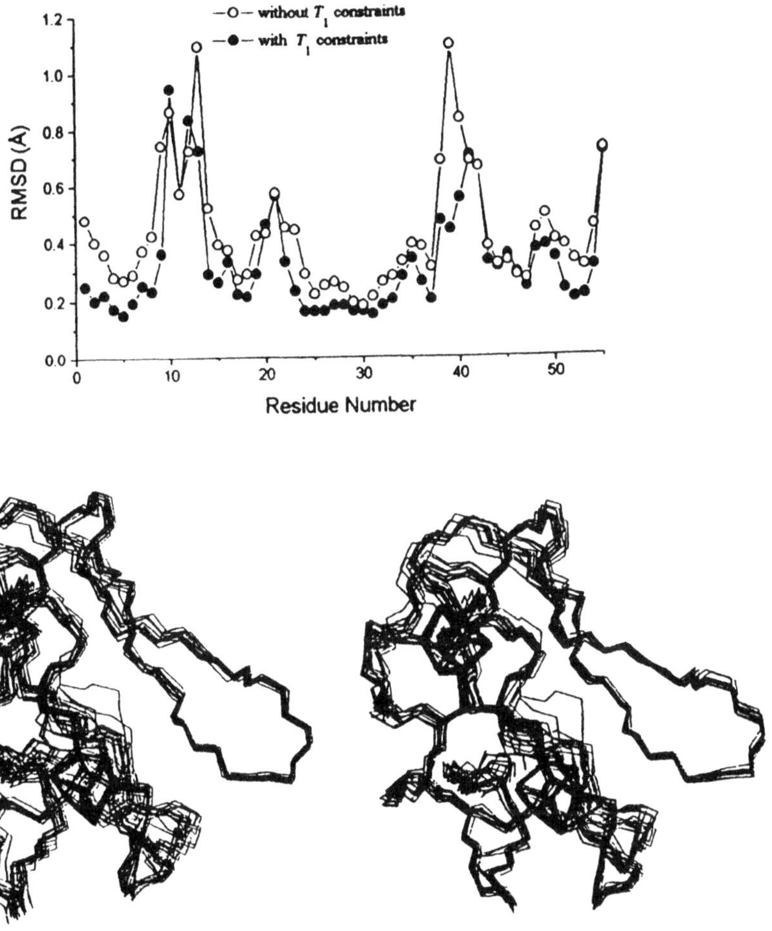

Fig. 2. Top: the RMSD of the backbone atoms obtained for the family of structures of the $2[Fe_4S_4]$ ferredoxin from *C. pasteurianum* with and without the use of T_1 constraints. Bottom: Plot of the backbone of the families obtained with (right) and without (left) T_1 constraints [26].

the Solomon contribution. Thus, the relative weight of the two contributions is irrelevant for structure determinations. However, as yet the systematic use of T_2's in structure calculations of proteins has never been attempted.

3. The hyperfine shift

The hyperfine shift is the difference between the chemical shift of a given nucleus in a paramagnetic system and its chemical shift in an otherwise identical diamagnetic system. In practice, the value of the diamagnetic contribution can be obtained from a system where the paramagnetic center has been substituted with a diamagnetic metal ion, from the apoprotein, or from the same protein with the same metal center in another (diamagnetic) oxidation state.

The hyperfine coupling between a nucleus and an electron is represented by a tensor **A** which couples the resonating nucleus **I** to the unpaired electron **S**. The corresponding hamiltonian is

$$\hat{H} = \mathbf{I} \cdot \mathbf{A} \cdot \mathbf{S} \tag{4}$$

The tensor **A** may be decomposed into the sum of two terms, which represent the Fermi contribution [28] and the dipolar contribution [29] to the hyperfine shift. The former is isotropic in space. Therefore its contribution to the hamiltonian (4) is described by:

$$\hat{H} = A\mathbf{I} \cdot \mathbf{S} \tag{5}$$

where A is a scalar constant expressing the Fermi coupling between the nucleus and the electron. The dipolar hyperfine shift is anisotropic, and is represented by a tensor, which, in turn, may be expressed as the sum of two tensors. One of the two is traceless, and, upon rotation (such as that of the molecule in solution) averages to zero, whereas the other has a non-zero trace, and thus does not vanish upon rotation. The latter contribution gives rise to the so-called pseudocontact contribution to the hyperfine shift in solution.

The Fermi (or contact) contribution to the hyperfine shift is caused by the delocalization of unpaired electron spin density through the bond network. With increasing number of σ bonds between the resonating nucleus and the metal ion, the contact contribution diminishes.

3.1 THE DIPOLAR SHIFT

The hyperfine shift of all protons separated from the metal ion by a high enough number of σ bonds (i.e. more than 5, if no π bonding is present) is essentially dipolar in origin. The dipolar shift is significant when the energy of the first excited levels is close to that of the ground state (i.e. ≤ 10 cm^{-1}) and when the spin-orbit coupling is large. These are the conditions for short electronic relaxation times, i.e. for tolerable line broadening of NMR signals. This implies that in the systems that are suitable for high resolution NMR, hyperfine shifts of dipolar origin are often observed. This is the case for lanthanides (except for Gd(III)), which are often used as paramagnetic probes for calcium [30,31], low spin iron(III), high spin iron(II), and high spin cobalt(II), which is a good probe for zinc(II) [32-34]. In the metal-centered point dipole-point dipole approximation the pseudo contact shift may be calculated from the following equation:

$$\delta^{pc} = \frac{1}{12\pi r_i^3}\left[\Delta\chi_{ax}\left(3n_i^2 - 1\right) + \frac{3}{2}\Delta\chi_{rh}\left(l_i^2 - m_i^2\right)\right] \qquad (6)$$

where $\Delta\chi_{ax}$ and $\Delta\chi_{rh}$ are the axial and the rhombic anisotropy of the magnetic susceptibility induced by the paramagnetic ion, r_i is the distance of the nucleus i from the metal ion, and l_i, m_i, and n_i are the direction cosines of the position vector ($\mathbf{r}_i$) of atom i with respect to the orthogonal reference system formed by the principal axes of the magnetic susceptibility tensor. The above equation contains five unknowns: $\Delta\chi_{ax}$, $\Delta\chi_{rh}$, and three direction cosines which relate the (unknown) axes of the molecular system (determined by the ligand field around the metal), to the coordinate system of the laboratory frame (which is set arbitrarily). If more than 5 pseudo contact shifts are known, and a molecular model is available, it is possible to obtain the values of the above 5 parameters through a fitting procedure [35-42]. The same shifts, together with the above 5 parameters and with all other structural constraints (NOE's, ^3J couplings, nuclear T_1's) are then used to generate a new structural model. Depending on the algorithm used for structure calculations, pseudo contact shifts may be used either to fold randomly generated structures [43], or to refine already folded structures [44,45]. The resulting structure must be consistent with all the constraints. For a protein of about 100 residues, it is expected that an accurate solution structure may be determined when hundreds of pseudo contact shifts, and 1000-3000 NOE's are available [46].

For distance geometry calculations, we have defined a target function for pseudo contact shifts [43]:

$$T.F. = \sum_i w_i \left[\Theta\left(\left|\delta^{pc}_{exp} - \delta^{pc}_{calc}\right|, toll\right)\right]^2 \quad \Theta(x, toll) = \begin{cases} 0 \text{ if } x < toll \\ (x - toll) \text{ if } x > toll \end{cases} \tag{7}$$

where the summation is over all pseudo contact shift constraints, w_i is the weight of the ith constraint, δpc_{exp} is the ith measured pseudocontact shift, δpc_{calc} is the corresponding calculated pseudo contact shift, *toll* is the tolerance given to the pseudo contact shift constraints. Calculated and experimental shifts are considered equal if the absolute value of their difference is smaller than the tolerance. The above target function is similar to the target function defined for upper distance limits (i.e. NOE's) in the program DYANA [47]:

$$T.F. = \sum_i w_i \left[\Theta\left(d_i - b_i, 0\right)\right]^2 \quad \Theta(x,0) = \begin{cases} 0 \text{ if } x < 0 \\ x \text{ if } x > 0 \end{cases} \tag{8}$$

where the summation is over all NOE constraints, w_i is the weight of the ith constraint, b_i is the ith upper distance limit, and d_i is the corresponding distance in the calculated structure. The target functions of pseudo contact shifts and NOE's are simultaneously minimized during calculations.

In analogy to the treatment of dipolar relaxation discussed above, delocalization of electron spin density to ligand orbitals with a spin-orbit coupling different from zero (e.g. p orbitals of a carbon atom), provides an anisotropic magnetic susceptibility tensor and therefore causes a breakdown of the model. This may be significant for heteronuclei, but it is seldom important for protons [9]. However, electron spin delocalization to the donor atoms of the ligands is never negligible. This implies that the hyperfine shift of the atoms closest to the paramagnetic metal center, which experience the largest pseudo contact shifts, may deviate from that based on the metal centered approximation. Thus it is sensible to impose a tolerance proportional to the shift itself. That is, a proton close to the metal center, which has a large shift, will have a large tolerance, and this avoids problems from the possible breakdown of the metal centered point dipole approximation. With this procedure, protons with a small pseudo contact shift have a very restrictive tolerance. However, this cannot be smaller than a certain threshold, which should account for possible differences in the diamagnetic shift between the paramagnetic protein under study and a diamagnetic model (e.g. due to minor structural changes or to a different hydrogen bonding network of amide protons). Therefore, the correct procedure is to set the tolerance proportional to the pseudo contact shift (e.g. 10%) but larger than a a fixed threshold (e.g. 0.2 ppm). In practice, the calibration of pseudo contact shifts can be regarded as satisfactory if the following conditions are met: (*i*) the violations from the pseudocontact shifts are minimized, (*ii*) the increase of NOE's target function (i.e. of NOE's violations) upon introduction of pseudo contact shift constraints is minimized, and (*iii*) where the pseudocontact shifts are effective, the family obtained with the combined use of NOE's and pseudo contact shifts is better defined than the family obtained with NOE's alone. These principles maybe applied both to distance geometry and to molecular dynamics. A module that adds pseudocontact shifts as restraints in molecular dynamics calculations [45] has been added to the SANDER module [48] of the AMBER 4.1 package [49]. An example of results obtained from distance geometry calculations that include pseudocontact shifts is shown in Fig. 3 [44].

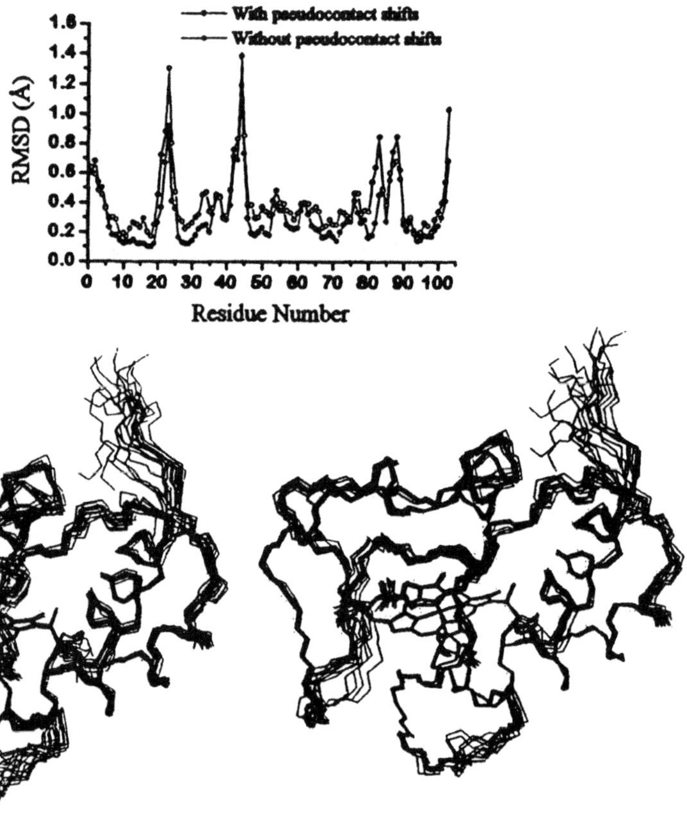

Fig. 3. Top: the RMSD of the backbone atoms obtained for the family of structures of the cyanide adduct of the Met80Ala mutant of cytochrome iso-1 from *S. cerevisiae* with and without the use of pseudocontact shift constraints. Bottom: Plot of the backbone of the families obtained with (right) and without (left) pseudocontact shifts [44].

3.2 THE CONTACT SHIFT

The contact shift is due to the presence of unpaired electron spin density in the 1s orbital of protons or in s orbitals for other nuclei. It is dependent on the coefficients of the atomic orbitals in the molecular orbitals and on spin polarization mechanisms. The former are related to the actual molecular structure. However, the understanding of the relationship between the contact shifts and the structural properties is still limited. For $NiNH_2CH_2$ moieties it has been reported that the contact shift of the protons depends on the cosine squared and the cosine of the Ni-N-C-H dihedral angle [50]. However, this relationship has never been used for structure determinations. In the case of iron-sulfur proteins the dependence of the contact shift on the Fe-S-C-H dihedral angle is as follows [51]:

$$\delta^{con} = a\,sin^2\theta + b\,cos\theta + c \tag{9}$$

where θ is the dihedral angle between the plane defined by the Sγ, the Cβ and the observed (Cα or Hβ) nuclei, and the Cβ-Sγ-Fe plane. The values of the parameters a, b, c are known [51,52]. The hyperfine shifts of Cα carbons can be adjusted to the same curve as the shifts of Hβ protons by simply scaling them [51]. Contact shifts have been used as structural constraints in the cases of reduced high potential iron sulfur proteins [24] and of oxidized Fe_4S_4 ferredoxins [53,54]. The dependence described by eqn. 9 has been attributed to contributions of p orbitals of the sulfur in the spin density function of the Cα and Hβ nuclei [51].

4. Concluding remarks

Molecular dynamics simulations may take advantage of the increased accuracy of the solution structure of paramagnetic metalloproteins due to the constraints discussed above. These are particularly significant in the neighborhood of metal centers, that is the more interesting regions of metalloproteins. Moreover, the use of paramagnetic constraints allows one to localize the position of the metal within the protein frame. This is important when the structure of the coordination sphere around the metal is unknown, as in the case of lanthanide-substituted calcium binding proteins [55]. As far as molecular modeling is involved, the accurate structural definition of the metal site helps to obtain reliable results in studies involving the docking of substrates or inhibitors.

The constraints for solution structure determinations based on the nuclear properties in the presence of paramagnetic centers in a protein are still far from being fully exploited. A general frame is provided in this article but the field is still open to new approaches. With an increasing number of examples becoming available, a standard procedure, comparable to that for diamagnetic proteins [4] is now emerging.

References

1. Davis, M.E. and McCammon, J.A. (1990) Electrostatics in Biomolecular Structure and Dynamics, *Chem. Rev.* **90**, 509-521.
2. Comba, P. and Hambley, T.W. (1995) *Molecular Modeling of Inorganic Compounds*, VCH, Weinheim, D.
3. Mouesca, J.M., Chen, J.L., Noodleman, L., Bashford, D. and Case, D.A. (1994) Density Functional/Poisson-Boltzmann Calculations of Redox Potentials for Iron-Sulfur Clusters, *J. Am. Chem. Soc.* **116**, 11898-11914.
4. Wüthrich, K. (1986) *NMR of Proteins and Nucleic Acids*, Wiley, New York.
5. Wüthrich, K. (1989) The development of nuclear magnetic resonance spectroscopy as a technique for protein structure determination, *Acc. Chem. Res.* **22**, 36-44.
6. (1973) *NMR of Paramagnetic Molecules*, Academic Press, New York.
7. Bertini, I. and Luchinat, C. (1986) *NMR of paramagnetic molecules in biological systems*, Benjamin/Cummings, Menlo Park, CA.

8. (1993) *Biological Magnetic Resonance, Vol. 12: NMR of Paramagnetic Molecules*, Plenum Press, New york.

9. Bertini, I. and Luchinat, C. (1996) *NMR of paramagnetic substances*, Coord.Chem.Rev. 150, Elsevier, Amsterdam.

10. Solomon, I. (1955) Relaxation Processes in a System of Two Spins, *Phys. Rev.* **99**, 559-565.

11. Solomon, I. and Bloembergen, N. (1956) Nuclear Magnetic Interactions in the HF Molecule, *J. Chem. Phys.* **25**, 261-266.

12. Guéron, M. (1975) Nuclear Relaxation in Macromolecules by Paramagnetic Ions: A Novel Mechanism, *J. Magn. Reson.* **19**, 58-66.

13. Vega, A.J. and Fiat, D. (1976) Nuclear Relaxation Processes of Paramagnetic Complexes. The Slow Motion Case, *Mol. Phys.* **31**, 347-362.

14. Bertini, I., Jonsson, B.-H., Luchinat, C., Pierattelli, R. and Vila, A.J. (1994) Strategies of signal assignments in paramagnetic metalloproteins. An NMR investigation of the thiocyanate adduct of the cobalt(II)-substituted human carbonic anhydrase II, *J. Magn. Reson. Ser. B* **104**, 230-239.

15. Abragam, A. (1961) *The Principles of Nuclear Magnetism*, Oxford University Press, Oxford.

16. Bax, A. (1982) *Two dimensional nuclear magnetic resonance in liquids*, Reidel, Dordrecht.

17. Shriver, J. (1992) Product operators and coherence transfer in multiple-pulse NMR experiments, *Concepts Magn. Reson.* **4**, 1-34.

18. Banci, L., Bertini, I., Eltis, L.D., Felli, I.C., Kastrau, D.H.W., Luchinat, C., Piccioli, M., Pierattelli, R. and Smith, M. (1994) The three dimensional structure in solution of the paramagnetic protein high-potential iron-sulfur protein I from *Ectothiorhodospira halophila* through nuclear magnetic resonance, *Eur. J. Biochem.* **225**, 715-725.

19. Banci, L., Bertini, I., Bren, K.L., Gray, H.B., Sompornpisut, P. and Turano, P. (1995) The three dimensional solution structure of the cyanide adduct of *Saccharomyces cerevisiae* Met80Ala-iso-1-cytochrome c. Identification of ligand-residue interactions in the distal heme cavity, *Biochemistry* **34**, 11385-11398.

20. Bertini, I., Luchinat, C. and Rosato, A. (1996) The solution structure of paramagnetic metalloproteins, *Prog. Biophys. Mol. Biol.* **66**, 43-80.

21. Banci, L., Bertini, I. and Luchinat, C. (1991) *Nuclear and electron relaxation. The magnetic nucleus-unpaired electron coupling in solution*, VCH, Weinheim.

22. Granot, J. (1982) Paramagnetic Relaxation in Dipolar-Coupled Homonuclear Spin Systems, *J. Magn. Reson.* **49**, 257-270.

23. La Mar, G.N. and de Ropp, J.S (1993) NMR Methodology for Paramagnetic Proteins, in Berliner, L.J. and Reuben, J. (eds) *Biological Magnetic Resonance, Vol. 12*, Plenum Press, New York, pp.1-78.

24. Bertini, I., Couture, M.M.J., Donaire, A., Eltis, L.D., Felli, I.C., Luchinat, C., Piccioli, M. and Rosato, A. (1996) The Solution Structure Refinement of the Paramagnetic Reduced HiPIP I from *Ectothiorhodospira halophila* by Using Stable Isotope Labeling and Nuclear Relaxation, *Eur. J. Biochem.* **241**, 440-452.

25. Huber, J.G., Moulis, J.-M. and Gaillard, J. (1996) Use of ^{1}H Longitudinal Relaxation Times in the Solution Structure of Paramagnetic Proteins. Application to [4Fe-4S] Proteins, *Biochemistry* **35**, 12705-12711.

26. Bertini, I., Donaire, A., Luchinat, C. and Rosato, A. (1997) Paramagnetic relaxation as a tool for solution structure determination: *Clostridium pasterianum* ferredoxin as an example, *Proteins: Structure,Function,and Genetics* in press

27. Ciurli, S., Cremonini, M.A., Kofod, P. and Luchinat, C. (1996) ^{1}H NMR of high potential iron-sulfur protein from the purple non-sulfur bacterium *Rhodoferax fermentans*, *Eur. J. Biochem.* **236**, 405-411.

28. McConnell, H.M. and Chesnut, D.B. (1958) Theory of isotropic hyperfine interactions in p-electron radicals, *J. Chem. Phys.* **28**, 107-117.

29. Kurland, R.J. and McGarvey, B.R. (1970) Isotropic NMR shifts in transition metal complexes: calculation of the Fermi contact and pseudocontact terms, *J. Magn. Reson.* **2**, 286-301.

30. Lee, L. and Sykes, B.D. (1983) Use of lanthanide-induced nuclear magnetic resonance shifts for determination of protein in solution: EF calcium binding site of Carp Parvalbumin, *Biochemistry* **22**, 4366-4373.

31. Shelling, J.G., Bjorson, M.E., Hodges, R.S., Taneja, A.K. and Sykes, B.D. (1984) Contact and dipolar contributions to lanthanide-induced NMR shifts of amino acid and peptide models for calcium binding sites in proteins, *J. Magn. Reson.* **57**, 99-114.

32. Bertini, I., Luchinat, C. and Scozzafava, A. (1977) Interaction of cobalt(II) bovine carbonic anhydrase with aniline, benzoate and anthranilate, *J. Am. Chem. Soc.* **99**, 581-584.

33. Banci, L., Dugad, L.B., La Mar, G.N., Keating, K.A., Luchinat, C. and Pierattelli, R. (1992) ^{1}H Nuclear Magnetic Resonance investigation of cobalt(II) substituted carbonic anhydrase, *Biophys. J.* **63**, 530-543.

34. Auld, D.S., Bertini, I., Donaire, A., Messori, L. and Moratal Mascarell, J.M. (1992) The pH dependent properties of cobalt(II) carboxypeptidase A inhibitor complexes, *Biochemistry* **31**, 3840-3846.

35. Williams, G., Clayden, N.J., Moore, G.R. and Williams, R.J.P. (1985) Comparison of the solution and crystal structures of mitochondrial cytochrome c. Analysis of paramagnetic shifts in the nuclear magnetic resonance spectrum of ferricytochrome c, *J. Mol. Biol.* **183**, 447-460.

36. Feng, Y.Q., Roder, H. and Englander, S.W. (1990) Redox dependent structure change and hyperfine nuclear magnetic resonance shifts in cytochrome c, *Biochemistry* **29**, 3494-3504.

37. Gao, Y., Boyd, J., Pielak, G.J. and Williams, R.J.P. (1991) Comparison of reduced and oxidized yeast iso-1-cytochrome c using proton paramagnetic shifts, *Biochemistry* **30**, 1928-1934.

38. Emerson, S.D. and La Mar, G.N. (1990) NMR determination of the orientation of the magnetic susceptibility tensor in cyanometmyoglobin: a new probe of steric tilt of bound ligand, *Biochemistry* **29**, 1556-1566.

39. Rajarathnam, K., La Mar, G.N., Chiu, M.L. and Sligar, S.G. (1992) Determination of the orientation of the magnetic axes of the cyano-metMb complexes of point mutants of myoglobin by solution 1H NMR: influence of HisE7-Gly and Arg CD3-Gly substitutions, *J. Am. Chem. Soc.* **114**, 9048-9058.

40. La Mar, G.N., Chen, Z.G., Vyas, K. and McPherson, A.D. (1995) An interpretative basis of the hyperfine shifts in cyanide-inhibited horseradish peroxidase on the magnetic axes and ligand tilt. Influence of substrate binding and extension to other peroxidases, *J. Am. Chem. Soc.* **117**, 411-419.

41. Banci, L., Bertini, I., Pierattelli, R., Tien, M. and Vila, A.J. (1995) Factoring of the hyperfine shifts in the cyanide adduct of lignin peroxidase from *P. chrysosporium*, *J. Am. Chem. Soc.* **117**, 8659-8667.

42. Gochin, M. and Roder, H. (1995) Protein Structure Refinement based on Paramagnetic NMR shifts. Applications to Wild-Type and mutants forms of Cytochrome c, *Protein Sci.* **4**, 296-305.

43. Banci, L., Bertini, I., Cremonini, M.A., Gori Savellini, G., Güntert, P. and Luchinat, C. (1997) *The pseudocontact shifts as structural constraints for solution structure determination of paramagnetic metalloproteins: a protocol for the Distance Geometry approach*, submitted.

44. Banci, L., Bertini, I., Bren, K.L., Cremonini, M.A., Gray, H.B., Luchinat, C. and Turano, P. (1996) The use of pseudocontact shifts to refine solution structures of paramagnetic metalloproteins: Met80Ala cyano-cytochrome *c* as an example, *JBIC* **1**, 117-126.

45. Banci, L., Bertini, I., Gori Savellini, G., Romagnoli, A., Turano, P., Cremonini, M.A., Luchinat, C. and Gray, H.B. (1997) The pseudocontact shifts as constraints for energy minimization and molecular dynamic calculations on solution structures of paramagnetic metalloproteins, *Proteins: Structure,Function,and Genetics* in press

46. Banci, L., Bertini, I., Gray, H.B., Luchinat, C., Reddig, T., Rosato, A. and Turano, P. (1998) Solution structure of oxidized horse heart cytochrome c, *Biochemistry* in press

47. Güntert, P., Mumenthaler, C. and Wüthrich, K. (1996) DYANA: torsion angle dynamics for structure calculation and automatic assignment of NOESY spectra, *XVIIth International Conference on Magnetic Resonance in Biological systems* (Abstract)

48. Pearlman, D.A. and Case, D.A. (1991) *SANDER: University of California, San Francisco,*

49. Pearlman, D.A., Case, D.A., Caldwell, J.W., Ross, W.S., Cheatham, T.E., Ferguson, D.M., Seibel, G.L., Singh, U.C., Weiner, P.K. and Kollman, P.A. (1995) *AMBER 4.1*, University of California, San Francisco.

50. Quaegebeur, J.P., Chachaty, C. and Yasukawa, T. (1979) Hyperfine couplings and ^{13}C relaxation in alkylamines coordinated to Ni(II) acetylacetonate, *Mol. Phys.* **37**, 409-424.

51. Bertini, I., Capozzi, F., Luchinat, C., Piccioli, M. and Vila, A.J. (1994) The Fe_4S_4 centers in ferredoxins studied through proton and carbon hyperfine coupling. Sequence specific assignments of cysteines in ferredoxins from *Clostridium acidi urici* and *Clostridium pasteurianum*, *J. Am. Chem. Soc.* **116**, 651-660.

52. Bertini, I., Ciurli, S. and Luchinat, C. (1995) The electronic structure of FeS centers in protein and models. A contribution to the understanding of their electron transfer properties, in *Structure and Bonding*, Springer-Verlag, Vol. 83, Berlin Heidelberg, pp.1-54.

53. Bertini, I., Donaire, A., Feinberg, B.A., Luchinat, C., Piccioli, M. and Yuan, H. (1995) Solution structure of the oxidized 2[Fe_4S_4] ferredoxin from *Clostridium pasteurianum*, *Eur. J. Biochem.* **232**, 192-205.

54. Wang, P.L., Donaire, A., Zhou, Z.H., Adams, M.W.W. and La Mar, G.N. (1996) Molecular model of the solution structure for the paramagnetic four-iron ferredoxin from the hyperthermophilic archaeon *Thermococcus litoralis*, *Biochemistry* **35**, 11319-11328.

55. Bentrop, D., Bertini, I., Cremonini, M.A., Forsén, S., Luchinat, C. and Malmendal, A. (1998) *The solution structure of the paramagnetic complex of the N-terminal domain of calmodulin with two* Ce^{3+} *ions by* ^1H *NMR*, submitted.

MODELING OF STRUCTURES AND MOLECULAR PROPERTIES OF TRANSITION METAL COMPOUNDS - TOWARD METALLOPROTEIN MODELING

PETER COMBA
Anorganisch-Chemisches Institut
Universität Heidelberg
Im Neuenheimer Feld 270
69120 Heidelberg
Germany

1. Introduction

Molecular modeling deals with the computational prediction of molecular structures and properties. These, i.e. stabilities, reactivities, thermodynamic and kinetic electron transfer parameters and spectroscopy (magnetic resonance, electronic and vibrational spectroscopy) are strongly correlated with molecular structural parameters and the environment: steric strain may lead to decreasing stabilities, bond stretching to the preference of certain substitution reaction paths, and structural variations may be accompanied by electron redistribution with possible consequences in various molecular properties. Thus, there are two parts of modeling: (i) the computation of the structure of an isolated molecule, a crystal lattice or a solvated species, i.e. the three-dimensional arrangement of atoms in space, and (ii) the computation of relevant molecular properties, based on structural parameters. The result of a modeling study may help to understand observed properties of a compound, to determine the structure of a molecule when more direct methods are not available, and to design new compounds with specified properties.

The Schrödinger equation, with many possible approximations, is the basis for quantum chemical modeling. Based on the Born-Oppenheimer approximation, the nuclear positions may be used as structural parameters to compute electronic properties, and an iterative optimization procedure may lead

L. Banci and P. Comba (eds.), Molecular Modeling and Dynamics of Bioinorganic Systems, 21–47.

to structural, electronic and other properties of the system in one step. Classical mechanical approaches such as force field calculations assume the arrangement of electrons to be fixed and the nuclear position and, depending on the approach used, steric strain and/or vibrational frequencies are the refined variables. The main result is an optimized structure, and most molecular properties need then to be computed with separate algorithms. There are various hybrid methods which use classical force field potentials with additional terms to model specific electronic effects, such as transition metal d-orbital effects [1a-c], electron redistribution due to π-bonding [1d], hybridization [1e] or coordination to metal ions [1f], and approaches where parts of a molecular structure are computed with quantum mechanical methods, whereas the rest is optimized with molecular mechanics [1g]. Some of these methods and corresponding applications are discussed in other chapters of this book. Here, I will mainly deal with classical force field based methods for structure refinement.

The indisputable advantages of using a quantum mechanical modeling tool or an empirical model with electronic terms included are that (i) this might be scientifically more satisfying, (ii) the model might be more generally applicable and (iii) the parameter field might be less complex. The fact that steric and electronic contributions are treated separately in molecular mechanics based modeling procedures may be an advantage in certain applications, and the high degree of parameterization may be a useful tool for obtaining accurate data. However, the major advantage of force field methods is that they are computationally less expensive. An important point here is that the accurate prediction of structures and properties of molecules in general, and biomolecules in particular, requires that large assemblies (e.g. metalloproteins) are optimized, and generally a large number of conformers needs to be screened. Also, for accurate and reliable predictions, parameter sets, basis functions and approximations need to be adjusted to as many experimentally characterized test molecules as possible. Thus, a thorough study may involve the optimization of hundreds or thousands of structures, and this is only possible with computationally inexpensive approaches.

2. The Basis of Molecular Modeling

The problem of understanding and predicting properties of chemical compounds and correlating their structures with stoichiometries and connectivities is to relate experimentally observable macroscopic properties to parameters on a molecular and submolecular level. This always involves a model, i.e. empirically derived laws, based on a pattern observed for certain properties of a large but limited group of compounds. Thus, a model itself is not based on true knowledge and the quality of the model is only related to the accuracy of the predictions within the well defined limits of application.

3. Structure Optimization

3.1. THEORETICAL BASIS

Molecular mechanics is an empirical method based on simple elements of theory. The assumption is that the positions of the atoms of a molecule are determined by forces between pairs and groups of three or more atoms (Fig. 1). The energies resulting from these forces and their sum, E_{total}, depend on the nuclear geometry, the functional form of the potentials and their parameterization. Minimization of the strain energy E_{total} by rearrangement of the nuclei, leads to an optimized structure and a value of the minimized strain energy.

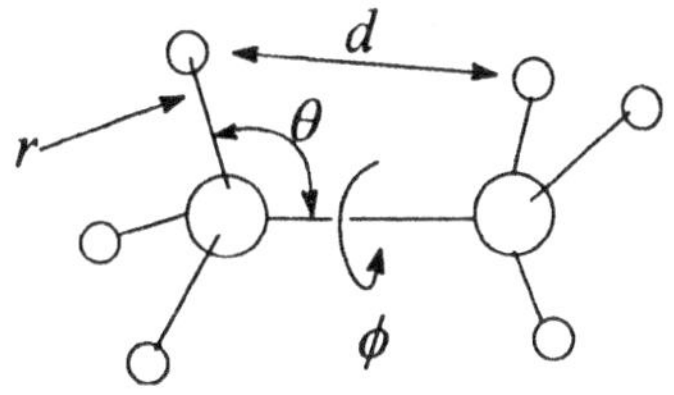

Fig. 1. The classical molecular mechanics concept

Typical potential energy functions for various potentials are given in equations 2 - 11.

$$E_{total} = \sum_{molecule} (E_b + E_\theta + E_\varnothing + E_{nb} + E_\sigma + \cdots) \quad (1)$$

$$E_b = \tfrac{1}{2} k_b (r_{ij} - r_o)^2 \quad (2)$$

$$E_b^{morse} = D(1 - \exp(-\alpha(r - r_o)))^2 - D \quad (3)$$

$$E_\theta = \tfrac{1}{2} k_\theta (\theta_{ijk} - \theta_o)^2 \quad (4)$$

$$E_\varnothing = \tfrac{1}{2} k_\varnothing (I + \cos(m(\varnothing_{ijkl} + \varnothing_{off}))) \quad (5)$$

$$E_{nb} = A \cdot e^{-Bd_{ij}} - C \cdot d_{ij}^{-6} \quad (6)$$

$$E_\delta = \tfrac{1}{2} k_\delta \delta^2 \quad (7)$$

$$E\varepsilon = \frac{q_i \quad q_j}{\varepsilon \quad d_{ij}} \quad (8)$$

$$E_{hb} = F \cdot d_{ij}^{-12} - G \cdot d_{ij}^{-10} \quad (9)$$

$$E_\theta^M = \tfrac{1}{8} k_{LML'} \sin^2(2\theta) \quad (10)$$

$$E\gamma = \tfrac{1}{2} k_\gamma (\gamma - \gamma_o) \quad (11)$$

A separation of the through bond interactions into bond stretching (E_b, eqns. 2 or 3), angle bending (E_θ, eqns. 4 or 10), torsional angle rotation ($E_\varnothing$, eqn. 5), out-of-plane bending (E_δ, eqn. 7) and various twist motions (E_γ, eqn. 11) is only possible, if these terms are not coupled [2]. Since the force constants for the corresponding potentials are very different, the local oscillators are largely decoupled. Some programs use cross-terms for cases where this is a poor approximation.

The number (eqn. 1) and types (eqns. 2 - 11) of potential energy functions are not the same in different programs. It is clear then that the parameter sets for the accurate computation of similar types of compounds may vary between different force fields. Molecular mechanics parameters are dependent on the functional forms and on the type of data to which they have been fitted (i.e. structural, thermodynamic and/or spectroscopic parameters; highly strained or unstrained compounds) and they are strongly correlated. Thus, these parameters are not transferable. These are typical properties of an interpolative approach, and that is what molecular mechanics is (Fig. 2) [2,3].

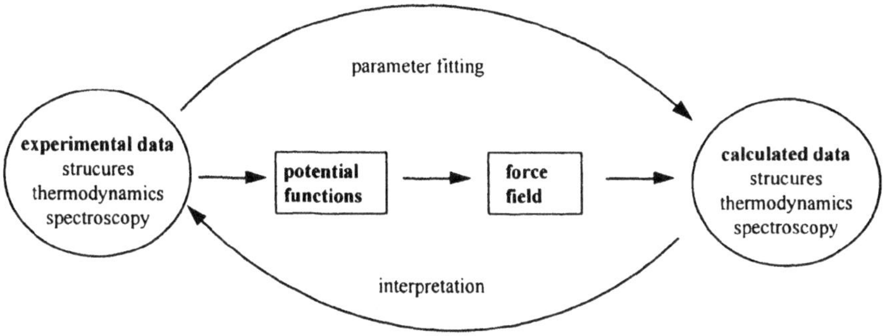

Fig. 2. The development of a force field

Therefore, there are no methodological limits to molecular mechanics in terms of the type of compound (purely organic or transition metal compounds) or of their size (small molecules, biomolecules, crystal lattices, solvates), see Fig. 3. Various approximations have been proposed and used to compute condensed phases and biomolecules [4], and the complications due to the variability of elements, oxidation states and electronic effects in transition metal compounds have now been solved to an extent where transition metal molecular modeling is becoming a routine tool [2,3,5].

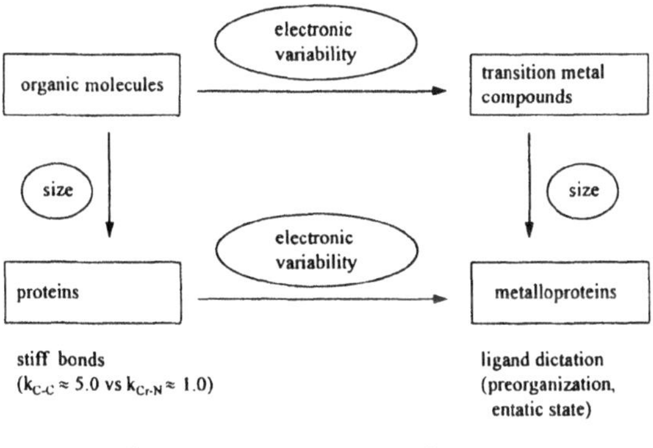

Fig. 3. The development of molecular mechanics for metalloproteins

The large number of transition metal ions and the broad range of oxidation states require a large number of parameter sets for the computation of coordination compounds. However, the major complication is due to the partially filled d-orbitals that are responsible for the widely varied structures with a large variety of coordination numbers and geometries. There still is an inherent pessimism that the molecular mechanics concept cannot work for transition metal compounds [6]. One of the complications is electron delocalization involving the metal center, the donor atoms and their neighbors.

This may lead to differences of bond strengths within the ligand backbone, between coordinated and metal-free ligands. Thus, there may be differences in the parameterization between metal-free and coordinated ligands, and this has been overlooked in most studies so far. From a recent analysis of this problem, it follows that, at least for many classical coordination compounds, the differences are so small that in many cases a separate parameterization of metal-free and coordinated ligands leads only to a minor improvement of the quality of the structural predictions (Table 1) [1f].

TABLE 1. Selected average bond distances for metal-free and coordinated ligands

	metal-free	coordinated
c- ***C* - *N*** - M	1.47	1.49
M - N - ***C* - *C*** -	1.51	1.50

Specific electronic effects that have been modeled with empirical functions include the directionality of d-orbitals, Jahn-Teller distortions, the trigonal twist of hexacoordinate compounds and the tetrahedral twist of square planar complexes [1a-c]. For other effects, such as structural variations due to π-bonding with metal d-orbitals and the trans-influence, generally applicable solutions have not been presented so far. This leads to rather complex parameterization schemes when the fully empirical approach of interpolations is applied. Of particular interest in this context are models involving ligand field terms (CLFSE/MM [1a]) and valence bond principles (VALBOND [1e,5b]); these are discussed in detail in other chapters.

For our own studies we use the MOMEC suit of programs [7] and force field [1b,1f,8] which is based on the potential energy functions given in eqns. 1 - 11. One of the main features is replacing the ligand-metal-ligand angle potential by 1,3-nonbonded interactions. For any given coordination number, the geometry of the chromophore is then mainly determined by ligand-ligand nonbonded effects. Thus, the geometry of the polyhedron (e.g. square planar vs. tetrahedral vs. trigonal pyramidal; trigonal bipyramidal vs. square pyramidal; octahedral vs. trigonal prismatic) does not need to be assumed and fixed, allowing the generally observed plasticity of coordination compounds. Also, this simplifies the parameterization enormously, since the only set of parameters needed for a particular metal-ligand arrangement is that for the metal-ligand stretching interaction [2,3].

Two additional functions have been introduced as small perturbations to mimic electronic effects: (i) the harmonic sine function (see eqn. 9) [1b]; this may be substituted by the usual harmonic function (eqn. 4) with θ_0 set to various specific values (e.g. 90°/180° for square planar and octahedral; 90°/120°/180° for trigonal bipyramidal); (ii) a twist angle potential for the angle between two planes [9]; this may be used to force the donor groups of a tetracoordinate compound to a plane (tetrahedral twist, Fig. 4a), those of a six-coordinate compound to the octahedral positions (Bailar twist, Fig. 4b) or the carbon atoms of a metallocene to the eclipsed or staggered positions (ring rotation of metallocene, Fig. 4c).

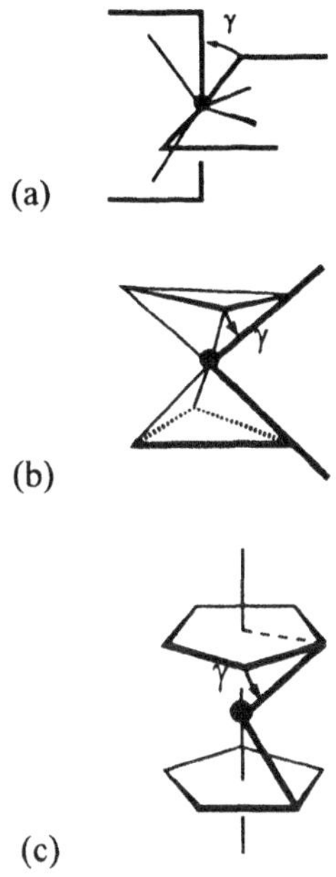

Fig. 4. Possible applications of the plane twist function

The advantages of the CLFSE/MM and the VALBOND models (see following chapters) are that the metal centered electronic effects are modeled directly, i.e. no specific functions, such as eqns. 10 and 11, are required to compute the coordination geometry, and this simplifies the parameterization of the force field. The advantage of our approach is that electronic and ligand centered steric effects may be separated and that the specific electronic functions may be constrained in order to map the conformational space.

3.2. APPLICATIONS

3.2.1. (μ-Peroxo)-dicopper(II) complexes

Recently, there has been considerable interest in the reversible binding of dioxygen by dicopper(I) complexes [10], relevant to arthropodal and molluscan hemocyanine active sites [11]. End-on (μ-peroxo)-dicopper(II) complexes are believed to be an important intermediate in the oxygenation, proceeding via oxidative addition to the dinuclear dicopper(I) center. Until recently, complete formation and satisfactory stability at room temperature of the μ-peroxo products of low molecular weight model compounds were a major problem. The entropic destabilization of the well studied $[CuL]_2O_2$ system (1) [12] was to be overcome by linking the two tripodal chromophores by an organic spacer group [13].

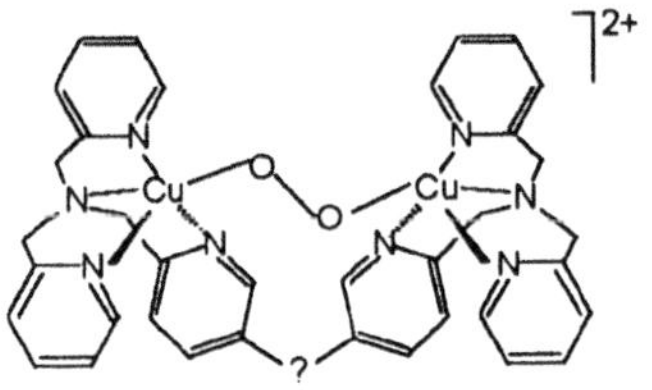

(1)

Initial studies with an ethyl bridge indicated that the favorable entropic effect was canceled by an unfavorable enthalpic effect, induced by steric strain. The optimization of the bridged ligand (see Chart below) involved the parameterization of the Cu-O-O-Cu fragment, and this was based on the only X-ray structure of end-on-(μ-peroxo)-dicopper(II) species available, i.e. that of (1) [12a,b] and vibrational spectroscopic studies of (1) [12c].

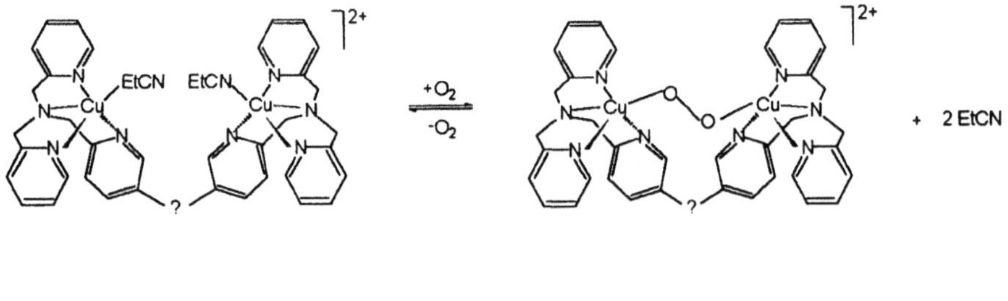

Also included in the development of the parameter set were structures of (μ-peroxo)-dicobalt(III) complexes [14]. The spacer group induced strain energies (see caption of Table 2), the rms of the chromophores and the structural parameters (see Table 2 and Fig. 5) indicate that, indeed, the ethyl-bridged (μ-peroxo)-dicopper(II) compound (2) is rather strained, and that the propane (3) and the ether (5) spacers lead to more relaxed products, while longer bridges (4) lead to considerable distortion.

TABLE 2. Strain energies and structural features of the calculated structures of the five (μ-peroxo)-dicopper(II) complexes [14]

Complex	(Spacer)	Spacer induced strain energy[a] [kJmol^{-1}]	RMS overlay of the chromophores[b]	Cu-Cu [Å]	Cu-O-O-Cu [°]	Tilt[c] [°]	Torsion [°]
(1)	(-)	0	0	4.42	180	0	60
(2)	(-CH_2-CH_2-)	9.94	0.402	4.42	175	24	35
(3)	(-CH_2-CH_2-CH_2-)	3.66	0.345	4.43	176	12	49
(4)	(-CH_2-CH_2-CH_2-CH_2-)	2.25	0.578	4.40	165	22	18
(5)	(-CH_2-O-CH_2-)	4.18	0.400	4.43	175	6	30

[a] Difference of the strain energy of the computed structures after replacing the spacer groups between the two chromophores by H-atoms, attached in ideal geometry at each of the aromatic carbon atoms, and of the strain energy of the computed structure without spacer. [b] RMS overlay of the $N_3CuO_2CuN_3$ fragment with the $N_3CuO_2CuN_3$ fragment of the calculated structure without spacer. [c] Angle between the calculated best planes through $Cu(N_{py})_3$. [d] Averaged torsion angle around a fictitious N_{py}-Cu-Cu-N_{py} fragment

Recent experimental studies involving (5) have confirmed these predictions, i.e. the corresponding (μ-peroxo)-dicopper(II) complex is stabilized with respect to (2) [13b].

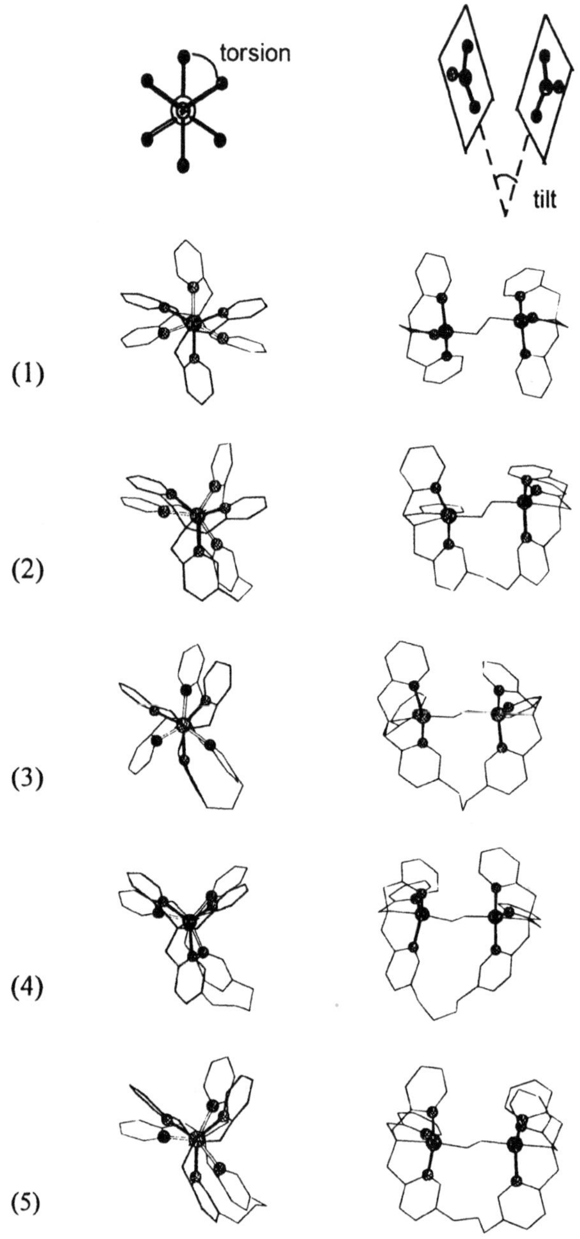

Fig. 5. Torsion and tilt of the five computed (μ-peroxo)-dicopper(II) complexes (reproduced with permission from [14])

3.2.2 *Rigid tetrahedrally preorganized ligands*

With tetrahedrally preorganized ligands, the well-known Irving-Williams series of divalent first row transition metal ions, with maximum stability for copper(II), may be reverted to obtain zinc(II) selectivity [15]. Also, with very rigid tetrahedral ligands, enforcing short metal-donor distances, the high ligand field might lead to unusual electronic properties and a possible stabilization of low spin electronic configurations. The substituted bispidine type ligands (6) were believed to fulfill this requirement.

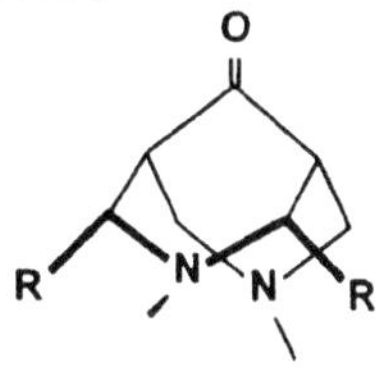

(6)

Strain energy minimized structures of metal complexes of (6) with R = pyridine or methylpyridine (Fig. 6) indicated that (i) cis-(6) and trans-(6) lead to very similar coordination geometries; (ii) ligands with 6-membered chelate rings involving the pyridine donors (R = methylpyridine) stabilize tetrahedral coordination polyhedra while those with 5-membered chelate rings (R = pyridine) lead to five- or six-coordinate complexes (Fig. 6); (iii) the ligand enforces rather short metal donor distances (see Fig. 7) [15d]. So far, no ligands leading to six-membered chelate rings have been isolated, but a cobalt(II) complex of a derivative of (6) with R = pyridine has been prepared [15d]. As predicted, with nitrate the complex is six-coordinate, and with the larger chloride ion it is five-coordinate.

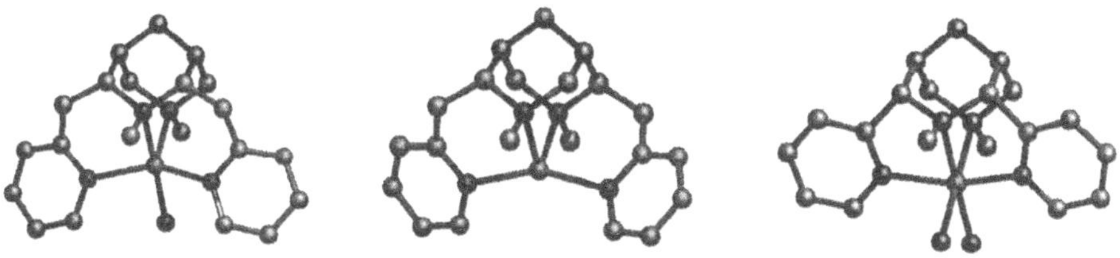

(a) (b) (c)

Fig. 6. Calculated structures of bispidine complexes; (a) 6-membered chelate rings, cis-substitution; (b) 6-membered chelate rings, trans-substitution; (c) 5-membered chelate rings, cis-substitution (see text) [15d]

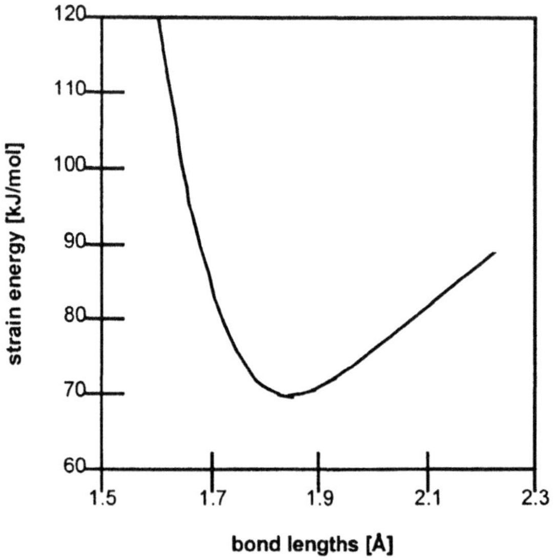

Fig. 7. Metal-donor dependent strain energy of structure (a) in Fig. 6 [15d]

The X-ray structure of the nitrato complex (Fig. 8) is in good agreement with the computed structure (Table 3) [15d]. This indicates that it is worthwhile to pursue the task of preparing a rigid ligand enforcing tetrahedral coordination, based on (6) with six-membered chelate rings involving the substituents R.

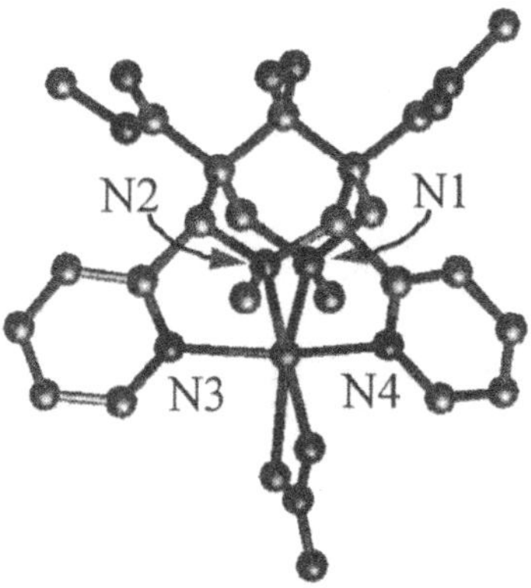

Fig. 8. The experimentally determined structure of a cobalt(II) bispidone complex (compare the computed structure (c) in Fig. 6) [15d]

Table 3. Comparison of observed and computed structural parameters of the cobalt(II) bispidone complex from Fig. 8 [15d]

bond distances [Å]	obs.	calc.
Co-N1	2.13	2.14
Co-N2	2.14	2.17
Co-N3	2.09	2.12
Co-N4	2.09	2.12
valence angles [deg]		
N3-Co-N4	150	150
N3-Co-N1	77	77
N3-Co-N2	94	96
N4-Co-N1	77	77
N4-Co-N2	99	97
N1-Co-N2	85	85

3.2.3 Linear metallocenes

A number of approaches have been published for the structural optimization of metallocenes [16], including an electrostatic model [16a,b], a fully bonded model (10-coordinate metal centers) [16c,d] and approaches using dummy atoms at the centroid of the Cp ring [16b,e,f]. So far, the most elegant method was that of an interactionless dummy atom at the centroid of the five-membered aromatic rings, i.e. the forces on the dummy atom are distributed among the carbon atoms [16f,g]. However, with the angular geometry around metal centers computed by 1,3-interactions (see 3.1) this elaborate technique is not necessary and dummy atoms are not needed. That is, the Cp rings are free to rotate even with all ten carbon atoms bound to the metal center. Furthermore, constraints on a plane-twist function may be used to drive the conformation from staggered to eclipsed (see Fig. 5c above). With this model and a preliminary parameterization, based on published force fields [16] and the usual van der Waals parameters [7,8], good agreement between experimental and computed structures was obtained [17]. The computed barrier to rotation of the Cp rings was close to 0 kJ mol^{-1}, and this is qualitatively in agreement with an observed value of ca. 4 kJ mol^{-1} (Table 4). The observed small barrier may be due to electronic or to solvent reorganization effects. Independent of the fundamental reason, the parameterization of the plane-twist potential may be used to model these.

TABLE 4. Preliminary results of ferrocene modeling

	Fe - C [Å]	Cp ... Cp [Å]	barrier [kJ/mol]
experiment	2.04	3.32	4
force field	2.04	3.34	0.1

4. Modeling of Molecular Properties

4.1. BASIC APPROACH

Some properties of transition metal compounds may, based on certain approximations, be directly related to the output (structures, strain energies and vibrational frequencies) of molecular mechanics calculations (Fig. 9). These include metal ion selectivities, conformational equilibria, which may be related to observable isomer distributions and stereo-selectivities, and vibrational spectroscopic data [2,3,18]. Also, parameters from magnetic resonance experiments are directly dependent on the relative orientation of electron and nuclear spins and may therefore be transformed into structural parameters and used in combination with strain energy minimized structures. NOE data and Karpus relations have been widely used for the structural refinement of biomolecules [19], and some examples have also been reported, where similar techniques have been used for low molecular weight transition metal compounds [2,3,18]. Chapters 1 and 9 in this book describe methods for the structure determination of paramagnetic metalloproteins by NMR techniques [20], and a combination of the simulation of EPR spectra of weakly coupled dinuclear S=½ / S=½ systems with molecular mechanics calculations (MM-EPR) has also been used for structure determination [2,3,18,21].

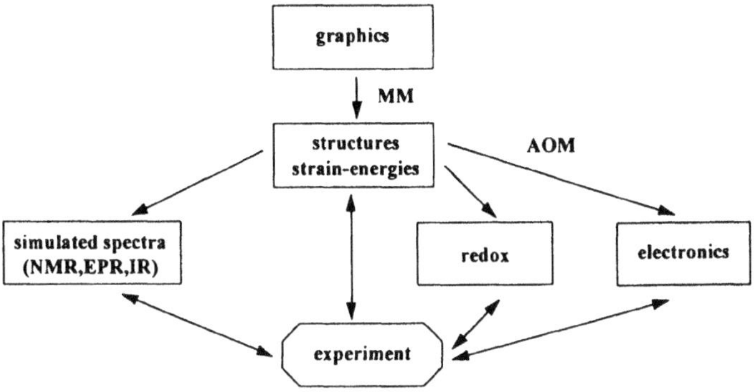

Fig.9. The determination of molecular structures using molecular mechanics in combination with the simulation of observable properties.

For other properties, such as redox potentials (MM-Redox) [2,3,22] and electronic spectra (MM-AOM) [2,3,18,23], specific algorithms are used to compute the properties from refined structures. All these techniques may be used for the design of new compounds or for structure refinement.

4.2. APPLICATIONS

4.2.1 Budotitane

Budotitane ([Ti(bzac)$_2$(OEt)$_2$]; bzac = 1-phenylbutane-1,3-dionate, OEt = ethoxide) is a potent anticancer drug [24], but its structure was not known experimentally. This is probably due to the lability of Ti(IV) compounds and the fact that, theoretically, five isomeric forms are possible (Fig. 10), i.e. no pure fractions may be isolated. NMR experiments indicated that only the three cis-isomers are stable in solution, and this observation was reproduced by force field calculations (Table 5) [25]. Results of similar quality were obtained for other Ti(IV), and for Co(III) complexes with acac-type ligands, for which X-ray data were in good agreement with the computed structures [25]. This, and the fact that a constant force field for the organic backbone is used, for which computed isomer ratios of Co(III) and hs Ni(II) compounds have been shown on a large data basis to be in good agreement with experimental data, leads to the conclusion that the computed structures are of high accuracy (Fig. 10).

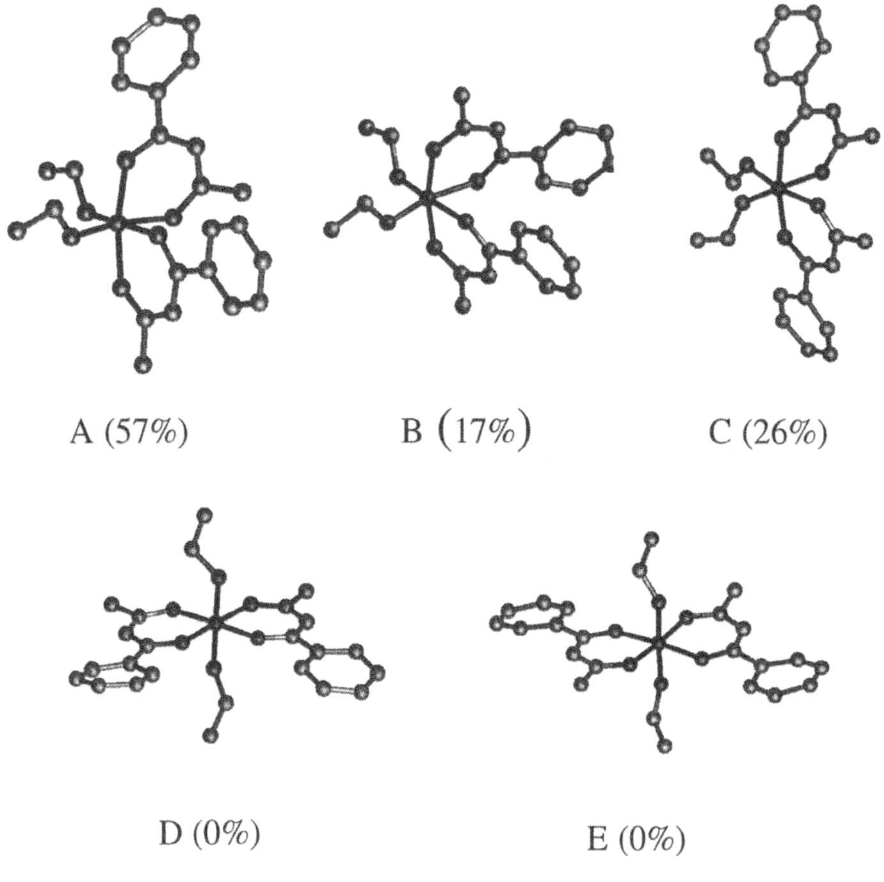

Fig. 10. Computed structures and isomer ratios of Budotitane; top: cis-isomers, bottom: trans-isomers [25]

TABLE 5. Computed and observed isomer ratios of Budotitane [25]

compound	isomer	calculated	observed
$[Ti(bzac)_2(OEt)_2]$	A	57	60
(budotitane)	B	17	19
	C	26	21
	D	0	0
	E	0	0

4.2.2 A Tyrosinase model compound

Tyrosinase ortho-hydroxylates phenols (tyrosine) to catechole. Similar to hemocyanine (see 3.2.1) it has two copper(I) centers, each bound to three histidine nitrogens and separated by 3.6 Å (type III copper site) [11,26]. A number of dinuclear copper model complexes that catalyze the oxidation of organic substrates have been reported [10], and the dicopper(I) compounds of dinucleating macrocyclic ligands with xylyl bridges (7) have been found to successfully model tyrosinase-type reactivity [27].

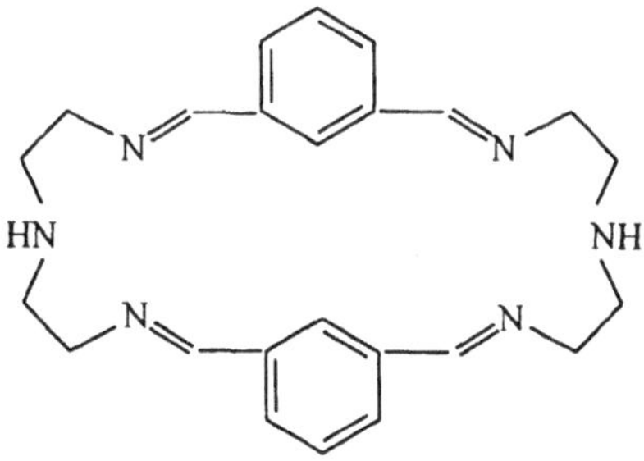

(7)

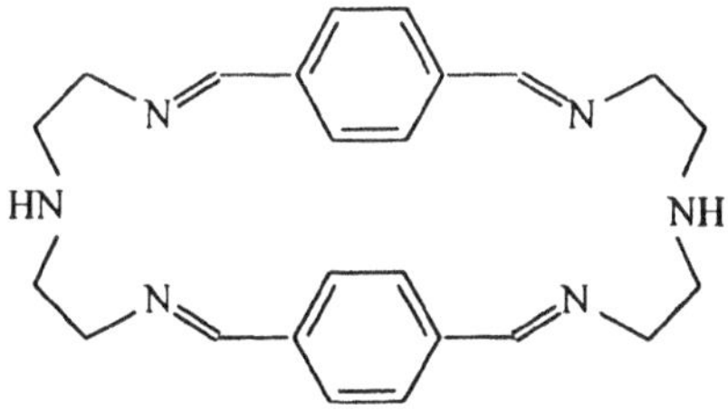

(8)

In fact, the hydroxylation of the aromatic spacers between the two copper sites is so fast and efficient that, so far, it has proven impossible to crystallize $[Cu_2(7)]^{2+}$. The dicopper complex of the isomer (8), however, is unreactive and its X-ray structure was solved [21c]. Molecular mechanics was used to model the two structures with the aim of understanding the strikingly different reactivities [21c]. The conformational search routine and the force field were carefully tested with $[Cu_2(8)]^{2+}$ and the dicopper(II) complexes of the reduced derivatives of (7) and (8), where the computed structures were used to simulate the weakly dipole-dipole coupled EPR spectra with the MM-EPR approach [21c]. The computed structure of $[Cu(X)_2(7)]^{2+}$ (see Fig. 11) is in good agreement with experimental data: the Cu···Cu distance of 3.4 Å is similar to that of tyrosinase (3.6 Å [11,26]) and that observed in the crystal structure of the oxygenated and the deoxygenated forms of hemocyanine (3.6 Å and 4.6 Å, respectively [11,26]), and it is different to that found for the unreactive isomeric form $[Cu_2(X)_2(8)]^{2+}$ (6.3 Å). The structural differences are also supported by electrochemical data which indicate that the copper centers in (7) are strongly coupled (two oxidation waves separated by 650 mV), suggesting a copper-copper distance smaller than 4.5 Å, while the copper centers in (8) are not coupled.

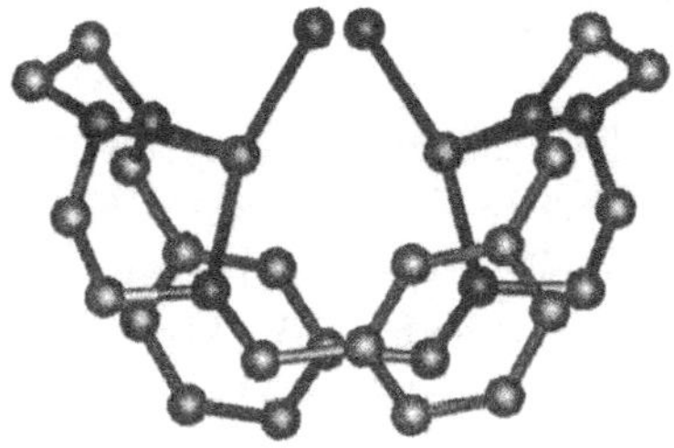

Fig. 11. Presentation of the lowest strain energy structure of $[Cu_2(X)_2(7)]^{2+}$ [21c]

4.2.3 Solution structures of a pair of stable hexaaminecobalt(III) conformers

The recent report of a pair of stable hexaaminecobalt(III) conformers of a new cage ligand (9), with strikingly different properties [28], has shown new ways of

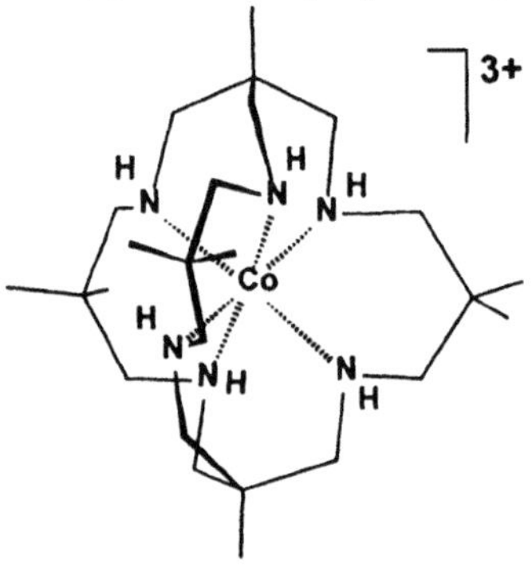

(9)

changing electronic and structural properties of metal complexes. One of the two species is yellow ($^1A_1 \rightarrow {}^1T_1$ at 20'800 cm^{-1}) with a redox potential of around zero Volts, as expected for this class of compounds. The other is blue

($^1A_1 \rightarrow {}^1T_1$ at 16'700 cm^{-1}) with E° = +8.84V (vs SHE). The structure of the "normal" yellow compound was solved by X-ray diffraction, but not that of the more exciting blue form. The experimental structure had the three six-membered chelate rings involving the 2,2-dimethylpropane bridges in skew-boat conformations, and these three chelate rings were oriented oblique (ob), with respect to the C_3-axis through cobalt and the two adamantane-type caps (see Fig. 12). Molecular mechanics indicated that another low energy conformation, with close to identical strain energy, had the above mentioned chelate rings in skew-boat conformation and oriented parallel (lel) to the C_3-axis, leading to an elongated cavity with significantly longer Co-N distances.

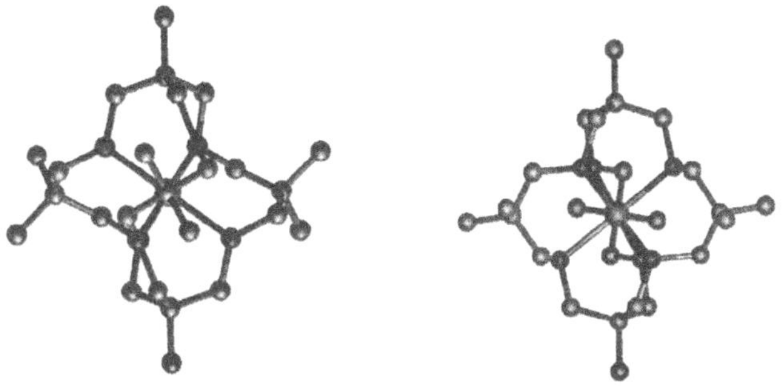

Fig. 12. Computed structures of the lel_3 (left) and ob_3 conformers of $[Co(9)]^{3+}$

Table 6. Observed and computed structural, electrochemical and spectroscopic properties of the ob_3 and lel_3 conformers of $[Co(9)]^{3+}$

		ob_3		lel_3	
		obs	calc	obs	calc
Co^{III} - N	[Å]	1.990	1.985		2.046
Co^{II} - N	[Å]		2.124	2.235	2.224
strain energy (Co^{III})	[kJ/mol]		279		282
strain energy (Co^{II})	[kJ/mol]		238		177
E°	[V]	0.0	-0.1	0.84	0.89
d-d absorption	[nm]	480	481	600	612
		350	344	370	416

Qualitatively, these structural changes are in agreement with the observations, i.e. a larger cavity will better accommodate cobalt(II) than cobalt(III), leading to a shift of the redox potential towards more positive values, and a bond elongation decreases the ligand field strength with a concommitent shift of the dd transitions to lower energy. Based on the refined structures of a series of low energy conformers, the electronic spectra and redox potentials were computed with methods and parameterization schemes reported elsewhere (MM-AOM [23], MM-REDOX [22]). These data are given in Table 6 [22a]. The good agreement between observed and calculated spectroscopic and electrochemical parameters and the nearly identical strain energy of the hexaaminecobalt(III) species allowed the assignment of the two structures shown in Fig. 12 to the yellow and the blue conformers.

5. The Challenge of Metalloproteins

The examples presented so far are some of our more recent results which demonstrate that (i) empirical force field calculations with a set of functions adapted to transition metal centers and a carefully tuned parameterization scheme are well suited to accurately predict structures and strain of transition metal compounds; (ii) there are simple empirical methods to compute molecular properties such as stabilities, redox potentials and spectra from known (observed or computed) structures; (iii) a combination of modeling structures and properties is a reliable tool for the design of new materials and for the determination of molecular structures.

In theory, the step from low molecular weight transition metal compounds to metalloproteins is straight forward, and a number of chapters in this book demonstrate that the tools and know-how are available to optimize proteins with empirical force field methods. For a number of reasons, we have chosen blue copper proteins as a first example. Our aim is to model structures and spectroscopic properties in order to design new mutants with specified structural and electronic properties and to determine their structure, once they have been isolated. The approach we want to use is much the same as that described so far, i.e. the computation of properties based on algorithms that use structural data obtained by molecular mechanics. These might involve the techniques described above or quantum chemical calculations (single point ab-initio or semi-empirical calculations), e.g. if charge transfer spectra instead of ligand field spectra are involved. There are a number of recent studies on the computation of spectra of blue copper proteins [29]. Hence, a combination of our experience in the area of coordination compounds with those of protein modeling and quantum chemistry of blue copper protein active sites is bound to be successful.

The amino acid sequences of the C-terminal loop with three of the four side chains coordinating to copper (see Fig. 13 for a model of a wt amicyanin active site) are listed in Table 7. It appears that a loop mutation should shed

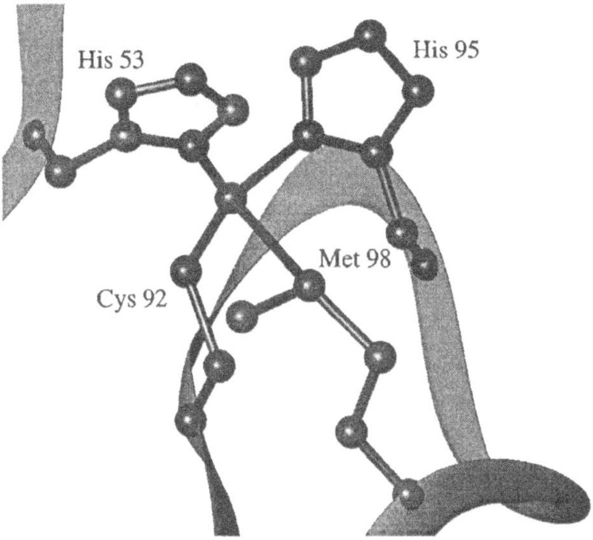

Fig. 13. Representation of the Cu site of amicyanin from paracoccus denitrificans

TABLE 7. Amino acid sequence of the C-terminal loop of some blue copper proteins

Protein	Amino acid sequence									
Amicyanin[1]	Cys	Thr	-	Pro	-	His	Pro	-	-	Phe	Met
Plastocyanin[2]	Cys	Ser	-	Pro	-	His	Gln	Gly	Ala	Gly	Met
Pseudoazurin[3]	Cys	Thr	-	Pro	-	His	Tyr	Ala	Met	Gly	Met
Azurin[4]	Cys	Thr	Phe	Pro	Gly	His	Ser	-	Ala	Leu	Met

[1] Thiobacillus versutus; [2] Populus nigra; [3] Alcaligenes faecalis S-6; [4] Pseudomonas aeruginosa

some light on the relative influences of the protein backbone and that of the C-terminal loop to the structure of the active site. The structural properties and their influence on the UV-vis and EPR spectra are outlined in Fig. 14. Two mutants of amicyanin, those with a plastocyanin [30] and with an azurin [31] C-terminal loop, have been isolated and characterized spectroscopically (Fig. 15).

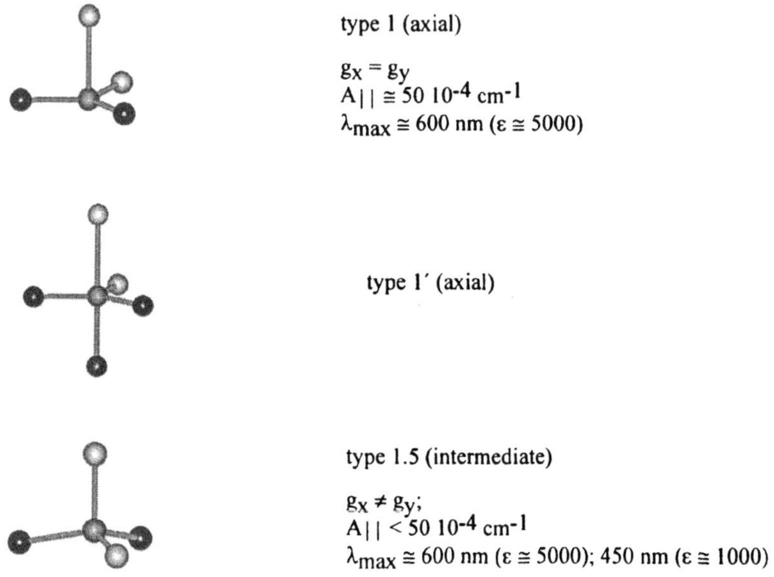

Fig. 14. Copper coordination geometries and properties of blue copper proteins

Both have spectroscopic properties that are intermediate between the two corresponding wild type proteins and indicate an increasing rhomicity of the active site with respect to the wild type amicyanin.

For our preliminary modeling studies we have parameterized a rigid geometry molecular mechanics module to optimized structures of blue copper proteins. The copper site and a limited number of adjacent amino acids, typically those of the C-terminal loop discussed above and some amino acids

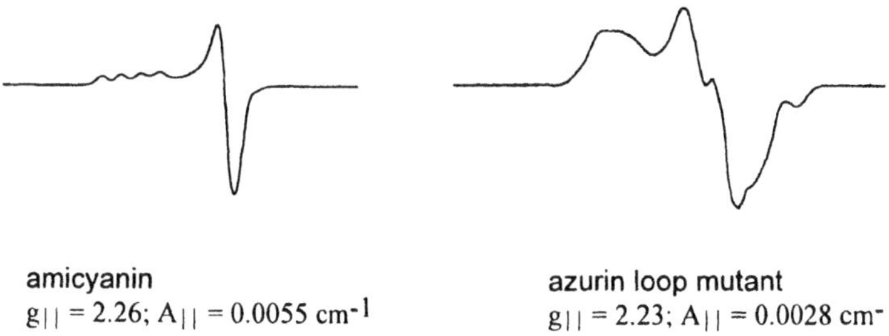

Fig. 15. X-band EPR spectra (77K) of the wilde type amicyanin and the amicyanin-azurin loop mutant

adjacent to the second His, were fully optimized, including van der Waals interactions to atoms in a shell of 3 Å around the optimized site. The remainder of the backbone was kept rigid. Force field parameters of the protein part were adapted from AMBER [4], parameters for the chromophore were those of our own force field [1b,1f,8], and fitted to published blue copper protein structures and model compounds. Preliminary results are shown in Table 8 [32].

TABLE 8. Selected observed and computed structural data of blue copper proteins

Amicyanin (Paracoccus denitrificans)			Plastocyanin (Populus nigra)		
Bond distance/ Out-of-plane	X-ray	Calculated	Bond distance/ Out-of-plane	X-ray	Calculated
Cu-S (Met98)	2.89Å	2.94Å	Cu-S (Met98)	2.82Å	2.84Å
Cu-S (Cys92)	2.15Å	2.17Å	Cu-S (Cys92)	2.07Å	2.12Å
Cu-N (His95)	2.00Å	1.98Å	Cu-N (His95)	2.06Å	2.08Å
Cu-N (His53)	1.95Å	1.98Å	Cu-N (His53)	1.91Å	2.08Å
Cu-$\{N_{His53}$-S_{Cys92}-$N_{His95}\}$	0.30Å	0.20Å	Cu-$\{N_{His53}$-S_{Cys92}-$N_{His95}\}$	0.36Å	0.33Å

Acknowledgment

Our studies are supported by the German Science Foundation (DFG), the Fonds of the Chemical Industry (FCI) and the VW-Stiftung. I am grateful for that, for the excellent work done by my coworkers whose names appear in the references and to Karin Stelzer for her help in preparing this manuscript.

References

1. a) Burton,V. J.; Deeth, R. J.; Kemp, C. M.; Gilbert, P. J. (1995) Molecular mechanics for coordination complexes: the impact of adding d-electron stabilisation energies, *J. Am. Chem. Soc.* **117**, 8407-8415; b) Comba, P.; Hambley, T. W.; Ströhle, M. (1995) The directionality of d-orbitals and molecular mechanics calculations of octahederal transition metal compounds, *Helv. Chim. Acta* **78**, 2042-2047; c) Comba, P.; Zimmer, M. (1994) Molecular mechanics and the Jahn-Teller effect, *Inorg. Chem.* **33**, 5368-5369; d) Warshel, A.; Karplus, M. (1972) Calculation of ground and excited state potential surfaces of conjugated molecules. I. Formulation and parametrization, *J. Am. Chem. Soc.* **94**, 5612-5625; e) Landis, C. R.; Cleveland, T.; Firman, T. K. (1995) Making sense of the shape of simple metal hydrides, *J. Am. Chem. Soc.* **117**, 1859-1860, f) Bol, J. E.; Buning, C.; Comba, P.; Ströhle, M.; Reedijk, J. (1997) Molecular mechanics modeling of the organic backbone of metal-free and coordinated ligands, *submitted*; g) Maseras, F.; Morokuma, K. (1995) IMOMM: A new integrated ab initio plus molecular mechanics geometry optimization scheme of equilibrium structures and transition states, *J. Comput. Chem.* **16**, 1170-1179.
2. Comba, P.; Hambley, T. W. (1995) *Molecular modeling of inorganic compounds*, VCH, Weinheim.
3. a) Comba, P. (1993) The relation between ligand structure, coordination stereochemistry and electronic and thermodynamic properties, *Coord. Chem. Rev.* **123**, 1-48; b) Comba, P. (1996) Modeling of structural and spectroscopic properties of transition metal compounds, In Gans W.; Amann, A; Boeyens, J. C. A., Eds. *Fundamental Priciples of Molecular Modeling*, Plenum Press, New York, 167-187.
4. a) Gavezzotti, A. (1997) *Theoretical aspects and computer modeling of the molecular solid state*, Wiley, New York, Chaper 4; b) Weiner, S. J.; Kollman, P. A.; Case, D. A.; Singh, C.; Ghio, C.; Profeta, S.; Alagona, G.; Weiner, P. (1984) A new force field for molecular mechanical simulation of nucleic acids and proteins, *J. Am. Chem. Soc.* **106**, 765-784; c) Cornell, W. D.; Cieplak, P.; Bayly, C. I.; Gould, I. R.; Merz Jr., K. M.; Ferguson, D. M.; Spellmeyer, D. C.; Fox, T.; Caldwell, J. W.; Kollman, P. A. (1995) A second generation force field for the simulation of proteins and nucleic acids, *J. Am. Chem. Soc.* **117**, 5179-5196.
5. a) Hay, B. P. (1993) Methods for molecular mechanics modeling of coordination compounds *Coord. Chem. Rev.* **126**, 177-236; b) Landis, C. R.; Root, D. M.; Cleveland, T. (1995) Molecular mechanics force fields for modeling inorganic and organometallic compounds In: Lipkowitz, K. B. and Boyd, D. B., Eds. *Reviews in Computational Chemistry*, VCH, Weinheim, Volume 6, 73-136.

6. Bersuker, I. (1996) *Electronic Structures and Properties of Transition Metal Compounds*, Wiley, New York.
7. Comba, P.; Hambley, T. W.; Lauer, G.; Okon, N. (1997) MOMEC97, a molecular modeling package for inorganic compounds, Lauer & Okon, Heidelberg, Germany, e-mail: CVS-HD@T-Online.de.
8. Bernhardt, P. V.; Comba, P. (1992) Molecular mechanics calculations of transition metal complexes, *Inorg. Chem.* **31**, 2638-2644.
9. Comba, P.; Lauer, G.; Okon, N. (1997), *in preperation*.
10. a) Kitajima, N.; Moro-oka, Y. (1994) Copper-dioxygen complexes. Inorganic and bioinorganic perspectives, *Chem. Rev.* **94**, 737-758; b) Karlin, K. D.; Tyeklár, Z., Eds. (1993) *Bioinorganic Chemistry of Copper*, Chapman & Hall, New York; c) Reedijk, J., Ed. (1993) *Bioinorganic Catalysis*, Marcel Dekker, New York; d) Vigato, P. A.; Tamburini, S.; Fenton, D. E. (1990) The activation of small molecules by dinuclear complexes of copper and other metals, *Coord.Chem.Rev.* **106**, 25-170; e) Sorrell, T. N. (1989) Synthetic models for binuclear copper proteins, *Tetrahedron* **45**, 3-68.
11. Magnus, K. A.; Ton-That, H.; Carpenter, J. E. (1994) Recent structural work on the oxygen transport protein hemocyanin, *Chem. Rev.* **94**, 727-735.
12. a) Jacobson, R. R.; Tyeklár, Z.; Farooq, A.; Karlin, K. D.; Liu. S.; Zubieta, J. (1988) A Cu_2-O_2 complex. Crystal structure and characterization of a reversible dioxygen binding system, *J. Am. Chem. Soc.* **110**, 3690-3692; b) Tyeklár, Z.; Jacobson, R. R.; Wei, N.; Murthy, N. N.; Zubieta, J.; Karlin, K. D. (1993) Reversible reaction of O_2 (and CO) with a copper(I) complex. X-ray structures of relevant mononuclear Cu(I) precursor adducts and the trans-(μ-1,2-peroxo)dicopper(II) product, *J. Am. Chem. Soc.* **115**, 2677-2689; c) Baldwin, M. J.; Ross, P. K.; Tyeklar, Z.; Karlin, K. D., Solomon, E. I. (1991) Spectroscopic and theoretical studies of an end-on peroxide-bridged coupled binuclear copper(II) model complex of relevance to the active sites in heomocyanin and tyrosinase, *J. Am. Chem. Soc.* **113**, 8671-8679.
13. a) Lee, D.-H.; Wei, N.; Murthy, N. N., Tyeklár, Z.; Karlin, K. D.; Kaderli, S.; Jung, B.; Zuberbühler, A. D. (1995) Reversible O_2 binding to a dinuclear copper(I) complex with linked tris(2-pyridylmethyl)amine units: kinetic-thermodynamic comparisons with mononuclear analogues, *J. Am. Chem. Soc.* **117**, 12498-12513; b) Karlin, K. D.; Lee, D. H.; Kaderli, S., Zuberbühler, A. D. (1997) Copper dioxygen complexes stable at ambient temperature: Optimization of ligand design and solvent, *J. Chem Soc., Chem. Commun.*, 475-476.
14. Comba, P.; Hilfenhaus, P.; Karlin, K. D. (1997) Modeling of end-on (μ-peroxo)dicopper(II) complexes, *Inorg. Chem.* **36**, 2309-2313.
15. a) Rawle, S. C.; Clarke, A. J.; Moore, P; Alcock, N. W. (1992) Ligands designed to impose tetrahedral coordination: A convenient route to aminoethyl and aminopropyl pendant arm derivatives of 1,5,9-triazacyclonanane, *J. Chem. Soc., Dalton Trans.*, 2755-2757; b) Alcock, N. W.; Benniston, A. C.; Moore, P.; Pike, G. A.; Rawle, S. C. (1991) Macrocyclic ligands designed to impose tetrahedral coordination: [1(3-dimethylaminopropyl)-1,5,9-triazacyclononane], L1, [1{2-(pyrrolidin-1-yl)ethyl}-1,5,9-triazacyclononane], L2, and their zinc(II) complexes *J. Chem. Soc., Chem. Comm.*, 706-708; c) Turonek, M. L.; Moore, P.; Clase, H. J.; Alcock, N. W. (1995) Synthesis and coordination chemistry of the pyridyl pendant-arm azamacrocycles 1-(2-pyridylmethyl)-1,5,9-triazacyclodocecane L^1 and 1-(2-pyridyl-2'-ethyl)-1,5,9-triazacyclododecane L^2, with nickel(II), copper(II) and zinc(II). Crystal structures of $[Ni(L^1)(O_2NO)]NO_3$ and $[Zn(L^2)(NO_3)_{3.67}Cl_{0.33}]$, *J. Chem. Soc., Dalton Trans.*, 3659-3666; d) Comba, P.; Nuber, B.; Ramlow, A. (1997) Design of a new type of very rigid tetradentate ligand, *J. Chem. Soc., Dalton* Trans., 347-352. e) Comba, P.; Goll, W.; Nuber, B.; Várnagy, K. (1997) Metal ion selective ligands: Design, synthesis and

complexation of bisamidobispyridyl ligand systems, *submitted*; f) Bates, G. B.; Parker, D.; Tasker, P. A. (1996) Synthesis and solution complexation behaviour of dimeric zinc-selective bis(benzimidazole) derivatives, *Chem. Soc. Perkin Trans. 2*, 1117-1125.

16. a) Menger, F. M. Sherrod, M. J. (1988) Docking calculations on ferrocene complexation with cyclodextrins, *J. Am. Chem. Soc.* **110**, 8606-8611; b) Davies, S. G.; Derome, A. E.; McNally, J. P. (1991) Conformational analysis and dynamics of the triphenylphosphine ligand in $[(\eta^5\text{-}C_5H_5)Fe(CO)(PPh_3)COCH_3]$, *J. Am. Chem. Soc.* **113**, 2854-2861; c) Rudzinski, J. M.; Osawa, E. (1992) Molecular Mechanics of bridged ferrocene derivatives: Conformational energy surfaces of [3]-, [4]- and [4_5] ferrocenophanes, *J. Phys. Org. Chem.* **5**, 382; d) Rudzinski, J. M.; Osawa, E. (1993) Conformational interconversion of [1.1] ferrocenophane: A molecular mechanics study, *J. Phys. Org. Chem.* **6**, 107; e) Blom, R.; Hammel, A.; Haaland, A.; Weidlein, J.; Timofeeva, T. V.; Struchkov, Y. T. (1993) Molecular structures of tris(methylcyclopentadienyl)-scandium and -ytterbium as studied by gas phase electron diffraction and molecular mechanics calculations: The scandium atom is too small to accommodate three pentahapto cyclopentadienyl rings, *J. Organomet. Chem.* **462**, 131-139; f) Doman, T. N.; Landis, C. R.; Bosnich, B. (1992) Molecular mechanics force fields for linear metallocenes, *J. Am. Chem. Soc.* **114**, 7264-7272; g) Mercandelli, P.; Sironi, A. (1996) Ligand stereochemistry of metal clusters containing π-bonded ligands, *J. Am. Chem. Soc.* **118**, 11548-11554.
17. Comba, P. (1997), *work in progress*.
18. Comba, P. (1994) Solution structures of coordination compounds, *Comm. Inorg. Chem.* **16**, 133-151.
19. a) Wüthrich, K. (1986) *NMR of proteins and nucleic acids*, Wiley, New York; b) Wüthrich, K. (1989) The development of nuclear magnetic resonance spectroscopy as a technique for protein structure determination, *Acc. Chem. Res.* **22**, 36-44.
20. a) Bertini, I.; Rosato, A. (1997) Solution structures of proteins containing paramagnetic metal ions, chapter 1 in this book; b) Banci, L.; Gori Savellini, G. (1997) Molecular dynamics calculations on metalloproteins, chapter 9 in this book.
21. a) Bernhardt, P. V.; Comba, P.; Hambley, T. W.; Massoud, S. S.; Stebler, S. (1992) Determination of solution structures of binuclear copper(II) complexes, *Inorg. Chem.* **31**, 2644-2650; b) Comba, P.; Hilfenhaus, P. (1995) One-step template synthesis and solution structures of bis(macrocyclic) octaamine dicopper(II) complexes, *J. Chem. Soc., Dalton Trans.*, 3269-3274. c) Comba, P.; Hambley, T. W.; Hilfenhaus, P.; Richens, D. T. (1996) Solid state and solution structures of two structurally related dicopper complexes with markedly different redox properties, *J. Chem. Soc., Dalton Trans.*, 533-539.
22. a) Comba, P.; Sickmüller, A. F. (1997) The solution structures of a pair of stable cobalt(III) hexaamine conformers, *Angew. Chem., Angew. Chem. Int. Ed. Engl., accepted*; b) Comba, P.; Sickmüller, A. F. (1997) Modeling of the redox properties of hexaaminecobalt(III/II) couples, *Inorg. Chem., accepted*; c) Comba, P.; Jakob, H. (1997) Reduction potentials of tetraaminecopper(II/I) couples, *submitted*.
23. a) Bernhardt, P. V.; Comba, P. (1993) Prediction and interpretation of electronic spectra of transition metal complexes via the combination of molecular mechanics and angular overlap model calculations, *Inorg. Chem.* **32**, 2798-2803.; b) Comba, P. (1994) Prediction and interpretation of EPR spectra of low spin iron(III) complexes with the MM-AOM method *Inorg. Chem.*, **33**, 4577-4583.; c) Comba, P.; Hambley, T. W.; Hitchman, M. A.; Stratemeier, H. (1995) Interpretation of electronic and EPR spectra of copper(II) amine complexes - A test of the MM-AOM method, *Inorg. Chem.* **34**, 3903-3911.
24. Keppler, B. K.; Friesen, C.; Vongerichten, H.; Vogel, E. (1993) Budititane, a new tumor-inhibiting titanium compound: Preclinical and clinical development, In Keppler, B. K. Ed. *Metal complexes in Cancer Chemotherapy*, VCH, Weinheim.

25. Comba, P.; Jakob, H.; Keppler, B. K., Nuber, B. (1994) Solution structures and isomer distributions of bis(ß-diketonato) complexes of titanium(IV) and cobalt(III), *Inorg. Chem.* **33**, 3396-3400.
26. Magnus, K. A.; Ton-That, H.; Carpenter, J. E. (1993) Three-dimensional structure of the oxygenated form of the hemocyanin subunit II of Limilus Polyphemus at atomic resolution, In Karlin, K. D., Tyeklar, Z., Eds. *Bioinorganic Chemistry of Copper*, Chapman & Hall, New York, 143-150.
27. Menif, R., Martell, A. E. (1989) Oxygen insertion by a new tyrosinase model binuclear Cu^I macrocyclic complex, *J. Chem. Soc., Chem. Commun.*, 1521-1523.
28. Geue, R. J.; Hanna, J.; Höhn, A.; Qin, C. J.; Ralph, S. F.; Sargeson, A. M.; Willis, A. C. (1997) Steric effects in redox reactions and electron transfer rates, In Isied, S. S., Ed. *Electron Transfer Reactions*, ACS Symposium Series, ACS, Washington, Chapter 8.
29. a) Solomon, E. I.; Baldwin, M. J.; Lowry, M. D. (1992) Electronic structures of active sites in copper proteins: Contributions to reactivity, *Chem. Rev.* **92**, 521-542; b) Ryde, U., Olsson, M. H. M.; Pierloot, K.; Roos, B. O. (1996) The cupric geometry of blue copper proteins is not strained, *J. Mol. Biol* **261**, 586-596; c) Pierloot, K.; DeKerpel, J. O. A.; Ryde, U.; Roos, B. O. (1997) Theoretical study of the electronic spectrum of plastocyanin, *J. Am. Chem. Soc.* **119**, 218-226; d) Larsson, S.; Broo, A.; Sjölin, L. (1995) Connection between structure, electronic spectrum and electron-transfer properties of blue copper proteins, *J. Phys. Chem.* **99**, 4860-4865.
30. Dennison, C.; Vijgenboom, E.; Hagen, W. R., Canters, G. W. (1996) Loop-directed mutagenesis converts amicyanin from *Thiobacyllus versutus* into a novel blue copper protein, *J. Am. Chem. Soc.* **118**, 7406.
31. Buning, C.; Canters, G. W.; Comba, P.; Dennison, C. (1997), *work in progress*.
32. Comba, P.; Melter, M. (1997), *work in progress*.

EXTENDING MOLECULAR MECHANICS METHODS TO THE DESCRIPTIONS OF TRANSITION METAL COMPLEXES AND BOND-MAKING AND -BREAKING PROCESSES

CLARK R. LANDIS, TIMOTHY K. FIRMAN, THOMAS CLEVELAND, AND DANIEL M. ROOT
Department of Chemistry
University of Wisconsin
1101 W. University Ave.
Madison, WI 53706
United States of America

1. Introduction

The advent of rapid molecular modeling techniques has revolutionized the use of computers in chemical research. One result of the success of molecular modeling is the feasability of asking increasingly intimate questions about the potential energy surfaces for molecules of escalating complexity. These developments place pressure on the development of new empirical force field methods, particularly in the areas of transition metal modeling and the modeling of chemical reactions. In this paper we review the application of valence bond concepts to the qualitative understanding of molecular geometries of transition metal complexes. Furthermore, we demonstrate how valence bond concepts can be extended to a molecular mechanics description of bond-breaking and -making processes.

We believe that progress in these force field developments requires the reformulation of potential energy functions according to a more fundamental bonding theory. Recently, we have shown that valence bond concepts form the basis of a molecular mechanics model (VALBOND) for normal valent molecules of the p-block [1], hypervalent molecules of the p-block [2], and some simple transition metal hydrides and alkyls [3]. In this paper we focus on how qualitative valence bond (VB) theory forms a unifying basis for understanding molecular shapes across the periodic table and provides a theoretical form from which powerful new molecular mechanics potential energy functions can be derived.

L. Banci and P. Comba (eds.), Molecular Modeling and Dynamics of Bioinorganic Systems, 49–75.

2. Qualitative Valence Bond Theory

At the heart of the qualitative VB theory for describing molecular shapes is the concept of hybridization. Despite the simplicity of hybrid orbitals, schemes for the localization of electron density such as Weinhold's NBO method [4-10] demonstrate that hybridized localized bonds are good descriptors of electron density distributions derived from electronic structure calculations for many molecules. For normal and hypovalent molecules of the p-block, approximate hybridizations derive directly from Lewis structures. Adjustments in hybridization following Bent's rule [11] provides a mechanism for rationalizing structural deviations from idealized sp, sp^2, and sp^3 derived geometries. Historically, the geometries of transition metal complexes and hypervalent main group compounds have been rationalized through the use of sp^md^n hybridization [12,13]. However, analyses of electron density distributions from high level ab initio computations, such as Magnusson's definitive study on hypervalency [14], do not support the participation of d orbitals in the hybridization of main group elements. Analogously, we have presented evidence that p-orbitals do not significantly participate in the formation of covalent bonds at transition metals [3].

2.1. RULES FOR APPROXIMATING BOND HYBRIDIZATIONS

Let us begin by defining relevant terms and summarizing the rules for estimating bond hybridizations. The terms hypovalent (electron deficient), hypervalent (electron surplus), and normal valent (or electron precise) are controversial. For our discussion we use the terms as follows: For central atoms from the p-block, only the valence s and p orbitals are accessible. We will consider a molecule in which all four valence orbitals are doubly or singly occupied to be normal valent. Accordingly, both CH_4 and triplet CH_2 are considered normal valent. If not all valence orbitals are occupied, e.g., monomeric BH_3 and singlet CH_2, then the molecule is considered hypovalent. Hypervalent molecules [15,16] have Lewis structures that overfill the available valence orbitals (e.g., XeF_2 and ClF_2). By these criteria, non-radical molecules with a p-block central atom are hypervalent if the electron count exceeds 8 electrons and hypovalent if the count is less than 8 electrons.

For the transition elements engaged in covalent bonding, only the valence s and the five d orbitals participate significantly in bonding. Normal valent molecules occupy each of these six valence orbitals; examples include WMe_6 and WH_6. Hypovalent, or electron deficient, metal complexes include YH_3, ZrH_4, and $TaMe_5$. Most common transition metal complexes are hypervalent. Examples include $[WMe_7]^{1-}$, $[PtH_4]^{2-}$, and

$ReMe_6$. Thus, non-radical molecules with a d-block central atom are hypervalent if the electron count exceeds 12 electrons and hypovalent if the count is less than 12 electrons.

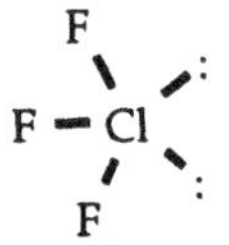
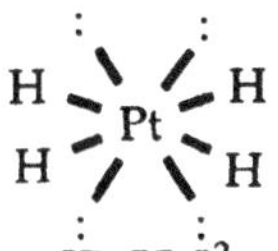
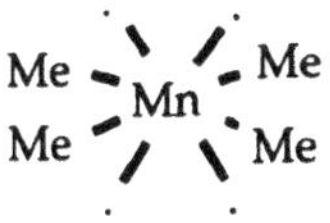

ClF_3	$[PtH_4]^{2-}$	$[MnMe_4]^-$
$10e^-$ singlet	$16e^-$ singlet	$12e^-$ quintet
5 orbitals filled	8 orbitals filled	8 orbitals filled

The following rules prescribe a simple scheme for counting electrons and assigning hybridizations.

Rule 1. The s-block and p-block elements form sp^n hybrids, whereas d-block elements form sd^n hybrids.

Here 'n' is defined as the number of occupied orbitals minus one. We will call such hybridizations the gross hybridization of the central atom. Gross central atom hybridizations of some normal valent and hypovalent molecules are: CH_4 (sp^3), triplet CH_2 (sp^3), BH_3 (sp^2), singlet CH_2 (sp^2), WMe_6 (sd^5), PtH_2 (sd^5), triplet WH_4 (sd^5), ZrH_4 (sd^3), TaH_5 (sd^4), and singlet WH_4 (sd^4).

Rule 2. For molecules with mixed ligands and/or multiple bonds, the distribution of p and d character among the hybrid orbitals depends on the relative electronegativities of the ligands (Bent's rule [11]) and the bond orders.

More specifically the following empirical corollaries yield good approximations to hybrid orbital preferences. (i) Lone pairs of the p-block elements prefer s-character. (ii) Lone pairs of the d-block elements have essentially pure d-character. (iii) Radical orbitals of the p-block elements have a greater preference for p-character than most ligand bonds. (iv) Radical orbitals of d-block elements have essentially pure d-character. (v) π-Bonds at p-block elements are pure p-character. (vi) Main group compounds follow Bent's rule.

Bent's rule states that "Atomic s character concentrates in orbitals directed toward electropositive substituents" [11]. Consider BF_2H. The gross hybridization at the boron is sp^2. According to Bent's rule this gross hybridization will distribute to place more s-character in the bond to the electropositive H ligand and, hence, more p-character in the B-F bonds. The geometry of BF_2H is consistent with approximately $sp^{2.1}$ hybridization in the B-F bonds and $sp^{1.8}$ hybridization in the B-H bond. The value of n in an sp^n hybrid represents the proportion of the squares of the p-coefficient to the s-coefficient in the hybrid orbital function. One interprets the $sp^{2.1}$ hybrid above as having 2.1 times as much p-character as it does s-character, or 68% p-character. Natural Bond Orbital (NBO) analyses of *ab initio* electron densities [8] consistently support the implications of Bent's rule.

Lone pairs can be considered as bonded to very electropositive ligands (a vacuum). In accordance with Bent's rule we expect high s-character lone pairs of main group atoms. For example, singlet CH_2 has a gross hybridization of sp^2. However the geometry is consistent with greater s-character ($sp^{0.580}$) in the lone pair relative to the C-H bonds ($sp^{4.445}$). For transition metals, lone pairs have no s-character and apparently violate Bent's rule [17]. For example, PtH_2 has gross hybridization of sd^5; use of four pure d-orbitals to accommodate the four lone pairs leaves the Pt-H bonds with sd hybridization. The apparent discrepancy between lone pairs on p-block and d-block elements can be traced to the relative energies of the valence orbitals; for the p-block the s-orbitals have lower one-electron energies than the p-orbitals whereas the transition metal d-orbitals generally lie at lower energies than those of the s-orbitals. Because lone pairs do not engage in bonding and experience large intra-pair repulsions, they adopt the character of the lower energy valence orbitals.

The hybridization preferences of main group radicals are subtle; usually they are predominantly p-orbital in character. For example, the geometry of triplet CH_2 is consistent with two $sp^{8.954}$ hybrids to the radicals and two $sp^{1.503}$ hybrids to the hydrogen atoms [1]. Similarly, the methyl radical has essentially pure p-character for the unpaired electron and sp^2 hybridization for the three C-H bonds. For transition metals, radicals prefer pure d-character to a good approximation. For example, triplet WH_4 has a gross hybridization of sd^5 with the two unpaired electrons occupying pure d-orbitals and with sd^3 hybridization of the four W-H bonds [18].

By symmetry, no s character can be used by π bonds. Therefore, π-bonds are pure p-character in main group elements and pure d-character in transition metals. For example, the carbons of ethylene have sp^3 gross hybridization. However, p-character is distributed such that one p-orbital is used for the π bond, while the three sigma bonds are approximately sp^2. We previously have published a detailed description of the application of these rules to main group elements, including a numeric algorithm for distributing hybridizations [19].

Rule 3. Strong ionic-covalent resonance stabilizes hypervalent centers; such resonance commonly maximizes at a linear arrangement of the electron pair bond and the ligand localized electron pair.

The simplest hypervalent molecule of the p-block is XeF_2. Because only four valence orbitals are available at Xe, a minimum first order VB description must include the two ionic resonance structures shown below.

$$F^{-} \quad {}^{+}Xe\text{—}F \longleftrightarrow F\text{—}Xe^{+} \quad F^{-}$$

2.2. VB THEORY AND THE VALBOND FORCE FIELD

Creating structures from the above rules requires knowledge of the shapes of the bond forming orbitals. We [1,2] recently have derived generalized hybrid orbital shape functions (Pauling referred to these as "strength functions") [20] for any arbitrary combination of valence s, p, and d atomic orbitals. Table 1 lists the angular preferences associated with common sp^n and sd^n hybridizations.

Unlike many simple qualitative bonding models, we have shown that VB-derived MM models not only survive demanding quantitative tests but uniquely permit the description of the complex shapes of hypervalent main group compounds and homoleptic transition metal alkyls and hydrides [1-3]. VALBOND is a rule-based MM force field that is generated from a user-sketched topology. From this topology, a Lewis structure is generated and "gross hybridizations" at the central atom (e.g., sp^3 for CF_2H_2) are derived. We have shown that these gross hybridizations can be refined by a generic "Bent's rule algorithm" (e.g. the C-F bonds of CF_2H_2 have greater p-character than the C-F bonds). From these hybridizations, the MM angle energy terms are set up using the generalized hybrid orbital strength functions shown below.

TABLE 1. Molecular Shapes Associated with Different Bond Hybridizations

Idealized Hybridization and Angles	Idealized Molecular Shape	Examples
sp^3 *109.5°*	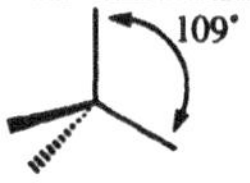 T_d	CH_4
sp^2 *120°*	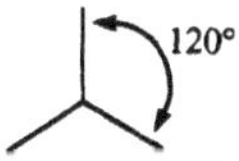 C_{3v}	BCl_3

sp *180°*	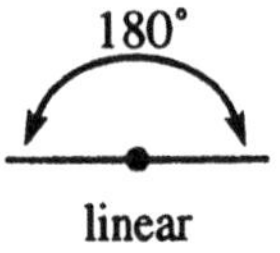	C_2H_2
sd^5 *117° and 63°*	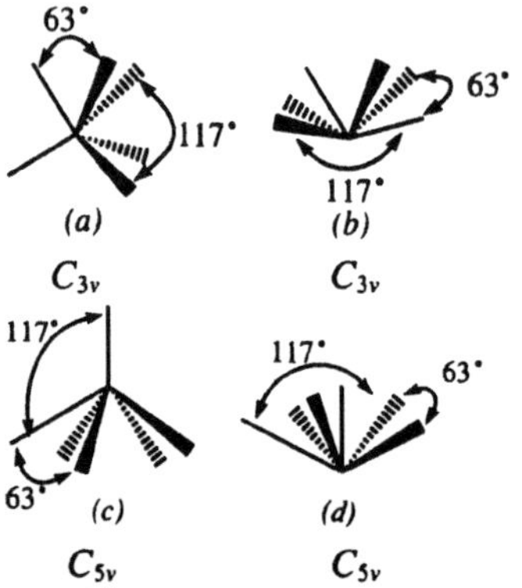	WH_6 [21,22], $[TcH_6]^+$ [3], $W(Me)_6$ [23,24]
sd^4 *114° and 66°*	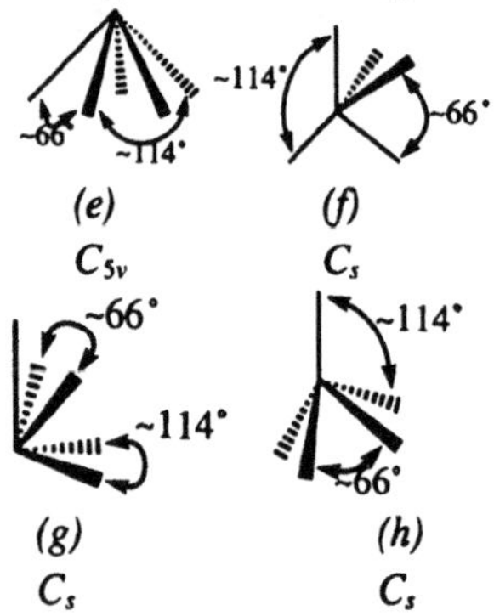	TaH_5 [3], ReH_5, $TaMe_5$ [25], $Ta(CH_2\text{-p-tol})_5$ [26]
sd^3 *109° and 71°*	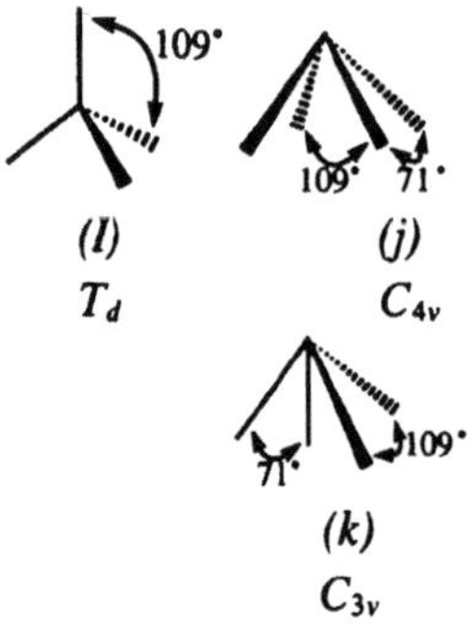	RuH_4, ZrH_4, $Os(Ph)_4$ [27]
sd^2 *90°*	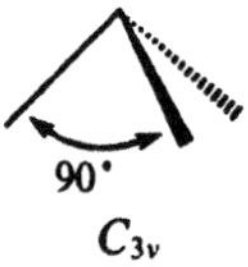	RhH_3 [3], $[ZrH_3]^+$ [3], $La(CH(SiMe_3)_2)_3$ [28], $Rh(Mes)_3$ [29]

sd	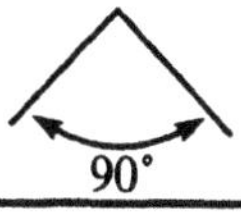	$Pt(Me)_2$, PtH_2 [30,31]
90°		

Consider two sp^md^n hybrids. For one orbital (hy_1) oriented along the z-axis, a second hybrid orbital hy_2 may be constructed in the xz plane that makes an angle α with the first (eq. 1 and 2). The Pauling-like strength functions can be expressed as a function of the interhybrid overlap integral, Δ, as shown in eq. 3. These functions lead to the VALBOND angle energy term of eqn. 6.

$$hy_1 = \sqrt{\frac{1}{1+m+n}}\left(1s + \sqrt{m}p_z + \sqrt{n}dz^2\right) \qquad (1)$$

$$hy_2 = \sqrt{\frac{1}{1+m+n}}\left(\begin{array}{c} 1s + \sqrt{m}\left(\cos\alpha p_z + \sin\alpha p_x\right) + \\ \sqrt{n}\left(\frac{1}{2}\left(3\cos^2\alpha - 1\right)dz^2 + \frac{\sqrt{3}}{2}\sin^2\alpha dx^2\text{-}y^2 + \sqrt{3}\sin\alpha\cos\alpha dxz\right)\end{array}\right) \qquad (2)$$

$$\Delta = \frac{1}{1+m+n}\left(1s + m\cos\alpha p_z + \frac{n}{2}\left(3\cos^2\alpha - 1\right)dz^2\right) \qquad (3)$$

$$S^{max} = \sqrt{\frac{1}{1+m+n}}\left(1 + \sqrt{3m} + \sqrt{5n}\right) \qquad (4)$$

$$S(\alpha) = S^{max}\sqrt{1 - \frac{1 + \sqrt{\left(1 - \Delta^2\right)}}{2}} \qquad (5)$$

$$E = k(S(\alpha) - S^{max}) \qquad (6)$$

Despite the use of rule-based parameterization, the VALBOND force field exhibits excellent accuracy and flexibility [1-3]. For example, bond angles and vibrational frequencies for a large selection of normal valent molecules from throughout the p-block are reproduced at MM3-level accuracies. Furthermore, the geometries of radical fragments (such as the methyl radical and singlet and triplet methylene) also are reproduced accurately without special parameterization.

In predicting the geometries of molecules containing lone pairs or radicals, we consider the hybridizations of only the bond-forming orbitals. This does not mean that lone pairs are stereochemically inactive; in the VALBOND model lone pairs and radicals affect geometries solely via their influence on the distribution of p and d character of the bond forming orbitals. For example, singlet CH_2 has C-H bond hybridizations of $sp^{4.443}$

[1]. These hybrid orbitals are orthogonal at about 103°, leading to the prediction of a 103° H-C-H bond angle.

One of the greatest successes of the qualitative VB theory outlined above has been its application to transition metal complexes for which covalent bonding dominates. We now examine some simple hybridizations associated with normal and hypovalent transition metal complexes and their geometric consequences.

2.2.1. *sd*^{5} *Bond Hybridization*

Hybridization considerations predict 12-electron WH_6 to adopt intriguingly complicated molecular shapes. Based on sd^5 bond hybridization, the best geometry for WH_6 is one in which all bond angles are either 63° or 117°. Four different molecular geometries accommodate the 63° and 117° H-W-H angular preferences; two of these have C_{3v} point group symmetry and two have C_{5v} (see Table 1). One may view these unusual structures as derived from an icosahedron. Picking any combination of six vertices such that no two vertices make a 180° angle generates the two C_{3v} and two C_{5v} structures. These structures have been previously analyzed by VALBOND [3].

Ab initio computations predict structures essentially identical to the hybridization predictions. Independent computations performed by Albright and coworkers [22,32] and by Schaefer and coworkers [21] lead to the conclusion that four local minima of similar energies exist. These local minima span the two C_{5v} and two C_{3v} geometries discussed above. They are not equal in energy, however. The two least crowded structures (Table 1, *a* and *c*) are preferred; they differ in energy by about 1 kcal/mol [22]. A combination of steric and electrostatic effects makes the other two less favorable, by about 8 kcal/mol for *b*, and 16 kcal/mol for *d* [22]. Using the VALBOND force field, we compute the following relative energies: *c* is lowest in energy, followed by *a*, which is less than 0.5 kcal/mol higher in energy; *b* is 4.0 kcal/mol higher in energy than *c*, and *d* is 5.0 kcal/mol higher.

Further support of the hybridization model has come from experimental and computational analyses of 12-electron $W(CH_3)_6$. One expects intermethyl steric effects to cause a preference for the least crowded of the four idealized minima. Indeed the crystallographic [24], electron diffraction [23], and ab initio results [33] for $W(CH_3)_6$ all are minor distortions of the less congested C_{3v} geometry, *a*.

None of the WH_6 minima corresponds to the octahedral VSEPR prediction. Indeed, ab initio computations suggest a cost of 142 kcal/mol for distortion to the octahedral geometry. Contrary to simple VSEPR ideas, ligand arrangements for stable structures of WH_6 need not be the most symmetric. Most dramatically, structures *b* and *d* place all of the ligands on one side of a plane normal to the principal symmetry axis that contains the tungsten.

Similar to the $W(CH_3)_6$ is the approximately trigonal prismatic, six-coordinate tris(o-xylidene)tungsten molecule [34]. The chelate C-W-C angles for this molecular compound are 74°, similar to the 73° angles of $W(CH_3)_6$, but the chelating ligands force the methylene hydrogens to align, increasing steric repulsion to distort the shape to trigonal prismatic. Distortion from an idealized C_{3v} geometry to a higher symmetry trigonal prism comes with little penalty; previous computations suggest a 3 kcal/mol barrier for distortion of $W(CH_3)_6$ through the trigonal prismatic transition state. Kaupp's recent paper finds a 6 kcal/mol barrier for the distortion [33].

2.2.2. *sd⁴ Bond Hybridization*

The sd^4 hybridization of 10-electron TaH_5 leads to idealized H-Ta-H bond angles of 66° and 114°. Four local minima have been found using VALBOND computations; two (*f* and *h*) belong to the C_s point group and are quite close to square pyramidal [3]. The remaining idealized structures are a pentagonal pyramid *e* (C_{5v}) and a distorted pentagonal pyramid *g* (C_s). Our Density Functional Theory (DFT) computations [35] led to at least one local minimum (there may be others that we have failed to locate) at a C_{4v} capped square pyramid minimum with bond angles of 77°, 118°, and 123°. Neither the ab initio structure nor any of the VALBOND structures resemble the VSEPR predicted trigonal bipyramidal structure.

2.2.3. *sd³ Bond Hybridization*

For sd^3 hybridization, the ideal angles are 71° and 109°. In addition to the tetrahedron three other structures are consistent with these angles. These structures have C_{4v} or C_{2v} point group symmetry. All of the idealized structures can be derived from arranging four ligands at the vertices of a cube such that no two ligands make a 180° bond angle. Consideration of interligand steric repulsion favors the tetrahedron as the most stable minimum. By the rules presented above, both 8-electron ZrH_4 and 12-electron RuH_4 have sd^3 bond hybrids. For RuH_4, two pairs of electrons are accommodated in pure d lone pairs. In addition to the tetrahedral geometry of RuH_4 a C_{4v} square pyramidal minimum *j*, with bond angles of 68° and 105°, was located and found to be lower in energy than the tetrahedral local minimum by 8.5 kcal/mol.

2.2.4. *sd² Bond Hybridization*

For both 6-electron $[ZrH_3]^+$ and 12-electron RhH_3, qualitative valence bond theory predicts sd^2 bond hybridization and a trigonal pyramidal structure with 90° bond angles. RhH_3 has three lone pairs of pure d-hybridization. A particularly interesting contrast with VSEPR results occurs in $[ZrH_3]^+$: despite the absence of lone pairs, a decidedly nonplanar structure is predicted. VSEPR theory predicts a trigonal planar structure. Our DFT calculations predict a trigonal pyramidal structure with 103° bond angles. $[ZrH_3]^+$ is used instead of YH_3 to lower the amount of hydridic character; ionic bonding will be discussed later in this paper. For RhH_3, our DFT results indicate a pyramidal structure with H-Rh-H angles of 84°.

2.2.5. *sd Bond Hybridization*

Like sd^2 bond hybridization, 90° degree angles are predicted for sd bond hybrids. Thus, both the 4-electron $[YH_2]^+$ and the 12-electron PtH_2 are predicted to exhibit strongly bent structures. DFT computations of both molecules exhibit bond angles near 90°: for PtH_2 the bond angle is 83°, and for $[YH_2]^+$ the bond angle is 103°. These results agree with the GVB computations of Low and Goddard [30,31] for PtH_2 which yielded an 82.5° angle. Similar GVB computations for $Pt(CH_3)_2$ yielded a 98° C-Pt-C bond angle. To the best of our knowledge, there are no experimental determinations of the structures of normal or hypovalent hydride or alkyl molecules containing sd bond hybridization.

2.3. WHY DOES HYBRIDIZATION WORK?

One test of simple qualitative models of bonding is consistency with more exact analyses of electron density distributions. We have relied primarily on Weinhold and coworkers' Natural Bond Orbital (NBO) analysis [10] as a general method for casting the results of a variety of types of electronic structure calculations into localized bonding terms. As shown in Table 2, simple transition metals are described well by Lewis structures involving covalent bonds and lone pairs. The average hybridizations of various transition metal complexes determined by NBO exhibit remarkable consistency with the simple rules stated here. For example, each metal forms sd^{n-1} hybrids where n is the sum of the bonds, lone pairs, and radicals; practically no p-character is used in these hybrids. We also find that transition metal-hydrogen bonds generally are covalent, although more significant departures from pure covalency are observed for the polar covalent bonds of the electropositive early transition metal hydrides.

TABLE 2. DFT(B3LYP) and NBO Results for Normal and Hypovalent Metal Hydrides

Compound (Relative Energy)	Optimized Geometry	NBO Bond Hybridizations and (Polarities)
$[YH_2]^+$ (C_{2v})	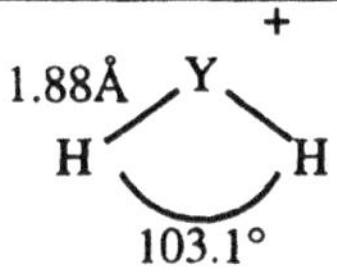	Y-H: $sd^{2.27}p^{0.03}$ (27% Y: 73% H)
PtH_2 (C_{2v})	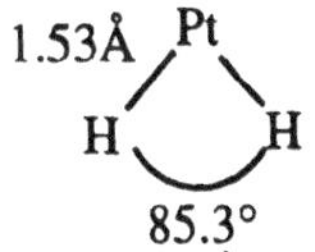	Pt-H: $sd^{1.25}p^{0.0}$ (55% Pt: 45% H)
ZrH_3^+ (C_{3v})	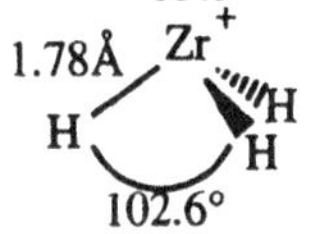	Zr-H: $sd^{2.45}p^{0.01}$ (38% Zr: 62% H)
RhH_3 (C_{3v})	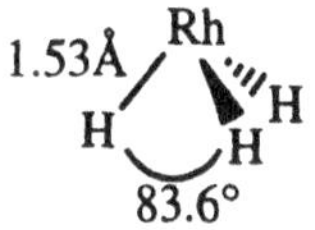	Rh-H: $sd^{2.00}p^{0.01}$ (54% Rh: 46% H)
ZrH_4 (T_d)	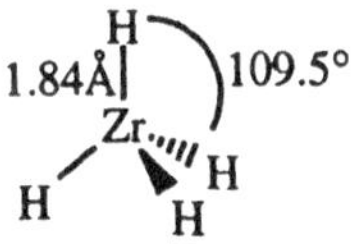	Zr-H: $sd^{2.97}p^{0.03}$ (32% Zr: 68% H)
RuH_4 (C_{4v}) *Rel. Energy:* *0 kcal/mol*	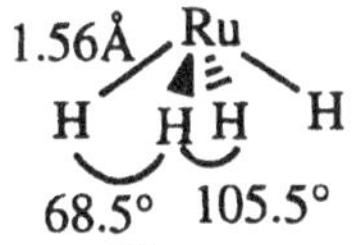	Ru-H: $sd^{2.99}p^{0.01}$ (55% Ru: 45% H)
RuH_4 (T_d) *Rel. Energy:* *+8.4 kcal/mol*	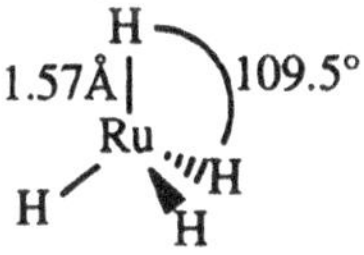	Ru-H: $sd^{3.00}p^{0.00}$ (53% Ru: 47% H)
TaH_5 (C_{4v})	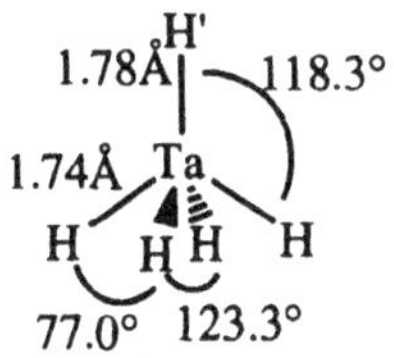	Ta-H: $sd^{4.11}p^{0.04}$ (39% Ta: 61% H') Ta-H': $sd^{3.47}p^{0.02}$ (34% Ta: 66% H) *Avg.:* $sd^{3.65}p^{0.04}$
ReH_5 (C_{5v}) *Rel. Energy:* *0 kcal/mol*	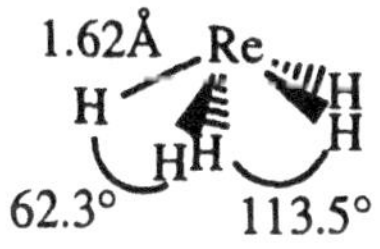	Re-H: $sd^{4.18}p^{0.01}$ (52% Re: 48% H)

ReH_5 (C_s) Rel. Energy: +1.2 kcal/mol	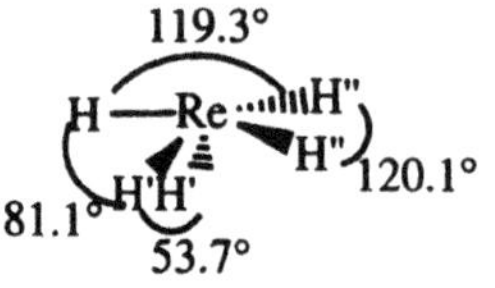 H'-Re-H'': 66.8°, 112.9° Re-H: 1.63Å Re-H': 1.61Å Re-H'': 1.65Å	Re-H: $sd^{3.34}p^{0.02}$ (48% Re: 52% H) Re-H': $sd^{4.26}p^{0.02}$ (48% Re: 52% H') Re-H'': $sd^{4.14}p^{0.01}$ (54% Re: 46% H'') *Avg.:* $sd^{4.02}p^{0.02}$
WH_6 (C_{3v}, *a*)	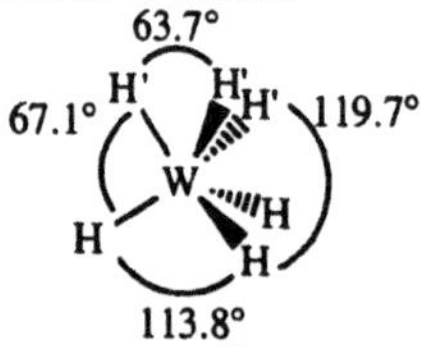 W-H: 1.71Å;W-H': 1.67Å	W-H: $sd^{4.08}p^{0.01}$ (43% W: 57% H) W-H': $sd^{6.27}p^{0.02}$ (50% W: 50% H') *Avg.:* $sd^{4.88}p^{0.02}$
WH_6 (C_{3v}, *b*)	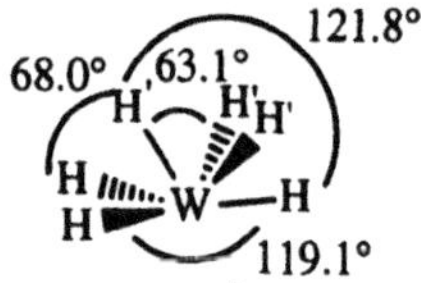 W-H: 1.70Å;W-H': 1.65Å	W-H: $sd^{4.48}p^{0.02}$ (44% W: 56% H) W-H': $sd^{5.58}p^{0.01}$ (53% W: 47% H') *Avg.:* $sd^{4.98}p^{0.02}$
WH_6 (C_{5v}, *c*)	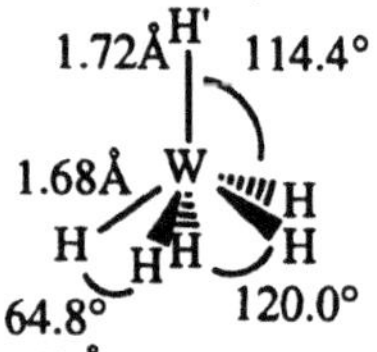	W-H: $sd^{5.43}p^{0.01}$ (48% W: 52% H) W-H': $sd^{3.45}p^{0.01}$ (41% W: 59% H') *Avg.:* $sd^{4.99}p^{0.01}$
WH_6 (C_{5v}, *d*)	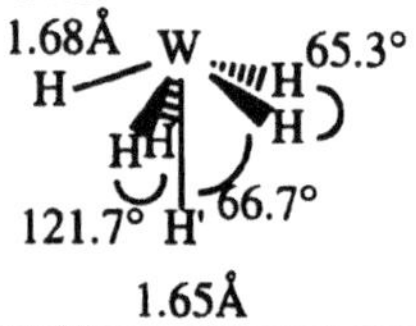	W-H: $sd^{5.10}p^{0.02}$ (49% W: 51% H) W-H': $sd^{4.45}p^{0.02}$ (58% W: 42% H') *Avg.:* $sd^{4.98}p^{0.02}$

As we have pointed out before [2], the hybridizations determined by NBO analysis and those used in VALBOND computations are not based on the same criteria. The hybridizations used in this qualitative VB theory represent idealized hybridizations that are derived inductively from observed geometries. In contrast, NBO hybridizations are based on manipulating density matrices in order to obtain maximum occupancy localized bond orbitals. As a result, NBO hybridizations for inequivalent M-H bonds are not equivalent. For example, NBO analysis of WH_6 in the preferred C_{3v} geometry yields two sets of W-H hybrids corresponding to the two sets of symmetry equivalent W-H bonds. One set has $sd^{6.27}p^{0.02}$ hybridization and the other $sd^{4.08}p^{0.01}$. We note the interpretation of NBO derived p-orbital populations has been questioned by Morokuma [36]. In contrast, our rules describe WH_6 as having six sd^5 bond orbitals. Nonetheless, the average hybridization of $sd^{4.98}p^{0.02}$ closely corresponds to sd^5. The general qualitative agreement between the two methods provides strong support for the simple rules of our qualitative VB method.

A less sophisticated method for examining bond orbital hybridization is Mulliken analysis of GVB bond pair orbitals. Although GVB bond orbitals are not completely localized and the interpretation of Mulliken populations can be misleading [37], Goddard and coworkers' [38] examinations of $[MH]^+$ support our conclusion that M-H bonds have little p character and extensive s-d hybridization.

3. Qualitative VB Theory and Hypervalent Molecules

Hypervalent molecules have Lewis structures with greater than four occupied valence orbitals for molecules of the p-block or greater than six occupied orbitals for molecules of the d block. Because the qualitative VB model restricts the number of bond-forming orbitals to four and six for p-block and d-block elements, respectively, ionic resonance configurations

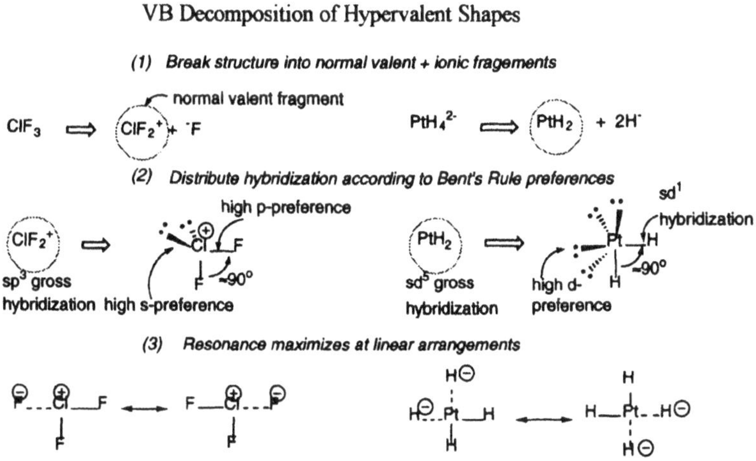

Fig. 1. A valence bond treatment of ClF_3 and PtH_4^{2-}.

must be invoked. To a good approximation each ionic configuration consists of a valence core fragment with all orbitals occupied and the "excess" electron pairs localized on ligands. The resonance of covalent and ionic structures was first used to describe the trans effect in square planar platinum structures by Yakshin [39] and Syrkin [40] in the 1940's.

Consider two simple examples from the p- and d-blocks. The 10-electron ClF_3 and 16-electron $[PtH_4]^{2-}$ each exceed the available orbital space (Fig. 1). The primary resonance configurations have the following forms: $[ClF_2]^+$ F^- and PtH_2 $2H^-$. The "core fragments" are $[ClF_2]^+$ and PtH_2, respectively. ClF_3 has three ionic configurations, corresponding to placement of the excess electron pair on each of the three ligands, whereas $[PtH_4]^{2-}$ has six unique configurations corresponding to PtH_2 $2H^-$. Because ClF_3 requires just one ligand-localized electron pair we consider it (as well as XeF_2, PF_5, and SF_4) to be "singly hypervalent". By analogy, $[PtH_4]^{2-}$, XeF_4, and SF_6, all of which have two excess electron pairs, are "doubly hypervalent".

The most stable geometry is that which maintains idealized bond angles for the filled valence core fragment while maximizing resonance stabilization among the ionic configurations. In general, maximum resonance occurs when the ligands bearing the excess electron pairs are diametrically opposed in the two configurations. The core fragments, $[ClF_2]^+$ and PtH_2, are bent to approximately 90°. For $[ClF_2]^+$ qualitative VB considerations suggest that the sp^3 gross hybridization should be distributed to give high s-character to the lone pairs and high p-character to the Cl-F bond hybrids. As a result of the high p-character of the bonds, $[ClF_2]^+$ should be bent to less than 109° in agreement with the DFT (B3LYP) computed value of 103°. Placing the ligands bearing the excess electron pair opposite one of the core fragment bonds yields the T-shape for ClF_3 and a square planar $[PtH_4]^{2-}$.

There are cases which are not adequately described by such linear resonances. A molecule with fewer covalent interactions than it has hypervalent, resonating ligands must form different shapes to accommodate the additional ligands. For example, 16-electron $Pt(PR_3)_3$ complexes are trigonal planar. There is a single covalent bond possible for the platinum, and it resonates with a lone pair on each of the other two ligands.

We previously have published the results of VALBOND molecular mechanics computations on hypervalent molecules of the main group. In this paper we summarize the application of these concepts to hypervalent transition metal complexes (see Table 3) and highlight the close structural analogies between hypervalent molecules of the p- and d-blocks.

TABLE 3. Comparison of p- and d- Block Hypervalent Centers

Shape	Dominant Resonance Structures	Examples
Linear *main group*	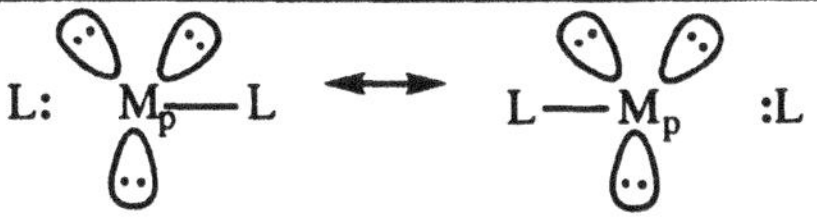	XeF_2 [41]
transition group	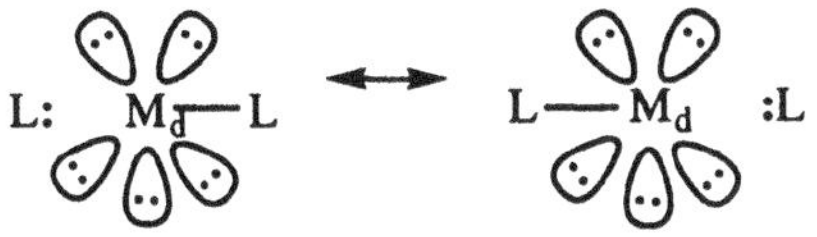	$[Au(CH_2PPh_3)_2]^+$ [42]
T-Shape *main group*	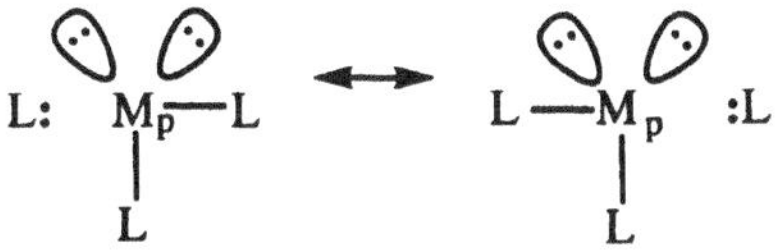	ClF_3 [41]
transition group	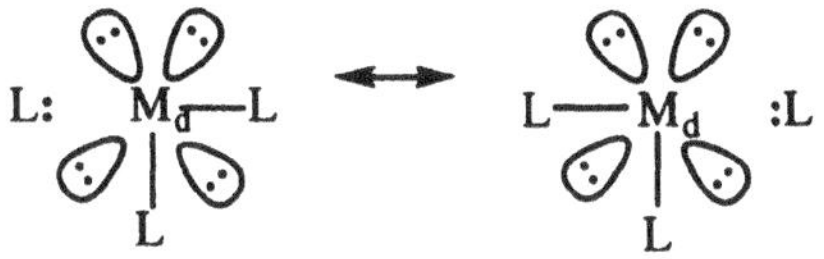	$[Rh(PPh_3)_3]^+$ [43]
See-Saw *main group*	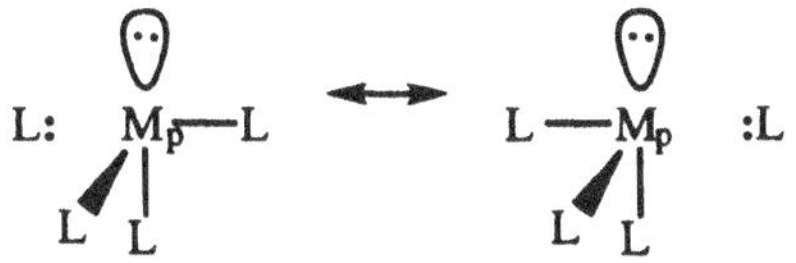	SF_4 [41]
transition group	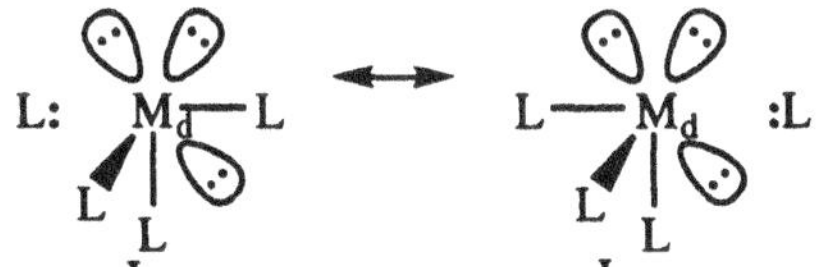	$[Fe(napthyl)_4]^{2-}$ [44]
Square Plane *main group*	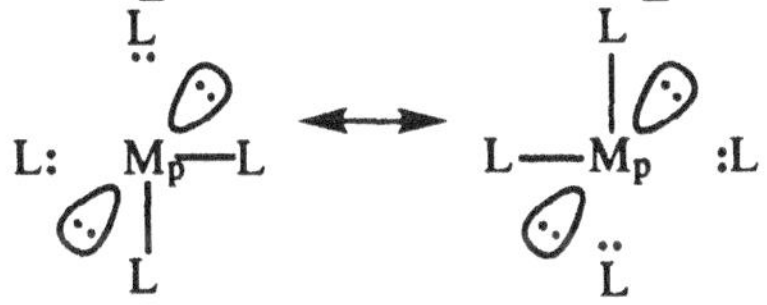	XeF_4 [41]
transition group	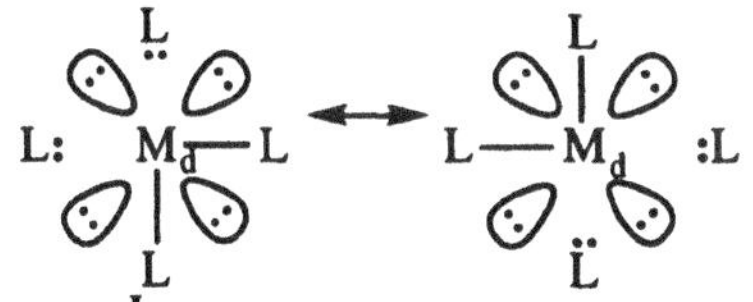	$[PtH_4]^{2-}$ [45]
Mono-vacant Octahedron *main group*	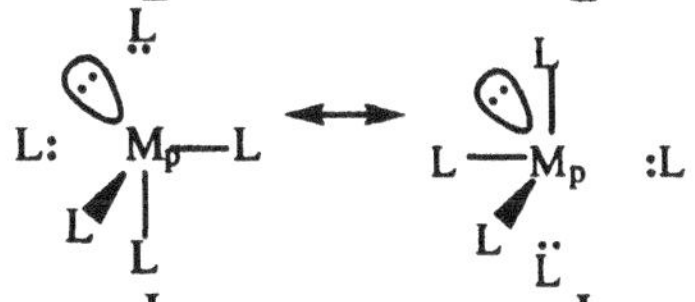	BrF_5 [41]
transition group	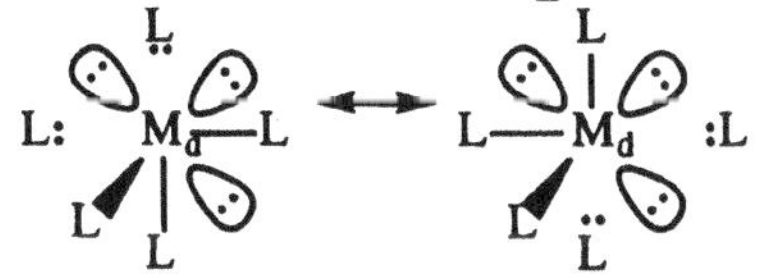	$[Rh(C_6F_5)_5]^{2-}$ [46]

Octahedron *main group*	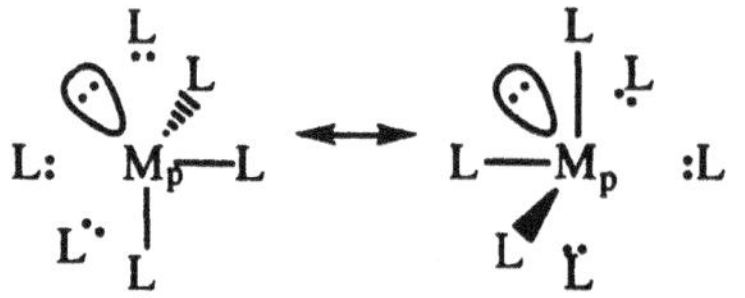	XeF_6 [41]
transition group	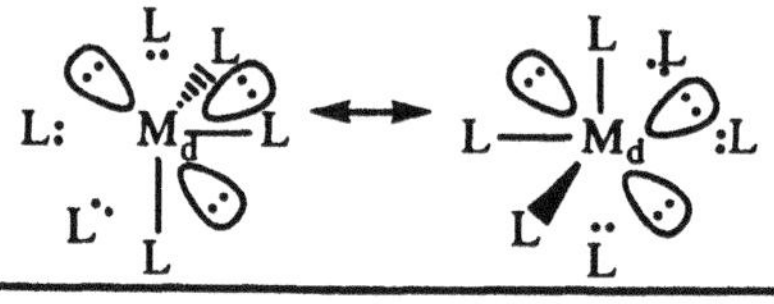	$[FeH_6]^{4-}$ [47]

3.1. HYPERVALENT MOLECULES WITH sd BOND HYBRIDIZATION: $[PtH_3]^-$, $[PtH_4]^{2-}$, AND $[Mn(CH_3)_4]^-$

As described above, a singly hypervalent molecule with sd bond hybridization such as $[PtH_3]^-$ forms a T-shape, similar to that of the main group compound ClF_3 [2]. The sd core fragment prefers a 90° bond angle, and the remaining ligand will be 180° from one of the bonds, maximizing hypervalent resonance (Table 3). We are unaware of any alkyl or hydride crystal structure which is analogous in electron count and number of ligands to $[PtH_3]^-$; the isoelectronic $[Rh(PPh_3)_3]^+$ is an analogue and has a T-shaped geometry [43].

Doubly hypervalent molecules with sd bond hybridization are common and span both high- and low-spin metal complexes. Some crystallographically characterized examples include: 12-electron $[MnMe_4]^-$, which has four unpaired electrons and a doubly hypervalent sd hybridized structure [48]; the 15-electron complexes $[Ir(C_6Cl_5)_4]^{2-}$ [46,49] and $[Pt(C_6Cl_5)_4]^-$ [50], with one unpaired electron; and the 16-electron, diamagnetic $[Pt(C_6Cl_5)_4]^{2-}$ [50], $[PtH_4]^{2-}$ [45], $[AuPh_4]^-$ [51], $[Au(C_6F_5)_4]^-$ [52], $[Ag(CF_3)_4]^-$ [53], and $[Cu(CF_3)_4]^-$ [54]. Maximum resonance stabilization occurs by placing each of the two ligands, bearing excess electron pairs, opposite metal-ligand bonds in the core fragments. As expected, each of these molecules adopts an approximately square-planar crystallographic structure (Table 3). Thus, VB theory provides an alternative to crystal field or ligand field theory rationalizations for the common square-planar geometry in 16-electron d^8 complexes. We note that VB theory does not predict the spin state of these complexes. Rather, VB theory leads to a structure after assuming a particular number of unpaired electrons. A particularly appealing aspect of the VB analysis of hypervalent molecules is the strong interconnection made between the bonding of square-planar, doubly hypervalent molecules of the d-block and those of the p-block such as XeF_4. Indeed, virtually every simple hypervalent main group compound has a transition metal cousin, related by both structure and electron count.

3.2. HYPERVALENT MOLECULES WITH sd^2 BOND HYBRIDIZATION: $[RhH_4]^-$, $[Rh(C_6F_5)_5]^{2-}$, AND $[FeH_6]^{4-}$

As described in the previous section, an sd^2 bond hybridization leads to trigonal pyramidal structure with angles of roughly 90°. Building upon this core fragment by the placement of ligands bearing excess electron pairs opposite to the core bonds yields the molecular geometry. A singly hypervalent molecule with sd^2 bond hybridization will generate a 'see-saw' geometry analogous to SF_4. For example, ab initio computations on the crystallographic structure of 14-electron $[Fe(nap)_4]^{2-}$ (nap = naphthyl) and the model compound 14-electron $[RhH_4]^-$ both exhibit this 'see-saw' shape, although the wide angle of the iron compound is only 118°, presumably due to steric repulsions among the very bulky naphthyl ligands [44]. The DFT minimum for $[RhH_4]^-$ is a very clear see-saw geometry, with H-Rh-H angles of 178°, 89°, and 85°.

Doubly hypervalent molecules with sd^2 bond hybridization form mono-vacant octahedra (square pyramidal structures with bond angles of approximately 90° and 180°), in analogy with the doubly hypervalent, five-coordinate molecules of the p-block such as 12-electron BrF_5. A crystallographically characterized example that contains no unpaired electrons is 16-electron $[Rh(C_6F_5)_5]^{2-}$ [49]. The 13-electron complex $[CrPh_5]^{2-}$ [55], for which the core fragment contains three unpaired electrons and three electron-pair bonds, also exhibits a mono-vacant octahedral structure.

Triply hypervalent molecules based on sd^2 hybridization form the familiar octahedral shape. This structure results from three 3c-4e$^-$ bonds, all at right angles to each other. XeF_6, which is slightly distorted from octahedral geometry, is a main group analogue. Crystallographically characterized 18-electron hydride and methyl compounds forming this shape include $[FeH_6]^{4-}$ [47], $[RhMe_6]^{3-}$ [56], and $[IrMe_6]^{3-}$ [56]. Open-shell 15-electron analogues that have been crystallographically characterized include $[CrMe_6]^{3-}$ [57] and $[MnMe_6]^{2-}$ [58].

3.3. HYPERVALENT MOLECULES WITH sd^4 BOND HYBRIDIZATION: $Re(CH_3)_6$ AND ReH_6

The simplest hypervalent molecule with an sd^4 bond hybridization is the 13-electron ReH_6 radical. With an 11-electron $[ReH_5]^+$ core fragment comprising five sd^4 bond hybrids and one pure d radical, purely qualitative considerations do not define the expected structure. As mentioned previously for 10-electron TaH_5, there is no geometry in which the idealized angles, corresponding to the core fragment sd^4 bond hybridization, can be achieved. The addition of one trans ligand arrangement for the hypervalent molecule exacerbates the problem. Qualitatively, one anticipates a structure with acute bond angles near 66° and obtuse angles opened from 114° due to maximization of resonance. DFT computations lead to the two minima (there may be others) shown below.

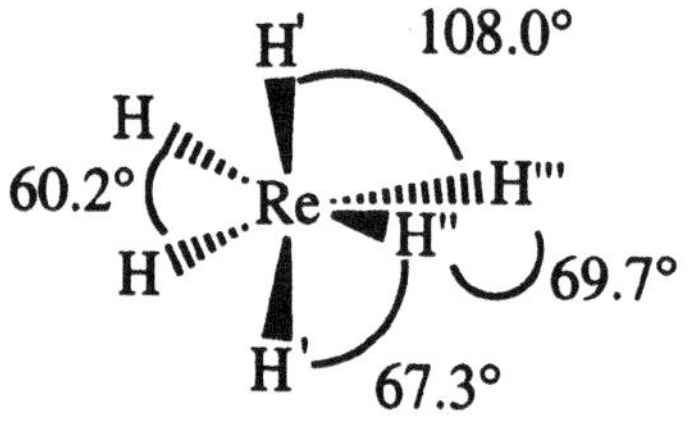

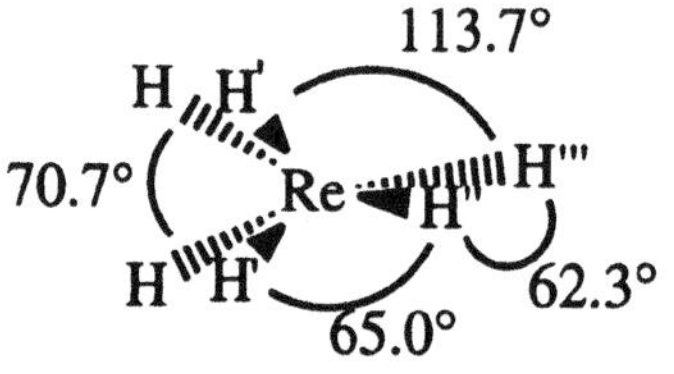

H'-Re-H': 104.7°
H-Re-H': 67.6°, 114.5°
H-Re-H'': 133.4°
H-Re-H''':137.0°
Re-H: 1.666 Å
Re-H': 1.614 Å
Re-H'': 1.635 Å
Re-H''': 1.686 Å
+0.2 kcal/mol

H'-Re-H': 74.0°
H-Re-H': 67.5°, 108.3°
H-Re-H'': 131.9°
H-Re-H''':136.5°
Re-H: 1.660 Å
Re-H': 1.606 Å
Re-H'': 1.640 Å
Re-H''': 1.697 Å
+0 kcal/mol

The primary difference between the two structures is the H'-Re-H' angle, which is obtuse in the left structure and acute in the right one. Although no experimental data for ReH_6 exists, Seppelt has reported the X-ray crystallographic structure of $Re(CH_3)_6$ [24]. This structure exhibits a geometry that is a slightly distorted trigonal prism with C-Re-C bond angles of ~86° and ~80° between C_3 symmetry related carbons.

3.4. HYPERVALENT MOLECULES WITH sd^5 BOND HYBRIDIZATION: $[WH_7]^-$ AND $[ReH_9]^{2-}$

Approximate structures for the sd^5 hybridized hypervalent compound $[WH_7]^-$ can be constructed from the normal valent WH_6 core. $[WH_7]^-$ is a singlet with a 14e$^-$ formal electron count, and so is singly hypervalent. Addition of H^- to either of the C_{5v} structures (*c* or *d* from Table 1) should lead to a strainless seven coordinate C_{5v} structure. In contrast, addition of H^- along the C_3 axis of the C_{3v} structures (*a* and *b* from Table 1) leads to C_{3v} structures that lack a linear hypervalent resonance stabilization. Thus, we anticipate a small amount of electronic strain in the C_{3v} structures.

We have found two DFT (B3LYP) geometries that are true minima; these correspond to the two C_{3v} structures described above. These two minima are very close in estimated energy, differing by less than 1 kcal/mol, so we are not certain which is the global minimum. In cases with large numbers of ligands there may be no single structure which can accommodate the idealized angles of the core fragment and the preferred linear arrangements of the hypervalent bonding components. In such cases of frustrated molecules, multiple resonance structures will be populated such that the lowest total energy is achieved. Interestingly, for $[WH_7]^-$ the "strainless" C_{5v} geometry is 9 kcal/mol higher in energy than the C_{3v} structures and is not a local minimum. Presumably, this is due to adverse steric and electrostatic interactions between the pentagonal array of pseudo-equatorial H's and the proximal axial H, as was found for WH_6 structure *d*.

Another example of a hypervalent sd^5 compound is the famous 18-electron nonahydride, $[ReH_9]^{2-}$ [59], which has D_{3h} symmetry. By our analysis this molecule is another example of a frustrated molecule. The core sd^5 fragment is roughly analogous to WH_6 structure *b*. (see below and Table 1) The bond angles are close to idealized for this fragment; 70° and 73° acute angles and 120° and 137° obtuse. However, this molecule cannot achieve idealized linear hypervalent resonance stabilization without engendering strain: placing the charge-bearing hydrogens trans to the acute trigonal cluster of H's would create unfavorable 47° bond angles between charge-bearing H's and the obtuse trigonal cluster.

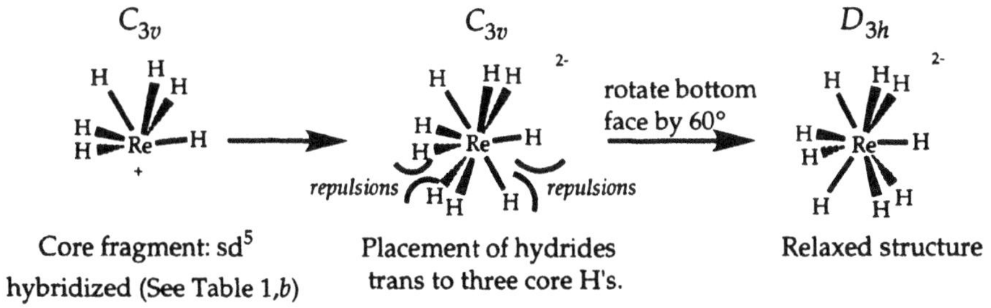

As a result, a more complicated mixture of resonance structures is populated. Like $[WH_7]^-$, $[ReH_9]^{2-}$ achieves maximum resonance without having linear bond angles. For ReH_9^{2-}, a low-energy distortion from a C_{3v} to D_{3h} geometry relieves the interligand repulsions with minimal disruption of bond hybridizations and resonance.

4. Force Field Description of Bond-Breaking and Bond-Making

Our work on modeling bond-making/-breaking processes begins with the simplest process: homolytic cleavage. Consider homolytic cleavage of methane to $\bullet CH_3 + \bullet H$. As a C-H bond is broken, geometry smoothly changes from a tetrahedron to a trigonal plane. This requires that the MM hybridizations smoothly change as a function of C-H bond order for an accurate description of the bond breaking process. We achieve this by dynamically mutating from a CH_4 MM topology to that of $\bullet CH_3 + \bullet H$.

The bond mutation approach makes use of the bond order concepts embodied in Parr and Johnston's BEBO method [60-63] to effect a continuous transformation between the CH_4 and $\bullet CH_3 + \bullet H$ potential energy formulations. The order of a bond at distance r is defined as the ratio of the extended Rydberg potential at r with that at the equilibrium distance, r_0, ($BO(r) = E^{Rydberg}(r) / E^{Rydberg}(r_0)$). In essence, the user enters a combined topology (shown below) which contains 4 C-H bonds and a bond (C-Rad) between C and a radical pseudoatom. A C-H bond and the C-Rad bond are identified as a mutating pair. As the C-H bond is stretched its bond order decreases and the C-Rad bond order increases such that the net bond order is conserved. The hybridizations associated with the C-H/C-Rad mutating bond pair are identical but evolve smoothly from sp^3 at the methane geometry to pure p at the methyl radical geometry. We thus create a very accurate potential energy surface [64] for a bond dissociation reaction as illustrated in Fig. 2.

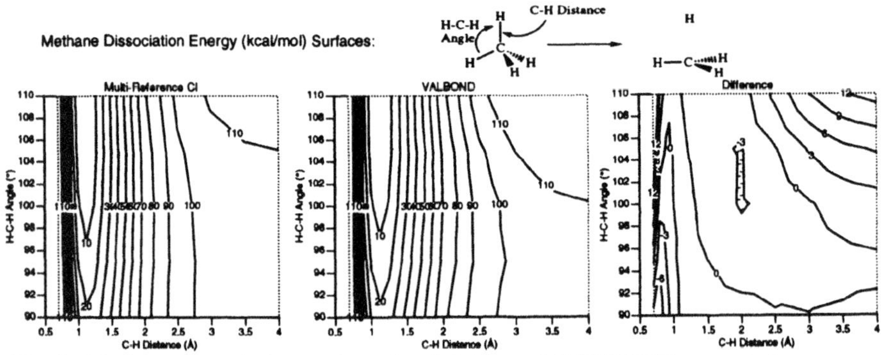

Fig. 2. Methane dissociation surfaces by ab initio and VALBOND computations

The bond mutation process is general. Consider (1) homolytic cleavage of WH_6 to yield $\bullet H + \bullet WH_5$ and (2) heterolytic cleavage of the Lewis acid-base adduct, $H_3P\text{-}BH_3$. As a W-H bond is stretched the hybridization of the intact W-H bonds change from sd^5 to sd^4 and the molecular shape smoothly transforms from the distorted trigonal prismatic (C_{3v}) to C_s geometries in accordance with ab initio [18] (GVB level) predictions (Fig. 3).

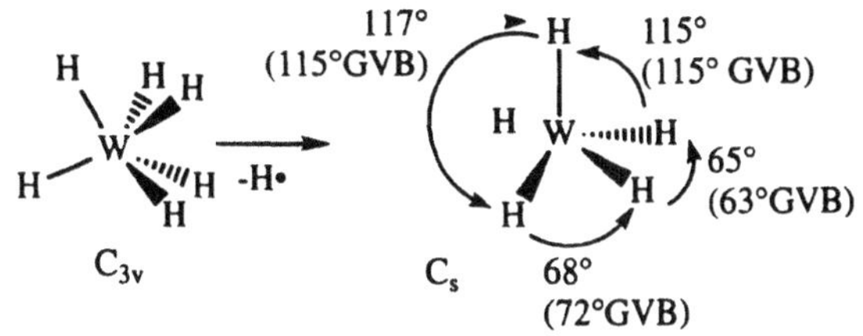

Fig. 3. Computed structural changes upon W-H dissociation.

For the H_3P-BH_3 adduct the mutation involves the P-B mutating with H_3P: and BH_3. As shown in Fig. 4, our molecular mechanics model [64] smoothly changes between the correct limiting geometries in a fashion that is consistent with ab initio computations.

Geometric Changes upon Cleaving the H_3P-BH_3 Bond

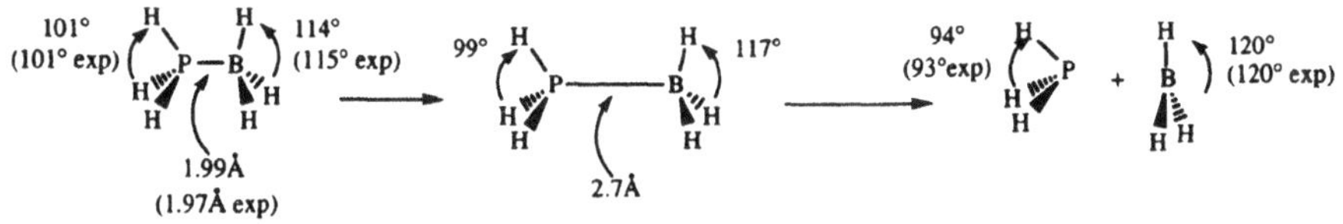

Fig. 4. Computed changes upon heterolytic cleavage of the P-B bond.

4.1. DESCRIBING BREAKABLE MULTIPLE BONDS IN A VALENCE BOND FORCE FIELD

Many of the most important processes in modern chemistry involve reactions of alkenes and other unsaturated substrates. Whether the reaction occurs in the absence of catalysts (e.g., Bergman cyclization, Cope rearrangement, Diels-Alder reaction, etc.) or the presence of catalysts (alkene hydrogenation, isomerization, polymerization, hydroformylation, metathesis, etc.) a challenge to modern force field developments is the generalized description of multiple bonds as they undergo bonding changes. Our previous VALBOND work demonstrated that a VB-perspective leads to good descriptions of geometries and vibrational frequencies of simple molecules. However, we have found deficiencies in these descriptions of unsaturated compounds, primarily the PES for bending motions of alkenes and alkynes are too shallow. Furthermore, bond-making and bond-breaking descriptions of unsaturated compounds require an improved MM description of π-bonding.

Consider ethene. A correct global potential energy surface must yield (1) a planar equilibrium geometry with a 1.34 Å C-C bond length and a 117° H-C-H bond angle (2) a stretched C-C bond length (1.48 Å) for ethene in which the planes of the two methenes are twisted to perpendicularity (3) a reduced barrier to rotation about the C-C linkage as the C-C bond length increases and (4) increase to a ca. 134° H-C-H bond angle as the C-C linkage cleaves to yield two triplet methenes. Our VB-consistent representation contains four C-H bonds, one C-C s bond, and a C-C π bond. The σ and π bonds have different equilibrium bond lengths (1.47 Å vs. 0 Å, respectively), bond dissociation energies (98 kcal/mol vs 600 kcal/mol), and different hybridization preferences (ca. sp^2 vs. pure p). For the dissociation of ethene into two triplet methylenes, as the C-C distance increases both the σ and π-bond mutate into two C centered radicals but at different rates. The π bond energy depends on both distance and the π-C-C-π dihedral angle: $E(r,\phi) = E_{\pi}^{Rydberg} * \cos^2 \phi_{\pi-C-C-\pi}$. Aspects of the resulting [64] global ethene potential energy surface are illustrated in Fig. 5. Note that our MM method closely reproduces all aspects of the GVB ab initio surface: as the C=C bond is twisted from $\phi=0°$ to $\phi=90^0$ the C-C bond stretches from its normal double bond length (1.34 Å) to a single bond length (1.48 Å); extension of the C-C bond results in a smooth decrease of the C-C rotation barrier; breaking of the C=C bond requires ca. 160 kcal/mol; the H-C-H bond angles change smoothly from 117° of ethene to 134° of triplet methene. As with all of our MM developments, analytic derivatives are carried through second derivatives and all surfaces are continuous.

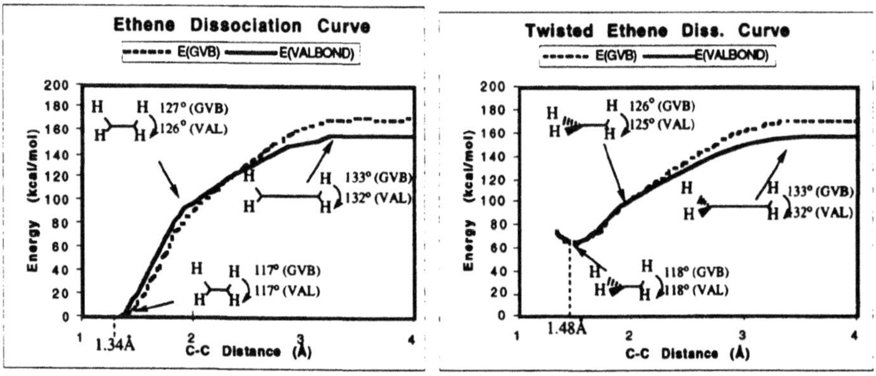

Fig. 5. Computed ethene dissociation profiles.

The ethene example readily carries over to other multiple bonds. For example, ethyne yields a linear molecule with a 1.24Å bond length. The changes in H-C-C bond angles and C-C distance of ethyne relative to ethene arises from the difference in the number of π-bonds, only; all parameters used in the computations are identical. We currently are exploring the extension of these simple yet powerful methods to all types of unsaturated compounds, including metal-carbon π-bonds.

5. Summary

Although the extension of force field methods to transition metal complexes and to bond-making and bond-breaking is fraught with difficulties, significant advancements result from the incorporation of valence bond perspectives into the formulation of the force field potential energy functions. Perhaps this is not surprising as Warshel and others previously have demonstrated that valence bond concepts are the key to creating robust methods for modeling biological processes. In addition to advancing modeling technology, VB perspectives offer unique insights into the forces controlling the shapes of molecules from across the periodic table, particularly when covalent bonding predominates.

References

1. Root, D. M., Landis, C. R., and Cleveland, T. (1993) Valence Bond Concepts Applied to the Molecular Mechanics Description of Molecular Shapes 1. Application to Non-hypervalent Molecules of the P-block. *Journal of the American Chemical Society* **115**, 4201-4209.
2. Cleveland, T., and Landis, C. R. (1996) Valence Bond Concepts Applied to the Molecular Mechanics Description of Molecular Shapes. 2. Applications to Hypervalent Molecules of the P-Block. *Journal of the American Chemical Society* **118**, 6020-6030.
3. Landis, C. R., Cleveland, T., and Firman, T. K. (1995) Making Sense of the Shapes of Simple Metal Hydrides. *Journal of the American Chemical Society* **117**, 1859-1860.
4. Foster, J. P., and Weinhold, F. (1980) Natural Hybrid Orbitals. *Journal of the American Chemical Society* **102**, 7211-7218.
5. Reed, A. E., Weinstock, R. B., and Weinhold, F. (1985) Natural Population Analysis. *Journal of Chemical Physics* **83**, 735-746.
6. Reed, A. E., and Weinhold, F. (1985) Natural Localized Molecular Orbitals. *J. Chem. Phys.* **83**, 1736-1740.
7. Reed, A. E., Weinhold, F., Curtiss, L. A., and Pochatko, D. J. (1986) Natural Bond Orbital Analysis of Molecular Ineractions: Theoretical Studies of Binary Complexes of HF, H_2O, NH_3, N_2, O_2, F_2, CO, and CO_2 with HF, H_2O, and NH_3. *Journal of Chemical Physics* **84**, 5687-5705.
8. Carpenter, J. E., and Weinhold, F. (1988) Transferability of Natural Bond Orbitals. *Journal of the American Chemical Society* **110**, 368-372.
9. Glendening, E. D., Badenhoop, J. K., Reed, A. E., Carpenter, J. E., and Weinhold, F. (1994) NBO 4.0. University of Wisconsin, Madison, WI.
10. Glendening, E. D., and Weinhold, F. (1994) Natural Resonance Theory. I. General Formalism. University of Wisconsin Theoretical Chemistry Institute.
11. Bent, H. (1961) An Appraisal of Valence-Bond Structures and Hybridization in Compounds of the First-Row Elements. *Chemical Reviews* **61**, 275-311.

12. Pauling, L. (1960) The Nature of the Chemical Bond. Cornell University, Ithaca.
13. Pauling, L. (1975) Valence-bond Theory of Compounds of Transition Metals. *Proceedings of the National Academy of Sciences of the United States of America* **72**, 4200-4202.
14. Magnusson, E. (1990) Hypercoordinate Molecules of Second-Row Elements: d Functions or d Orbitals? *Journal of the American Chemical Society* **112**, 7940-7951.
15. Musher, J. (1969) The Chemistry of Hypervalent Molecules. *Angewandte Chemie (International edition in English)* **8**, 54-68.
16. Musher, J. L. (1972) The Role of Nonbonding Orbitals, s Mixing, and d-Orbital Participation in Hypervalent Bonding. *Journal of the American Chemical Society* **94**, 1370-1371.
17. Jonas, V., Boehme, C., and Frenking, G. (1996) Bent's Rule and the Structure of Transition Metal Compounds. *Inorganic Chemistry* **35**, 2097-2099.
18. Firman, T. K., and Landis, C. R. unpublished results.
19. Landis, C. R., Root, D. M., and Cleveland, T. (1995) Molecular Mechanics Force Fields for Modeling Inorganic and Organometallic Compounds. VCH Publishers, New York. 73-136.
20. Pauling, L. (1931) The Nature of the Chemical Bond. Application of Results Obtained from the Quantum Mechanics and from a Theory of Paramagnetic Susceptibility to the Structure of Molecules. *Journal of the American Chemical Society* **58**, 1367-1400.
21. Shen, M., Schaefer, H. F., and Partridge, H. (1992) Tungsten Hexahydride (WH_6). An Equilibrium Geometry Far from Octahedral. *Journal of Chemical Physics* **98**, 508-521.
22. Kang, S. K., Tang, H., and Albright, T. A. (1993) Structures for d^0 ML_6 and ML_5 Complexes. *Journal of the American Chemical Society* **115**, 1971-1981.
23. Haaland, A., Hammel, A., Rypdal, K., and Volden, H. V. (1990) Coordination Geometry of Gaseous Hexamethyltungsten: Not Octahedral. *Journal of the American Chemical Society* **112**, 4547-4549.
24. Pfennig, V., and Seppelt, K. (1996) Crystal and Molecular-Structures of Hexamethyltungsten and Hexamethylrhenium. *Science* **271**, 626-628.
25. Pulham, C., Haaland, A., Hammel, A., Rypdal, K., Verne, H. P., and Volden, H. V. (1992) The Structures of Pentamethyltantalum and -Antimony: One Square Pyramid and One Trigonal Bipyramid. *Angewandte Chemie (International edition in English)* **31**, 1464-1467.
26. Piersol, C. J., Profilet, R. D., Fanwick, P. E., and Rothwell, I. P. (1993) Crystal and Molecular Structure of Penta(4-methylbenzyl)tantalum, $[Ta(CH_2C_6H_4\text{-}4Me)_5]$. *Polyhedron* **12**, 1779-1783.
27. Stavropoulos, P., Savage, P. D., Tooze, R. P., Wilkinson, G., Hussain, B., Motevalli, M., and Hursthouse, M. B. (1987) The Synthesis and X-Ray Structures of Homoleptic Tetrahedral Aryls of Osmium(IV) and of Cyclohexyls of

Ruthenium(IV), Osmium(IV), and Chromium(IV). *J. Chem. Soc., Dalton Trans.* , 557.

28. Hitchcock, P. B., Lappert, M. F., Smith, R. G., Bartlett, R. A., and Power, P. P. (1988) Synthesis and Structural Characterization of the First Neutral Homoleptic Lanthanide Metal Alkyls: [LnR_3] [Ln=La or Sm, R=$CH(SiMe_3)_2$]. *Journal of the Chemical Society - Chemical Communications* , 1007-1008.
29. Hay-Motherwell, R. S., Hussain-Bates, B., Hursthouse, M. B., and Wilkinson, G. (1990) Synthesis and X-ray Crystal Structure of Trimesitylrhodium(III). *Journal of the Chemical Society - Chemical Communications* , 1242-1243.
30. Low, J. J., and Goddard, W. A. (1984) Theoretical Studies of Oxidative Addition and Reductive Elimination: $H_2 + Pt(PH_3)_2 \rightarrow Pt(H)_2(PH_3)_2$. *Journal of the American Chemical Society* **106**, 6928-6937.
31. Low, J. J., and Goddard, W. A. (1986) Theoretical Studies of Oxidative Addition and Reductive Elimination. 2. Reductive Coupling of H-H, H-C, and C-C Bonds from Palladium and Platinum Complexes. *Organometallics* **5**, 609-622.
32. Kang, S. K., Albright, T. A., and Eisenstein, O. (1989) The Structures of d^0 ML_6 Complexes. *Inorganic Chemistry* **28**, 1611-1613.
33. Kaupp, M. (1996) The Structure of Hexamethyltungsten, $W(CH_3)_6$- Distorted Trigonal Prismatic with C-3 Symmetry. *Journal of the American Chemical Society* **118**, 3018-3024.
34. Lappert, M. F., Raston, C. L., Skelton, B. W., and White, A. H. (1981) A Metallotricycle: Synthesis, one-electron Reduction, and Crystal Structure of the Thermally Robust Complex Tris-o-xylidenetungsten(VI), [$W(CH_2C_6H_4CH_2\text{-o})_3$]. *Journal of the Chemical Society - Chemical Communications* , 485-486.
35. Landis, C., Firman, T. K., and Cleveland, T., unpublished results.
36. Maseras, F., and Morokuma, K. (1992) Application of the natural population analysis to transition-metal complexes. Should the empty metal p orbitals be included in the valence space? *Chemical Physics Letters* **195**, 500-504.
37. Hehre, W. J., Radom, L., Schleyer, P. v. R., and Pople, J. A. (1986) Ab Initio Molecular Orbital Theory. John Wiley & Sons, New York. 548 pp.
38. Schilling, J. B., Goddard, W. A., and Beauchamp, J. L. (1987) Theoretical Studies of Transition Metal Hydrides. 3. SrH^+ through CdH^+. *Journal of the American Chemical Society* **109**, 5565-5583.
39. Yakshin, M. M. (1941) Dielectric Polarization of Complex Compounds of Platinum. *Compt. Rend. (Doklady) l'Acad. Sci. U.R.S.S.* **32**, 555-557.
40. Syrkin, Y. K. (1948) Theory of Cis and Trans substitutions in Complex Compounds of Platinum. *Bull. Acad. Sci. U.R.S.S., Classe Sci. Chim.* , 69-82.
41. Gillespie, R. J., and Hargittai, I. (1991) The VSEPR Model of Molecular Geometry. Allyn and Bacon, Boston.
42. Cerrada, E., Gimeno, M. C., Laguna, A., Laguna, M., Orera, V., and Jones, P. G. (1996) Charge-transfer salts with mononuclear and dinuclear ylide gold(I) complexes: x-ray structure of [$Au(CH_2PPh_3)_2$](TCNQ)(TCNQ = 7,7',8,8'-tetracyanoquinodimethane). *J. Organomet. Chem.* **506**, 203.

43. Yared, Y. W., Miles, S. L., Bau, R., and Reed, C. A. (1977) The First Formally Three-Coordinate d^8 Complex: Tris(triphenylphosphine)rhodium(I) Perchlorate and Its Novel Structure. *Journal of the American Chemical Society* **99**, 7076.
44. Bazhenova, T. A., Lobkovskaya, R. M., Shibaeva, R. P., Shilova, A. K., Gruselle, M., Leny, G., and Deschamps, E. (1983) Structure of the Napthyl Iron(II) Complex Formed in the Reaction of $FeCl_3$ and $C_{10}H_7Li$. *Journal of Organometallic Chemistry* **244**, 375-382.
45. Bronger, W., Müller, P., Schmitz, D., and Spittank, H. (1984) Synthese und Struktur von Na_2PtH_4, einem ternären Hydrid mit quadratisch planaren PtH_4^{2-}-Baugruppen. *Zeitschrift Anorganishe Allgermaine Chemie* **516**, 35-41.
46. Garcia, M. P., Jiminez, M. V., Oro, L. A., Lahoz, F. J., Tiripicchio, M. C., and Tiripicchio, A. (1993) A Homoleptic Mononuclear Iridium(II) Organometallic Complex: Synthesis and X-ray Structure of $[Ir(C_6Cl_5)_4]^{2-}$. *Organometallics* **12**, 4660-4663.
47. Bau, R., Ho, D. M., and Gibbins, S. G. (1981) The Binary Metal Hydrido Anion Fe_H6^{4-}: An X-ray Structural Characterization. *Journal of the American Chemical Society* **103**, 4960-4962.
48. Morris, R. J., and Girolami, G. S. (1991) High-Valent Organomanganese Chemistry. 1. Synthesis and Characterization of Manganese(III) and -(IV) Alkyls. *Organometallics* **10**, 792.
49. Garcia, M. P., Oro, L. A., and Lahoz, F. J. (1988) Novel Anionic Aryl Complexes of Rh(III): $[Rh(C_6F_5)_5]^{2-}$ and $[Rh(C_6F_5)_4(CO)]$-. *Angewandte Chemie (International edition in English)* **27**, 1700-1701.
50. Usón, R., Forniés, J., Tomás, M., Menjón, B., Sünkel, K., and Bau, R. (1984) The First Mononuclear Pt(III) Complex. Molecular Structures of $(NBu_4)[Pt(C_6Cl_5)_4]$ and of its Parent Compound $(NBu_4)_2[Pt(C_6Cl_5)_4]$. *Journal of the Chemical Society - Chemical Communications* , 751-752.
51. Markwell, A. J. (1985) The Crystal and Molecular Structure of Tetrabutylammonium Tetraphenylaurate(III). *Journal of Organometallic Chemistry* **293**, 257.
52. Murray, H. H., Fackler, J. P. J., Porter, L. C., Briggs, D. A., Guerra, M. A., and Lagow, R. J. (1987) Synthesis and X-ray Crystal Structures of $[Au(CH_2)_2PPh_2]_2(CF_3)_2$, $[Au(CH_2)PPh_2]_2(C_6F_5)_2$, and $[PPN][Au(C_6F_5)_4\}$: Two Dinuclear Gold(II) Ylide Complexes Containing Alkyl and Aryl Ligands and a Tetrakis(pentafluorophenyl)aurate(III) Anion Complex. *Inorganic Chemistry* **26**, 357.
53. Geiser, U., Schlueter, J. A., Williams, J. M., Naumann, D., and Roy, T. (1995) Anion Disorder in the 115-118 K Structures of the Organic Superconductors κ_L-$(BEDT\text{-}TTF)_2Ag(CF_3)_4(C_2H_3Cl_3)$ [BEDT-TTF = 3,4;3',4'- Bis(ethylenedithio)-2,2',5,5'-tetrathiafulfalene]. *Acta Cryst. B (Str. Sci.)* **51**, 789.
54. Geiser, U., Schlueter, J. A., Williams, J. M., Naumann, D., and Roy, T. (1995) $CuCF_3^{4-}$ *Acta Cryst. B (Str. Sci.)* **51**, 789.

55. Müller, E., Krause, J., and Schmiedeknecht, K. (1972) Strukturuntersuchungen an Organochromverbindungen. III. Röntgenstrukturanalytische und Spektroskopische Untersuchungen an $Na_2Cr(Ph)_5$. *Journal of Organometallic Chemistry* **44**, 127-140.
56. Hay-Motherwell, R. S., Wilkinson, G., Hussain, B., and Hursthouse, M. B. (1989) Homoleptic Methyl Compounds of Rhodium and Iridium(III). X-ray Crystal Structures of Tetramethylethylenediamine Lithium hexamethyl-rhodate(III) and -iridate(III). *Journal of the Chemical Society - Chemical Communications* , 1436-1437.
57. Krausse, J., and Marx, G. (1974) Strukturuntersuchungen an organochrom-verbindungenV*. IR UND Röntgen-Strukturanalyse des $Li_3Cr(CH_3)_6{\bullet}{}_3C_4H_8O_2$. *Journal of Organometallic Chemistry* **65**, 215-222.
58. Morris, R. J., and Girolami, G. S. (1988) . *Journal of the American Chemical Society* **110**, 6245-6252.
59. Abrahams, S. C., Ginsberg, A. P., and Knox, K. (1963) Transition Metal-Hydrogen Compounds. II. The Crystal and Molecular Structure of potassium Rhenium Hydride, K_2ReH_9. *Inorganic Chemistry* **3**, 558-567.
60. Truhlar, D. G. (1972) Test of Bond-Order Methods for Predicting the Position of the Minimum-Energy Path for Hydrogen Atom Transfer Reactions. *Journal of the American Chemical Society* **94**, 7584-7586.
61. Johnston, H. S., and Parr, C. (1963) . *Journal of the American Chemical Society* **94**, 2544.
62. Truhlar, D. G., and Parr, C. A. (1971) . *Journal of Physical Chemistry* **75**, 1844.
63. Truhlar, D. G., Steckler, R., and Gordon, M. (1987) Potential Energy Surfaces for Polyatomic Reaction Dynamics. *Chemical Reviews* **87**, 217-236.
64. Root, D. M., and Landis, C. R. unpublished results.

A NOVEL MOLECULAR MECHANICS STRATEGY FOR TRANSITION METALS BOUND TO BIOLOGICAL MOLECULES

ROBERT J. DEETH*, IAN J. MUNSLOW AND VERONICA J. PAGET
Inorganic Computational Chemistry Group
Department of Chemistry
University of Warwick
Coventry CV4 7AL, UK

1. Introduction

Transition metals bound to biological molecules are frequently associated with unusual ligand sets and low symmetry coordination environments [1]. Additional complications arise for certain species, such as the d^9 Cu(II) ion, where the electronic effects from the d-electrons already result in highly variable and 'plastic' coordination even in simple complexes. These features pose serious problems for conventional molecular modelling [2].

Ideally, the molecular mechanics (MM) simulation of the metal centre should be independent of coordination number and ligand type and should be capable of treating the d-electron energies explicitly. This is achieved [3] by three modifications of the MM methodology normally employed for molecules comprising light atoms. Firstly, all

L. Banci and P. Comba (eds.), Molecular Modeling and Dynamics of Bioinorganic Systems, 77–103.

L-M-L angle and L-M-L-A torsion terms in the force field (FF) are set to zero. Secondly, explicit 1,3 L-L van der Waals and electrostatic and 1,2 M-L electrostatic interactions are enabled. Thirdly, and most importantly, the d-electron stabilisation energy contribution is explicitly incorporated by adding to the conventional strain energy expression a generalised ligand field theory term. The latter is based on the cellular ligand field (CLF) model of Gerloch, Harding and Woolley [4].

The power of this hybrid cellular ligand field stabilisation energy/molecular mechanics (CLFSE/MM) approach in coordination chemistry has been demonstrated [3,5,6] by a series of applications on Cu(II), Ni(II) and Co(III) species. Some notable firsts include i) the automatic generation of Jahn-Teller elongated CuL_6 species, ii) a single set of FF parameters for four-, five- and six-coordinate Cu(II) complexes and iii) a single FF which simultaneously reproduces the geometries of both high- and low-spin Ni(II) complexes.

This paper presents extensions designed for application to TMs binding to biological molecules. On the one hand, FF parameters for metal-amino acid complexes have been developed and, in particular, the structures of four-coordinate CuL_2 (L = amino acid) complexes are discussed. On the other hand, the concept of Structure and Energy Tuned (STET) MM for metal complexes is developed. Here, the unique features of the CLFSE/MM method are exploited in conjunction with *ab initio* Density Functional Theory (DFT) to permit modelling both ground and Transition State structures and energies using a single set of FF parameters. The prospects for the *de novo* design of new metal species capable of specific Pt-DNA cross-linking interactions are explored.

2. Theoretical Background

The nature of this volume removes the need for an elaborate description of Molecular Mechanics (MM). Instead, a brief general outline can be given followed by a more detailed description of those features which distinguish our approach from more conventional MM schemes.

2.1. MOLECULAR MECHANICS

The basic expression for the strain energy, E_{tot}, comprises the usual terms for bond stretching, E_{str}, angle bending, E_{bend}, torsional twisting, E_{tor}, van der Waals interactions, E_{vdw}, and electrostatics, E_C [3]. In addition, E_{tot} is augmented with a metal-centred d-electron stabilisation energy term based on the CLF† model, the CLFSE.

$$E_{tot} = \Sigma E_{str} + \Sigma E_{bend} + \Sigma E_{tor} + \Sigma E_{vdw} + \Sigma E_C + CLFSE \quad (1)$$

Apart from the CLFSE, the following modifications are applied. (i) E_{str} is computed using a Morse function; (ii) the force constants for all E_{bend} terms centred on the metal are set to zero; (iii) the force constants for all E_{tor} terms involving rotation about a metal-ligand bond are set to zero and (iv) explicit 1,3 van der Waals and Coulomb interactions between ligand donor atoms and 1,2 M-L Coulomb interactions are permitted.

The first modification is prompted by the desire to model six-coordinate Cu^{II} complexes. Even for a $[CuL_6]$ species with six chemically identical ligands, the Jahn-Teller distortion results in much longer axial than equatorial bond lengths [7]. In order to use a common set of Cu-L parameters for both situations, a Hooke's Law E_{str} term is less appropriate.

Fig. 1 shows the difference between a Morse function and the harmonic oscillator expression for a hypothetical M-L bond with a length of 2.3 Å, a dissociation energy of 100 kcal mol^{-1} and a negligible interaction by about 3 Å. The deviation remains small until the bond lengthens by around 0.1 Å. Above 2.4 Å, the discrepancy

---

† The CLF model shares a number of features with the Angular Overlap Model (AOM) and was originally designated the ligand field version of the AOM. See refs. [4], [12] and [13] for details.

rapidly increases reaching 100 kcal mol^{-1} around 2.6 Å. For bond lengths less than the ideal value, the difference is even more rapid but in the opposite sense. In summary, the Morse function facilitates bond lengthening but resists bond shortening.

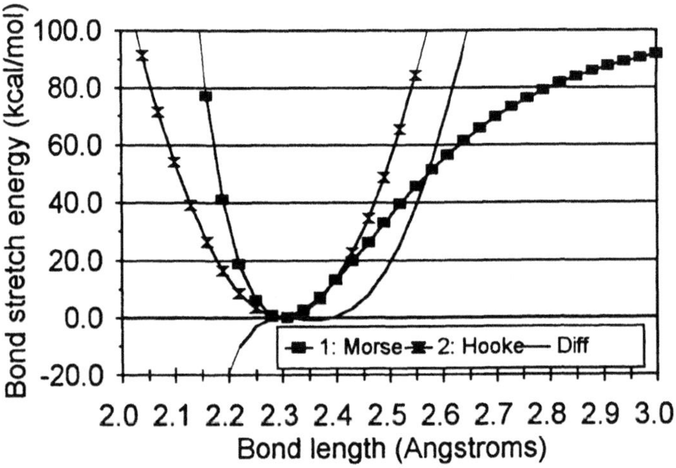

Fig. 1. A comparison of E_{str} terms.

The second and fourth modifications were in response to the so-called 'unique labelling problem' [8] wherein a given L-M-L angle in a complex may adopt several very different values. For example, in octahedral complexes, the cis L-M-L angles are 90° while the trans angles for the same set of atoms are 180°. This multiplicity does not occur in carbon chemistry and each coordination number can be treated with a single angle bend parameter. For metal complexes, the 'normal' approach has to be modified.

Solutions to this problem have been developed [9] which still retain explicit L-M-L E_{bend} terms. Alternatively, one can recognise from simple models like VSEPR theory [10] and the extensive work of Kepert [11], that, in the absence of special electronic effects, the basic angular geometry of molecules can be described in terms of ligand-ligand repulsions. The latter can be incorporated into MM by permitting 1,3 non-bonding interactions between ligand donor atoms. These would normally be subsumed within E_{bend}. We have also found it advantageous to include 1,3 L-L and 1,2 M-L interactions.

The third modification is simply to reduce the degree of parameterisation.

2.2. STEREOCHEMICAL ACTIVITY OF D ELECTRONS AND LIGAND FIELD STABILISATION ENERGIES

As alluded to above, molecular shape can largely be described in terms of repulsions *in the absence of special electronic effects*. This latter point justifies the inclusion of the CLFSE term in eq. 1. Coordination chemistry provides many examples of irregular geometries which can be traced to the 'stereochemical activity' of the underlying incomplete d shell [12]. Jahn-Teller distortions are particularly famous examples of a rather more general phenomenon which also encompasses, for example, the tendency for d^3 and low-spin d^6 complexes to be octahedral and for low-spin d^8 species to be planar.

The structural preferences displayed by certain d configurations are often rationalised in terms of the ligand field stabilisation energy (LFSE). When the d-orbital energies for a particular geometry and spin state are evaluated relative to a global barycentre, those arrangements which maximise the LFSE, like octahedral d^3 and low spin d^6, are found to be the most stable.

The Jahn-Teller effect in six-coordinate d^9 species can likewise be explained using d-electron stabilisation (Fig. 2) [7]. Relative to the octahedral precursor, the total d-electron energy is lowered by a tetragonal distortion which splits the e_g orbitals. To first order, the energy change of each e_g component is equal but opposite. However, the lower orbital is doubly occupied while the upper one contains a single electron. The energy difference ΔE_{JT} drives the distortion.

Low-spin d^8 species are notionally subject to the same forces although some additional mechanism is required initially to induce the low-spin state. However, once this occurs, the same tetragonal distortion becomes favoured as for the d^9 case but with twice the driving force. One presumes the distortion ultimately becomes so large that the axial ligands are lost [7].

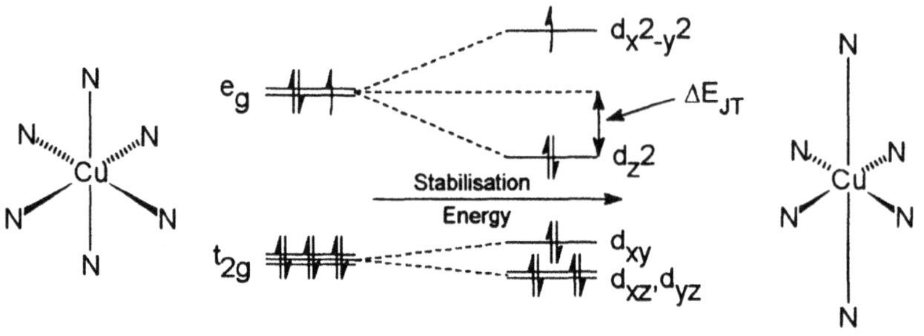

Fig. 2. Diagram illustrating the ligand field stabilisation energy for a tetragonal distortion of an octahedral d^9 Cu(II) species.

The stereochemical effects of the d-electrons do not always alter the symmetry. Trigonal bipyramidal complexes are a good example [12]. Bromo complexes of Me_6tren (Me_6tren = $N(CH_2CH_2NMe_2)_3$) are known for several divalent first row metals

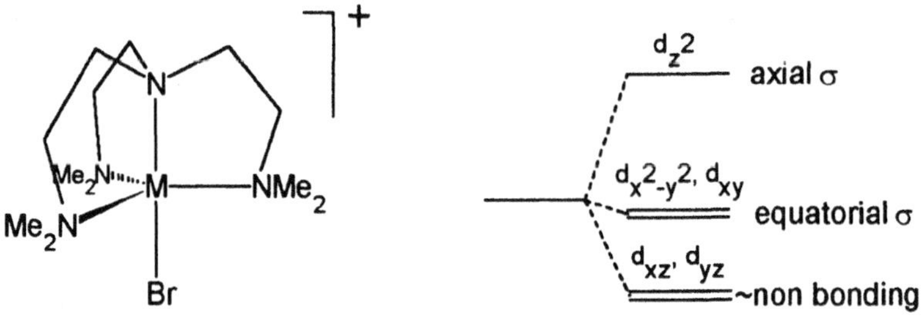

Fig. 3. Schematic representation of the structure of $[M(Me_6tren)Br]^+$ complexes and the associated d-orbital splitting pattern.

(Fig. 3). The M-L interactions in these species are dominated by the σ-bonding amines. In tbp symmetry, the d-orbitals divide into three sets - d_{z^2}, $d_{x^2-y^2}/d_{xy}$ and d_{xz}/d_{yz} - which can conveniently be characterised in terms of axial σ-bonding, equatorial σ-bonding and non-bonding, respectively. The weighted average of the four M-N distances increases slightly from Co^{II} to Zn^{II} (Table 1) but, superimposed on this general trend, are variations which can be traced back to the underlying arrangements of the d-electrons in the σ-bonding sets.

At zinc, all the d_σ functions are symmetrically occupied (i.e. full) and the axial bond lengths are longer than the equatorial contacts as expected from the simple models applied in main group chemistry [10,11]. For high-spin d^7 Co^{II}, these σ functions are also symmetrically occupied (i.e. half filled) and again longer axial bonds arise. At Ni^{II}, there is now an extra electron in the $d_{x^2-y^2}/d_{xy}$ pair which repels the equatorial ligands while the axial contacts shorten somewhat to compensate. The result is actually a longer $Ni\text{-}N_{eq}$ than $Ni\text{-}N_{ax}$ distance. This effect is enhanced further a d^9 Cu^{II} where now the hole in d_{z^2} leads to a pronounced axial compression.

TABLE 1. Selected structural parameters for $[M(Me_6tren)Br]^+$ complexes. (See ref. [12].)

M-N lengths (Å)	Co^{II}	Ni^{II}	Cu^{II}	Zn^{II}
d_{av}	2.10	2.12	2.12	2.13
d_{ax}	2.15	2.10	2.07	2.19
d_{eq}	2.08	2.13	2.14	2.11
$d_{ax} - d_{eq}$	+0.07	-0.03	-0.07	+0.08

2.3. IMPLEMENTING THE CELLULAR LIGAND FIELD STABILISATION ENERGY

The power of the CLF model is that each M-L interaction is treated as a local perturbation of the d orbitals. The complete ligand field potential is constructed by summing all these local contributions [4]. There are no symmetry constraints inherent in this procedure so that the CLF can be used for any coordination number and geometry.

Each M-L bond is associated with local energy parameters describing individual σ and π interactions, e_σ and e_π respectively (Fig. 4). For simplicity, a linear dependence of e_σ on the M-L distance is used [3] and e_π is ignored [6]. Given a molecular structure, the M-L distances are computed to yield the relevant e_σ values while the angular geometry is processed via tensor operator theory to give appropriate matrices relating the local perturbations to the global reference frame. Using Gerloch's Master equation M.2 [13], a complete one-electron ligand field matrix is constructed and diagonalised to give the d-orbital energies. The latter obey a crystal field barycentre. This is a relatively time-consuming process since there is no simple analytical expression for the CLFSE derivatives with respect to atomic displacements and these quantities are evaluated by finite difference. Fortunately, only the metal and ligand donor atoms need to be considered.

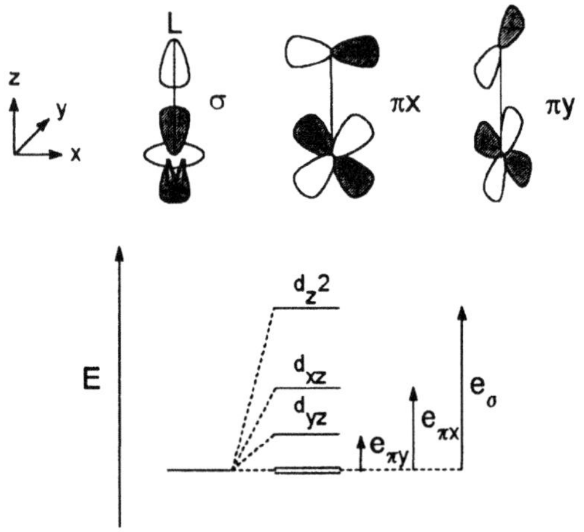

Fig. 4. Cellular ligand field e parameter definitions (π-donor assumed).

Routines for calculating the CLFSE and its derivatives have been added to existing MM software [14]. Most of the force field parameters for the 'organic' parts remain unchanged but parameters describing M-L interactions often require significant adjustments. In particular, E_{str} terms for M-L bonds developed by other workers are *not* transferable since the CLFSE makes a significant contribution. However, equivalent parameters can easily be developed as shown, for example, for Co^{III}-amine complexes (Fig. 5).

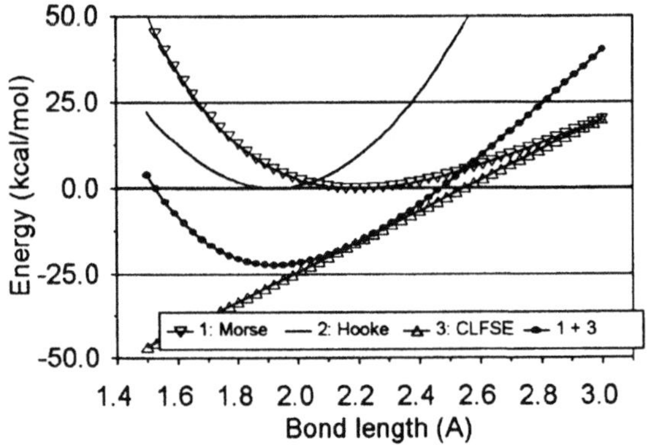

Fig. 5. Comparison of bond stretch terms for a Co(III)-N interaction.

The CLFSE for an octahedral low-spin d^6 complex is $-12/5\Delta_{oct}$. Within the CLF formalism, $\Delta_{oct} = 3e_\sigma$ and thus, the contribution from to the CLFSE from a single ligand is $-6/5e_\sigma$. The latter is expressed as eq. 2:

$$e_\sigma = a_0 + a_1 r \qquad (2)$$

where r is the M-L bond length and a_0 and a_1 are empirical constants chosen to yield approximately the same e_σ values as derived from the ligand field analysis of the d-d spectra of known complexes. Taking the CLFSE contribution and a Morse function, we can rapidly define parameters such that the sum of the CLFSE and Morse contributions matches curves derived from published MM parameters for Co(III)-N bonds [15]. In the example above, the Co-N force constant for the Hooke's law expression was taken as 1.7 mdyn $Å^{-1}$ (245 kcal mol^{-1} $Å^{-2}$) with an ideal bond length of 1.925 Å. Note, in particular, that the effect of the CLFSE is to give a minimum very much shorter than the 'ideal' Morse function bond length of 2.36 Å. Note also, that the CLFSE/MM parameters can be tuned to yield a bond length of 2.15 Å as computed using DFT for a hypothetical $^5T_{2g}$ $[Co(NH_3)_6]^{3+}$ complex [16]. The conclusion here is that the CLFSE/MM model can fairly easily mimic the results from existing conventional treatments. Data for some representative $[CoN_6]^{3+}$ species are given in Table 2.

TABLE 2: Comparison of selected geometrical parameters for six-coordinate Co(III)-amine complexes. (Key: bold = CLFSE/MM, italic = experimental (see ref. [6]) normal = conventional MM [15]. en = 1,2-ethylenediamine, tn = 1,3-propylenediamine, dptn = di-3-aminopropylamine)

Parameter	$[Co(NH_3)_6]^{3+}$	$[Co(en)_3]^{3+}$	$[Co(tn)_3]^{3+}$	$[Co(dptn)_2]^{3+}$
Co-N	**1.96** *1.96* 1.96	**1.96** *1.96* 1.96	**1.99** *1.98* 1.99	**1.99** *2.00* 1.97
				2.02 *2.04* 2.04
cis N-Co-N	**90.0** *90.0* 90.0	**90.0** *90.6* 88.0	**90.0** *90.0* 94.7	**90.0** *90.1* 94.5
trans N-Co-N	**179.8** *179.9* -	**176.0** *175.6* -	**178.6** *177.8* -	**176.9** *174.6* -

3. Applications

Many of the examples mentioned thus far could be accommodated within a conventional MM scheme as is no doubt described elsewhere in this volume. The utility of the CLFSE/MM method is the generality of the d-electron energy term which greatly facilitates the development of FF parameters that apply to far wider ranges of complexes.

Applications of the CLFSE/MM model have so far focused mainly on amine and imine complexes of Cu^{II}, Ni^{II} and Co^{III}. Amine ligands are electronically simple and form only σ bonds to metals. The contribution to the CLFSE from each M-L interaction is thus described by a single e_σ parameter. Imines are more biologically relevant and can also, in principle, engage in M-L π bonding. However, given our main aim is to model molecular structure, in practice it turns out that good geometries can be computed without including e_π components [6].

3.1. THE JAHN-TELLER EFFECT IN COPPER(II) COMPLEXES

Perhaps the most significant feature of the CLFSE/MM approach is its ability to model radically different M-L bond lengths with a single set of M-L parameters. The CLFSE/MM scheme appears to be unique in this regard. The large tetragonal elongations seen in six-coordinate Cu^{II} complexes have already been mentioned and some representative examples are shown in Fig. 6. If all the ligands are not the same, the model automatically selects the lowest energy path for distortion [5].

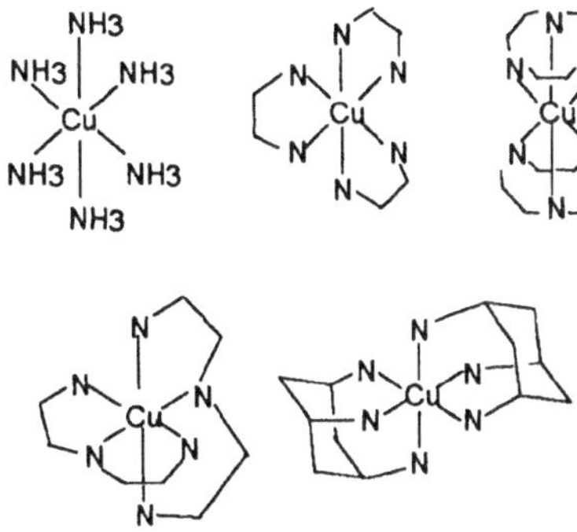

Fig. 6. Structural diagrams for CuL_6 complexes with saturated amine ligands. For clarity, hydrogens have been omitted from the alkylated amines.

TABLE 3. Comparison between CLFSE/MM (bold) and experimental (italic) geometrical parameters for the complexes shown in Fig. 6 proceeding from left to right and top to bottom. (The comparison for $[Cu(NH_3)_6]^{2+}$ should be considered as illustrative only. See ref. [5] for details.)

Complex	$[Cu(NH_3)_6]^{2+}$	$[Cu(en)_3]^{2+}$	$[Cu(ten)_2]^{2+}$	$[Cu(dien)_2]^{2+}$	$[Cu(tach)_2]^{2+}$
$Cu\text{-}N_{eq}(av)$	**2.04** *2.15*	**1.99** *2.04*	**2.05** *2.06*	**2.06** *2.07*	**2.01** *2.07*
$Cu\text{-}N_{ax}(av)$	**2.48** *2.45*	**2.46** *2.41*	**2.29** *2.32*	**2.41** *2.41*	**2.28** *2.35*
cis N-Cu-N(av)	**90** *90*	**80** *81*	**81** *81*	**80** *81*	**89** *97*
trans N-Cu-N (av)	**180** *180*	**170** *169*	**177** *178*	**164** *166*	**180** *180*

3.2. HIGH- AND LOW-SPIN NICKEL(II) SPECIES

Another good example is the geometry change accompanying the different spin states for d^8 Ni^{II} species. The Ni-N bonds in low-spin Ni amines are about 0.2Å shorter than in their high spin analogues (Fig. 7 and Table 4). This change is modelled using a single set of Ni-N parameters by simply changing the d orbital occupations to reflect the desired spin state [3].

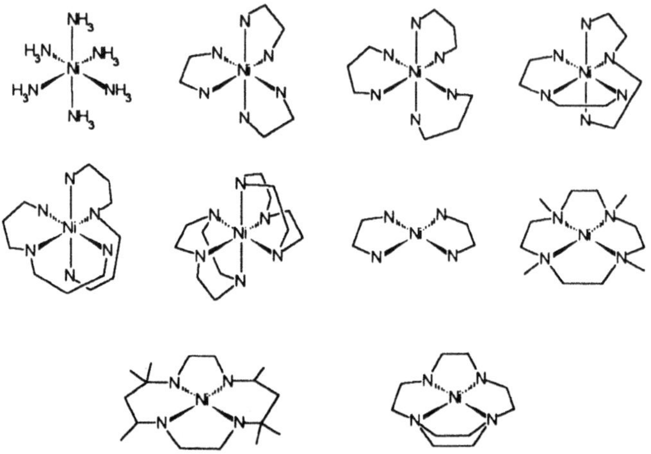

Fig. 7. Structural diagrams for NiN_4 and NiN_6 amine complexes.

TABLE 4: Comparison between CLFSE/MM (bold) and experimental (italic) geometrical parameters for the Ni complexes shown in Fig. 7 proceeding from left to right and top to bottom.

Complex	Ni-N	*cis* N-Ni-N	*trans* N-Ni-N
$[Ni(NH_3)_6]^{2+}$	**2.12** *2.13*	**90** *90*	**180** *180*
$[Ni(en)_3]^{2+}$	**2.11** *2.13*	**81** *82*	**172** *172*
$[Ni(tn)_3]^{2+}$	**2.12** *2.15*	**86** *87*	**175** *176*
$[Ni(dien)_2]^{2+}$	**2.04** *2.06*	**82** *82*	**168** *167*
	2.14 *2.15*		
$[Ni(dptn)_2]^{2+}$	**2.06** *2.15*	**90** *90*	**175** *176*
	2.15 *2.23*		
$[Ni(tcn)_2]^{2+}$	**2.08** *2.11*	**82** *83*	**177** *177*
$[Ni(en)_2]^{2+}$	**1.93** *1.92*	**87** *86*	**180** *180*
$[Ni(tmc)_2]^{2+}$	**1.97** *1.98*	**91** *91*	**167** *169*
$[Ni(meso14ane)]^{2+}$	**1.97** *1.96*	**90** *90*	**180** *180*
$[Ni(12ane)]^{2+}$	**1.84** *1.87*	**90** *90*	**169** *169*

The Ni-N parameters were derived from existing conventional treatments [15]. Of course, the latter required independent FFs - one for high-spin and one for low-spin nickel(II). The requirement only for a single set of CLFSE/MM FF parameters opens up other exciting possibilities. For example, in principle, we could modify the program to allow for non-integral orbital occupations and thereby follow certain types of spin-crossover behaviour directly.

3.3. FLEXIBILITY IN PENTACOORDINATE COMPLEXES

Five-coordinate complexes are known to be very flexible with only a small energy difference between the limiting tbp and square pyramidal (sqp) stereochemistries [17]. However, the tbp to sqp conversion for Cu^{II} species has severe structural consequences. (See Fig. 8.) As we have already seen, tbp Cu^{II} complexes are distorted albeit with respect to bond lengths rather than symmetry. Similarly, sqp Cu^{II} species are distorted inasmuch as the bond to the apical ligand is elongated. Thus, a nitrogen ligand at an axial tbp site would have a Cu-N distance of around 1.95 Å while at the apical sqp position the distance would be around 2.3 - 2.4 Å. Since the underlying cause of these changes is the stereochemical activity of the d electrons, the CLFSE automatically models both symmetries with a single FF [3].

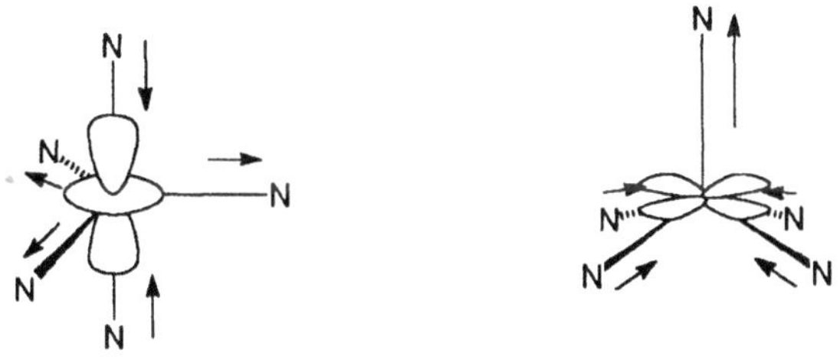

Fig. 8. Illustration of the geometrical consequences of the stereochemical activity of the singly occupied d-orbital in d^9 trigonal bipyramidal (left) and square pyramidal (right) stereochemistries.

The energy difference between the limiting geometries for $[Cu(papd)]^{2+}$ (papd = 2,5,8,11,14-pentaazapentadecane), for example, is calculated [3] to be only a few kcal mol^{-1}. This is of the order of crystal packing forces so it is not surprising that the observed solid-state structure is a sensitive function of the crystalline environment. Indeed, Hathaway [18] has followed the structural change from near tbp to near sqp for a range of $[Cu(bipy)_2X]^{n+}$ species and attempted to map out the pseudorotation pathway connecting them.

This presents difficulties for molecular modelling since the parameterisation is based on solid state data and, therefore, averages the environmental effects and cannot account for specific interactions between the complex and its environment unless the environment is explicitly included in the calculations. We must therefore take great care when interpreting both modelling and experimental results for such 'plastic' species.

3.4. COPPER(II) BIS-OXAZOLINE COMPLEXES

Copper(II) oxazoline complexes of the type shown schematically in Fig. 9a are active catalysts for the enantioselective Diels-Alder reaction between acrylimide and cyclopentadiene. Davies *et al.* [19] suggest a correlation between the ligand bite angle and the enantio- and diastereoselectivity but were only able to test this indirectly by computing the angle at the bridging carbon in the parent free ligands. However, there is no guarantee that this correlation will be mirrored by the actual N-Cu-N bite angle in the active Cu catalysts. This problem is ideally suited to the CLFSE/MM scheme. Crystal structures of the diaqua and dichloro complexes are available for FF parameterisation.

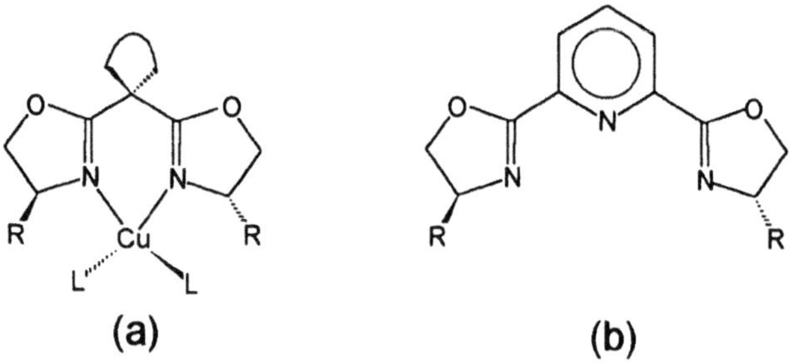

Fig. 9. (a) Schematic representation of $[Cu(oxazoline)_2L_2]^{n+}$ complexes and (b) the pybox ligand.

Both show pronounced tetrahedral distortions which is a particularly challenging feature for molecular modelling. Moreover, there are several related pyridine-bis(oxazoline) (pybox) systems (Fig. 9b) - a four-coordinate aqua, a five-coordinate dichloro and a six-coordinate bis-pybox complex. The goal was to develop a single unified FF capable of modelling all five molecules.

As described previously [6], there are three ways of overcoming the inherent tendency of the CLFSE to enforce planar coordination in $[CuL_4]$ species: (i) very large e_π values (ii) very large van der Waals repulsion terms or (iii) ligand-ligand electrostatic interactions. The latter is certainly the most appealing, especially as we had already anticipated the need for electrostatics. Guided by the results for other imine systems, FF parameters for both $[Cu\{CH_2(oxaz)_2\}(H_2O)_2]^{2+}$ and $[Cu\{CH_2(oxaz)_2\}Cl_2]$ were developed. This initial FF, which included M-L and L-L electrostatic interactions, was augmented and modified to accommodate the pybox systems as well and, after a degree of trial and error, a reasonable set of FF parameters was obtained capable of reproducing the structures of all five complexes with acceptable accuracy (Fig. 10) [20].

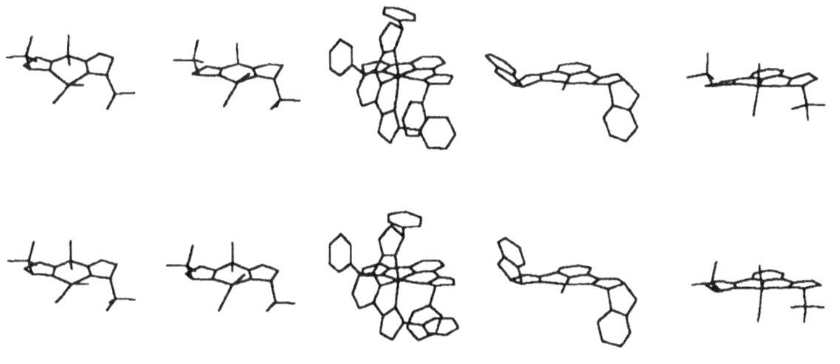

1 2 3 4 5

Fig. 10. Observed (upper) and calculated (lower) structures for some Cu(II) oxazoline complexes. 1: $[Cu\{Me_2C(^tBu\text{-}oxaz)_2\}(H_2O)_2]^{2+}$; 2: $[Cu\{Me_2C(^tBu\text{-}oxaz)_2\}Cl_2]$; 3: $[Cu(Ph\text{-}pybox)_2]^{2+}$; 4: $[Cu(In\text{-}pybox)(H_2O)]^{2+}$; 5: $[Cu(^tBu\text{-}pybox)Cl_2]$. (Ph = phenyl, In = indole)

Overall, the agreement between theory and experiment is good. Bond lengths and angles are reproduced to within about 0.03Å and 4° respectively. The tetrahedral twists for 1 and 2 are correctly predicted as is the slight twist of the $CuCl_2$ plane relative to the pybox plane in 4. Complex 5 has the observed Jahn-Teller elongated structure and 3 is planar as observed. Note that 3 has triflate oxygen atoms positioned along the notional z-axis at about 2.45Å from the Cu centre (Fig. 11). Ligand Field studies [21] suggest the effect from such distant groups would be minimal which was confirmed by CLFSE/MM calculations. Placing one or both triflates on the z-axis does not significantly alter the structure of the rest of the molecule.

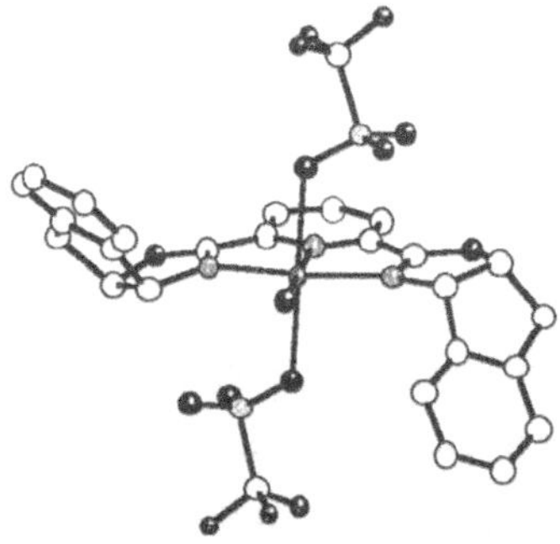

Fig. 11. Calculated structure for [Cu(In-pybox)(H_2O)(triflate)$_2$] showing elongated axial coordination.

This FF was then used to study the relationship between the bite angles and enantioselectivity in the bis-oxazoline complexes. It turns out that the correlation between the free-ligand bridge-head angle, Φ, and the enantioselectivity is mirrored by a similar correlation between Φ and the N-Cu-N bite angle in the complexes (Table 5).

TABLE 5: Correlation between bridgehead C-C-C angles (see Fig. 8a), calculated N-Cu-N bite angles and observed enantiomeric excesses.

Ligand	Φ (free ligand)	Φ (in complex)	N-Cu-N	% e.e.	endo:exo
In-cyclopropyl	111	121	97.6	96.3	44:1
In-cyclobutyl	108	116	96.3	92.0	38:1
In-cyclopentyl	106	113	95.6	89.5	37:1
In-cyclohexyl	104	111	94.9	83.0	26:1

3.5. COPPER(II) AMINO-ACID COMPLEXES

As emphasised repeatedly in this paper, the inherent flexibility of Cu^{II} species often leads to irregular geometries. For tetracoordinate complexes, the observed geometry is often intermediate between the extremes of planar and tetrahedral as in, for example, $[CuCl_4]^{2-}$ which, in the dicesium salt, is a flattened tetrahedron† (Fig. 12) [22]. A

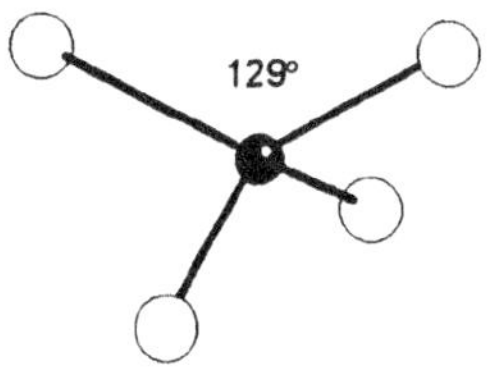

Fig. 12. The structure of the $[CuCl_4]^{2-}$ ion in Cs_2CuCl_4.

planar geometry is favoured by the d-electron stabilisation energy while a tetrahedral structure is favoured by ligand-ligand repulsions. The energy balance is quite subtle. In $[CuCl_4]^{2-}$, the energy of the vibration which distorts the square planar D_{4h} geometry towards the flattened tetrahedral D_{2d} structure is only about 50 cm^{-1} [23].

Bis-amino acid complexes of Cu^{II} display a range of geometries from flattened tetrahedral to nearly planar [24]. These structures were determined relative to the solid state and we may therefore wonder what effect, if any, crystal packing has on the observed structure. Indeed, others have attempted to model these complexes using a more conventional MM scheme and concluded [24] that certain $[CuL_2]$ complexes are quite distorted *in vacuo* but much more planar in the solid state. Our experience leads us to query this conclusion.

Firstly, an obvious and critical test is to use some suitably high level, preferably an *ab initio* Quantum Mechanical model, to verify the *in vacuo* structure of a representative complex. Density Functional Theory (DFT) is the method of choice for such calculations. Accordingly, we selected $[Cu(L\text{-but})_2]$ (L-but = L-2-aminobutyric acid) for study since this was suggested [24] as having the largest *in vacuo* distortion. As can be seen in Fig. 13, the computed *in vacuo* DFT structure around the Cu centre agrees remarkably well with the solid state structure with an rms deviation for the

† A planar geometry can occur for $[CuCl_4]^{2-}$ with a suitable countercation but we have not yet extended our MM modelling to include the lattice environment. *In vacuo* DFT calculations show that in the absence of environmental effects, the flattened tetrahedral structure is the lowest energy conformation.

CuO_2N_2 set of only 0.025 Å. Indeed, we have consistently found that DFT calculations on TM complexes reproduce well experimental condensed-phase data [25]. For [Cu(L-but)$_2$] there are some deviations associated with the ethyl substituents and these could possibly be attributed to crystal packing or to local minima associated with different conformations. However, the conclusion proposed by Sabolovic and Rasmussen [24] that 'intermolecular interactions maintain the planarity of the Cu coordination polyhedron' appears to be an artefact of their MM model.[†]

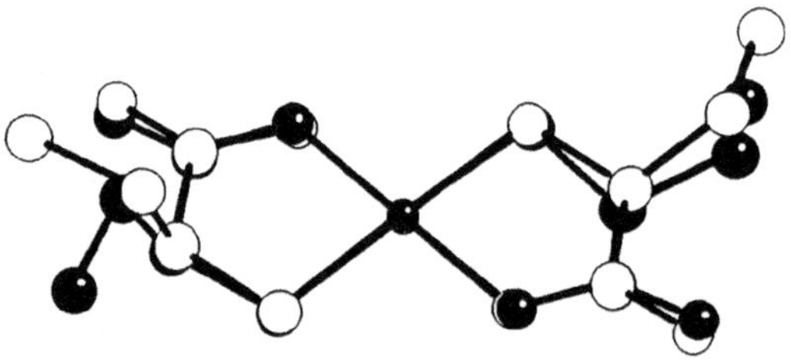

Fig. 13. Comparison of the observed (white) and DFT calculated (black) structures for [Cu(L-but)$_2$]. The coordinates of CuO_2N_2 atoms have been fitted by least squares.

In contrast, there is no difficulty in generating CLFSE/MM FF parameters which reproduce the observed solid state structures of these Cu amino acid species (Fig. 14 and Table 6) [26]. The force field is based on our previous work [3, 5, 6] and includes metal-ligand and ligand-ligand electrostatic interactions.

Indeed, the whole point about MM is that its basic interpolative nature should ensure that it is always be possible to reproduce a desired feature (e.g. molecular structure) for a more or less closely related series of molecules based on whatever experimental values of that feature (e.g. X-ray crystal structure analyses) are available [2]. Pathological behaviour therefore indicates either that the molecule in question is far removed from those used to establish the force field parameterisation or that there is some intrinsic problem with the model itself.

[†] Note that new results obtained by Sabolovic *et al.* which were presented at this meeting are now in substantial agreement with our results.

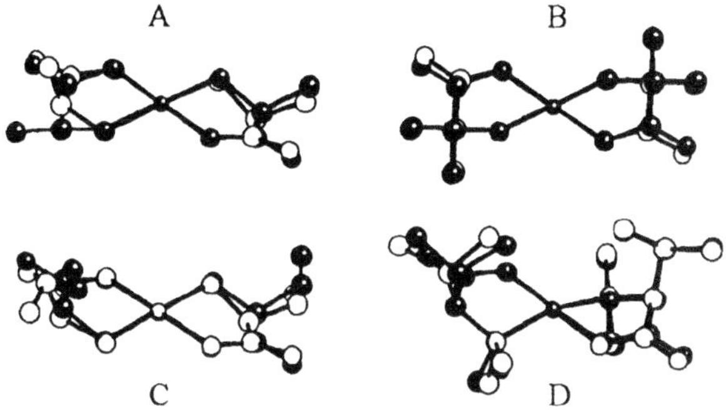

Fig. 14. Comparison of the observed (white) and CLF/MM calculated (black) structures for some selected bis-amino acid Cu(II) complexes. A: [Cu(L-2-aminopropanoate)$_2$]; B: [Cu(2-amino-2-methylpropanoate)$_2$]; C: [Cu(L-2-aminobutanoate)$_2$]; D: [Cu(L-2-dimethylamino-3-methylbutanoate)$_2$]

TABLE 6: Comparison of selected geometrical parameters (Å and °) for the complexes in Fig. 14. Bold = CLF/MM, italic = experimental

Complex	Cu-O	Cu-N	O-Cu-N	O-Cu-O	N-Cu-N
A	**1.92** *1.95*	**2.00** *1.98*	**80** *84*	**180** *179*	**179** *175*
B	**1.92** *1.95*	**2.00** *1.98*	**80** *84*	**180** *180*	**178** *180*
C	**1.92** *1.96*	**2.00** *1.99*	**80** *84*	**180** *179*	**177** *175*
D	**1.91** *1.90*	**2.00** *2.02*	**80** *85*	**176** *176*	**155** *161*

4. New Developments

The CLFSE/MM model embodies a number of novel features which are uniquely suited to study problems in coordination chemistry. For decades, LFSE arguments have been used to try to rationalise all manner of experiments from heats of hydration [27] through to activation energies for associative and dissociative substitution in six-coordinate complexes [28]. The latter is especially exciting since it suggests a way in to modelling reaction mechanism within a MM framework.

4.1 THE TRANS INFLUENCE IN PLANAR D^8 COMPLEXES

The trans influence is a well known phenomenon in the coordination chemistry of planar d^8 systems and relates to variations observed in the M-L bond lengths of ground state species which depend on the nature of the trans ligand [27]. In some cases, these effects can be quite pronounced [29].

Modelling the trans influence is another exacting challenge for MM. The desire is, once again, to develop a single set of FF parameters for a given M-L interaction which can cope with a range of M-L distances. Given that the trans influence is often described in terms of the competing electronic effects of the various ligands, the CLFSE/MM scheme, with its built-in d-electron term, should be ideally suited to this problem [30].

Our first strategy was as follows. The Cambridge Structural Database was searched for suitable planar Pt(II) complexes from which a table of 'ideal' Pt-L bond lengths was derived describing both homoleptic species and the trans influence from the other ligands in the test set. CLFSE/MM calculations were then carried out for 'bare ligand' species - i.e. hypothetical five atom complexes involving the Pt and the four ligand donor atoms.

Treating the homoleptic PtL_4 complexes first, series of suitable combinations of bond stretch parameters were developed which gave the required Pt-L distance but had varying degrees of curvature at the minimum. The hope was that a ligand with a steeper curve would tend to retain its homoleptic Pt-L distance and oblige the trans ligand to adjust its bond length to compensate.

Unfortunately, the problem is not this straightforward. The d-orbitals are centrosymmetric which means that individual perturbations from centrosymmetrically-related ligands cannot be separated. This effect has been well documented in applications of the CLF model. For the CLFSE/MM approach, it results in an inability to control mutually *trans* M-L bond lengths independently.

To overcome this problem, ligand-ligand bond stretch terms were defined for all possible combinations of donor atoms in the test set. The 'ideal' bond length is set at the largest contact observed in the test data and, after some experimentation, a suitable set of FF parameters was derived capable of reproducing virtually all the desired bond length variations to within about 0.02 Å (Table 7).

TABLE 7. Comparison of CLF/MM calculated bond lengths (bold) for 'bare ligand' Pt complexes with averaged experimental values (italic) derived from the Cambridge Structural Database.

Ligand donor set	Pt-Cl	Pt-S	Pt-P
Cl_4	**2.30** *2.30*		
Cl_3S	**2.30**[a] *2.30*	**2.25** *2.26*	
	2.32[b] *2.32*		
trans-Cl_2S_2	**2.30** *2.30*	**2.31** *2.31*	
cis-Cl_2S_2	**2.32** *2.32*	**2.27** *2.26*	
S_3Cl	**2.31** *2.33*	**2.31**[b] *2.32*	
		2.28[a] *2.27*	
S_4		**2.32** *2.32*	
P_4			**2.32** *2.32*
cis-Cl_2P_2	**2.36** *2.35*		**2.25** *2.23*
cis-Cl_2SP	**2.30**[b] *2.33*	**2.27** *2.27*	**2.22** *2.23*
	2.37[c] *2.35*		

[a] *trans* to Cl; [b] *trans* to S; [c] *trans* to P

4.2. STRUCTURE AND ENERGY TUNED MOLECULAR MECHANICS

The final topic in this section also arose out of parallel work on modelling exchange/substitution processes for d^8 systems [31,32]. Ligand exchange is a fundamental process in chemistry especially if the exchanging ligand is a solvent molecule. We have been using DFT to compute the mechanism of ligand exchange at the planar d^8 centres Pd(II) and Pt(II).

Three intriguing features emerged from these calculations. First, the *in vacuo* DFT optimised geometries are in good agreement with available experimental data. Second, a tbp model structure is an excellent energetic and structural model for the true transition state† and shows the expected axial compression due to the stereochemical activity of the d electrons. Thirdly, accurate activation enthalpies were computed based on the *in vacuo* DFT energies plus a solvation energy correction based on the Born model [33] and using the DFT structure. Evidently, the computed structure for the transition state is reasonable despite the absence of any explicit solvent in the DFT calculation.

These DFT results provide a basis for parameterising a CLFSE/MM study. However, not only can the CLFSE/MM treatment be designed to yield the correct structure, we can also arrange for the calculated activation energy to match the DFT value as well. The CLFSE/MM approach is then STructure and Energy Tuned (STET).

A complete STET approach requires a transition state searching algorithm to be implemented within our modelling software. This has not yet been completed but it is nevertheless possible to demonstrate that the basic idea is sound.

DFT calculations on chloride exchange on $[PdCl_4]^{2-}$ show [32] that a tbp structure is a good approximation to the true C_{2v} transition state (Fig. 15). The $[PdCl_5]^{3-}$ transition state model has D_{3h} symmetry and analytical expressions for the CLFSE, E_{str}, E_{vdw} and E_C can be derived in terms of the axial and equatorial Pd-Cl contacts. The expressions can conveniently be fed into a spreadsheet where the minimum configuration can be located using 3-dimensional plotting routines.

---

† The transition state, incidentally, can never posess *rigourous* D_{3h} symmetry since this would result in a degenerate transition state vector. See discussion in ref. [32].

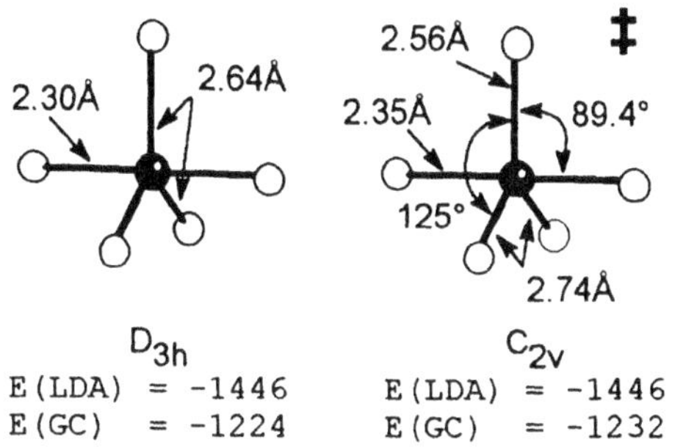

Fig. 15. Calculated (DFT) structures and binding energies for D_{3h} $[PdCl_5]^{3-}$ and the true C_{2v} Transition State for chloride exchange.

The result is that the CLFSE/MM model can simultaneously reproduce the *in vacuo* DFT structure of $[PdCl_5]^{3-}$ and the theoretical activation energy (Table 8). The latter is some 140 kcal mol^{-1} since, in a vacuum, there are large electrostatic repulsions associated with bringing a chloride anion up to $[PdCl_4]^{2-}$. The effects are strongly attenuated in solution. The Born model corrections suggest a true activation energy nearer 12 kcal mol^{-1} which is consistent with values found for similar processes. In contrast, the experimental activation energy for water exchange at $[Pd(H_2O)_4]^{2+}$ [34] is also 12 kcal mol^{-1} but the electrostatic effects are much smaller - so much so that the *in vacuo* DFT calculations [12] suggest that the near tbp $[Pd(H_2O)_5]^{2+}$ transition state model has a lower energy than the starting materials. By modifying the Pd and O charges, this feature can also be accommodated within the STET scheme (Table 8).

TABLE 8: Comparison of *in vacuo* structures (Å) and energies (kcal mol^{-1}) calculated for model ligand exchange processes at Pd(II) centres. (STET parameter values for $[PdCl_n]^{2-n}$: D_0 100 kcal mol^{-1}, α 0.8, b_0 2.49 Å; a_0 36400 cm^{-1}, a_1 -10600 cm^{-1}; ρ(Pd) +0.47, ρ(Cl) -0.67; $[Pd(H_2O)_n]^{2+}$: D_0 100 kcal mol^{-1}, α 0.7, b_0 2.38 Å; a_0 30000 cm^{-1}, a_1 -11000 cm^{-1}; ρ(Pd) +1.2, ρ(O) -0.2; Standard van der Waals parameters used for Cl and O.)

L	Pd-L		Pd-L_{ax}		Pd-L_{eq}		$E^{\ddagger}$	
	STET	DFT	STET	DFT	STET	DFT	STET	DFT
Cl	2.33	2.31	2.28	2.30	2.62	2.64	141	143
H_2O	2.03	2.01	1.96	1.98	2.26	2.28	-6	-7

The results from this idealised treatment have been confirmed by complete CLFSE/MM calculations with the proviso that the Pd-L_{ax} distances have to be constrained to a set of selected values while the rest of the molecule is minimised. The strain energy for each geometry is then recalculated (without optimisation) and the results plotted to determine the minimum energy configuration.

In conjunction with full transition state optimisation, this approach offers many exciting possibilities. Of special relevance to this meeting is the prospect of modelling the detailed mechanism of Pt binding to DNA [35]. Much of this binding is under kinetic control implying that the nature of the transition state is crucial to selectivity. Previous modelling studies [36] have been unable to treat the activated complex in any rigorous way. The STET approach represents a unique method for a more complete study of this important problem.

5. Conclusions

The combination of conventional Molecular Mechanics with a term for treating explicitly the stabilisation energy associated with incompletely filled d-orbitals provides a powerful tool for modelling the structures of coordination complexes. This CLFSE/MM approach is especially suited to studying metals in biological systems since no symmetry constraints are required to handle the often unusual geometries and ligand sets.

Some highlights of the CLFSE/MM model include:

- six-coordinate Cu^{II} complexes automatically show Jahn-Teller elongations
- four coordinate CuN_4 centres are automatically planar
- a single set of FF parameters treats 4, 5 and 6 coordinate Cu^{II} species
- a single set of FF parameters treats both high and low-spin d^8 Ni^{II} complexes

Further applications described above illustrate how the model can easily be extended to biologically relevant complexes. The implicit ability to treat the stereochemical activity of the d-electrons offers possibilities for modelling reaction mechanisms. In particular, the FF parameters can be tuned to provide a single set which simultaneously models both ground and transition states for ligand exchange processes at Pt(II) centres. This may pave the way for the *de novo* design of new Pt-based anti-cancer agents.

Acknowledgements

We acknowledge the support of the Engineering and Physical Sciences Research Council for the provision of a studentship to VJP.

References

1. da Silva, J. J. R. F., Williams, R. J. P. (1991) The Biological Chemistry of the Elements: The Inorganic Chemistry of Life, Oxford University Press, New York, USA.

2. Comba, P., Hambley, T. W. (1995) Molecular Modeling of Inorganic Compounds, VCH, Weinheim, Germany.
3. Burton, V. J., Deeth, R. J., Kemp, C. M., Gilbert, P.J. (1995) Molecular mechanics for coordination complexes: the impact of adding d-electron stabilisation energies, *J. Amer. Chem. Soc.* **117**, 8407-8415.
4. Gerloch, M., Harding, J. D., Woolley, R. G. (1981) The context and application of ligand-field theory, *Struct. Bonding (Berlin)* **46**, 1-46.
5. Burton, V. J., Deeth, R. J. (1995) Molecular modelling for copper(II) centres, *Chem. Commun.* 573-574.
6. Deeth, R. J., Paget, V. J. (1997) Molecular mechanics calculations on imine and mixed-ligand systems of Co^{III}, Ni^{II} and Cu^{II}, *J. Chem. Soc., Dalton Trans.* 537-546.
7. Deeth, R. J., Hitchman, M. A. (1986) Factors influencing Jahn-Teller distortions in six-coordinate copper(II) and low-spin nickel(II) complexes, *Inorg. Chem.* **25**, 1225-1233.
8. Wiesemann, F., Teipel, S., Krebs, B., Hoeweler, U. (1994) Force-field calculations on the structures of transition-metal complexes 1. Application to copper(II) complexes in square-planar coordination, *Inorg. Chem.* **33**, 1891-1898.
9. Allured, V. S., Kelly, C. M., Landis, C. R. (1991) Shapes empirical force-field - new treatment of angular potentials and its application to square-planar transition-metal complexes, *J. Am. Chem. Soc.* **113**, 1-12.
10. Gillespie, R. J., Nyholm, R. S. (1957) Inorganic stereochemistry, *Quart. Rev.* **11**, 339-380.
11. Kepert, D. L. (1987) Comprehensive Coordination Chemistry, Vol. 1, Pergamon Press., Oxford, UK.
12. Deeth, R. J., Gerloch, M. (1985) Stereochemical activity of d shells in trigonal-bipyramidal complexes of the first transition series, *Inorg. Chem.* **24**, 4490-4493.
13. Gerloch, M. (1983) Magnetism and Ligand Field Analysis, Cambridge University Press, New York, USA.
14. The DOMMINO program (D-Orbitals in Molecular Mechanics of INOrganic complexes) is based on in-house MM software, kindly provided by Dr D. J. Osguthorpe, Molecular Graphics Unit, University of Bath, UK, modified to include the CLFSE term.
15. Hancock, R. D. (1989) Molecular mechanics calculations as a tool in coordination chemistry, *Prog. Inorg. Chem.* **37**, 187-292.
16. Deeth, R. J. *unpublished results.*
17. Gillespie, R. J. (1963) The stereochemistry of five-coordination. Part II. Transition elements, *J. Chem. Soc.* 4679-4685.
18. Harrison, W. D., Kennedy, D. M., Power, M., Sheahan, R., Hathaway, B. J. (1981) A structural profile of the bis(2,2'-bipyridyl)monochlorocopper(II) cation - crystal structures of bis(2,2'-bipyridyl)monochloro-copper(II) perchlorate and the nitrate trihydrate, *J. Chem. Soc., Dalton Trans.* 1556-1564.
19. Davies, I. W., Gerena, L., Castonguay, L., Senanayake, C. H., Larsen, R. D., Verheoven, T. R., Reider, P. J. (1996) The influence of ligand bite angle on the enantioselectivity of copper(II)-catalyzed Diels-Alder reactions, *Chem. Commun.* 1753-1754.
20. Deeth, R. J., Davies, I. W. *in preparation.*
21. Deeth, R. J., Gerloch, M. (1984) Redirected ligand field analysis 1. Ligand fields of coordination voids and "semi-coordination" in copper(II) complexes, *Inorg. Chem.* **23**, 3846-3853.
22. McGinnety, J. A. (1972) Cesium tetrachlorocuprate. Structure, crystal forces, and charge distribution, *J. Amer. Chem. Soc.* **94**, 8406-8413.
23. Hitchman, M. A., Cassidy, P. J. (1979) Polarized crystal spectrum of bis(methylphenethylammonium)tetrachlorocuprate(II): analysis of the energies, vibrational fine structure, and temperature dependence of the "d-d" transitions of the planar $CuCl_4^{2-}$ ion, *Inorg. Chem.* **18**, 1745-1754.

24. Sabolovic, J., Rasmussen, K. (1995) In vacuo and in crystal molecular-mechanical modeling of copper(II) complexes with amino-acids, *Inorg. Chem.* **34**, 1221-1232.

25. Bray, M. R., Deeth, R. J., Paget, V. J., Sheen, P. D. (1997) The relative performance of the local density approximation and gradient corrected density functional theory for computing metal-ligand distances in Werner-type and organometallic complexes, *Int. J. Quant. Chem.* **61**, 85-91.

26. Deeth, R. J., Munslow, I. J. *in preparation.*

27. Cotton, F. A., Wilkinson, G. (1988) Advanced Inorganic Chemistry (5th ed.), Wiley-Interscience, New York, USA.

28. Jordan, R. B. (1991) Reaction Mechanisms of Inorganic and Organometallic Systems, Oxford University Press, New York, USA.

29. Loevqvist, K (1996) A crystallographic study of platinum(II) complexes, PhD Thesis, University of Lund, Sweden.

30. Paget, V. J. (1996) The modelling of transition metal centres, PhD Thesis, University of Bath, UK.

31. Deeth, R. J., Elding, L. I. (1996) Theoretical modelling of water exchange on $[Pd(H_2O)_4]^{2+}$, $[Pt(H_2O)_4]^{2+}$ and *trans*-$[Pt(H_2O)_2Cl_2]$, *Inorg. Chem.* **35**, 5019-5026.

32. Deeth, R. J. (1996) Transition state symmetries and theoretical activation enthalpies for chloride exchange at planar $[PdCl_4]^{2-}$, *Chem. Phys. Lett.* **261**, 45-50.

33. Dasent, W. E. (1982) Inorganic Energetics, 2nd ed., Cambridge University Press, Cambridge, UK.

34. Elmroth, S., Bugarcic, Z., Elding, L.I. (1992) High-pressure stopped-flow study of kinetics and mechanism for complex-formation reactions of tetraaquapalladium(II) and tetraaquaplatinum(II) with thioethers in aqueous-solution, *Inorg. Chem.* **31**, 3551-1554.

35. Fouchet, M. H., Guittet, E., Cognet, J. E. A., Kozelka, J., Gauthier, C., Le Bret, M., Zimmermann, K., Chottard, J. C. (1997) Structure of a nonanucleotide duplex cross-linked by cisplatin at an ApG sequence, *J. Biol. Inorg. Chem.* **2**, 83-92.

36. Kozelka, J., Petsko, G. A., Lippard, S. J., Quigley, G. J. (1985) Molecular mechanics calculations on cis-$[Pt(NH_3)_2(D(GpG)]$ adducts in 2 oligonucleotide duplexes, *J. Amer. Chem. Soc.* **107**, 4079-4081; Kozelka, J., Petsko, G. A., Quigley, G. J., Lippard, S. J. (1986) High-salt and low-salt models for kinked adducts of cis-diamminedichloroplatinum(II) with oligonucleotide duplexes, *Inorg. Chem.* **25**, 1075-1077.

COMPUTATIONAL ANALYSIS OF INORGANIC AND BIO-INORGANIC NICKEL COMPLEXES

CSILLA CSIKI, KAREN M. NORENBERG, CHRISTINA M. SHOEMAKER AND MARC ZIMMER*

Connecticut College

Chemistry Department

New London, CT 06320

U.S.A.

1. Introduction

The idea of using a mathematical model, based on the ball and spring concept, to describe the geometry of molecules, as is done in molecular mechanics (MM), has been in the literature for more than 50 years. However its use was fairly rare until the advent of relatively inexpensive workstations which has led to a large proliferation of MM programs and to the everyday use and acceptance of MM methods, especially in organic and bioorganic chemistry. Bioinorganic and inorganic MM calculations are less common and their use has been limited by a number of factors.

- There are no commercial MM programs with reliable force field parameters for all the transition metals. Organic force fields need to contain parameters for relatively few atoms (C,H,N,O,S,Cl...). A general bioinorganic force field would have to contain parameters for a number of different transition metals in all their oxidation and spin states, in addition to the organic parameters.
- It is difficult to parameterize a bioinorganic MM force field from solid state structures as there are often insufficient high resolution crystal structures of analogous molecules.
- Deriving force constants from infra-red spectra is not a trivial procedure, because metal-ligand vibrational frequencies are low and their analysis is

L. Banci and P. Comba (eds.), Molecular Modeling and Dynamics of Bioinorganic Systems, 105–129.

often complicated by strong electronic effects, as well as by mixing with ligand vibrational modes [1].

- Ab initio calculations of metal centers are not straight forward, and are of limited use in parameterization.
- Partial charges around metal ions are not easily obtained, and though they are often ignored in inorganic MM, their importance in bioinorganic MM is still being debated [2].

In this paper I would like to use our research on nickel complexes to show how we are dealing with these problems. I will start with our analysis of nickel complexes containing six-membered rings, progress to discussions of larger sixteen membered macrocyclic complexes and finally present our initial modeling of the active site of urease.

1.1. NICKEL

Nickel is found in many coordination compounds and in the last 15 years it has been found to be an essential component of four enzymes, urease, carbon monoxide dehydrogenase, methyl coenzyme-M reductase and hydrogenase [3]. The computational analysis of nickel complexes is complicated by the fact that nickel is a d^8 system, which can adopt high-spin and low-spin forms, and that these have been found in a variety of coordination geometries. High-spin nickel(II) can adopt six-coordinate (octahedral), five-coordinate (trigonal-bipyramidal (tbp) and square-pyramidal (spy)), and four-coordinate (tetrahedral) geometries, while low-spin complexes have been found in five coordinate (tbp,spy) and four-coordinate (square-planar) geometries.

Numerous molecular mechanical (MM) analyses of nickel complexes have been published, in this paragraph we will list just a few of them. Comba [4], Hancock [5,6] and Lindoy [7] have all developed parameters for nickel(II) and used them for the analysis of small coordination compounds. Drew has modeled the adsorption of tetraaza[14]annulenes and their nickel complexes by a carbon surface [8], Shelnutt has published numerous papers dealing with the non planar deformations of nickel porphyrins [9], and we have analyzed and proposed a mechanism for the action of methyl coenzyme-M reductase [10]. Comba has used the angular overlap model together with MM to predict and interpret electronic spectra of transition metal complexes including nickel [11], and Houk has modeled organometallic nickel complexes [12].

2. Six-membered nickel(II) rings [13]

2.1. INTRODUCTION

The Cambridge Structural Database (CSD) contains the coordinates of over 150,000 crystal structures [14]. This wealth of information has barely been tapped. We have searched the CSD for all the complexes having the substructure shown below, where X is N, O, P or S and all carbon and ligating atoms are sp^3 hybridized. Many papers, and even some books have been written about the conformations of cyclohexane and its organic derivatives. In contrast very little attention has been paid to the conformation of six-membered rings in inorganic systems. Hancock has published many interesting papers in which he has analyzed the stability and bite-size of five and six-membered rings [5]. However, he only considered the six-membered chair conformation. We have recently studied, in the first inorganic cluster analysis to be published, the conformations of cobalt(III) propanediamine and its derivatives [15]. We have expanded this study to complexes containing nickel(II) in saturated six-membered ring systems to establish

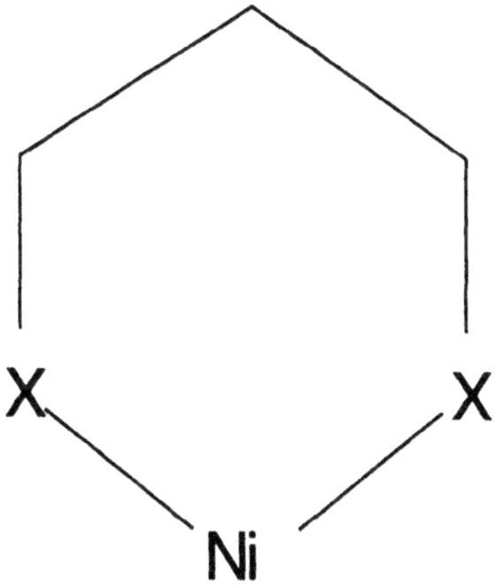

- whether cluster analysis can be used to find significant groupings in small inorganic molecules;
- whether the spin state of the metal ion has an effect on the conformation of the six-membered rings;
- whether having ligating atoms other than nitrogen in the six-membered ring has any conformational effect;

- whether the conformations of the six-membered rings are influenced by the coordination geometry of the metal ion;
- whether inorganic molecular mechanical calculations of molecules involving six-membered rings could be done by minimizing all the possible combinations of chair, twist boats and boats, or whether a conformational search had to be undertaken in which the internal dihedral angles of the six-membered rings were varied. In the vast majority of inorganic molecular mechanical analyses no conformational searches are undertaken. In fact in a recent review of bioinorganic molecular mechanics it was found that less than 10 of the 126 studies being reviewed, undertook a conformational search [16]. Instead, it is common practice to enter all probable structures individually and minimize them separately, or to minimize the crystal structure and ignore all other possible conformations. The possible structures are generated by drawing all possible combinations of the chair, boat and twist-boat conformers for six membered rings, and λ and δ conformers for five membered rings. Isomers and stereoisomers are handled in the same way [17].

2.2. RESULTS AND DISCUSSION

There are 220 nickel(II) complexes with the six-membered substructure shown on the previous page. Some of these complexes have more than one six-membered ring resulting in 433 ring systems, which were all clustered with the Xcluster program [18].

Clustering algorithms attempt to find groupings such that the similarities within the groups are significantly greater than those between groups. There are many different methods of clustering data. Xcluster is an agglomerative, hierarchical single-link clustering method. The method starts with separate conformations and unites them into clusters in a stepwise procedure (agglomerative) by joining pairs of conformations that are closest in conformational space (single-link), if the pair of closest conformations are in adjacent clusters the clusters are fused to form a new larger cluster (hierarchical).

It was found that the best clustering was obtained when all the conformations were divided into 7 different clusters. The abundance (%), conformation and the dihedral angles of the average conformation in each cluster are given in Table 1. The table and overlaps of the members of each cluster (not shown) confirm that cluster analysis can be used to separate the conformations of small inorganic systems.

TABLE 1: The abundance (%), conformational identity and dihedral angles (°) of the average conformation of each cluster.

Cluster #	*φ1*	*φ2*	*φ3*	*φ4*	*φ5*	*φ6*	*Conf.*	*%*
1	-40.0	39.8	-56.5	68.3	-68.0	56.1	chair	83.4
2	-10.9	20.1	-26.5	70.8	-62.3	11.1	sofa	0.3
3	-58.5	57.1	-8.8	60.5	58.9	10.7	boat	10.3
4	-44.0	19.6	14.6	51.2	-81.3	32.1	λ twist-boat	2.4
5	-17.4	-40.4	62.6	-49.5	-23.6	65.5	λ twist-boat	0.8
6	35.9	23.8	-67.0	45.3	26.0	-48.9	δ twist-boat	2.9
7	6.9	-6.9	-4.9	-0.1	0.6	4.4	flat	0.3

Due to the Ni-N distance being much longer than the organic bonds in the six-membered ring, the conformations are not symmetric. This is the reason the average chair conformation has dihedral angles of -40°, 40°, -56°, 68°, -68° and 56° rather than the -60°, 60°, -60°, 60°, -60°, 60° expected on the basis of cyclohexanes structure. A molecular mechanical analysis of nickel(II) propanediamine $[Ni(pn)(H_2O)_4]^{2+}$ revealed that its lowest energy conformation is a chair that is 16.1kJ/mol lower in energy than the twist-boat. The flat, boat and sofa conformations are not energy minima of nickel(II) propane diamine.

The cluster analysis showed (Table 1) that most of the rings are in the chair conformation, this is not unexpected because the chair is the lowest energy conformation. However, it is surprising to find 10% of the rings in the boat conformation since boats are not even an energy minimum conformation. Clearly, the intermolecular interactions and ligand topologies of the individual complexes that were ignored in the minimization of nickel(II) propanediamine are responsible for the six-membered rings adopting the high energy boat conformation. An inspection of the complexes which contain six-membered rings in a boat conformation revealed that they are all due to the six-membered ring being part of a bicyclic system as shown on the left bottom or due to the rings being 'bound back' to the metal ion as shown on the left top. The sofa conformation is found in a nickel complex that contains three other six-membered rings, two of which are chairs and one is a twist boat. A molecular mechanical conformational search of this compound [19] revealed that this was the lowest energy conformation and that non-bonded repulsions from adjacent six-membered rings forced one of the six-membered rings to adopt a sofa conformation.

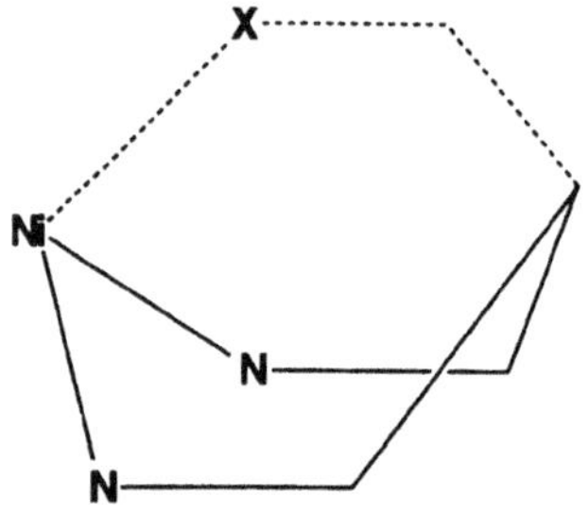

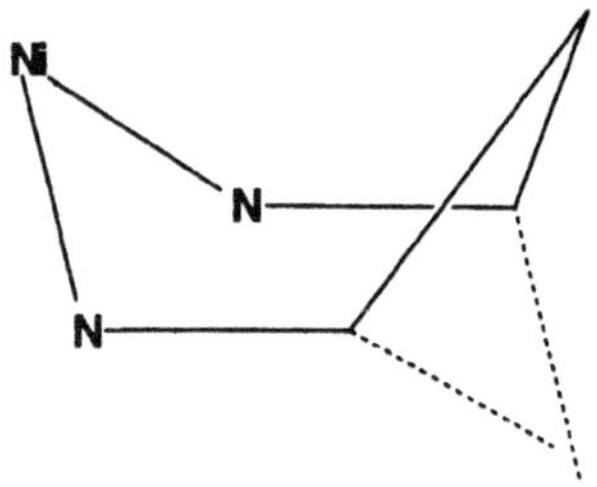

No correlation was found between the Ni-N distance, which is a function of the spin state, and the cluster membership or the Cremer-Pople parameters. It was also found that there was no correlation between the coordination geometry around the metal ion or the nature of the metal ligating atoms and the conformation of the six-membered rings.

One of the aims of this analysis was to establish whether inorganic molecular mechanical calculations of molecules involving six-membered rings could be done by minimizing all the possible combinations of chair, twist boats and boats, or whether a conformational search had to be undertaken in which the internal dihedral angles of the six-membered rings were varied. Since the chair conformation is the global minimum conformation we assumed that, if there were any problems with using ideal six-membered ring conformations as starting conformations, these problems would occur in the complexes in which the rings adopt higher energy non-chair conformations. We therefore analyzed all complexes containing at least one non-chair six-membered ring, to see whether the solid state structure would be obtained upon minimizing from ideal starting conformations. Starting geometries were drawn and the six-membered rings were constrained to the idealized chair, boat or twist-boat nearest to the conformation observed in the solid state. The structures were minimized before releasing the constraints and reminimizing. In all cases this procedure found the solid state minimum, in fact the boat conformation was not a required starting conformation, the crystal structure conformation was always obtained if only the twist-boats and chairs were used as starting conformations.

Although the solid state structure was always in the energy minimum closest to the starting geometry, it wasn't always in the global energy minimum. We found that both copper(II) and nickel(II) sixteen membered macrocycles adopted non-global energy minima and therefore analyzed them in more detail.

3. Metal incorporation into sixteen-membered macrocycles.

3.1. INTRODUCTION

Complexed sixteen membered macrocycles can adopt four different configurations depending on the stereochemistry around the ligating atoms. When a metal ion is bound to 1,5,9,13-tetraazacyclohexadecane, which is commonly abbreviated 16aneN_4, the nitrogen atoms are each bound to one hydrogen. This hydrogen can be above (+) or below (-) the tetraaza plane as shown below. The four different configurations arising from this are the ++++, +++-, ++-- and +-+- configuration. The same configurations can occur in 16aneS_4 macrocycles where the lonepairs on the sulfur atom can be above and below the tetrathio plane.

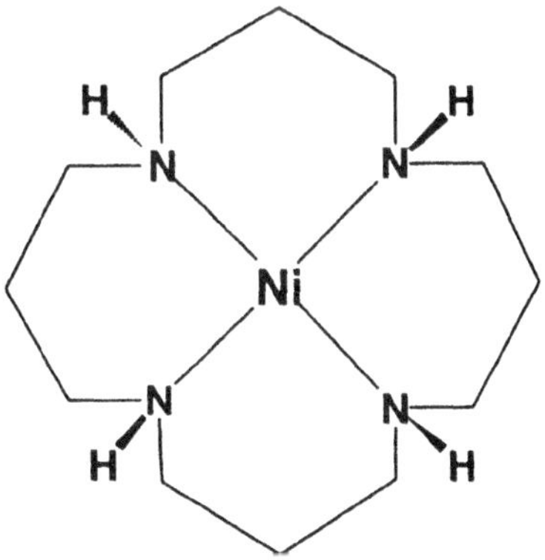

Ni(16aneN_4) +++-

3.2. RESULTS AND DISCUSSION

The solid state 16aneN_4 nickel(II) structure is square planar and has the +,+,+,- configuration with its four six-membered rings in chair, chair, δ twist-boat and λ twist-boat conformations. Minimizing nickel(II) 16aneN_4 from a conformation with the rings in idealized chair, chair, δ twist-boat and a λ twist-boat conformations resulted in the conformation observed in the crystal structure. However, a thorough configurational/conformational search revealed that the +-+- configuration is the most stable square planar Ni(II)16aneN_4 configuration. This configuration has the four six-membered rings in alternating λ and δ twist boat conformations, see Table 2. It is 1.23 kJ/mol lower in energy than the +++- configuration, that is observed in the solid state structure. This energy difference can easily be accounted for by packing forces.

Sixteen membered tetrathio macrocycles exhibit similar behavior. The crystal structure of nickel 16aneS_4 has a ++-- configuration with its six-membered rings in alternating chair and twist-boat conformations. Our molecular mechanical analyses have shown that this is the lowest energy conformation available to the ++-- configuration but that, just like for 16aneN_4, there is a configuration that is lower in energy than the solid state configuration. Table 2 shows that the ++++ configuration with all the six-membered rings in the chair conformation is 18.7 kJ/mol lower in energy than the ++-- configuration. This is a significantly large energy difference that cannot attributed to packing forces.

Table 2: All the configurations of nickel(II) 16aneN_4 and 16aneS_4 with the conformations of the six-membered rings and the relative strain energy in kJ/mol.

Macro.	Config	Ring 1	Ring 2	Ring 3	Ring 4	Rel. Energy*
16aneN_4	++++	Chair	λ-tb	Chair	Sofa	-267.98
	+++-	Chair	Chair	δ-tb	λ-tb	-280.35
	++--	Chair	λ-tb	Chair	δ-tb	-279.86
	+-+-	δ-tb	λ-tb	δ-tb	λ-tb	-281.58
16aneS_4	++++	Chair	Chair	Chair	Chair	-603.00
	+++-	Chair	Chair	δ-tb	λ-tb	-581.27

++--	Chair	λ-tb	Chair	δ-tb	-584.33
+-+-	λ-tb	δ-tb	λ-tb	δ-tb	-536.08

(* the energies differ from Table 3 because these calculations were carried out using Hooke's law type bond angle deformations, while those in Table 3 were obtained with a "points on a sphere" model. The same minimum structures and trends are obtained using both methods.)

The energies of the different configurations for square planar 16aneS_4 complexes were obtained as a function of the metal-ligand distance using the method of Hancock [20]. Similar results were obtained when using Wiberg's method [21]. The ++++ configuration, which has all its six-membered rings in the chair conformation, is the lowest energy configuration for metal-sulfur distances greater than 2.15Å. A CSD search found square planar 16aneS_4 complexes with Cu [22], Rh[23], Ni[29], Mo[24], Fe[25] Hg[26] and Cd [42] metal ions. The Cu, Rh, Ni, Fe and Cd complexes all adopt the higher energy ++-- configuration, and only Mo adopts the global minimum structure. The solid state structure of the mercury 16aneS_4 complex is in the +-+- configuration, which is over 90 kJ/mol higher energy than the ++++ configuration at a metal-sulfur distance of 2.60Å. Presumably this high energy configuration is due to the metal incorporation process and is observed in the solid state because there is no mechanism of configurational interconversion, or if interconversion can occur it is slower than the crystallization process. A very likely penultimate intermediate in the metal incorporation that could be responsible for the observed solid state configuration is a cis or folded intermediate, as shown in the mechanism on the next page.

We therefore calculated the strain energy of all the configurations of folded 16aneS4. The results are shown in Table 3. These results might explain why the planar ++++ configuration that is the global minimum configuration is not adopted in the solid state. If the mechanism proposed here, or any other mechanism involving a folded penultimate structure, is correct then the ++++ planar configuration is not found because the ++++ folded intermediate is extremely strained.

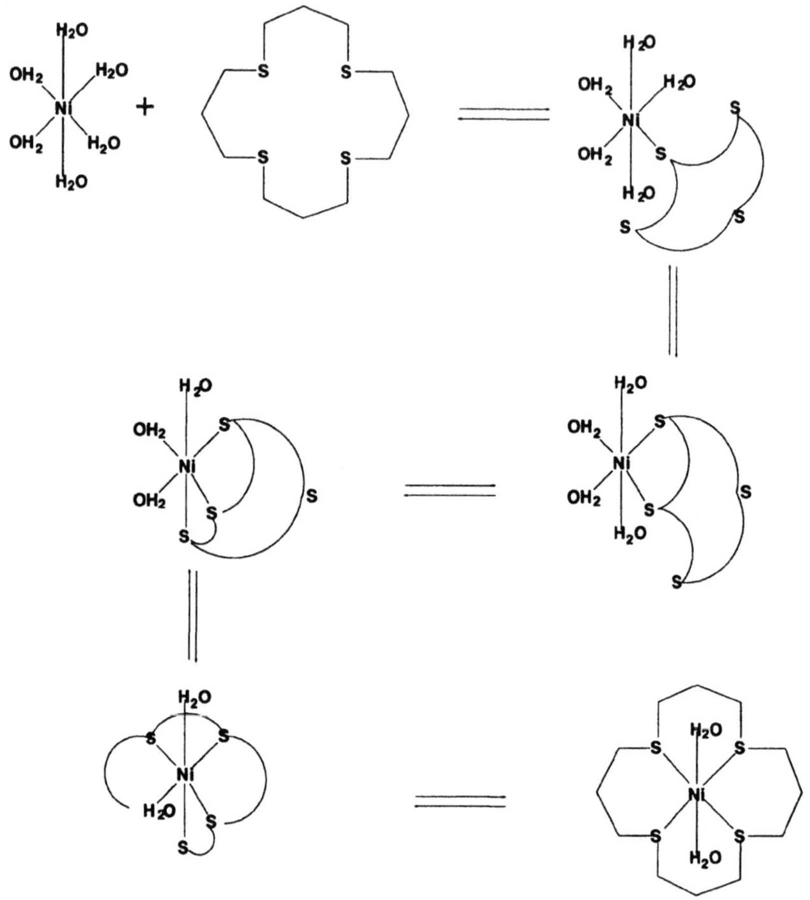

Table 3: Relative energies (kJ/mol) of all the configurations of nickel(II) 16aneS4 in both the planar and the folded conformation.

	++++	+++-	++--	+-+-
Planar	86.97	112.85	102.47	142.28
Folded	236.72	161.37	122.90	107.21

A +-+- or ++-- intermediate folded structure, which would result in a +-+- or ++-- planar configuration is much more likely. If the conversion barrier between the different configurations is high this would explain why the solid state structures of Cu, Rh, Ni, Fe, Cd and Hg adopt square planar and octahedral +-+- and ++-- configurations, although they are higher in energy than the ++++ planar configuration. We think it is very important to consider the mechanism of metal incorporation when designing a chelating agent.

4. Urease

4.1. INTRODUCTION

Urease catalyzes the hydrolysis of urea to ammonia and carbamate, which is unstable and subsequently degrades to produce a second ammonia molecule and carbonic acid.

$$H_2N\overset{O}{\overset{\|}{C}}NH_2 + H_2O \longrightarrow NH_3 + H_2N\overset{O}{\overset{\|}{C}}OH$$

$$H_2N\overset{O}{\overset{\|}{C}}OH + H_2O \longrightarrow NH_3 + H_2CO_3$$

The catalyzed reaction is 10^{14} times the rate of the uncatalyzed reaction in which urea breaks down to ammonia and cyanic acid.

$$H_2N\overset{O}{\overset{\|}{C}}NH_2 \longrightarrow NH_3 + HOCN$$

Urease is found in bacteria, fungi and higher plants. It is of importance in both agriculture and in medicine. In agriculture urease allows microorganisms to use urea as a source of nitrogen [27], however urea-based fertilizers are rapidly broken by urease resulting in phytopathicity and loss of volatilized nitrogen. Medicinally urease activity has been shown to be important in the development of urinary stones, acutc pyloncphritis, peptic ulceration and urinary catheter obstruction. The first enzyme ever to be crystallized was urease and for many years textbooks used urease as an example of an enzyme that could catalyzed its reactions without a coenzyme or metal ion. However in 1975 it was reported that urease contains nickel and that in fact the active site probably contained two nickel ions [28]. Fifty years after first being crystallized the solid state structure of urease was finally solved in 1995 [29]. The enzyme has three bi-nickel sites located in a tightly packed trimer of trimers. One of the nickel ions, labeled Ni-1, is pseudotetrahedrally coordinated by His246, His 272, Lys217 and a weakly coordinated water. Ni-2 is in a distorted trigonal bipyramidal or distorted square pyramidal geometry and is coordinated to Lys217, His134, His136, Asp360 and a water molecule. The two nickel ions are bridged by a carbamylated lysine, this explains why carbon dioxide is required for the in vitro incorporation of nickel [30]. The structure of the apoenzyme has been solved and is nearly identical to active holoenzymes except in the active site where the nickel ions and the carbamate are replaced by water [31]. In the apoenzyme Lys217 is not carbamylated, even in the presence of carbon dioxide. Nickel binding and carbamylation seem to be related processes. The crystal structures of three inactive urease mutants have also been solved. Replacing His134, which coordinates Ni-2, with an Ala results in a mononickel metallocenter. The remaining nickel changed from pseudotetrahedral to octahedral geometry by coordinating additional water molecules. A comparison of the H134A mutant with the wild-type urease shows very little difference, except for the changes mentioned and in residues 314 to 331 that comprise a flap which covers the metallocenter [32]. In the H320A mutant a solvent water molecule that is primarily associated with Ni-2 in the wild-type urease, binds Ni-1 as well and assumes a bridging role. The 10^5 fold lower activity of H320A could be due to the bridging water molecule taking the coordination site of the urea. The major structural differences between H219A and the wild-type urease are not in the active site, but at the flap. Possibly this prevents the migration of the urea molecule to the bi-nickel active site [32]. The structures of the wild-type binickel urease, the apoenzyme, and the three mutants are all remarkably similar. It is surprising that Zn^{2+}, Cu^{2+} and Co^{2+}, which are found in the active site of other hydrolytic enzymes are not functional in urease [33]. Urease requires nickel, perhaps this is a consequence of the extreme preorganization of the apoenzyme and its fairly complex metal incorporation process, as well as nickel's ability to adopt 4,5 and 6 coordinate geometries.

4.2. COMPUTATIONAL DESIGN OF A UREASE MODEL SYSTEM [34].

Structural and functional model compounds are often synthesized to model the active site of a protein [35]. To date choosing a ligand to achieve some desired structure in a bioinorganic model compound has largely been a matter of trial and error. Many efforts are being undertaken to make this process more efficient, the most promising of these efforts are in the field of combinatorial chemistry [36]. We will present a computational method for the rational design of model compounds with specific structural properties, and we will show how the method has been used to propose a series of potential urease mimics.

The first step in designing a model system is to establish a list of properties the model is to mimic. In order to prevent duplicating published work and to use as much of the available structural data as possible, the Cambridge Structural Database (CSD) [37] is searched for any structures that have properties similar or identical to those desired for the mimic. The CSD is used because it is a vastly underutilized resource that contains more than 150,000 crystal structures, and can easily and efficiently be searched for structural motifs.

The graphic "building" modules of modeling programs are used to patch together the structural units obtained in the CSD search or to modify the ligands so that they have the desired properties of the model compound. The steric feasibility of the designed complexes are established by molecular mechanical methods, and if specific electronic properties are required, quantum mechanical methods can be used to screen potential models for those having the most desired properties. Molecular mechanics (MM) has often been used to analyze bio-inorganic systems [16], however it has not been used to design model systems. In the remainder of this section we will show how this method has been used to design a family of urease model systems.

A mechanism for action of urease has been proposed on the basis of its structure [38].We would like to design a model compound of the complex enclosed in the dashed box. The properties that we were looking for in our model was that it would be easy to synthesize, contain two nickel ions, and have a cavity between the two metal ions that is large enough to contain urea bound in the manner shown. It would be a structural model of the intermediate and not necessarily an active model of urease.

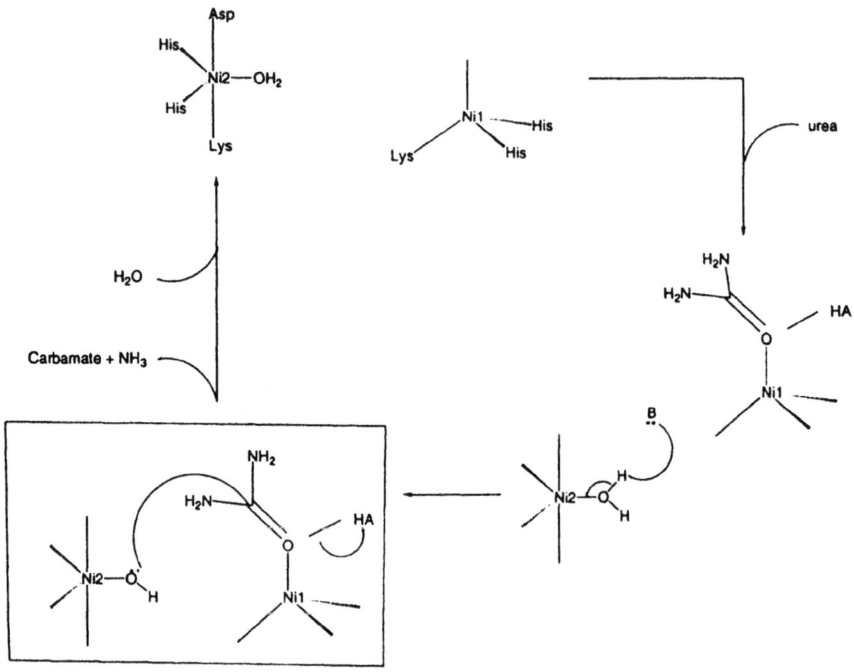

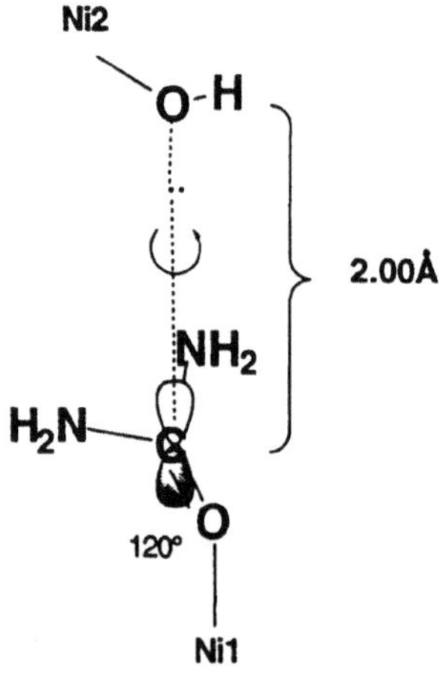

Two approaches can be taken to designing the model. A bridging ligand can be used to keep the two nickels about 3.5Å apart as is observed in crystal structure, or a more rigid ligand system can be used that will hold the two metal ions in a way that allows for the formation of the intermediate shown in the proposed mechanism. We have chosen the latter approach.

The minimum and maximum Ni-Ni distance desirable for a model system were determined by drawing an ideal intermediate as shown on the left. Monitoring the Ni-Ni distance while rotating the O-C-O-Ni2 dihedral angle reveals that this distance should be between 3.16Å and 5.36Å if the ideal geometry shown is assumed.

A CSD search for structures containing two nickel ions bridged by a RCO_2 group found 61 complexes. Fourteen percent of these structures had Ni-Ni distances of less than 2.80Å, all these complexes had a nickel-nickel bond. It was also found that all structures with Ni-Ni distances of between 2.8 and 3.6Å were bridged not only by a RCO_2 group but by an additional group like a water molecule. Complexes, like urease, with no Ni-Ni bonds and no additional bridging groups have Ni-Ni distances longer than 3.7Å.

Therefore the criteria we used for designing our urease mimic were that it should have a ligand framework that will hold the two nickel ions at the calculated distance (between 3.00 and 5.80Å), and coordinate the metals through oxygens and nitrogens, as these make up the coordination sphere of the metals in the enzyme. We searched the database for all the transition metal complexes fulfilling the criteria. The reasoning behind this decision was that it should be relatively simple to take a ligand system that met all the requirements for a urease model system, but had the wrong metals, and synthesize it with nickel.

Ninety structures were found in the CSD that fulfilled all the requirements and had a metal to metal distance of between 3.00 and 5.80Å. Forty percent of the hits were copper complexes and fifteen percent contained nickel. The most interesting hits were a family of substituted 1,4,8,11-tetrakis(2-pyridylmethyl)-1,4,8,11-tetraazacyclotetradecane macrocycles which hold the metal ions in place as

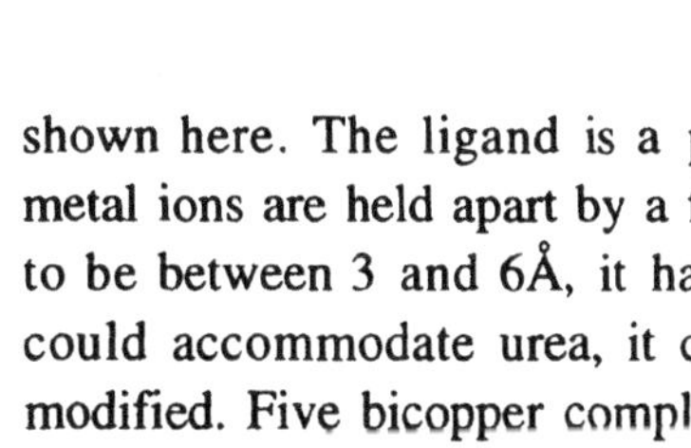

shown here. The ligand is a potential urease structural model because the metal ions are held apart by a fairly flexible framework that allows the metals to be between 3 and 6Å, it has an empty cavity between the metals which could accommodate urea, it can be synthesized in two steps and is easily modified. Five bicopper complexes with L^1 were found with Cu--Cu distances of between 3.993 and 5.741Å. No nickel complexes of the ligand were found, however there is no reason why nickel should not bind to the ligand in a similar fashion to copper.

In order to ensure that the cavity between the metal ions does not collapse and that there is room for urea coordination a molecular dynamics simulation of the binickel complex was run. The crystal structure of one of the bi-copper complexes, in which the coppers were computationally replaced with nickel ions, was used as a starting geometry. The nickel-nickel distance was monitored as a function of time. Although the ligand is fairly flexible, the Ni-Ni distance only varied from 4.40 to 5.40Å, it did not undergo a major conformational change, nor did it collapse.

Another possible shortcoming of the structural model could be that it adopts a low energy conformation that is not a viable structural urease model, and that this conformation was not observed in the molecular dynamics simulation because it is separated from the nickel 'mutated' structure by a large energy hill. In order to ensure that no other low energy conformations exist we undertook a Monte Carlo dihedral conformational search of the complex. All conformations found within 25kJ/mol of the lowest energy conformation are viable urease mimics.

We are currently contemplating actually going into the laboratory and synthesizing these model compounds.

4.3. MOLECULAR MECHANICAL ANALYSIS OF THE ACTIVE SITE OF UREASE.

The structure and mechanism of urease is well suited to a molecular mechanical and dynamical analysis because enough is known about the structure of urease that one has a very good starting point for simulations, however there are still many questions that need to be answered about this important enzyme. The crystal structures of the apoenzyme, the wild-type and the three mutant forms are all very similar. The five structures only vary at the active site and at the flap that covers the nickel ions. Urease is inactive in the crystalline form and gains some, but not all, its activity when the crystals are ground up. The most likely explanation for this observation is that the flap cannot open in the crystalline state thereby preventing the urea from entering the active site [39]. We have taken this, perhaps a little optimistically, as a sign that the enzyme does not undergo major conformational changes and that we can safely concentrate on our computational analysis of the active site.

4.3.1. *Nickel Parameters.*

While there are a number of force fields that have been parameterized for protein simulations, most of them do not include parameters for metal ions. Modeling urease presents difficulties because the wild-type enzyme has two nickel ions, one four-coordinate and the other five-coordinate, while the mononickel mutant has a six-coordinate nickel. Therefore parameters for nickel in octahedral, trigonal bipyramidal, square pyramidal and tetrahedral geometry are required to model the different forms of urease. Parameters for the carbamylated lysine which bridges the two nickel ions in the wild-type enzyme are also required.

In order to have a single set of metal parameters to simulate all coordination geometries we have modified the AMBER* force field as implemented in MacroModel v5.5 [40], it uses a 6,12 Lennard Jones hydrogen bonding treatment [41] and an improved protein backbone parameter set [42]. Hancock's nickel(II) high spin octahedral parameters [43] that were derived for the MOLBLD-3 program were used as the basis for our parameters. They were modified by solid state structure intrapolation [44], of octahedral, trigonal bipyramidal, square pyramidal and tetrahedral nickel(II) crystal structures obtained from the CSD. In order to have a single set of parameters for all coordination geometries the typically used Hooke's law type bond angle description was replaced with a points on a sphere model [17]. The parameters used are given in Table 4.

Table 4: Nickel parameters added to the AMBER* force field in MacroModel v5.5. A points on a sphere model was used to simulate bond angle deformations around the nickel ion. (MacroModel descriptors used)

Bonds

Bond Type	Bond length (Å)	Force constant (kcal/mol)	Bond moment
Ni - N3	2.1000	110.00	-0.0270
Ni - N2	2.0800	110.00	-0.3000
Ni - O3	2.0764	110.00	-0.0270
Ni - O2	2.0440	110.00	0.0270
Ni - Cl	2.4170	90.00	-0.400

Bond Angles

Bond Angle	Angle (°)	Force constant (kcal/mol)
Ni - N3 - H0	109.50	15.00
Ni - N3 - C3	109.50	15.00
Ni - N2 * C2	120.00	15.00
Ni - 03 - C0	109.50	15.00
Ni - 03 - H0	109.50	15.00
Ni - 03 - C0	109.50	15.00

Torsions: All dihedral angles around the nickel ions were set at 0.00

Improper Torsions

	V1/2 (kcal/mol)	V2/2 (kcal/mol)	V3/2 (kcal/mol)
C2*O2*O2*N3	0.000	20.000	0.000

Non-bonded interactions:

	r_o (Å)	ε (kcal/mol)
Ni	2.300	0.1700

Special substructure: (carbamylated lysine)

Ni-O2*C2*O2-Ni

C2*O2	1.24Å	510.00 kcal/mol
O2*C2*O2	126°	80.00 kcal/mol

The parameters given in Table 4 were used to minimize the structures listed in Table 5. The differences between the metal-ligand distances in the solid state structure and the minimized structures are small and compare well with those obtained with other nickel force fields. The differences between the ligand-metal-ligand angles of the crystal structures and the minimized structures are slightly larger. We find this acceptable because our parameters allow us to minimize all coordination geometries of nickel with the same set of parameters.

Table 5: Differences between the crystal structures and minimized structures.

CSD Refcode **Octahedral**	Mean metal-ligand bond length differences (Å)	Mean ligand-metal-ligand bond angle differences (°)
ACENNI	0.124	9.70
ACLBNI	0.169	3.37
ANIENP	0.258	0.00
AQACNJ	0.085	0.00
AQRLB	0.108	0.00
CIHGEF	0.249	2.17
COHWUR10	0.050	8.10
DAIMNI	0.139	4.33
FUGMID	0.166	9.73
HIMZNI	0.094	0.79
HISTNI	0.154	6.10
PEKPAW	0.166	9.73
PEXNAH	0.118	4.51
PHYANI	0.072	5.57
VOFZUL	0.075	5.33
YAHGAP	0.062	0.693

CSD Refcode	Mean metal-ligand bond length differences (Å)	Mean ligand-metal-ligand bond angle differences (°)
Five coordinate		
DEBFUL	0.093	8.95
DOCVIA	0.126	3.00
DOHXED	0.072	6.50
HALBEB	0.053	9.97
MOPAON	0.0247	5.32
NICEAS10	0.306	14.5
SALDNI	0.457	6.00
Tetrahedral		
BURCEW	0.144	2.30
BURCIA	0.171	2.35
BURCOG	0.174	2.53
CESWEC	0.111	14.1
CPIANI	0.186	0.90
DMPLNI	0.084	14.9
FELJIP	0.184	6.21
FEWTUW	0.287	11.1
GLYXNI	0.168	3.18
ISALNI	0.140	9.20
KEHXIE	0.128	2.35
RCO_2 bridges		
CAENNI	0.055	6.67
CONNEY	0.054	9.25
CPRENI10	0.035	7.03
GEGDUR	0.021	4.53

KAMVOJ	0.040	4.07
KUVLOC	0.0675	12.6

4.3.2. Molecular Mechanics Analysis of the Active Site of Wild-Type Urease.

The coordinates of the crystal structure of the wild-type urease were obtained from the Brookhaven Protein Databank (refcode = 1kau). Since the nickel ions are contained in the α subunit of the trimer, the β and γ subunits were deleted. All water molecules except the one coordinated to Ni-2 were removed, hydrogen atoms were added where required and a water was added to Ni-1 so that it adopted a tetrahedral coordination geometry. The structure was minimzed "in vacuo" until the derivative convergence criterion of 0.05kJ/mol was reached. All atoms within 12Å of the two nickel ions were minimized in the calculations and the non-bonded interactions of all these atoms with the atoms between 12 and 15Å from the metals were considered. However the position of the atoms in this outer sphere were fixed as were their internal coordinates. All atoms further than 15Å from the nickel ions were ignored. The minimized structure obtained using our parameters and the approximations enumerated above was significantly different from the solid state structure. Most of the deviations occurred around the metal ions which separated from 3.5Å to 4.30Å. In view of our CSD finding that an additional bridging group was present in all carbonyl bridged nickel complexes with Ni--Ni distances of between 2.8 and 3.6Å, we reminimized the urease solid state structure with a water molecule bridging Ni1 and Ni2. The minimized structure showed remarkable agreement with the solid state structure, indicating that Ni1 and Ni2 are bridged by a water molecule that is weakly coordinated to Ni1. This bridging water molecule was observed in the crystal structure of the H320A mutant [31]. It has been assumed that the 10^5 lower activity observed in the H320A mutant is due to the bridging water taking up the coordination site of the substrate. It seems that this bridging also occurs in the native enzyme.

We are currently using molecular dynamics to investigate the movement of the flap that covers the active site, the internal waters and the amino acids thought to be involved in catalysis. We are also doing calculations with urea in the active site.

5. Experimental

The Cambridge Structure Database (CSD) V5.11 was searched for all nickel(II) six-membered rings having the sub-structure shown in the text. Structures having only sp^3 hybridized carbons and ligating atoms in the six-membered rings were saved. Version 5.11 of the CSD was released in April 1996 and contains 152,464 crystal structures. All the hits were stored and converted to MacroModel format using the csdconv program. All atoms that were not part of the six-membered ring were removed with the draw function in MacroModel [40]. Atoms were renumbered so that all six-membered rings were numbered in the same way, and placed into a masterfile for cluster analysis. The Xcluster [18] program was used for cluster analysis. Proximity matrices were obtained by determining the pairwise distances between six-membered rings using (i) the rms differences between corresponding dihedral angles and (ii) the rms displacement between pairs of identically numbered atoms after optimal rigid-body superimposition. The rings were symmetry adapted as described previously [15].

The correlation/covariance spreadsheet of Vista 2.0 in the CSD was used to establish whether any correlations between structural parameters were present.

MacroModel v5.5 was used for all the molecular mechanical analyses. For the analysis of the six-membered ring systems and the sixteen membered macrocycles the MM2* parameter set was used with some additional parameters. Hancock's [43] and Lindoy's [7] nickel, nitrogen parameters were used and the following nickel(II) sulfur parameters were added Ni-S (r° = 2.42Å, k = 0.68mdyn/Å), S-Ni-S (θ° = 90° or 180°, k = 0.50mdyn/rad^2) and Ni-S-C (θ° = 109.5°, k = 0.50mdyn/rad^2). As is usually the case in inorganic molecular mechanics the torsions around the metal ion were taken as zero. Molecular dynamics runs were used to sample the conformational space around the different 16aneN$_4$ and 16aneS$_4$ configurations. One nanosecond simulations at 1000K were sampled 100 times and the sampled structures were minimized to find low energy conformations.

In order to model the active site of urease the AMBER* parameters in MacroModel were modified as described in the text.

References

1 Nakamoto, K. *Infrared and Raman Spectra of Inorganic and Coordination Compounds*, 3rd ed.; Wiley: New York, **1978**.

2 Vedani, A.; Dunitz, J.D. *J. Am. Chem. Soc.* **1985**, *107*, 7653.

Hoops S.C., Anderson, K.W., and Merz, K.M. Jr. (1991) Force field design for metalloproteins, *J. Am. Chem. Soc.* **113**, 8262-8270, and other papers in this book.

3. Halcrow M.A., and Christou G. (1994) Biomimetic chemistry of nickel, *Chem. Rev.* **94**, 2421-2481.

4. Comba, P. (1993) The relationship between ligand structures, coordination stereochemistry, and electronic and thermodynamic properties, *Coord. Chem. Rev.* **123**, 1-48. Comba, P and Hambley, T.W. (1995) *Molecular Modeling of Inorganic Compounds*, VCH, Weinheim, Germany.

5. Martell, A.E., Hancock, R.D., and Motekaitis, R.J. (1994) Factors affecting stabilities of chelate, macrocyclic and macrobicyclic complexes in solution, *Coord. Chem. Rev.* **133**, 39-65. Hancock, R. D. (1990) Molecular mechanics calculations and metal ion recognition, *Acc. Chem. Res.* **23**, 253-257.

6 Hancock, R.D., Dobson, S.M., Evers, A., Wade, P., Ngwenya, M.P., Boeyens, J.C.A., and Wainwright, K.P. (1988) More rigid macrocyclic ligands that show metal ion size-based selectivity. A crystallographic, molecular mechanics, and formation constant study of the complexes of bridged cyclen, *J. Am. Chem. Soc.* **110**, 2788-2794.

7 Adam, K.R., Antolovich, M., Brigden, L.G., and Lindoy, L.F. (1991) Comparative molecular mechanics study of the low-spin nickel(II) complexes of an extended series of tetraaza macrocycles, *J. Am. Chem. Soc.* **113**, 3346-3351.

8 Drew, M.G.B., Jutson, N.J., Mitchell, P.C.H., Potter, R.J., and Thompsett, D. (1993) Experimental and computer modelling studies of carbon-supported metal complexes, *J. Chem. Soc, Faraday Trans.* **89**, 3963-3973.

9 Song, X.-Z., Jentzen, W., Jia, S.-L., Jaquinod, L., Nurco, D.J., Medforth, C.J., Smith, K.M., and Shelnutt, J.A. (1996) Representation of nonplanar structures of nickel(II) 5,15-disubstituted porphyrins in terms of displacements along the lowest-frequency normal coordinates of the macrocycle, *J. Am. Chem. Soc.* **118**, 12975-12988.

10 Zimmer, M., and Crabtree, R.H. (1990) Bending of the reduced porphyrin of factor F430 can accomodate a trigonal-bipyramidal geometry at nickel: A conformational analysis of this nickel-containing tetrapyrrole, in relation to archaebacterial methanogenesis. *J. Am. Chem. Soc.* **112**, 1062-1066. Zimmer, M. (1993) Empirical force field analysis of the revised structure of coenzyme F430. Epimerization and geometry of the corphinoid tetrapyrrole, *J. Biomol. Struct. & Dyn.* **11**, 203-214.

11 Bernhardt, P.V., and Comba, P. (1993) Prediction and interpretation of electronic spectra of transition metal complexes via the combination of molecular mechanics and angular overlap model calculations, *Inorg. Chem.* **32**, 2798-2803.

12 Gugelchuk, M.M., and Houk, K.N. (1994) Stereoselective organometallic reactions; A force field study of pi-allyl intermediates in nickel(0)-catalyzed cycloadditions, *J. Am. Chem. Soc.* **116**, 330-339.

13 Norenberg, K.N., Shoemaker, C.M., and Zimmer, M. Molecular mechanics and cluster analysis of nickel(II) six-membered rings. *J. Chem. Soc., Dalton Trans.* in press.

14 Allen, F.H., and Kennard, O. (1993) 3D search and research using the Cambridge Structural Database, *Chem. Design Automation News* **8** 31-37.

15 DaCruz, M.F., and Zimmer, M. (1996) Cluster and molecular mechanical analysis of the conformation of all six-membered cobalt(III) diamine rings in the Cambridge Structure Database, *Inorg. Chem.* **35**,

2872-2877.

16 Zimmer, M. (1995) Bioinorganic molecular mechanics, *Chem. Rev.* **95**, 2629-2649.

17 Hay, B. P. (1993) Methods for molecular mechanics modeling of coordination compounds, *Coord. Chem. Rev.* **126**, 177-236.

18 Shenkin, P.S., and McDonald, D.Q. (1994) Cluster analysis of molecular conformations, *J. Comp. Chem.* **15**, 899-916.

19 Rawle, S.C., Clarke, A.J., Moore, P., and Alcock, N.W. (1992) Ligands designed to impose tetrahedral co-ordination; a convenient route to aminoethyl and aminopropyl pendant arm derivatives of 1,5,9-triazacyclododecane, *J. Chem. Soc., Dalton Trans.* **1992**, 2755-2757. CSD name = pakvus.

20 Hancock, R.D. (1989) Molecular Mechanics, *Prog. Inorg. Chem.* **37** 187- 291

21 Wiberg K.B. and Boyd R.H. (1972) Application of strain energy minimization to the dynamics of conformational changes, *J. Am. Chem. Soc.* **94**, 8426-8430.: Hancock, R.D., Drew, M.G.B., and Yates, P.C. (1986) The close equivalence to earlier reported methods of a recently reported method of calculating hole sizes in macrocyclic ligands, *J. Chem. Soc., Dalton Trans.* **1986**, 2505-2507.

22 V.B. Pett, L.L. Diaddario, E.R. Dockal, P.W. Corfield, C. Ceccarelli, M.D. Glick, L.A. Ochrymowycz and D.B. Rorabacher, Ring size effects on the structure of macrocyclic ligand complexes: copper(II) complexes with 12-16-membered cyclic tetrathia ethers *Inorg. Chem.* **22**, 1983, 3661-3670.

23 Blake, A.J., Reid G., and Schroeder, M. (1989) Platinum metal thioether macrocyclic complexes: Synthesis, electrochemistry, and single-crystal X-ray structures of cis-$[RhCl_2L^2]PF_6$ and trans-$[RhCl_2L^3]PF_6$ L^2 =1,4,8,11-tetrathiacyclotetradecane, L^3 = 1,5,9,13-tetrathiacyclohexadecane, *J. Chem. Soc., Dalton Trans.* **1989**, 1675.

24. Desimone, R.E., Cragel, J.,Jr., Ilsley, W.H. and Glick, M.D. (1979) Structural chemistry of molybdenum complexes of cyclic polythiaethers:the crystal and molecular structure of ethoxido-oxo-bis(1,5,9,13-tetrathiacyclohexadecane)-.mu.-oxodimolybdenum(IV) trifluoromethanesulfonate hydrate, *J. Coord. Chem.* **9**, 167-175.

25. Hills, A., Hughes, D.L., Jimenez-Tenorio, M., Leigh, G.J., Houlton, A. and Silver, J.A. (1989) Large Moessbauer quadrupole splittings in high-spin iron(II) complexes: the structure of diiodo-1,5,9,13-tetrathiacyclohexadecaneiron(II) (or 1,5,9,13-tetrathiacyclohexadecaneiron(II) diiodide) *J. Chem. Soc., Chem. Comm.* **1989**, 1774-1775.

26 Setzer, W.N., Tang, Y., Grant, G.J., and VanDerveer D.G. (1991) Synthesis and X-ray Crystal Structures of Heavy-Metal Complexes of 1,5,9,13-Tetrathiacyclohexadecane, *Inorg. Chem.* **30**, 3652-3656.

27 Mobley, H.L.T. and Hausinger, R.P. (1989) Microbial ureases: significance, regulation, and molecular characterization *Microbiol. Rev.* **53**, 85-108.

28 Dixon, N.E., Gazzola, C., Blakeley, R.L., and Zerner, B. (1975) Jack bean urease (EC3.5.1.5). A metalloenzyme. A simple biological role for nickel? *J. Am. Chem. Soc.* **97**, 4131-4132.

29 Jabri, E., Carr, M.B., Hausinger, R.P., and Karplus, P.A. (1995) The crystal structure of urease from *Klebsiella aerogenes, Science* **268**, 998-1004.

30 Park, I.-S., and Hausinger, R.P. (1995) Requirement of carbon dioxide for in vitro assembly of the urease nickel metallocenter. *Science* **267**, 1156-1158.

31 Jabri, E., and Karplus, P.A. (1996) Structures of the *Klebsiella aerogenes*, urease apoenzyme and two acive-site mutants, *Biochemistry* **35**, 10616-10626.

32 Park, I.-S., Michel, L.O., Pearson, M.A., Jabri, E., Karplus, P.A., Wang, S., Dong, J., Scott, R.A. Koehler, B.P., Johnson, M.K., and Hausinger, R.P. (1996) Characterization of the mononickel metallocenter in H134A mutant urease, *J. Bio. Chem.* **271**, 18632-18637.

33 Park, I.-S., and Hausinger, R.P. (1996) Metal ion interactions with urease and ureD-urease apoproteins, *Biochemistry* **35**, 5345-5352.

34 Csiki, C., and Zimmer, M. (in press) Computational Design of Biomimetic Compounds: Urease an example, *J. Mol. Struct.*

35 Lippard S.J., and Berg J.M. (1994) Principles in Bioinorganic Chemistry, University Science Books: Mill Valley, CA.

36 Francis M.B., Finney N.S., and Jacobsen E.N. (1996) Combinatorial approach to the discovery of novel coordination complexes, *J. Am. Chem. Soc.* **118,** 8983-8984.

37 Allen F.H., and Kennard O. (1991) The development of versions 3 and 4 of the Cambridge Structural Database System, *J. Chem. Inf. Comput. Sci.* **37**, 187-204.

38 Lippard, S.J. (1995) At last - the crystal structure of urease, *Science* **268**, 996-997.

39 Moncrief, M.B.C., Hom, L.G., Jabri, E., Karplus, P.A., Hausinger, R.P. (1995) Urease activity in the crystalline state, *Protein Science* **4**, 2234-2236.

40. Mohamadi, F., Richards, N.G.F., Guida, W.C. Liskamp, R., Lipton, M., Caulfield, C., Chang, C., Hendrickson, T. and Still, W.C. (1990) MacroModel-An integrated software system for modeling organic and bioorganic molecules using molecular mechanics *J. Comp. Chem.* **11** 440-467.

41. Ferguson, D.M., and Kollman P.A. (1991) Can the Lennard-Jones 6-12 function replace the 10-12 form in molecular mechanics calculations? *J. Comp. Chem.* **12** 620-626.

42. McDonald, Q. and Still, W.C. (1992) AMBER* torsional parameters for the peptide backbone *Tetrahedron Lett.* **33**, 7743-7746.

43 Hancock, R.D., Dobson, S.M., Evers, A., Wade, P.W., Ngwenya, M.P., Boeyens, J.C.A., and Wainwright, K.P. (1989) More rigid macrocyclic ligands that show metal ion size-based selectivity. A crystallographic, molecular mechanics, and formation constant study of the complexes of bridged cyclen, *J. Am. Chem. Soc.* **110**, 2788-2794.

44. Comba, P., and Zimmer, M. (1996) Inorganic molecular mechanics, *J. Chem. Ed.* **73**, 108-110.

MOLECULAR MODELING OF PLATINUM COMPLEXES WITH OLIGONUCLEOTIDES:
Methodological Lessons and Structural Insights

MIGUEL-ANGEL ELIZONDO-RIOJAS§, FLORENCE GONNET, PASCALE AUGE-BARRERE-MAZOUAT, FREDERIC ALLAIN, JACQUELINE BERGES#, ROGER ATTIAS, JEAN-CLAUDE CHOTTARD, and JIŘÍ KOZELKA*

Laboratoire de Chimie et Biochimie Pharmacologiques et Toxicologiques, URA CNRS 400, 45, rue des Saints-Pères, 75270 Paris Cedex 06, FRANCE

§ On leave from *Centro Universitario Contra el Cancer (CUCC), Hospital Universitario "Dr. José Eleuterio Gonzalez", Universidad Autónoma de Nuevo León, Monterrey, N. L., MEXICO*

Laboratoire de Chimie Théorique, Université Pierre et Marie Curie, 4, place Jussieu, 75005 Paris, FRANCE

1. Introduction

The fortuitous discovery of the antitumor properties of *cis*-$[PtCl_2(NH_3)_2]$ ("cisplatin") by Rosenberg in 1969 has demonstrated that a potentially toxic heavy metal complex can be useful in medicine [1]. It was subsequently shown that the antitumor activity is related to the drug's binding to DNA (for a review, see [2]), and platinum-DNA interactions have since become a hotspot of bioinorganic research.

L. Banci and P. Comba (eds.), Molecular Modeling and Dynamics of Bioinorganic Systems, 131–160.

Cisplatin is a bifunctional DNA-binding agent and thus capable of crosslinking different DNA residues. The sites with the highest affinity for Pt(II) are the N7 atoms of the guanines [3]. The most frequent adducts found in cisplatin-treated DNA are depicted in Fig. 1. The binding of cisplatin to DNA is believed to act as a signal giving rise to a complex cellular response, which, in susceptible cells, activates the programmed cell death (apoptosis) [4,5]. So far it has not been conclusively demonstrated which of the different cisplatin-DNA-adducts is responsible for the induction of apoptosis.

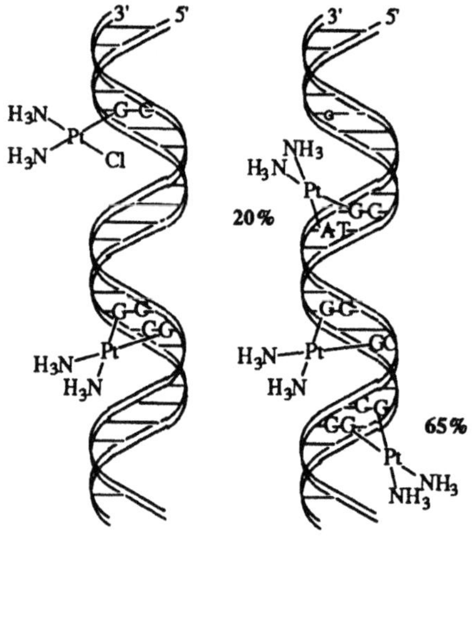

Fig. 1. Schematic representation of cisplatin-DNA adducts.

A correlation between the level of the intrastrand GG and AG chelates, i.e., the two major cisplatin-DNA-adducts, formed in the lymphocyte DNA of cisplatin-treated patients, and the response to the drug [6] has indicated an implication- causal or non-causal- of these adducts in the antitumor activity. In procaryotic systems, the biological effects, such as mutagenesis [7,8], inhibition of DNA polymerases [9], and repair of specific DNA adducts [10,11], have been quantified. These studies have highlighted the differences between the biological effects of the different adducts. For instance, the AG intrastrand chelate was shown to be significantly more mutagenic than the GG intrastrand adduct [7,8]. Based on the hypothesis that the biological effects of the platinum-DNA adducts are related to specific modifications of the helical DNA structure, several groups have studied platinum-oligonucleotide complexes as structural models for the different Pt-DNA adducts (reviewed by Kozelka [12] and Hambley [13]; see also [14,15]).

The principal aims of this contribution are to make the reader acquainted with the structural chemistry of platinum-DNA complexes and to point at methodological problems which arise from modeling a heavy metal complex incorporated into the polyanionic chain of a nucleic acid. Section 2 summarizes our NMR-guided molecular modeling work on platinum-oligonucleotide adducts featuring specific GG and AG intrastrand crosslinks, with emphasis on the structural differences between these two adduct types. Section 3 presents our recent molecular dynamics (MD) simulations of the dinucleotide complex *cis*-[$Pt(NH_3)_2${d(TpG)}]. In section 4, specific difficulties related to the development of force field parameters are discussed. A short concluding section closes our contribution.

2. NMR-guided modeling of *cis*-$Pt(NH_3)_2$-adducts of double-stranded oligonucleotides: GG versus AG intrastrand chelates

After the identification of the GG intrastrand chelate as the major adduct formed between *cis*-[$PtCl_2(NH_3)_2$] and DNA [16,17], the question arose whether the GG crosslink causes a local or a global distortion of the DNA double helix. Kozelka et al. have developed AMBER-based models for locally [18,19] ("model A" in Fig. 2) as well as globally [19,20] ("model C" in Fig. 2) distorted oligonucleotide duplexes. The two model types were calculated to be nearly equienergetic; thus, the molecular mechanics energy did not allow a commitment to one structure. Strategies to overcome this "multiple minima problem" include: i) improving the force field and hence the accuracy of the energy calculation; ii) correlation of the different minimum-energy-structures with experimental data. In this section, the results from molecular modeling correlated with NMR data will be discussed. For the development of force field parameters, the reader is referred to Section 4.

Since model A and model C have displayed several specific local structural differences such as different sugar puckers or particular interatomic distances, it appeared challenging to carry out an extended NMR study of a platinated oligonucleotide and to correlate the structural information with molecular models of both types. Such a study was accomplished by Herman et al. [21] who concluded that the double helix of the duplex oligonucleotide d(GCCG*G*ATCGC)-d(GCGATCCGGC) (**1**) crosslinked at the G*G* sequence by $\{cis\text{-}Pt(NH_3)_2\}^{2+}$ is kinked rather than linear. Lippard, Crothers and co-workers came to a similar conclusion based on gel electrophoresis measurements on multimers of site-specifically platinated oligonucleotides [22,23]. Their method proved faster than the lengthy NMR/molecular modeling analysis. Moreover, it allows for an estimate of the kink angle and its direction, provided that a structurally characterized standard exists. On the other hand, NMR-guided molecular models can yield structural details at the atomic level which gel electrophoresis techniques cannot provide. Thus, the study of the GG-platinated decamer **1** has shown that the structure has to be described in terms of an equilibrium between two major base arrangements, i.e., those in model C and in the slightly different model E (Fig. 2) [21,24].

Recently, Fouchet et al. have analyzed the DNA duplex d(CTCA*G*CCTC)-d(GAGGCTGAG) (**2**), crosslinked at the A*G* sequence by $\{cis\text{-}Pt(NH_3)_2\}^{2+}$ [25], and compared its structure with that of the GG-platinated decamer **1**. Similar modifications of the electrophoresis mobility [23,26] and a similar recognition by antibodies [27] had suggested that GG- and AG-crosslinked duplexes globally resemble each other. On the other hand, different reactivities towards chemical probes had indicated local structural differences between the two adduct types. In agreement with these expectations, the global NMR features of **1** and **2** resemble each other; nevertheless, a detailed comparison revealed some important differences. For instance, the deviations from the standard B-DNA-like NOE connectivity, linking H1' protons of one nucleotide with the base (H6 or H8) protons of the 3'-neighbor [28] are slightly different in **1** and **2** (Fig. 3): whereas in **1**, this connectivity is interrupted at two sites within the unplatinated strand, i.e., between the two cytidines complementary to the platinated G*G* dinucleotide (C16,C17) as well as between the 3'-cytidine (C17) and its 3'-neighbor (G18), in **2**, there is only one interruption within the unplatinated strand, that between the thymidine complementary to A* (T15) and its 3'-neighbor (G16). This is in agreement with the cytosine C17 of **1** oscillating between two stacked positions (marked as dashed bars in Fig. 4), whereas for **2**, the observation of a H1'-H6 NOE between T15 and C14 suggests that T15 (marked as cross-hatched bar in Fig. 4) remains stacked on its 5'-neighbor. In other words, we suggest that **2** is represented by a purely model E-like structure, in contrast to the C $\rightleftharpoons$ E equilibrium assigned to **1**. This assignment is further corroborated by other observations such as a more severely perturbed base-pairing of the platinated A*T pair of **2** compared to that of the corresponding G*C pair of **1**, which becomes manifest in a large upfield shift of the thymine imino proton.

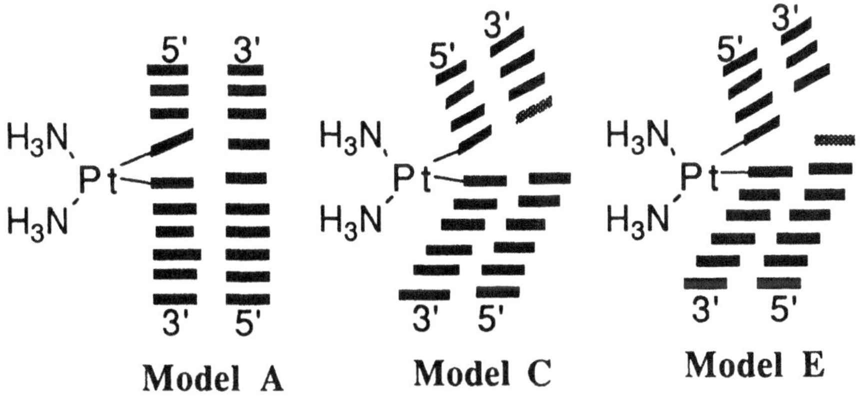

Fig. 2. Orientation of the nucleobases in the different models for GG-crosslinked oligonucleotides.

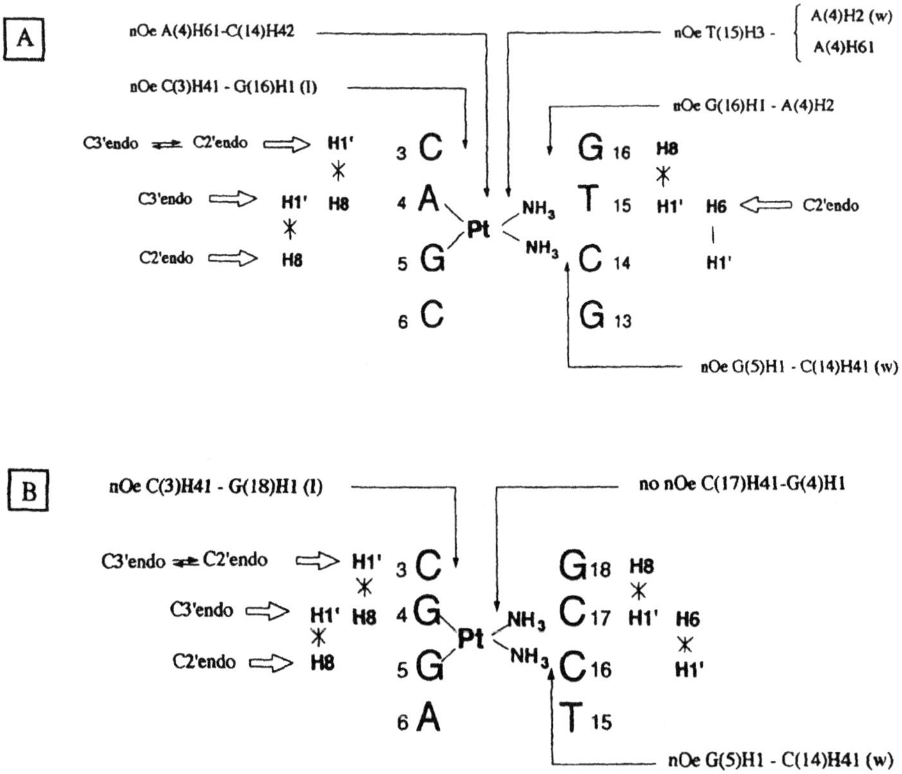

Fig. 3. Schematic summary of the characteristic NMR features observed for the platinated duplexes **1** (A) and **2** (B). I = intense; w = weak. (Reproduced from [25])

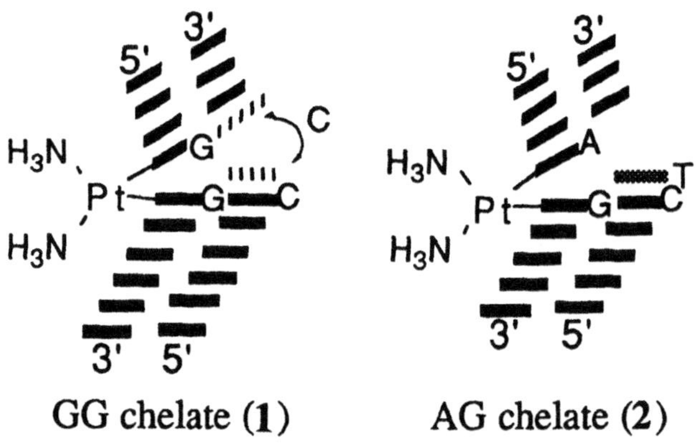

Fig. 4. Schematic view highlighting the principal structural difference between the platinated duplexes **1** and **2**, as inferred from NMR/molecular modeling analyses [21,25].

Our conclusion that the structural difference between GG- and AG-chelates of $\{cis\text{-}Pt(NH_3)_2\}^{2+}$ within DNA duplexes mainly concerns the *unplatinated* strand could be related to the finding reported by a dutch group that the recognition protein complex UvrAB of the E.Coli repair system UvrABC binds to the convex side of the platinum-induced kink, i.e., making contacts with the unplatinated strand at the site opposite the platinated dinucleotide [29]. It is therefore tempting to speculate that the slight structural difference between GG- and AG-chelates may give rise to a different recognition by UvrAB and possibly to a different repair of the lesions. In fact, AG crosslinks of the cisplatin derivatives $[PtCl_2(en)]$ and $[PtCl_2(dach)]$ (en = 1,2-diaminoethane; dach = *trans*-1,2-diaminocyclohexane) are indeed more effectively removed from DNA by UvrABC than GG crosslinks [10]. This differential repair could be related to the different mutagenic properties of the two lesions (see discussion in [25]).

3. Conformational analysis of *cis*-$[Pt(NH_3)_2\{d(TpG)\}]$ using NMR and MD-simulations

cis-$[Pt(NH_3)_2\{d(TpG)\}]$ (**3**) is a dinucleotide complex in which platinum crosslinks the N7 atom of guanine and the (deprotonated) N3 atom of thymine (Fig. 5). The main features of this complex are analogous to those of the *cis*-$[Pt(NH_3)_2\{d(GpG)\}]^+$ and *cis*-$[Pt(NH_3)_2\{d(ApG)\}]^+$ complexes studied previously [31,32], but the different angle subtended by the Pt-N3 bond with he glycosidic N1-C1' bond in thymidine, compared with that between the Pt-N7 and N9-C1' bonds in guanosine or adenosine, let us expect somewhat different structural behaviour.

The downfield part of the ^1H NMR spectrum of **3** in D_2O (Fig. 6) indicates the presence of two conformers which exchange slowly on the NMR time scale: a major conformer (75%; δ_{H6} = 7.18; δ_{H8} = 8.30 ppm) and a minor conformer (25%; δ_{H6} = 7.69; δ_{H8} = 8.13 ppm). An NMR analysis [33] revealed that both conformers have N(Thy)/S(Gua) sugar puckers and an *anti* orientation with respect to the guanosine glycosidic bond, and differ in the thymidine glycosidic bond orientation, which is *syn* in the major conformer and *anti* in the minor. This suggested that the two

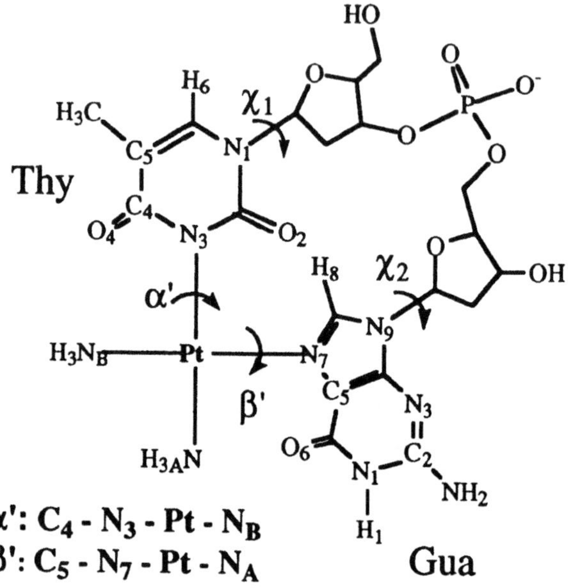

Fig. 5. Sructural formula of *cis*-[Pt(NH_3)$_2${d(TpG)}], the numbering of atoms and definition of the torsion angles α', β', χ_1, and χ_2.

conformers are rotamers, differing in their base orientations. However, from the NMR data alone, it was not possible to determine the base positions more specifically, and it appeared therefore challenging to try to obtain this information from molecular modeling.

In order to delineate the main conformational domains, we first calculated an energy map for the core complex *cis*-[Pt(NH_3)$_2$(Thy)-(Gua)]$^+$, as a function of the Pt-N(base) torsion angles α' and β' (defined in Fig. 5). The map (Fig. 7) features four low-energy zones, two corresponding to head-to-head and two to head-to-tail conformations. The main task consisted in assigning to each of the two rotamers a corresponding low-energy zone.

In a first series of experiments, we carried out 630 ps MD-runs at 350 K for each of the four low-energy domains of the complex *cis*-[Pt(NH_3)$_2$(Thy)(Gua)]$^+$. In a second series, four analogous runs were performed for the dinucleotide complex **3**. The first series was used to check that the conformational sampling with respect to α' and β' followed the calculated energy surface shown in Fig. 7. Fig. 8a demonstrates that this was indeed the case. For practical reasons, the calculations of the energy level curves (Fig. 7) used the BIOSYM software and the ESFF force field. The molecular dynamics simulations (Fig. 8), on the other hand, were done with the AMBER 3.0 program and force field. The similarity between Figs. 7 and 8a suggests that the energy surface defined by α' and β' is not sensitive to the choice of a particular force field.

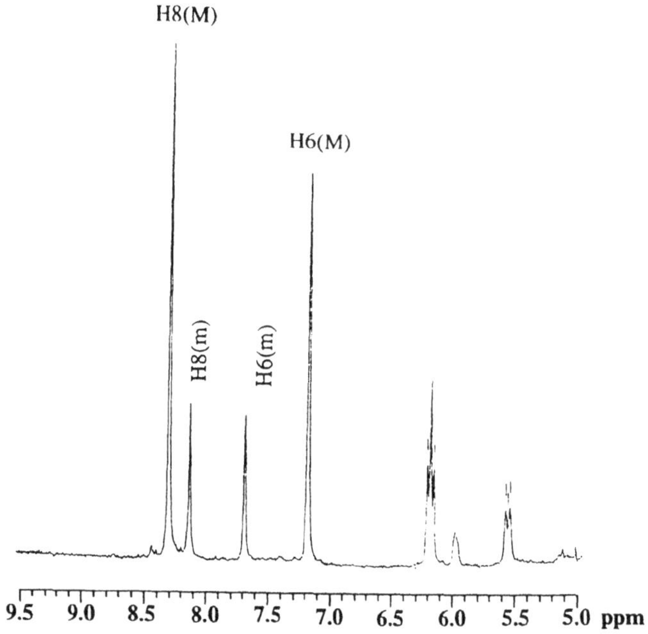

Fig. 6. Downfield part of the ^{1}H NMR spectrum of *cis*-[Pt(NH_3)$_2${d(TpG)}] (3) in D_2O.

The conformational sampling map obtained from the four MD runs carried out for the dinucleotide complex 3 (Fig. 8b) shows that the

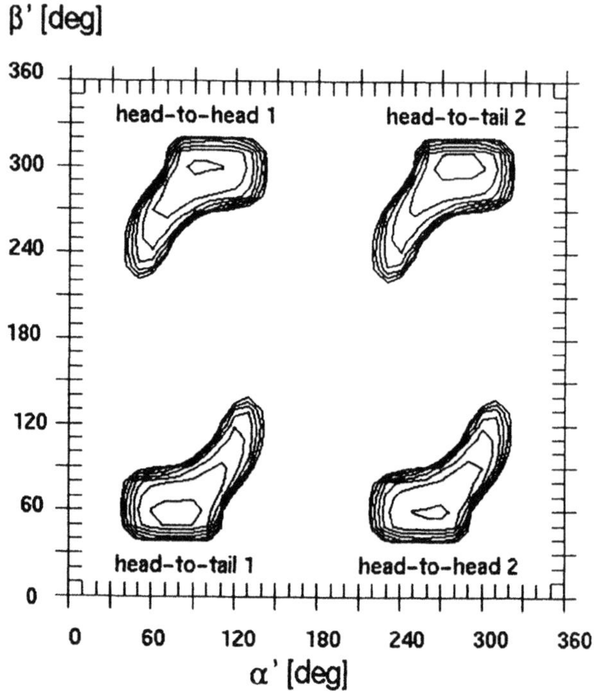

Fig. 7. Energy map of *cis*-[Pt(NH_3)$_2$(Thy)(Gua)]$^+$, as a function of the torsion angles α' and β' defined in Fig. 5, calculated using the ESFF force-field of BIOSYM [34]. The lowest level curve is 0.35 kcal/mol above the global energy minimum; the next six level curves shown are at intervals of 1.5 kcal/mol.

conformational space available to the nucleobase complex *cis*-[Pt(NH_3)$_2$-(Thy)(Gua)]$^+$ is reduced due to the sugar-phosphate-link between thymine and guanine. Fig. 9 presents the trajectories of the deoxyribose torsion angles δ_1 and δ_2, and the glycosidic torsion angles χ_1 and χ_2 for the same four simulations. These plots indicate that the main conformations belonging to the four low-energy zones are as follows:

Domain	α'	β'	χ(T)	χ(G)	Sugar pucker	
					T	G
HH1	+	-	*syn*	*anti*	N	S
HH2	-	+	*anti*	*anti*	N	S
HT1	+	+	*syn*	*syn*	N	N/S
HT2	-	-	*anti*	*syn*	N	N/S

Therefore, the experimentally determined conformations T(*syn*/N)-G(*anti*/S) for the major rotamer, and T(*anti*/N)-G(*anti*/S) for the minor rotamer, can be unambiguously assigned to the HH1 and HH2 domains. Fig. 10 shows views of energy-minimized structures representing the preponderant conformations of both rotamers. It can be recognized that in the minor rotamer, the *anti* orientation of the thymidine deoxyribose places the ring oxygen close to the H6 thymine proton, in a position where it causes deshielding of the proton. Conversely, in the major rotamer, the thymidine ring oxygen is farther away from the base and closer to C6 than to H6; thus, it has a shielding effect on H6. This is apparently the major effect responsible for the different H6 chemical shifts in the two rotamers (Fig. 6). The ring currents play a minor role in *cis*-[Pt(NH_3)$_2${d(TpG)}]. The ring current of thymine is about six times weaker than that of the five-membered ring of guanine or adenine [38], and the H6 atom is far away from the guanine rings and is therefore only marginally affected. Our calculations of the ring current effects indeed indicate that in the two head-to-head low-energy zones, the ring current effects of T on H8(G) and of G on H6(T) remain below 0.07 ppm [35]. This is at variance with the complexes *cis*-[Pt(NH_3)$_2${d(GpG)}]$^+$ and *cis*-[Pt(NH_3)$_2${d(ApG)}]$^+$, where the major influence that affects the chemical shifts of the base protons arises from the ring current effects [31,32].

All the presented MD simulations were carried out without explicit representation of solvent molecules. The dielectric screening was accounted for by a distance-dependent ε of 4r, and the 1-4-nonbonded interactions were reduced by a factor of 2. Throughout the calculations, we used weak V_6 torsional constraints of the type $k_d/2[1+\cos(n\phi-\gamma)]$, with $k_d/2 = 1$ kcal/mol and $\gamma = 180°$, imposed on all H-N-Pt-N_{cis} dihedrals, (no constraints were imposed on the H-N-Pt-N_{trans} dihedrals, since these are ill-defined due to the linear N-Pt-N_{trans} bond angle). Without these constraints,

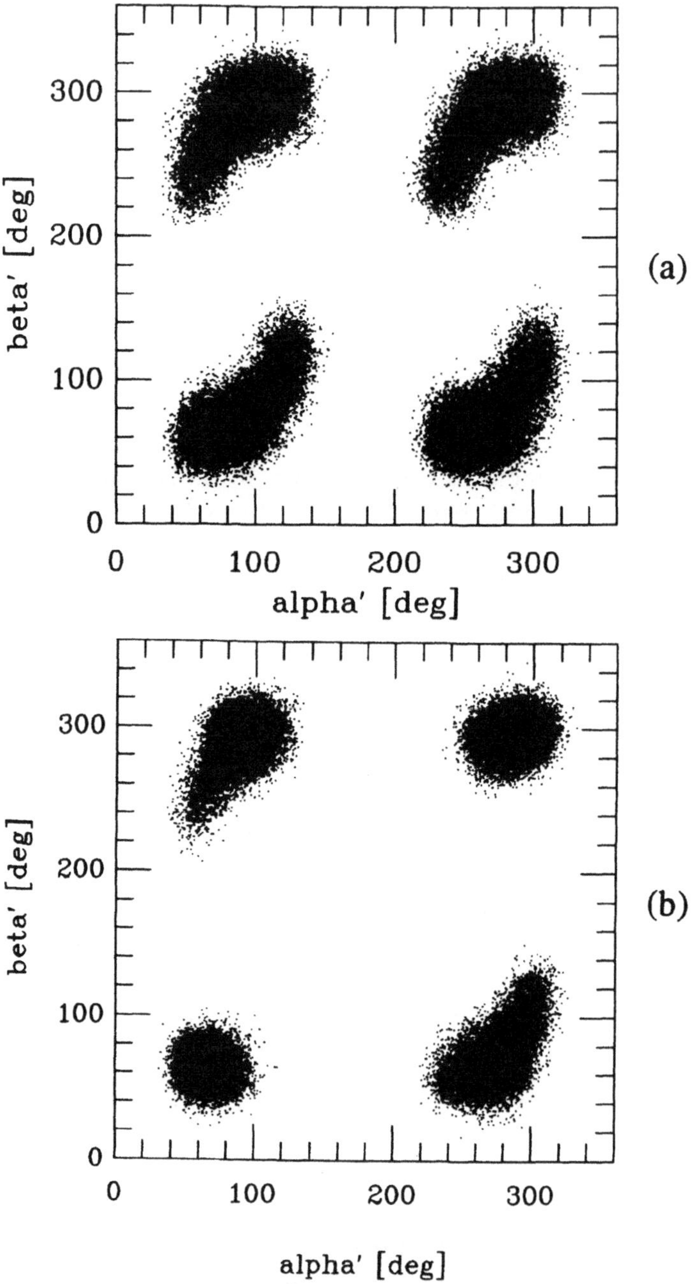

Fig. 8. Conformational sampling resulting from four independent 630 ps MD runs (one for each low-energy domain), at constant temperature of 350 K, for *cis*-$[Pt(NH_3)_2(Thy)(Gua)]^+$ (a), and for **3** (b). The simulations [35] were carried out using the AMBER 3.0 MOLDYN module [36] coupled to the MORCAD package [37].

we observed induction of a fast rotation of one of the two NH_3 groups about the Pt-N bond. This rotation absorbed so much kinetic energy (the simulations were carried out at constant temperature), that the vibrations were almost frozen, leading to a drop in potential energy, and thus to a decrease in total energy. The extreme slowing down of vibrational movements was clearly manifest in the conformational sampling (Fig. 11): The populated conformational space was restrained to a narrow zone close to the energy minima (compare with the energy map in Fig. 7). The weak torsional constraints imposed on the H-N-Pt-N_{cis} dihedrals were thought to simulate the damping of such rotations by solvent molecules. Some participants of the workshop questioned our approach and suggested that the fast NH_3 group rotation was an artifact arising from an inappropriate reassignment of velocity vectors by the temperature-maintaining algorithm [39]. We are currently checking this and trying to understand the origin of this artifact.

4. Evaluation of force field parameters for platinum-DNA complexes

4.1. *AB INITIO* CALCULATIONS: WHICH LEVEL OF THEORY?

Quantum chemical *ab initio* calculations are frequently used to determine force field parameters. For instance, the electrostatic potential calculated from the *ab initio* wave function is commonly employed in order to define atomic charges. A least-squares routine iterates the atomic charges to optimize the fit between the Coulomb potential based on the monopole charges and the *ab initio* electrostatic potential, both calculated for a finite number of space points set around the molecule. The GAUSSIAN94 package [40] supports two such routines [41,42]. Obviously, the meaningfullness of the optimized atomic charges depends primarily on the quality of the *ab initio* electron density. It is well-known that the latter depends on the choice of the orbital basis set as well as on the inclusion or neglect of electron correlation [43,44].

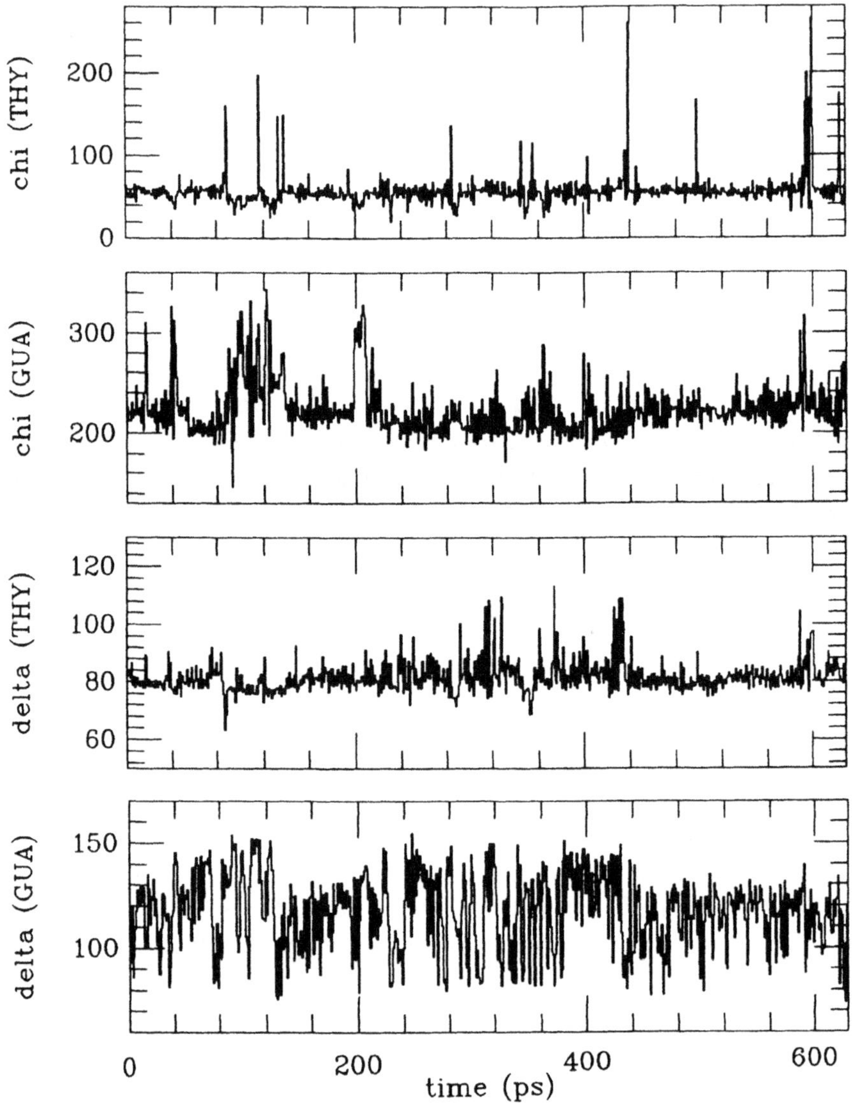

Fig. 9. MD trajectories for the torsion angles δ_1, δ_2, χ_1, and χ_2 of **3** resulting from 630 ps MD runs at constant temperature of 350 K. The simulations [35] were carried out using the AMBER 3.0 MOLDYN module [36] coupled to the MORCAD package [37]. a) Domain head-to-head 1 (0°<α'<180°; 180°<β'<360°).

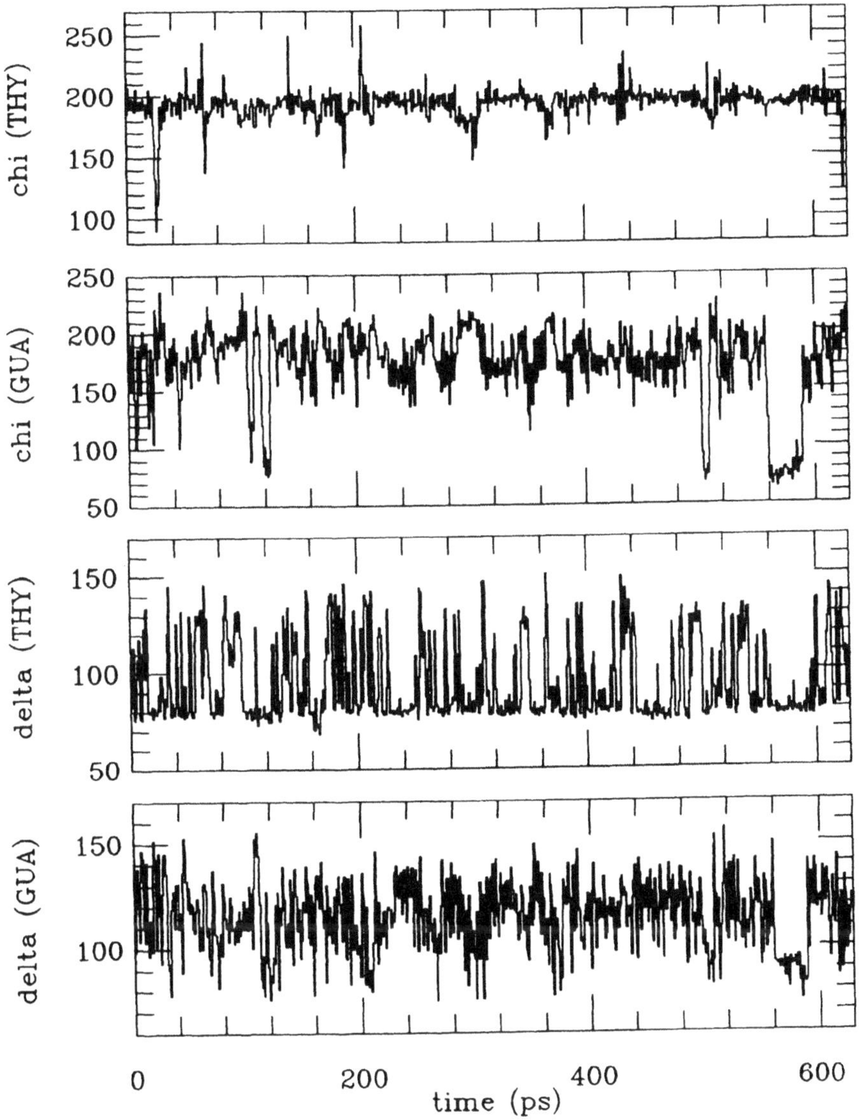

Fig. 9 (continued). MD trajectories for the torsion angles δ_1, δ_2, χ_1, and χ_2 of **3**, resulting from 630 ps MD runs at constant temperature of 350 K. b) Domain head-to-head 2 (180°<α'<360°; 0°<β'<180°).

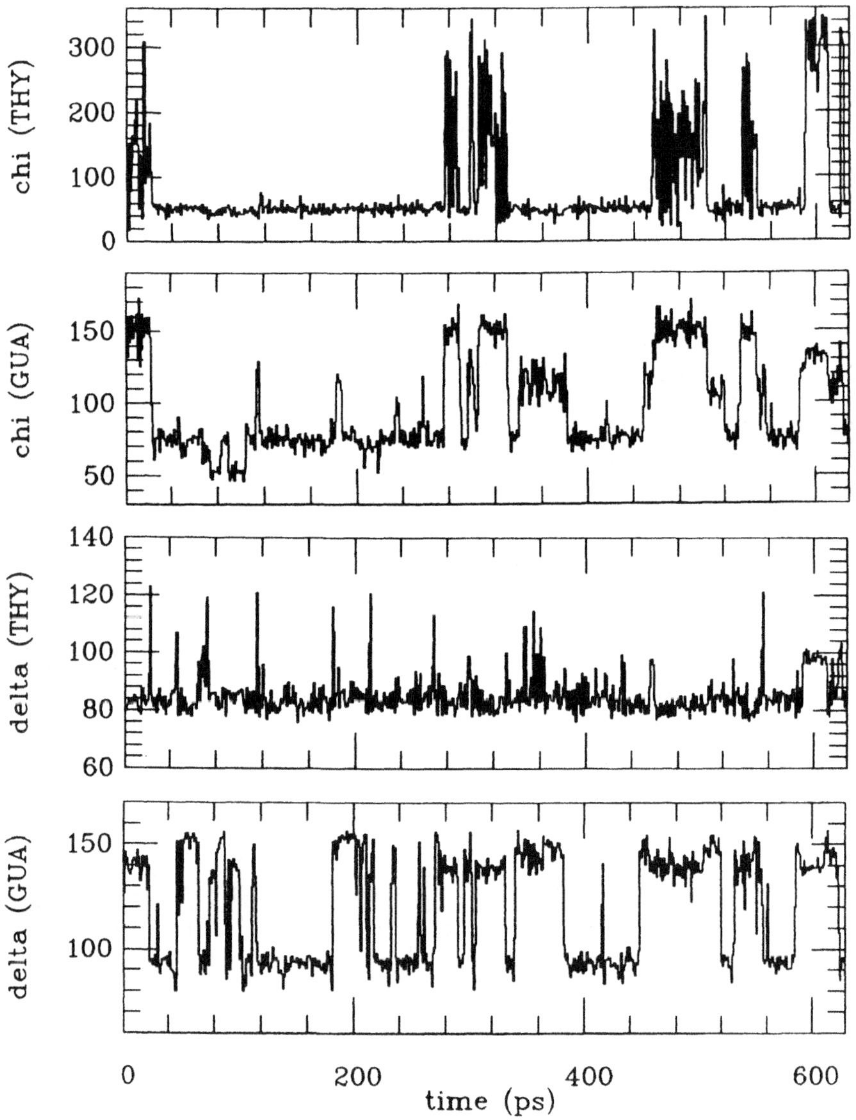

Fig. 9 (continued). MD trajectories for the torsion angles δ_1, δ_2, χ_1, and χ_2 of **3**, resulting from 630 ps MD runs at constant temperature of 350 K. c) Domain head-to-tail 1 ($0° < \alpha' < 180°$; $0° < \beta' < 180°$).

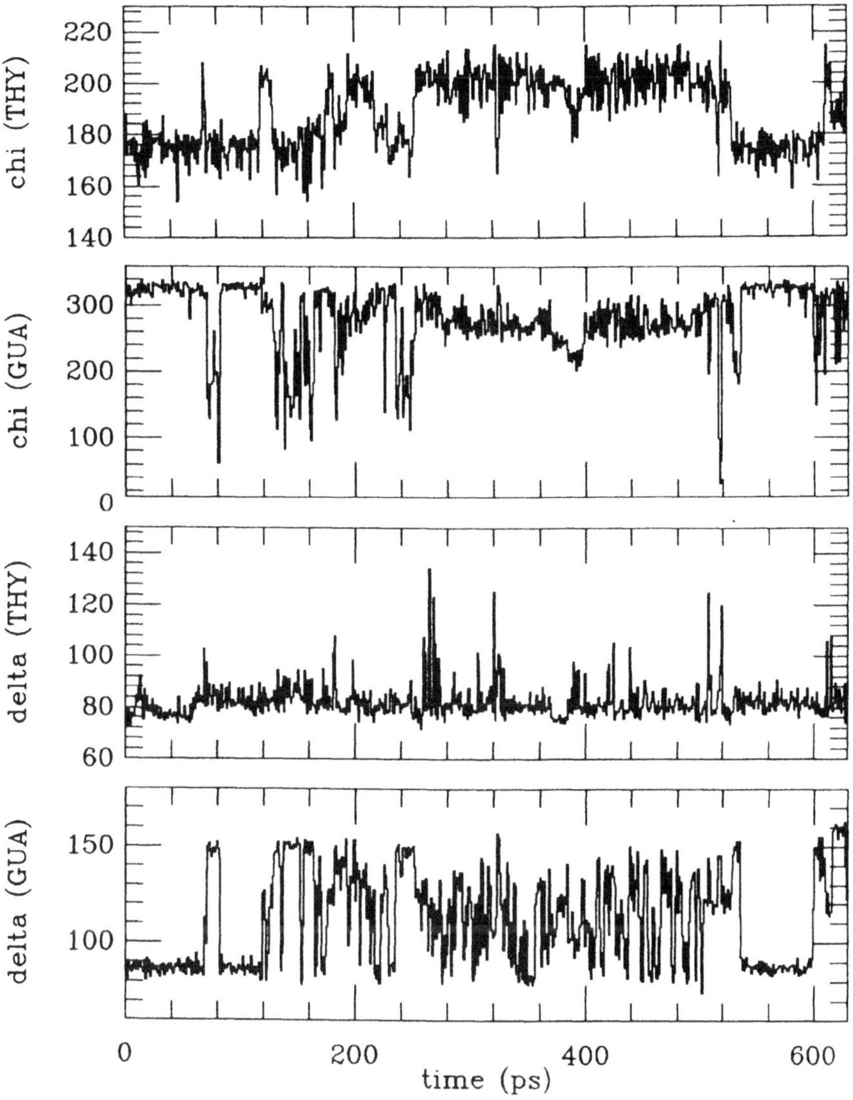

Fig. 9 (continued). MD trajectories for the torsion angles δ_1, δ_2, χ_1, and χ_2 of **3** resulting from 630 ps MD runs at constant temperature of 350 K. d) Domain head-to-tail 2 (180°<α'<360°; 180°<β'<360°).

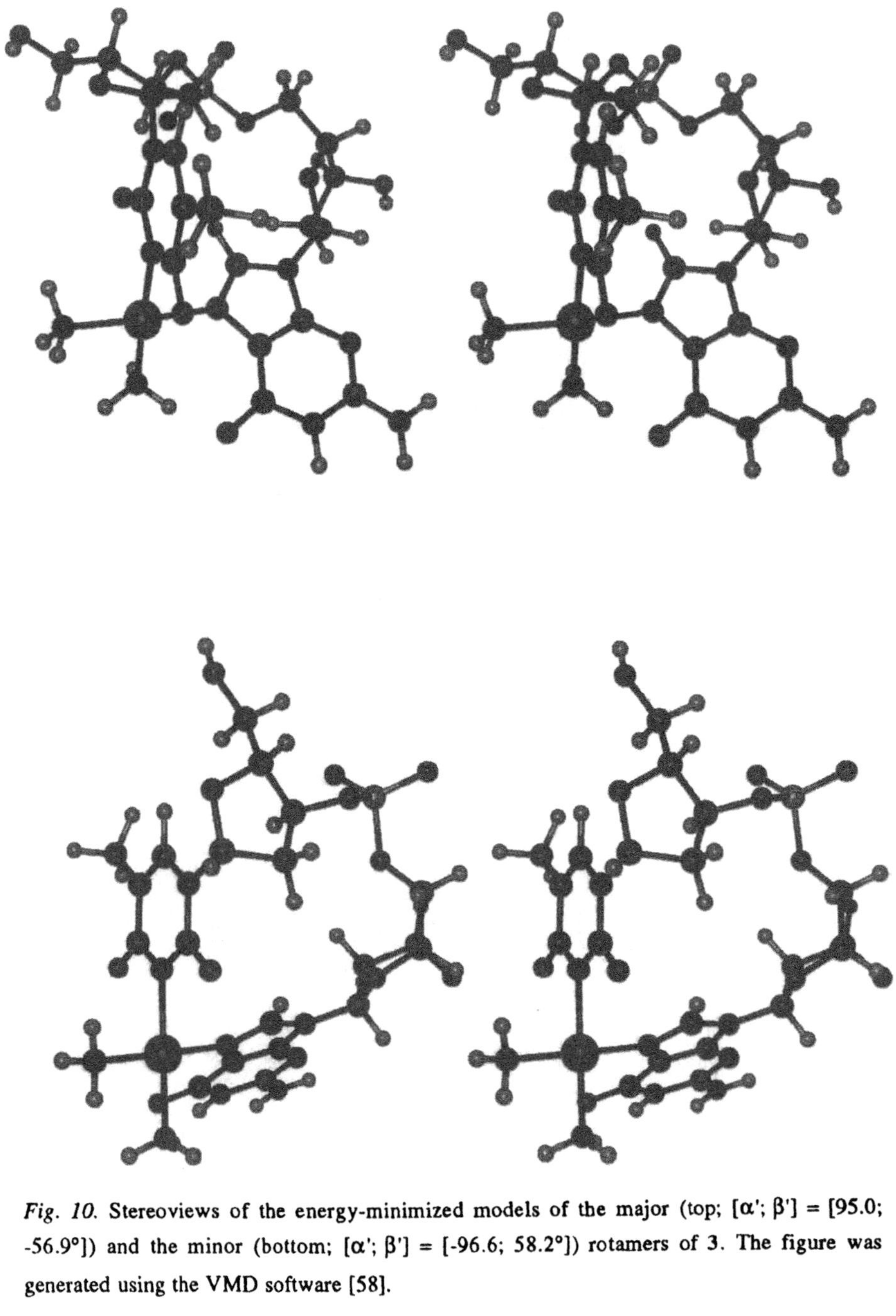

Fig. 10. Stereoviews of the energy-minimized models of the major (top; [α'; β'] = [95.0; -56.9°]) and the minor (bottom; [α'; β'] = [-96.6; 58.2°]) rotamers of 3. The figure was generated using the VMD software [58].

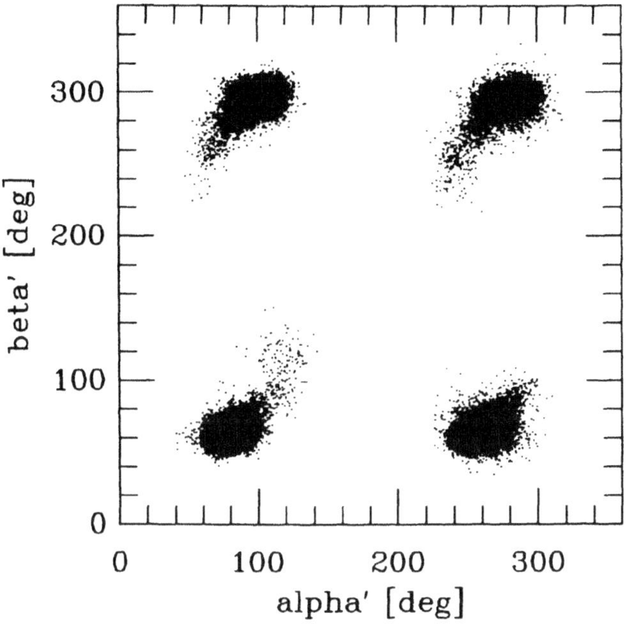

Fig. 11. Conformational sampling resulting from four independent 630 ps MD runs (one for each low-energy domain), at constant temperature of 350 K, for *cis*-[Pt(NH_3)$_2$(Thy)(Gua)]$^+$, with unconstrained rotations of the NH_3 groups. The simulations [35] were carried out using the AMBER 3.0 MOLDYN module [36] coupled to the MORCAD package [37].

For a complex of a third-row transition element such as platinum, electron correlation is expected to play a significant role. In order to quantify the impact of the basis set and of the inclusion of electron correlation on the calculated electron density and molecular geometry, we carried out a series of *ab initio* calculations for the complex *cis*-

$[PtCl_2(PMe_3)_2]$ at different levels of theory. The solid-state structure has been determined by two groups [45,46], and the dipole moment is expected to be similar to that of the triethylphosphine analog *cis*-$[PtCl_2(PEt_3)_2]$, for which two determinations (in benzene solution) are available [47,48]. A complex *cis*-$[PtCl_2(PR_3)_2]$ is particularly suitable for a test, since the high dipole moment (10.9 D for the triethyl derivative) mainly arises from the electron transfer trough the Pt-Cl and Pt-P bonds, and the calculated μ value therefore indicates how well the polarity of the coordinative bonds is described. In addition, the two different bond lengths, Pt-Cl (2.37 Å) and Pt-P (2.24 Å), reflect the different *trans*-influences of the Cl^- and PR_3 ligands [49].

Table 1 lists the dipole moments calculated at different levels of theory and using different basis sets. For the platinum atom, three sets of effective core potentials (ECP) were employed: the Los Alamos potentials LANL1 [51] and LANL2 [52] implemented in GAUSSIAN94, and the pseudopotential derived by the Dolg group [53], whose pseudo-orbital basis set was reoptimized by Pelissier and augmented by two uncontracted f functions with the exponents 0.5 and 1.5 (basis designated DP*) [54]. The double zeta pseudo-orbital basis set corresponding to the LANL2 potential, LANL2DZ, which docs not contain polarization functions, was augmented in one part of the calculations (basis LANL2DZ*) by a single f function with an exponent of 0.78 [55]. The LANL1 pseudopotential replaces all but the *5d* and *6s* electrons, the LANL2 and DP ECP's leave the *5s*, *5p*, *5d*, and *6s* orbitals in the valence space.

From Table 1, it emerges: i) The dipole moment is overestimated in all calculations. ii) The calculated dipole moment approaches the experimental value upon extension of the orbital basis set; addition of polarization functions has a significant effect. iii) Inclusion of electron correlation (MP2, DFT/B3LYP) dramatically improves the result. iv) With the same basis set, Møller-Plesset second order perturbation (MP2) and density-functional theory (DFT) calculations using the hybrid functional B3LYP [40] yield similar dipole moments, the DFT values being slightly closer to the experimental dipole moment.

Table 2 displays the Pt-Cl and Pt-P bond lengths, obtained from two-parameter geometry optimizations at different levels of theory. These bond lengths have to be compared with the experimental values: Pt-Cl (2.374±0.009 Å) and Pt-P (2.243±0.008 Å) [45,46]. We observe that all

TABLE 1. Calculated dipole moments for *cis*-[$PtCl_2(PMe_3)_2$][a] using Hartree-Fock (HF), Møller-Plesset second order perturbation (MP2) and density funcional theory DFT(B3LYP) calculations [50]. See text for the basis set codes.

Basis set	Basis set		Dipole moment [D]	
(Pt)	(P, Cl, C, H)	HF[b]	MP2[c]	DFT(B3LYP)[b]
LANL1MB	LANL1MB	16.55		
LANL1DZ	LANL1DZ	14.24	12.88	13.30
LANL2DZ	LANL2DZ	14.26	12.75	12.66
LANL2DZ*	LANL2DZ	14.15	12.73	12.59
LANL2DZ*	3-21G*	13.62	12.65	12.32
LANL2DZ*	6-31G**	13.67	12.45	12.25
DP*	3-21G*	13.43	12.45	12.23
DP*	6-31G	13.98	12.70	
DP*	6-31G*	13.42	12.33	
DP*	6-31G**	13.42	12.16	12.09
DP*	6-31^{++}G**	13.32		11.91

[a] Averaged experimental geometry from [45]; C-H, 1.10 Å; P-C-H, 109.87° (values for free PMe_3 [30]); Pt-P-C-H dihedrals optimized using MP2/LANL1DZ.

[b] calculated as the expectation value $< r >$

[c] calculated using the Z-vector method [44]

calculations overestimate the bond distances. More seriously, the HF calculations assign almost identical values to the Pt-Cl and Pt-P bonds or even invert their order. Again, generally, expanding the basis set and inclusion of electron correlation alleviates the disagreement with the experimental values.

The calculations of dipole moments as well as the geometry optimizations indicate that Hartree-Fock calculations on compounds of third-transition-row-elements such as Pt are likely to yield highly erroneous

TABLE 2. Pt-Cl and Pt-P bond lengths optimized for *cis*-[$PtCl_2(PMe_3)_2$] using Hartree-Fock (HF) and density funcional theory DFT(B3LYP) calculations[a] [50].

Basis set	Basis set	Distance [Å]			
(Pt)	(P, Cl, C, H)	HF		DFT(B3LYP)	
		Pt-Cl	Pt-P	Pt-Cl	Pt-P
LANL1MB	LANL1MB	2.47	2.64		
LANL1DZ	LANL1DZ	2.461	2.438	2.497	2.475
LANL2DZ	LANL2DZ	2.428	2.428	2.432	2.387
LANL2DZ*	LANL2DZ	2.415	2.419	2.418	2.375
LANL2DZ	3-21G*	2.394	2.347	2.393	2.302
LANL2DZ	6-31G**	2.401	2.364	2.403	2.320
DP	3-21G*	2.365	2.320	2.377	2.295
DP	6-31G**	2.371	2.350	2.386	2.314

[a] Starting from the same averaged experimental geometry as in Table 1, the Pt-Cl and Pt-P bonds were optimized as the single parameters.

results. Inclusion of electron correlation and the use of a sufficiently flexible basis set including polarization functions seem indispensable for a meaningful description of the electon density. DFT/B3LYP calculations using the DP* basis set for Pt and the 6-31G** basis for the light atoms appears as a reasonable model chemistry, yielding satisfactory agreement with the experiment at acceptable computer time cost (DFT/B3LYP calculations required, on the average, 2-3 times more cpu time than HF calculations but 2-4 times less cpu time than MP2 calculations using the same basis set).

4.2. DETERMINATION OF FORCE-FIELD PARAMETERS FOR Pt(II)-NUCLEOBASE-BINDING USING DFT CALCULATIONS ON THE MODEL COMPLEX $[Pt(NH_3)_3(N\text{-Methylimidazole})]^{2+}$

Based on the results of Section 4.1, we used DFT/B3LYP calculations with the DP*/6-31G** basis to determine some key force field parameters for platinum binding to nucleobases. In the example presented here, we examined the energetics of the bending of the Pt-N(base) bond out of the base plane (angle ψ in Fig. 12). The calculations were done on the model complex $[Pt(NH_3)_3(N\text{-methylimidazole})]^{2+}$. We used N-methylimidazole as a non-bulky model for a purine base, in order to keep the contribution of nonbonded ligand-ligand interactions to the total energy at a minimum.

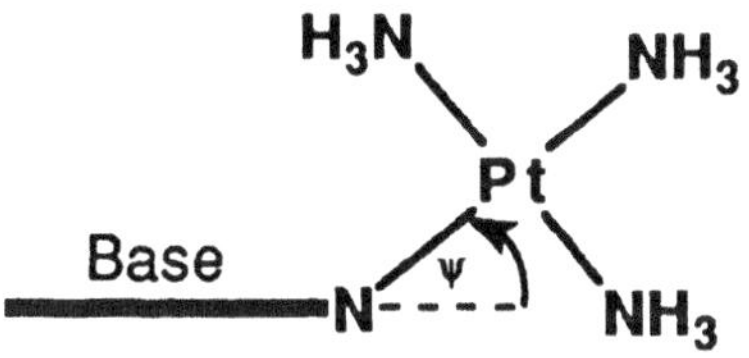

Fig. 12. Angle ψ defining the deviation of the Pt-N(base) bond from the base plane.

Fig. 13 shows the *ab initio* energy as a function of the angle ψ, together with the AMBER molecular mechanics energy, consisting of bond-bending, 6-12 Lennard-Jones, and Coulomb terms. ESP charges [42] for the Coulomb energy were obtained from a DFT/B3LYP calculation using the DP*/6-31G** basis sets.

The decrease of overlap between the lone-pair of the ligating N atom of the base and the Pt acceptor orbitals (*6s, 5d*) upon increasing ψ is expected to follow a cosine function, such as $\{1\text{-}\cos(2\psi)\}$. The decrease of Pt-N bond energy, if assumed proportional to the overlap [56], should follow the same curve. It would therefore seem appropriate to parameterize this out-of-plane bending with a classical V_2 dihedral term applied to the

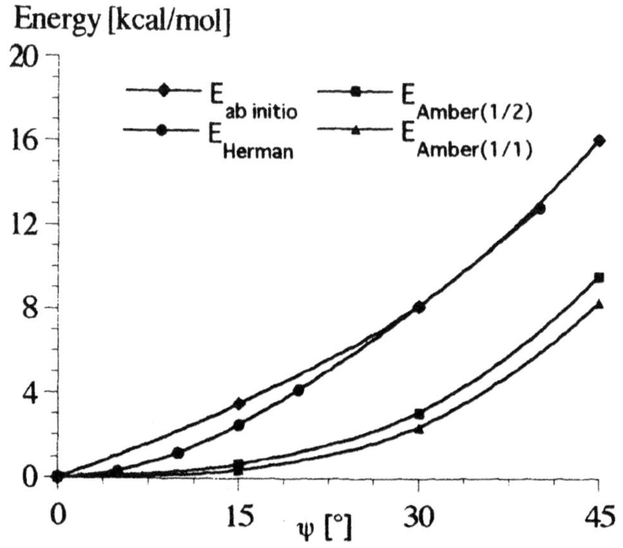

Fig. 13. Dependence of the energy on the angle ψ (Fig. 12) calculated for $[Pt(NH_3)_3(N\text{-methylimidazole})]^{2+}$ ($E_{ab\ initio}$, E_{Amber}) or $[Pt(NH_3)_3(Gua)]^{2+}$ (E_{Herman}). $E_{ab\ initio}$ was calculated using the DFT/B3LYP functional implemented in GAUSSIAN94 [40], with the DP* basis on Pt and 6-31G** on C, N, and H. E_{Amber} is the sum of bond-bending, Lennard-Jones, and Coulomb energy calculated with the AMBER force-field. (1/2) and (1/1) refer to the coefficients reducing the 1-4 electrostatic forces. E_{Herman} is the sum of the four improper dihedral terms described in the text and shown in Fig. 14.

Pt-N-C-X dihedral angles. However, in strained structures, such as platinum chelates of adjacent guanines within duplex oligonucleotides, this parameterization leads to artifactual puckering of the five-membered rings [21]. For Pt-purine binding, we have therefore proposed an alternative parameterization using four improper $V_1(\gamma = 0)$ dihedral terms: Pt-C5-C2-N7 ($k_d/2$ = 4.7 kcal/mol), Pt-C5-C2-C8 ($k_d/2$ = 4.7 kcal/mol), Pt-C8-C5-C4 ($k_d/2$ = 30.4 kcal/mol), and Pt-C8-C6-C5 ($k_d/2$ = 27.7 kcal/mol). The connectivities defining these four dihedrals are depicted in Fig. 14. The idea behind this parameterization is as follows: The dihedral terms Pt-A-B-X will tend to keep the four atoms in the base plane. Since the atoms Pt and

X are in *trans*-orientation, deviation of Pt from the base plane will "push" the atom X to the opposite side of the plane. This will counter-balance the effect of the stretching and bond-bending terms, which will "drag" the atoms of the five-membered ring towards the platinum atom. The $k_d/2$ values were calibrated by Herman et al. [21] in order to fit a HF calculation on $[Pt(NH_3)_3(pyrimidine)]^{2+}$ [57], which was the only calculation available at that time. As shown in Fig. 13, the *ab initio* energy (DFT/B3LYP for $[Pt(NH_3)_3(N\text{-methylimidazole})]^{2+}$) and the sum of four dihedrals calculated for a Pt-purine complex are astonishingly similar. For ψ angles

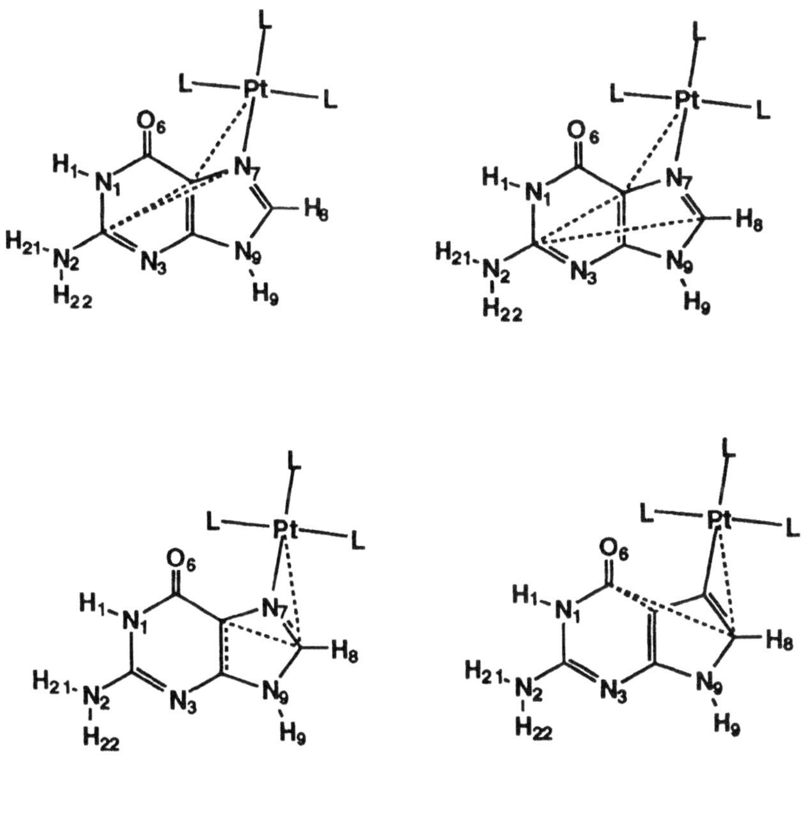

Fig 14. Improper dihedral angles used to parameterize the bending of the Pt atom out of the guanine plane [21].

smaller than 20°, where the nonbonded and angle-bending terms can be neglected (Fig. 13), the Herman parameterization thus describes adequately the *ab initio* energy. At stronger deviations of the Pt atom from the base plane (ψ > 20°), the sum of nonbonded, angle-bending, and Herman dihedral energy will lead to an overestimation of the total energy.

5. Conclusion

Structural studies on metal complexes with nucleic acids have to deal with problems inherent to nucleic acid structures as well as with those characteristic of transition metal compounds. For the modeling of platinum complexes with oligonucleotides, our approach has been to use AMBER, a force field developed and tested for the modeling of nucleic acids, extended by data parameterizing the platinum coordination sphere, which rely upon *ab initio* calculations on one hand, and on experimental observables on the other. In this account, we have presented molecular mechanics and dynamics calculations combined with NMR analyses. In addition, we have pointed out some specific problems related to the modeling of platinum-DNA complexes. Any comments from the readers will be greatly appreciated.

Acknowledgement

We are indebted to Drs. Jean-Marie Teuler and Roger Savinelli for technical assistence with *ab initio* calculations, and to Dr. Jean-Pierre Daudey for the communication of the DP* basis set. A generous loan of platinum complexes from Johnson-Matthey, financial support from the ARC (Contract N° 2042) and Ligue Nationale Française contre le Cancer, as well as computer time from the IDRIS center of the Centre National de la Recherche Scientifique are gratefully acknowledged.

References

1. Rosenberg, B., Van Camp, L., Trosko, J. E., and Mansoui, V. H. (1969) Platinum Compounds: a new Class of Potent Antitumour Agents, *Nature* **222,** 385-386.

2. Chu, G. (1994) Cellular Responses to Cisplatin, *J. Biol. Chem.* **269,** 787-790.

3. Martin, R. B. (1996) Dichotomy of Metal Ion Binding to N1 and N7 of Purines, *Met. Ions Biol. Syst.* **32,** 61-89.

4. Hickman, J. A., Potten, C. S., Merritt, A. J., and Fisher, T. C. (1994) Apoptosis and cancer chemotherapy, *Phil. Trans. R. Soc. Lond. B* **345,** 319-325.

5. Eastman, A. (1996) The Interaction of Cisplatin With Signal Transduction Pathways and the Regulation of Apoptosis, in Pinedo H. M. and J. H. Schornagel (eds.), *Platinum and Other Metal Coordination Compounds in Cancer Chemotherapy,* Plenum Press, New York, pp. 283-291.

6. Reed, E., Ozols, R. F., Tarone, R., Yuspa, S. T., and Poirier, M. C. (1987) Platinum-DNA adducts in leucocytes DNA correlate with disease response in ovarian cancer patients receiving platinum-based chemotherapy, *Proc. Natl. Acad. Sci. USA* **84,** 5024-5028.

7. Bradley, L. J. N., Yarema, K. J., Lippard, S. J., and Essigman, J. M. (1993) Mutagenicity and Genotoxicity of the Major DNA Adduct of the Antitumor Drug cis-Diamminedichloroplatinum(II), *Biochemistry* **32,** 982-988.

8. Burnouf, D., Daune, M., and Fuchs, R. P. P. (1987) Spectrum of cisplatin-induced mutations in *Escherichia coli, Proc. Natl. Acad. Sci. USA* **84,** 3758-3762.

9. Comess, K. M., Burstyn, J. N., Essigmann, J. M., and Lippard, S. J. (1992) Replication Inhibition and Translesion Synthesis on Templates Containing Site-Specifically Placed cis-Diamminedichloroplatinum(II) DNA Adducts, *Biochemistry* **31,** 3975-3990.

10. Page, J. D., Husain, I., Sancar, A., and Chaney, S. G. (1990) Effect of the Diaminocyclohexane Carrier Ligand on Platinum Adduct Formation, Repair, and Lethality, *Biochemistry* **29,** 1016-1024.

11. Zamble, D. B., Mu, D., Reardon, J. T., Sancar, A., and Lippard, S. J. (1996) Repair of Cisplatin-DNA Adducts by the Mammalian Excision Nuclease, *Biochemistry* **35,** 10004-10013.

12. Kozelka, J. (1996) Molecular Modeling of Transition Metal Complexes with Nucleic Acids and Their Constituents, *Met. Ions Biol. Syst.* **33,** 1-28.

13. Hambley, T. W. (1992) What Can Be Learnt from Computer-Generated Models of Interactions Between DNA and Pt(II) Based Anti-Cancer Drugs?, *Comments Inorg. Chem.* **14,** 1-26.

14. Yang, D., van Boom, S. S. G. E., Reedijk, J., van Boom, J. H., and Wang, A. H.-J. (1995) Structure and Isomerization of an Intrastrand Cisplatin-Cross-Linked Octamer DNA Duplex by NMR Analysis, *Biochemistry* **34,** 12912-12920.

15. Huang, H., Zhu, L., Reid, B. R., Drobny, G. P., and Hopkins, P. B. (1995) Solution Structure of a Cisplatin-Induced DNA Interstrand Cross-Link, *Science* **270,** 1842-1845.

16. Fichtinger-Schepman, A. M. J., Lohman, P. H. M., and Reedijk, J. (1982) Detection and Quantification of Adducts Formed upon Interaction od Diamminedichloroplatinum(II) with DNA, by Anion-exchange Chromatography after Enzymatic Degradation., *Nucleic Acids Res.* **10,** 5345-5356.

17. Eastman, A. (1983) Characterization of the Adducts Produced in DNA by cis-Diamminedichloroplatinum(II) and cis-Dichloro(ethylenediamine)platinum(II), *Biochemistry* **22,** 3927-3933.

18. Kozelka, J., Petsko, G. A., and Lippard, S. J. (1985) Molecular Mechanics Calculations on *cis*-[Pt(NH_3)$_2$d(GpG)] Adducts in Two Oligonucleotide Duplexes, *J. Am. Chem. Soc.* **107,** 4079- 4081.

19. Kozelka, J., Archer, S., Petsko, G. A., Lippard, S. J., and Quigley, G. J. (1987) Molecular Mechanics Modeling of Oligonucleotide Adducts of the Antitumor Drug *cis*-Diamminedichloroplatinum(II), *Biopolymers* **26,** 1245-1271.

20. Kozelka, J., Petsko, G. A., Quigley, G. J., and Lippard, S. J. (1986) Hight-Salt and Low-Salt Models for Kinked Adducts of *cis*-Diamminedichloroplatinum(II) with Oligonucleotide Duplexes, *Inorg. Chem.* **25,** 1075-1077.

21. Herman, F., Kozelka, J., Stoven, V., Guittet, E., Girault, J.-P., Huynh-Dinh, T., Igolen, J., Lallemand, J.-Y., and Chottard, J.-C. (1990) A d(GpG)-Platinated Decanucleotide Duplex Is Kinked - An Extended NMR and Molecular Mechanics Study, *Eur. J. Biochem.* **194,** 119-133.

22. Rice, J. A., Crothers, D. M., Pinto, A. L., and Lippard, S. J. (1988) The Major Adduct of the Antitumor Drug cis-DDP with DNA Bends the Duplex by 40°C Toward the Major Groove, *Proc. Natl. Acad. Sci. USA* **85,** 4158-4161.

23. Bellon, S. F. and Lippard, S. J. (1990) Bending Studies of DNA Site-Specifically Modified by Cisplatin, Trans-Diamminedichloroplatinum(II) and Cis-[Pt(NH_3)$_2$(N3-Cytosine)Cl]$^+$, *Biophys. Chem.* **35,** 179-188.

24. Kozelka, J. and Chottard, J.-C. (1990) How Does Cisplatin Alter DNA Structure? A Molecular Mechanics Study on Double-Stranded Oligonucleotides, *Biophys. Chem.* **35,** 165-178.

25. Fouchet, M.-H., Guittet, E., Cognet, J. A. H., Kozelka, J., Gauthier, C., Le Bret, M., Zimmermann, K., and Chottard, J.-C. (1997) Structure of a nonanucleotide duplex cross-linked by cisplatin at an ApG sequence, *JBIC* **2,** 83-92.

26. Schwartz, A., Marrot, L., and Leng, M. (1989) Conformation of DNA Modified at a d(GG) or a d(AG) Site by the Antitumor Drug *cis*-Diamminedichloroplatinum(II), *Biochemistry* **28,** 7975-7979.

27. Sundquist, W. I., Lippard, S. J., and Stollar, B. D. (1987) Monoclonal antibodies to DNA modified with *cis*- or *trans*- diamminedichloroplatinum(II), *Proc. Natl. Acad. Sci. USA* **84,** 8225-8229.

28. Scheek, R. M., Boelens, R., Russo, N., van Boom, J. H., and Kaptein, R. (1984) Sequential Resonance Assignements in 1H NMR Spectra of Oligonucleotides by Two-Dimensional NMR Spectroscopy, *Biochemistry* **23**, 1371-1376.

29. Visse, R., de Ruijter, M., Brouwer, J., Brandsma, J. A., and van de Putte, P. (1991) Uvr Excision Repair Protein Complex of Escherichia Coli Binds to the Convex Side of a Cisplatin-Induced Kink in the DNA, *J. Biol. Chem.* **266**, 7609-7617.

30. Lide D. R. Jr., Mann D. E. (1958) Microwave Spectra of Molecules Exhibiting Internal Rotation. IV. Isobutane, Tertiary Butyl Fluoride, and Trimethylphosphine, *J. Chem. Phys.* **28**, 914-920.

31. Kozelka, J., Fouchet, M. H., and Chottard, J. C. (1992) H8 Chemical shifts in Oligonucleotides Cross-Linked at a GpG Sequence by cis-$Pt(NH_3)_2^{2+}$- A Clue to the Adduct Structure, *Eur. J. Biochem.* **205**, 895-906.

32. Schröder, G., Kozelka, J., Sabat, M., Fouchet, M.-H., Beyerle-Pfnür, R., and Lippert, B. (1996) Model of the Second Most Abundant Cisplatin-DNA Cross-Link, *Inorg. Chem.* **35**, 1647-1652.

33. Gonnet, F. (1993) Etudes cinétique et structurale de la platination des séquences GG d'oligonucléotides modèles de l'ADN, PhD Thesis, Université Paris VI.

34. DISCOVER program version 95.0 (1996) BIOSYM Technologies, San Diego, USA.

35. Elizondo-Riojas, M.-A., Gonnet, F., Chottard, J.-C., and Kozelka, J. (1997) Conformational analysis of *cis*-[$Pt(NH_3)_2${d(TpG)}] using ^{1}H NMR and molecular dynamics simulations, unpublished results.

36. AMBER program Version 3.0 (1986) University of California, San Francisco, USA.

37. Le Bret, M., Gabarro, J., Gilbert, J. C., and Lemarechal, C. (1991) MORCAD, an object-oriented molecular modeling package running on IBM/RS6000 and SGI 4Dxxx work stations, *J. Chim. Phys.* **88**, 2489-2496.

38. Giessner-Prettre, C. and Pullman, B. (1987) Quantum mechanical calculations of NMR chemical shifts in nucleic acids, *Q. Rev. Biophysics* **20**, 113-172.

39. Merz, K. M., Kollman, P. A., and Knapp, E.-W. (1997) discussion at the NATO workshop.

40. a) GAUSSIAN 94 program, Revision C.3 (1995) Gaussian, Inc., Pittsburgh, USA;
b) Foresman, J. B. and Frisch, A. (1996) *Exploring Chemistry with Electronic Structure Methods*, Gaussian, Inc., Pittsburgh.

41. Chirlian, L. E. and Francl, M. M. (1987) Atomic Charges Derived from Electrostatic Potentials: A Detailed Study, *J. Comput. Chem.* **8**, 894-905.

42. Besler, B. H., Merz, K. M., Jr., and Kollman, P. A. (1990) Atomic Charges Derived from Semiempirical Methods, *J. Comput. Chem.* **11**, 431-439.

43. Bachrach, S. M. (1994) Population Analysis and Electron Densities from Quantum Mechanics, *Rev. Comput. Chem.* **5**, 171-227.

44. Wiberg, K. B., Hadad, C. M., LePage, T. J., Breneman, C. M., and Frich, M. J. (1992) Analysis of the Effect of Electron correlation on Charge Density Distributions, *J. Phys. Chem.* **96**, 671-679.

45. Messmer, G. G., Amma, E. L., and Ibers, J. A. (1967) Crystal Structure of *cis*-$Pt[P(CH_3)_3]_2Cl_2$, *Inorg. Chem.* **6**, 725-730.

46. Del Pra, A. and Zanotti, G. (1979) Crystal Structure Refinement of *cis*-$Pt[P(CH_3)_3]_2Cl_2$, *Cryst. Struct. Commun.* **8**, 737-742.

47. Jensen, K. A. (1936) Dipolmessungen an isomeren Platokomplexen. II., *Z. anorg. allg. Chem.* **229**, 225-251.

48. Chatt, J. and Wilkins, R. G. (1952) The Nature of the Co-ordinate Link. Part VI. A Comparison of Equilibria between *cis*- and *trans*-$(MEt_3)_2PtCl_2$, where M = P, As, and Sb, *J. Chem. Soc.* 4300-4306.

49. Appleton, T. G., Clark, H. C., and Manzer, L. E. (1973) The *trans*-influence: its measurements and significance, *Coord. Chem. Rev.* **10**, 335-422.

50. Kozelka, J. and Bergès, J. (1997) unpublished results.

51. Hay, P. J. and Wadt, W. R. (1985) Ab initio effective core potentials for molecular calculations. Potentials for the transition metal atoms Sc to Hg, *J. Chem. Phys.* **82**, 270-282.

52. Hay, P. J. and Wadt, W. R. (1985) Ab initio effective core potentials for molecular calculations. Potentials for K to Au including the outermost core orbitals, *J. Chem. Phys.* **82**, 299-310.

53. Andrae, D., Häussermann, U., Dolg, M., Stoll, H., and Preuss, H. (1990) Energy-adjusted *ab initio* pseudopotentials for the second and third row transition elements, *Theor. Chim. Acta* **77**, 123-141.

54. Pelissier, M. (1996) personal communication.

55. Irikura, K. K. and Goddard, W. A. I. (1994) Energetics of Third-Row Transition Metal Methylidene Ions MCH^+ (M = La, Hf, Ta, W, Re, Os, Ir, Pt, Au), *J. Am. Chem. Soc.* **116**, 8733-8740.

56. Larsen, E. and La Mar, G. N. (1974) The Angular Overlap Model. How to use it and why, *J. Chem. Ed.* **51**, 633-640.

57. Miller, K. J., Taylor, E. R., Bash, H., Krauss, M., and Stevens, W. J. (1985) A Theoretical Model for the Binding of Cis-$Pt(NH_3)_2^{+2}$ to DNA, *J. Biomol. Struct. Dyn.* **2**, 1157-1171.

58. VMD program version 1.2b1 (1997) A. Dalke, W. Humphrey, J. Ulrich, Theoretical Biophysics Group, University of Illinois and Beckman Institute, Urbana, USA.

METAL CATIONS IN BIOLOGICAL SYSTEMS: MODELING METAL IONS IN IONOPHORES AND DNA

PETER KOLLMAN
Department of Pharmaceutical Chemistry
University of California, San Francisco
SF, CA 94143-0446

1. Introduction and Perspective

Studying complex molecules of organic and biological interest usually demands the employment of molecular mechanical methods. Unlike first principle quantum mechanical electronic structure methods, molecular mechanical methods (e.g. eq. 1) require that chemical bonds in the system be specified and these are present throughout the calculation. This is straightforward for the atoms that make up an organic molecule, but the situation is more complex for metal cations, which interact with electron rich centers. These interactions usually have considerable ionic <u>and</u> covalent character, whereas the typical molecular mechanical model requires that one choose a fully non-covalent or fully covalent interaction. A non-covalent model employs only the van der Waals and electrostatic interactions between metal cation and the other atoms [1], or if non-additive effects are included in the model, also polarization effects [2] and, perhaps, three body exchange [3]. It appears that this approach is reasonable for mainly electrostatic interactions such as those involving alkali ions with "hard" Lewis bases such as N and O.

L. Banci and P. Comba (eds.), Molecular Modeling and Dynamics of Bioinorganic Systems, 161–166.

$$V=\sum_{bonds}\frac{1}{2}K_r\left(r-r_{eq}\right)^2+\sum_{angles}\frac{1}{2}K_\theta\left(\theta-\theta_{eq}\right)^2+$$
$$\sum_{dihedrals}\frac{V_n}{2}[1+\cos(n\theta-\gamma)]+\sum_{nonbonds}\left[\frac{A_{ij}}{R_{ij}^{12}}-\frac{B_{ij}}{R_{ij}^{6}}+\frac{q_iq_j}{\varepsilon R_{ij}}\right] \tag{1}$$

At the other extreme are strongly covalent interactions such as those involving Pt with nitrogen lone pairs [4]. In such cases, one uses explicit bonds (eq. 1) and angle terms involving these interactions, the latter to keep the N-Pt-N near to the canonical values of 90° or 180°. The covalent model allows a more precise representation of the bond geometry around the metal, with the disadvantage that the energetics of breaking of the metal-ligand bond cannot generally be simulated with a "standard" covalent molecular mechanical approach.

Nonetheless, just as there is a large variability in functional forms of molecular mechanical models, varying from the simple models of eq.1 [5] to those that include anharmonic effects and cross terms involving bonds, angles, dihedrals [6], there can be many variants in covalent models involving metal cations, such as (a) allowing bond dissociation using Morse like functions; (b) including non-bonded interactions between next neighbor [1-3] interactions; (c) making all dihedral energies around a metal center zero [7]. The basic premise of molecular mechanical models is to derive/calibrate them on simple systems and then use them on more complex systems; as long as the model has the capability of modeling the phenomena of interest, such an approach can be considered.

2. Applications

Here we describe some examples of modeling systems involving metal cations, focusing mainly on our work in this area.

Modeling Fe-porphyrin, as it occurs both in modeling systems and in myoglobin and hemoglobin is inevitably done with a covalent model, because only such a model is likely to give the correct bonding geometry of the heme center. In our studies [8,9], we also used a bonded model to describe Fe-CO and FeO_2 interactions, but the focus of our modeling was to calculate the relative free energies of the environment on the heme CO vs heme-O_2 systems via non-covalent interactions. Briefly recapitulating, it was known that CO ligated to an isolated Fe-porphyrin 10^4 more tightly than O_2. However, in various more complex organic models, and in myoglobin, this ratio was considerably reduced. So the goal of our free energy calculations was to see if the stronger non covalent interaction of the environment with Fe-O_2 than Fe-CO could quantitatively reproduce the reduction of differential free energy of binding from ~5.6 kcal/mole (10^4) favoring CO to its smaller values in the model systems and myoglobin. This was successfully accomplished, not only in simpler models [8] but also in myoglobin [9].

At the other extreme has been our model of alkali ions interacting with crowns, calixerenes, calixspherands and spherands [10]. Ours was the first to study cation-crown interactions, and we used a non-bonded model [1]. The usefulness of this approach has stood the test of time to date. A more extensive review of our work in this area is presented elsewhere [10], but a few brief highlights are in order. In our 1982 study, we were able to reproduce/rationalize the different structural preferences of 18-crown-6 when free (Ci), complexed with $Na^{+(1)}$ or when complexed with K^+, Rb^+, Cs^+ (D_{3d}) using energy minimization. In the period prior to 1990, due to computer limitations, only inclusion of solvent and molecular dynamics/free energy calculations could be done. Nonetheless, relative free energies could sometimes be estimated with simple cluster models for solvent.

The work of Damewood on Fe^{+2} and Fe^{+3} [11], Aqvist and Warshel on a number of divalent cations [12] and Aqvist on alkali metal cations [13] were important tests of the non-covalent's model ability to reproduce aqueous solvation free energies and these studies showed this could be done. The Aqvist study provided a set of cation-water parameters which were particularly useful for free energy calculations on host guest systems, since they were calibrated to reproduce the aqueous solvation free energy exactly.

We have carried out free energy calculations that reproduced the relative free energies of various alkali cations for hexaspherands [14], octaspherands [15,16] and open chain analogues of 18-crown-6 vs 18-crown-6 itself [17]. In all cases, we could simulate experiment semi-quantitatively and it is likely that fully non-additive models might offer an improvement. So the challenge for the future is to see what is needed for truly accurate quantitative modeling of such systems.

An exciting new area of simulation is the study of DNA double helices [18] and the modeling of the A→B transitions in such systems [19]. Our standard model of DNA decamer duplexes has been to include 18 Na^+ cations in the model which, together with the polyanionic DNA (-18), lead to a neutral system. Such systems also include thousands of H_2O

in a periodic box and a particle mesh Ewald (PME) representation of the long range electrostatics. We have shown that replaces 100 H_2O in the cell by an additional 50 Na+ and 50 Cl- corresponding to a 1M solution of NaCl has little effect on the results with only neutralizing cations [20]. Similarly, not including cations at all, and making the change on each atom fractionally more positive by 18/N, where N, the number of atoms in the system is ~1000) has little effect on the trajectories [20].

On the other hand, we have studied DNA in a solution including $Co(NH_3)_6^{+3}$, with the Co-N represented covalently, but the interaction of the cation, with the rest of the system represented non-covalently, using a DNA duplex sequence that has been shown to undergo B→A transition in the presence of $Co(NH_3)_6^{+3}$. The calculations were successful in simulating this metal driven process [20].

3. Summary

There are various ways to model metal ions within molecular mechanical models and we heard some promising new ones at this meeting presented, among others, by Deeth and Landis. This important area will continue to see much research effort, given the importance of these cations in structure/function relationships in organic and biochemical systems.

References

1. G. Wipff, P. Weiner, and P.A. Kollman. (1982) A Molecular Mechanism Study of 18-Crown-6, its Open Chain Analog and Their Complexes with Alkali Cations.*J. Amer. Chem. Soc.* **104**, 3249-3258
2. J. Caldwell, L.X. Dang, and P.A. Kollman. Implementation of Nonadditive Intermolecular Potentials by Use of Molecular Dynamics: Development of a Water-Water Potential and Water-Ion Cluster Interactions. (1990) *J. Amer. Chem. Soc.* **112**, 9144-9147.
3. L. Dang, J. Rice, J. Caldwell, and P. Kollman. (1991) The Effect of Water Models on the Interaction of the Sodium-Chloride Ion Pair in Water. *J. Amer. Chem. Soc.* **113**, 2481-2486 .
4. Kozelka, R. Savinelli, G. Berthier, J. Flament and R. Lavery, (1993) Force Field For Platinum Binding To Adenine. *J. Comp. Chem.* **14**, 45-53.
5. W.D. Cornell, P. Cieplak, C.I. Bayly, I.R. Gould, K.M. Merz, Jr., D.M. Ferguson, D.C. Spellmeyer, T. Fox, J.W. Caldwell and P.A. Kollman.A Second Generation Force Field for the Simulation of

Proteins and Nucleic Acids. (1995) *J. Amer. Chem. Soc.* **117**, 5179-5196

6. N.L. Allinger, K.S. Chen, and J.H. Lii, (1996) An Improved Force Field (MM4) For Saturated Hydrocarbons.*J. Comp. Chem.* **17**, 642-668 .
7. D. Ferguson and D. Raber, Molecular Mechanics Calculations Of Several Lanthanide Complexes - An Application Of The Random Incremental Pulse Search.*J. Comp. Chem.* **11**, 1061-1071.
8. M.A. Lopez and P.A. Kollman. (1989) Application of Molecular Dynamics and Free Energy Perturbation Methods to Metalloporphyrin-Ligand Systems I: CO and Dioxygen Binding to Four Heme Systems. *J. Amer. Chem. Soc.* **111**, 6212-6222 (1989).
9. M.A. Lopez and P.A. Kollman. (1993) Application Of Molecular Dynamics And Free Energy Perturbation Methods To Metalloporphyrin-Ligand Systems .2. CO And Dioxygen Binding To Myoglobin. *Protein Science*, **2**, 1975-1986 .
10. P. Kollman. (1995) The Nature of Molecular Recognition: Examples from Host/Guest Chemistry. In *Computational Approaches in Supramolecular Chemistry,* G. Wipff, ed, pp 175-181, Kluwer Academic Publishers, The Netherlands
11. J.J. Urban and J.E. Damewood. (1990) Accurate Free Energy Perturbation Calculation Of The Hydration Free Energies Of Fe-3+ And Fe-2+. *J. Chem. Soc. Chem. Comm.* **22**, 1636-1638
12. J. Aqvist and A. Warshel, (1990) Free Energy Relationships In Metalloenzyme-Catalyzed Reactions - Calculations Of The Effects Of Metal Ion Substitutions In Staphylococcal Nuclease. J. Amer. Chem. Soc. **112**, 2860-2868.
13. J. Aqvist, (1990) Ion Water Interaction Potentials Derived From Free Energy Perturbation Simulations.*J. Phys. Chem.* **94**, 8021 .
14. Y. Sun, J.W. Caldwell and P. Kollman. (1995) Molecular Dynamics and Free Energy Perturbation Study of Spherand Complexation with Metal Ions Employing Additive and Non-Additive Force Fields.*J. Phys. Chem.* **99**, 10081-10085
15. C.I. Bayly and P.A. Kollman (1994) Molecular Dynamics And Free Energy Calculations On The Peculiar Bimodal Alkali Ion Selectivity Of An 8-Subunit Cavitand. J. Amer. Chem. Soc. **116** , 697-703 .
16. B.E. Thomas and P.A. Kollman (1994) Free Energy Perturbation Calculations Of The Relative Binding Affinities Of An 8- Subunit Cavitand For Alkali Ions In Methanol.*J. Amer. Chem, Soc.* **116**, 3449-3452.
17. Y. Sun and P.A Kollman. (1995) The Quantitation and Nature of the Macrocyclic Effect: K^+ Complexation with 18-crown-6 and Pentaglyme. *J. Amer. Chem. Soc.***117**, 3599-3604 .
18. T.E. Cheatham, III, J.L. Miller, T. Fox, T.A. Darden and P.A. Kollman, (1995) Molecular Dynamincs Simulations on Solvated

Biomolecular Systems: The Particle Mesh Ewald Method Leads to Stable Trajectories of DNA, RNA and Proteins.*J. Amer. Chem. Soc.* **117**, 4193-4194

19. T.E. Cheatham III and P.A. Kollman, (1996) Observation of the A-DNA to B-DNA Transition During Unrestrained Molecular Dynamics in Aqueous Solution *J. Mol. Bio*l. **259**, 434-444 .
20. T.E. Cheatham III and P.A. Kollman, unpublished.

THE ROLE OF Ca^{2+} IN THE BINDING OF CARBOHYDRATES TO C-TYPE LECTINS AS REVEALED BY MOLECULAR MECHANICS AND MOLECULAR DYNAMICS CALCULATIONS

CLAUS-WILHELM VON DER LIETH
German Cancer Research Center
Spectroscopic Department,
Im Neuenheimer Feld 280,
69120 Heidelberg
Germany

1. Introduction

Lectins constitute a structurally diverse class of proteins or glycoproteins which are characterized by their ability to bind carbohydrates with considerable specificity [1-3]. They are found in various organisms, ranging from viruses, bacteria and plants to humans. There is increasing evidence that carbohydrate-lectin interactions have a fundamental role in cell adhesion processes [4,5], cell proliferation [6] , organogenesis and human pathology. Lectins merely bind but do not process carbohydrates. In contrast to, for example, antibodies, which can also bind carbohydrates, lectins are produced constitutively and not as a result of an external stimulus [7,8]. Lectins have been grouped into classes of discrete families based on homologies in their primary structures [9]. Although the number of reported animal lectins continues to increase, a recent classification [9] indicates that most fall into one of five major groups: the Ca^{2+}-dependent (C-type) lectins, the galectins (galactose binding proteins), the mannose-6-phosphate-binding (P-type) lectins, and the immunoglobolin-like (I-type) lectins including sialoadhesins and L-type lectins, related in sequence to the leguminous plant lectins. While the overall architecture of the lectins widely varies, carbohydrate-binding activity can often be assigned to one part of the structure, called a carbohydrate recognition domain (CRD). C-type CRDs are present in a diverse array of protein structures which have been found in serum, extracellular matrix and membranes of animals. These proteins share a common primary structural motif, consisting of 14 invariant and 18 highly conserved residues in the CRD region. Seven structural categories of C-type lectins have been defined [10]. Among these the mannose-binding proteins, the collectins, the asialoglycoproteins and the selectin cell adhesion molecules have been investigated in-

L. Banci and P. Comba (eds.), Molecular Modeling and Dynamics of Bioinorganic Systems, 167–190.

tensively. In particular the selectins have attracted considerable scientific interest since they mediate the initial phase of adhesion between leukocytes and the endothelia [11,12].

The various C-type lectins show a Ca^{2+}-dependent binding to several different carbohydrates, including mannose, glucose, fucose, galactose and several of their glucosamine and galactosamine derivatives [13,14]. Lectins are often characterized by their ability to bind mono- or sometimes disaccharides. The interactions of individual saccharides are normally of low affinity and their dissociation constants are generally in the mM range [15]. Nature seems to use low-affinity carbohydrate-protein interactions to mediate a vast array of biologically important functions [16].

Progress in the understanding of the biological function of lectins on an atomic level has been made during the last decade through a combination of techniques such as X-ray crystallography [17-19], site-directed mutagenesis [20-22], NMR-studies, various ligand binding experiments [13,14] and molecular modeling [23-25].

The results reveal that one of the two calcium ions which are bound to the CRD forms the nucleus of the sugar binding site, whereby two vicinal, equatorial hydroxyl groups of the carbohydrate act as coordinating ligands for the Ca^{2+}. Four of the five amino acid side chains which coordinate with the calcium ion can also form hydrogen bonds to the hydroxyl groups, which can have both donor and acceptor functions. The binding affinities of several mono-, di-, and polysaccharides have been determined [13]. Additionally, a variety of mutants of mannose-binding protein-A (MBP-A) and MBP-C have been created to study how different amino acids influence the process of binding [26-29] or to enhance the binding affinity of certain monosaccharides.

Computer simulation techniques can potentially provide a detailed view of lectin-carbohydrate interactions [30,31]. The aim of this study is to evaluate if and how classical, force field based methods are able to describe the geometry of various carbohydrate C-type lectin complexes. Molecular dynamics (MD) simulations should provide insights into the dynamic behavior of these complexes, i.e., information which is not available by experimental methods. Force-field-based calculations of proteins containing highly charged ionic species such as Ca^{2+} ions require an appropriate description of the protein-metal ion complex. Therefore, the validity of the force field and of the computational model used, as well as the selected protein model have to be evaluated carefully. Another goal of this study is to determine how well a simple model of the binding site can mimic the essential structural features of the binding process and can produce qualitative insights into the ligand binding.

2. Modeling Approaches

2.1. AVAILABLE X-RAY STRUCTURES

A limited number of high-resolution (below 2.0 Å) X-ray crystal structures have been determined for C-type lectins complexed with either monosaccharides or oligosaccharides (see Table 1). Saccharides with vicinal pair of equatorial hydroxyl groups such as those found at the 3 and 4 positions of mannose show the ability to bind to calcium-dependent lectins. The Brookhaven Protein Dat Bank currently contains structures of the serum collectins, the mannose-binding protein (MBP-A), complexed with a mannose-rich oligosaccharide, and mutants thereof, complexed with β-methyl galactose and N-acetyl galactosamine. Moreover, it contains the rat liver mannose-binding protein (MBP-C), complexed with different monosaccharides and one unliganded structure of the human MBP-C. Selectins-like mutants of MBP-A in the unliganded form and as complexes with three different Lewis^X derivatives have been deposited, as has a solved structure of E-selectin in the unliganded form. Additionally, the Brookhaven Protein Data Bank contains two theoretical models of E-selectin and one of P-selectin, obtained by knowledge-based homology modeling.

The collectins share a common arrangement of structural domains. A cysteine-rich domain at the amino terminus is followed by an α-helical collagenous domain and a so-called neck region which connects the collagenous domain to the C-terminal CRD. The individual monomers assemble into homotrimers (see Fig. 1), which further associate into larger oligomers. Serum MBP-A consists of hexamers of the trimeric ensembles. MBP-C has been reported to form smaller oligomers, probably consisting of two associated trimers. The older X-ray structure determinations were based on crystals of the 115-unit C-terminal peptide of rat MBP-A as the minimum fragment that exhibits carbohydrate-recognition activity, since the quality of the crystals of complete protein was not sufficient for high-resolution X-ray studies [37]. The CRD fragments alone formed dimers in solution as well as in crystals. Whereas complex sugar ligands are needed to grow the crystal of MBP-A, MBP-C does not require sugar ligands to crystallize [28]. Thus, MBP-C complexes provided a system for studying a variety of lectin carbohydrate complexes.

The CRD consists of 115 to 130 amino acids [10], containing 14 invariant and 18 highly conserved amino acid residues. All solved X-ray structures contain 2 or 3 Ca^{2+} ions in the CRD. One of the required Ca^{2+} coordinates directly with the sugar ligand and thus is involved in the binding process. Among the structurally characterized lectins this function of a Ca^{2+} at the binding site is unique. In other lectins metal ions are used to stabilize the binding site and fix the positions of the amino acids that interact with the sugar ligands [38].

TABLE 1: C-type lectins with X-ray structures in the Brookhaven Protein Data Bank

PDB code	Bound Sugar	Species	Ref.	Resol. [Å]	R-value	Remark
MPB-A						
1MSB	(unliganded)	rat	[32]	2.3	0.176	
2MSB	α-D-Man		[33]	1.7	0.174	Complexed with Man6
1YTT	(unliganded)	rat	[58]	1.8	0.185	
MPB-C						
1RDI	α-1-O-Me-L-Fuc	rat,eng.	[28]	1.8	0.196	Dimer
1RDJ	β-1-O-Me-L-Fuc	rat,eng.	[28]	1.8	0.194	Dimer
1RDK	D-Gal	rat,eng.	[28]	1.8	0.265	Dimer
1RDL	α-1-O-Me-D-Man	rat,eng.	[28]	1.8	0.203	Dimer 0.2 M
1RDM	α-1-O-Me-D-Man	rat,eng.	[28]	1.9	0.191	Dimer 1.3 M
1RDN	α-1-O-Me-D-GlcNAc	rat,eng.	[28]	1.8	0.203	Dimer
1RDO	(unliganded)	rat,eng.	[28]	2.5	0.250	Dimer
MBP-A-Gal						
1AFA	β-1-O-Me-D-Gal	rat,eng.	[29]	2.0	0,234	Trimer
1AFB	β-D-GalNAc	rat,eng.	[29]	1.9	0.217	Trimer
1AFD	(unliganded)	rat,eng.	[29]	2.0	0.234	Trimer
Selectin-like Mutant of MBP-A						
1KMB	(unliganded)	rat,eng	[34]	2.1	0.265	Trimer
2KMB	3'-SiaLewisX	rat,eng	[34]	2.0	0.247	Trimer
3KMB	3'sulfoLewisX	rat,eng	[34]	1.95	0.254	Trimer
4KMB	4'sulfoLewisX	rat,eng	[34]	2.0	0.272	Trimer
Selectin						
1ESL	(unliganded)	human	[35]	2.0	0.173	
1KDA	unliganded (E-Selectin)	human	[23]			2MSB as template
1KDJ	unliganded (P-Selectin)	human	[36]			2MSB as template

The second Ca^{2+} which is present in all solved structures fulfills a structural function for C-type lectins. The calcium of the structural site is at a distance of about 8.4 Å from the calcium of the binding site. The location of a third Ca^{2+}, which is not present in all crystals, is not well defined. When present, it is only weakly attached to the protein surface in the vicinity of the structural calcium. Therefore, it was concluded that the third site may arise from adventitious binding of excess Ca^{2+} present in the crystallization medium [33].

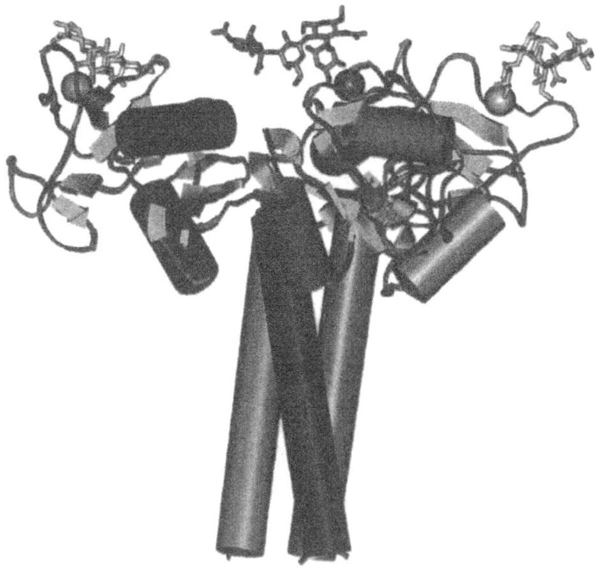

Fig. 1: Representation of the crystal structure of a selectin-like mutant of MBP-A, complexed with sialylated Lewis^x oligosaccharides (PDB entry 2KBM.pdb) [34]. The protein crystallizes as a trimer. Only the Ca^{2+}-ions (spheres) involved in the binding process located close to the fucose (stick representation) are shown. The secondary structural motifs of the protein backbone are cylinders for α helices, arrows for β–strands and ribbons for random coil regions; water molecules are not shown.

The calcium ion at each site is coordinated to 8 oxygen atoms. Fig. 2 gives a graphical representation of the geometry of the binding site. The binding calcium is coordinated to the two oxygen atoms of the sugar, the carbonyl oxygen of the side chains of Gln-190, Glu-198, Asp-192, Asn-210 and Asp-211 and the amid carbonyl atom of Asp-211. The structural calcium it is complexed by one water molecule, both carboxyl oxygen atoms of Asp-116, Glu-170, the carbonyl oxygen atoms of the side chains of Asn-193 and Asn-199 as well as the amid carbonyl of Glu-198. Since two different oxygen atoms of Glu-189 are involved at the two sites, it may be concluded that the structural calcium is important for maintaining the structural integrity of the binding site and to positioning certain amino acids for optimal binding geometry.

The appendix lists the calcium-oxygen distances for both sites for various experimentally determined structures of MBP-A, MBP-C and MBP-A mutants which recognize galactose and the Lewis^{X} analogues. These data indicate that the geometry at each site is well conserved in all structures. Not surprisingly, the distances for the various carbohydrates show relatively large variation. Although this interaction is a prerequisite for the binding of various carbohydrates to Ca^{2+}-dependent lectins, other structural features such as diffrences in the residues near the binding site may influence the binding affinities.

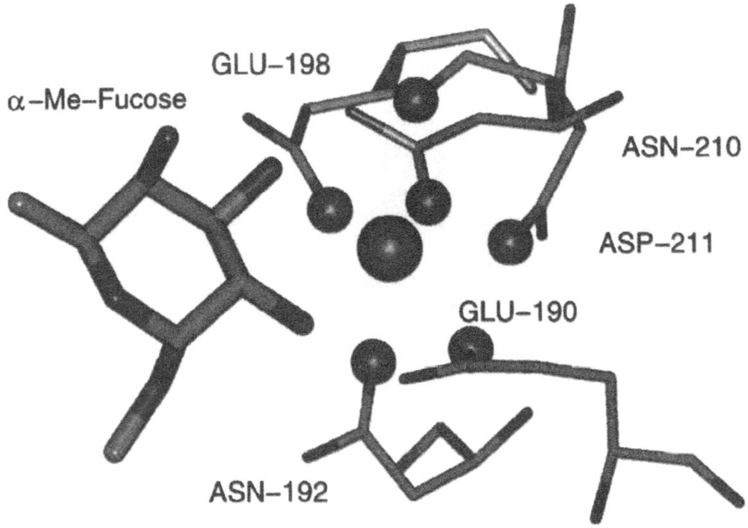

Fig. 2: Coordination geometry at the bindings site of a C-type lectin. The crystal structure of MBP-C coordinated by α-Me-fucose (PDB-entry 1RDJ) is shown. The calcium ion (large sphere) is coordinated to 6 oxygen atoms of the lectin (small spheres) and two oxygen atoms of the carbohydrate ligand.

2.2. SOFTWARE AND HARDWARE

Manipulation of the structures and analysis of the trajectories of the MD-simulations were performed using various modules (InsightII, Biopolymer, Analysis) of the MSI/ BIOSYM software. Force field calculations were carried out with the standard AMBER [39,40] or the CVFF [41,42] parameterization in DISCOVER 2.9.5.. The semi-empirical quantummechanical calculations to derive point charges for the Ca^{2+} and the active site were performed with the MOPAC 5.0 program included in the InsightII-module. For MD-calculations the parallel processing version of DISCOVER 2.9.5., implemented on an IBM-SP2, computer was used with a maximum of 4 processors. Interactive manipulations and visualization of the various molecules were performed on a SGI Indigo^2 with a R4400 processor and two graphics engines.

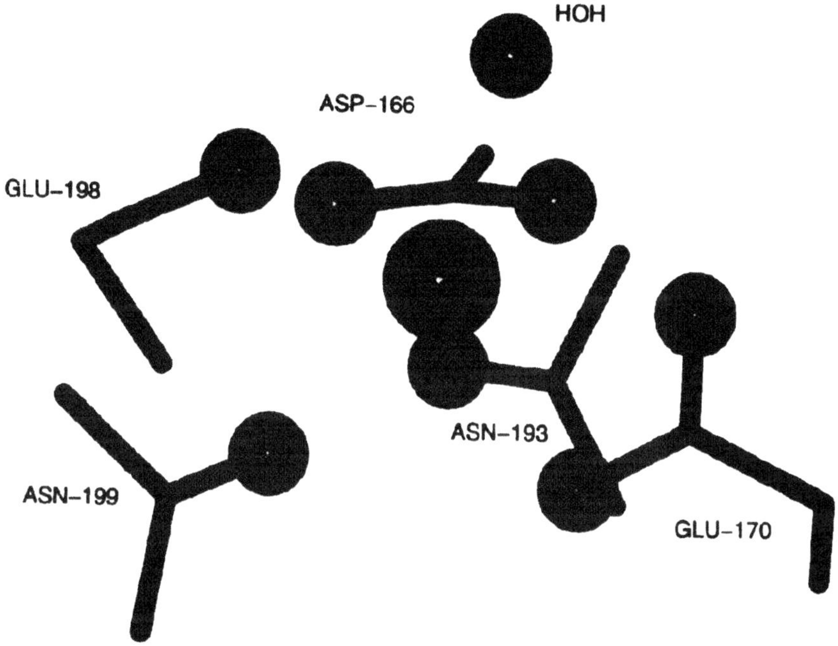

Fig. 3: Coordination geometry at the structural calcium site in C-type lectins. The crystal structure of MBP-C (PDB-entry 1RDJ) is taken as an example. The calcium ion (large sphere) is coordinated to 7 oxygen atoms (small spheres) of the lectin and one water molecule.

2.3. COMPUTATIONAL MODEL

Force-field-based calculations of proteins containing highly charged ionic species such as Ca^{2+} require an appropriate description of the protein-ion complex. Two different approaches have been described which are referred to as *bonded* models and *nonbonded* models. In the *bonded* models explicit bond and angle terms are introduced into the potential energy function to account for the interaction between the ion and the coordinating atoms [43-46]. The alternative *nonbonded* models only include van der Waals and electrostatic terms. These should be appropriate for Ca^{2+}, Zn^{2+}, Mg^{2+} and Cd^{2+} which have a closed shell electronic configuration and thus no crystal field splitting. The main difficulty is to find an appropriate description for the strong electrostatic interactions of the divalent ions. Early attempts of *nonbonded* models in molecular mechanics calculations of proteins resulted in coordinations geometries in disagreement with experimental data [47,48]. In a more recent study, Stote and Karplus [49] showed that the inclusion of long-range electrostatic interactions up to 15 Å is essential to obtain structures in agreement with experimental results. Most of the other simulation studies described used cutoffs in the range of 8 to 10 Å. Stote and Karplus included the long-range electrostatic interactions via the "extended electrostatics" model based on a multipole expansion of distant charges [50]. Comparing *bonded* and *nonbonded* models, Stote and Karplus concluded that a *nonbonded* model is physically more valid and should be the preferred choice. The use of a *bonded* model is only necessary in cases where the geometry of an essentially rigid coordination site must be preserved.

Following the advice of Stote and Karplus a *nonbonded* model was used for the Ca^{2+}-dependent lectins. The standard Lennard-Jones potential for the van der Waals interactions and the Coulomb potential for the electrostatic terms as implemented in many force fields were used. The AMBER and the consistent valence force field (CVFF) provide parameters for van der Waals interactions of Ca^{2+}. The parameterization of the electrostatic term strongly depends on the charge at the Ca-ion, the choice of the dielectric constant and the model of truncation for the nonbonded interactions. In the case of Zn^{2+}-containing proteins, early studies as well as more recent work of Stote and Karplus assumed a formal 2+ charge on the zinc ion. In combination with the AMBER force field a procedure to derive MNDO-based charges for the active site has been used [51]. In this study the point charges of the active site of MBP-A were calculated with PM3. The resultant charge on the calcium ion was close to +0.8. To evaluate the influence of the point charge of the calcium ion for the stability of the complex, three different MD simulations were performed, using a point charge of +2.0, +0.4 and +0.8. Here only the results which were obtained with the consistent valence force field will be reported.

2.4. PROTEIN SIMULATIONS

The high-resolution (1.8 Å) X-ray structure of MBP-C complexed with β-1-O-Me-L-fucose (PDB entry 1RDJ) was selected to evaluate the parameterization of the *nonbonded* model of Ca^{2+}-dependent lectins. The protein crystallizes as a dimer. Fig. 4 gives a simplified representation of its structure.

Each monomer consists of 110 residues, 2 calcium ions and the ligand. 505 water molecules are present in the X-ray structure and were included in the simulation. The missing H-atoms were added using the ADD-H-atoms facility in InsightII at a pH of 7 to give a total of 4312 atoms.

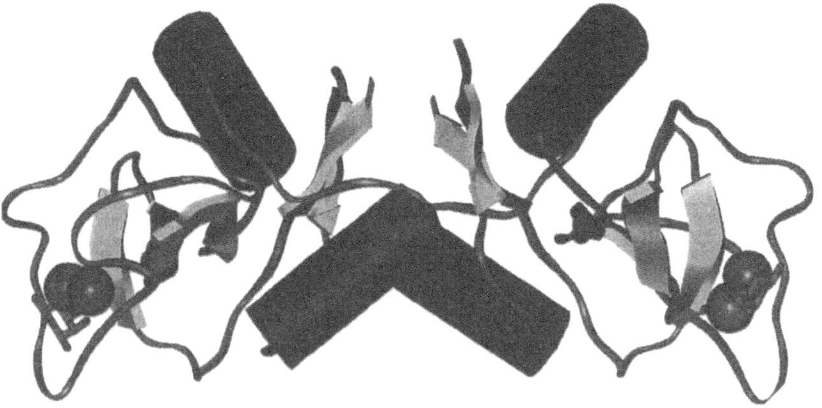

Fig. 4: X-ray structure of MBP-C complexed with β–1–O–Me-L-fucose (PDB entry 1RDJ). The protein crystallizes as a dimer. The Ca^{2+} involved in the binding process as well as the structural Ca^{2+} are shown as spheres. The fucose is given as a stick representation. The water molecules are not displayed.

The potential type for each atom was assigned using the automatic procedure in InsightII. The calcium ions which were not assigned automatically had to be assigned manually. Point charges for each atom were added with the automatic charge assignment facility in InsightII. The charge of the calcium ion was set to the values discussed above. A distance-dependent dielectric constant of 1 and a double cutoff of 13 and 15 Å have been applied for all MD simulations. The integration time step was 1 fs. The backbone atoms of all amino acid residues were tethered with a soft force constant of 20 kcal/$Å^2$ to avoid conformational changes in the vicinity of the structural calcium which occurred after about 100 ps when no constraints were applied. An energy minimization (1000 steps, steepest descent algorithm) was used to remove short van der Waals contacts and strained bond angles before starting the MD simulations. An equilibration period of 50 ps was followed by a production time of 100 ps. The simulations were carried out at a temperature of 300 K. The coordinates were stored after every 250 steps of integration (0.25 ps). The various trajectories were examined with the ANALYSIS option of InsightII.

2.5. MODEL OF THE ACTIVE SITE

A simple model of the binding site which mimics its essential structural characteristics was created for the study of the interactions of the various ligands in different orientations and for a number of mutants. All residues within a sphere of 5 Å around the ligand and the residues which form the complex with the binding Ca^{2+} were included in this simple model. For the analysis of MBP-A complexes the models were generated from the X-ray structure of the PDB entry 2MSB and for the MBP-C complexes the X-ray structure of 1RDI was used. The backbone atoms of all residues as well as the calcium ion were fixed. The two carbohydrate oxygen atoms which interact with the calcium ion were tethered using a weak force constant of 10 kcal/Å^2. No constraints were applied to the side chains of the lectin and the remaining atoms of the various ligands. The carbohydrate geometries were either extracted from the corresponding complex with MBP-C or MBP-A or taken from the monosaccharide library of the program SWEET [52]. Different complexes were generated using the SUPERPOSITION option of the BIOPOLYMER module by interactively selecting the corresponding oxygen atoms and at least three carbon atoms of the sugar ring. Mutations of the binding site were accomplished with the aid of the PROTEIN/REPLACE option of the BIOPOLYMER module. These models were minimized (conjugate gradient, convergence of 0.0001 kcal). Then, the nonbonded energies between the lectin and the ligand were calculated with the ENCLOSE option of the stand-alone version of DISCOVER. The program ANA_INTERACT (C.W.v.d.Lieth, unpublished) extracted the relevant information from the output files and converted it to files readable by EXCEL (Mirosoft).

3. Results and Discussion

3.1. MOLECULAR DYNAMICS SIMULATIONS

Molecular modeling of structures and dynamics of a metalloprotein is a more complex problem than modeling a protein without highly charged metal ions. Therefore, the validity of the force field, the computational model and the selected protein model have to be evaluated carefully. In this study the distances between the Ca^{2+} and the various coordinating oxygen atoms of the lectin (see Table 2) have been used as a measure of the validity of the chosen parameterization and models. Fig. 5 displays the trajectories of a simulation with a point charge of +0.8 for the calcium ion. All Ca-O distances exhibit some fluctuations around their mean values, but, with the exception of the distance involving Ca-OD1:Asn-210, do not show any jumps or shifts to other orientations over the duration of the simulation.

The mean values of the corresponding distances in 31 experimentally determined structures (see Appendix) were used as a reference for analyzing the simulation results. Of the three values tested a point charge of +0.8 for calcium gave the best agreement with the experimental results. The standard deviations

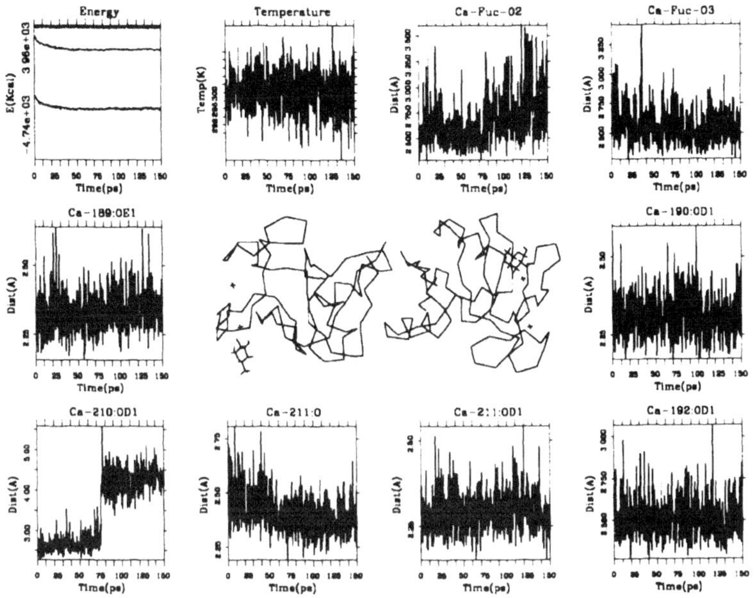

Fig. 5 Trajectories of Ca-O distances (simulation with a point charge +0.8 for the calcium ion). In the middle, a simple stick representation of the α-carbons of MBP-C, the calcium ions (cross) and the ligand fucose are shown

of the mean Ca^{2+}-oxygen distances for the binding site as well as their minimal and maximal values were similar to the experimental values (see Appendix).The Ca^{2+}-oxygen distances for the bound carbohydrate were close to experimental results. They were 0.1 to 0.3 Å longer than those involving the amino acid residues of the lectin and showed a higher flexibility. The Ca^{2+} to Glu:OE1 distance was in all three simulations 0.2 Å shorter than the experimental values. Reorientation of the Asn side chain has been investigated in more detail by analyzing various other trajectories. Asn-210 is located at the surface of the lectin in the vicinity of the binding site for the carbohydrates. From our studies it follows (see Fig. 6) that conformational reorientation occurs due to formation of a new hydrogen bond between the OD1 oxygen of Asn and OH3 of fucose with simultaneous reorientation of the OE1 oxygen of Glu-198 coordinated with the calcium ion. Thus, the coordination number of the calcium remains 8. The same coupled reorientation was also found in the simulation with a point charge of +0.4. However, in contrast to the simulation with a point charge of +0.8 the hydrogen bonding pattern between Asn-210 and fucose also changed significantly.

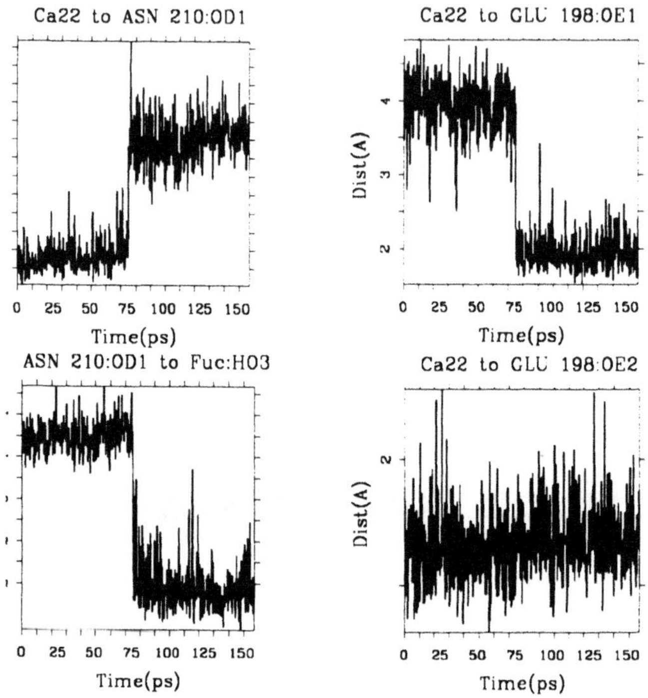

Fig. 6. Analysis of the reorientation of the ASN side chain. Four trajectories for distances within the ligand binding site are shown for the simulation with a point charge of +0.8 assigned to the calcium ion.

The new hydrogen bond between the OH3 of fucose and OD1 of Asn-210 causes a reorientation of fucose, leading to a structure so that only O3 of fucose had a short distance to Ca^{2+}. The dominant interaction for the O2 of fucose was then a hydrogen bond with Glu:OE2-190. This hydrogen bond was found for both orientations but it was stronger in the second one. The tendency of both oxygen atoms of the Glu-198 and Asp-211 carboxylate to coordinate the Ca^{2+} was very pronounced in the simulation with a point charge of +2.0 at the calcium ion and required a reorientation of the OD1 of Asn-210 and the carbonyl oxygen of the main chain of Asp-211. Thus, the coordination number of Ca^{2+} still remains 8. The reorientation of various side chains is due to the high charge of +2.0 which overestimates the electrostatic interactions between polar groups. Surprisingly, the orientation of the fucose and its binding mode to Ca^{2+} are not significantly effected by these reorientations.

In general it can be concluded that the chosen force field, simulation model and the parameterization are able to reproduce the structural details of the binding site of Ca^{2+}-dependent lectins within the experimentally reported variations. In particular, the mode of binding between the carbohydrate and the Ca^{2+} of the binding site are quite well described by the chosen setup.

TABLE 2: Mean values, standard deviations, minmium and maximum values for the distances between the calcium ion and the coordinated oxygen atoms from MD simulations with different point charges for the calcium ion.

		MFU	MFU	GLU	ASN	GLU	ASN	ASP	ASP	GLU	ASP
	residue			190	192	198	210	211	211	198	211
charge	atom	O2	O3	OE1	OD1	OE2	OD1	O	OD1	OE2	OD2
0.4	**mean**	**3.55**	**2.55**	**2.34**	**2.60**	**2.35**	**3.24**	**2.41**	**2.29**		
	std.dev	0.72	0.15	0.08	0.17	0.07	0.64	0.09	0.07		
	min	2.31	2.22	2.16	2.27	2.17	2.35	2.20	2.14		
	max	4.58	3.27	2.70	3.82	2.58	5.11	2.81	2.55		
0.8	**mean**	**2.67**	**2.56**	**2.31**	**2.51**	**2.33**	**3.44**	**2.40**	**2.29**		
	std.dev.	0.25	0.15	0.07	0.11	0.07	0.83	0.09	0.06		
	min	2.29	2.29	2.17	2.27	2.16	2.27	2.19	2.15		
	max	3.60	3.42	2.60	3.07	2.64	5.59	2.81	2.54		
2.0	**mean**	**2.65**	**2.63**	**2.33**	**2.44**	**2.34**	**4.68**	**2.59**	**3.36**	**2.43**	**2.44**
	std.dev.	0.35	0.32	0.07	0.10	0.08	0.66	0.20	0.97	0.18	0.15
	min	2.31	2.25	2.18	2.22	2.16	2.44	2.27	2.12	2.17	2.21
	max	3.67	4.28	2.59	2.91	2.64	6.23	3.87	4.62	3.89	4.03

3.2. INTERACTION ANALYSIS FOR VARIOUS COMPLEXES

The determination of more than 10 high-resolution 3D structures of Ca^{2+}-dependent lectins, complexed with various saccharides, has revealed the same basic binding characteristics for all complexes: two vicinal hydroxyl groups of the carbohydrate coordinate directly with the Ca^{2+} and form the nucleus of the sugar binding site. Although this fundamental interaction is very similar for all known structures, other orientations and considerable differences in the binding affinities have been reported. The reason for these findings is that other residues in the vicinity of the binding site contribute to the process of binding. The surprise was great when X-ray structure data revealed without any doubt that mannose binds in different orientations when complexed to MBP-A and MBP-C [28]. The mannose ligand is rotated by 180 degrees around the midpoint of the carbon_3/carbon_4 bond so that the oxygen_3 takes the position of oxygen_4 and vice versa (see Fig. 6).

Table 3 shows that the total interaction energy for mannose complexed to MBP-A is very similar to that for α-Me-mannoside complexd to MBP-C. The energetic contribution from Ca^{2+} is aslo nearly the same for the two complexes and on the order of one third of the total interaction energy. The individual contributions from the amnino acids side chain show a greater range of variation. The substitution of valine for histidine at position 189 in MBP-C

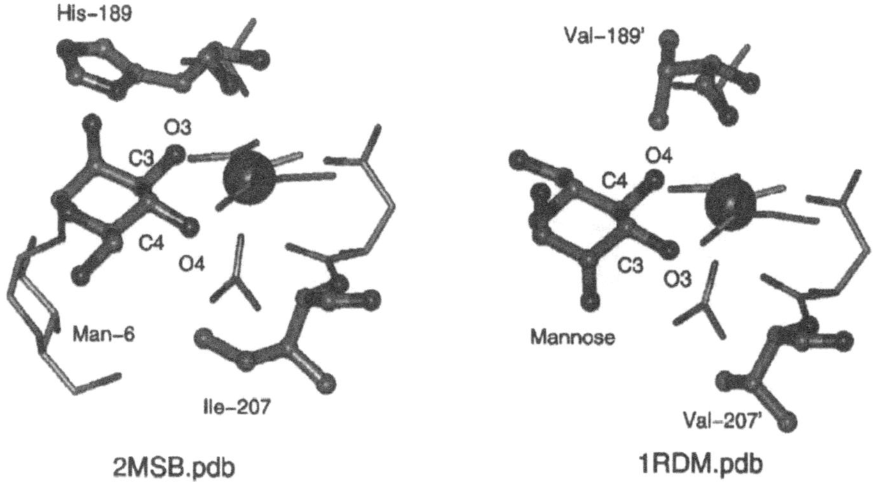

Fig. 7: Comparison of the two orientations of mannose in the crystal structure of MBP-A (normal orientation) and MBP-C (inverted orientation). MBP-A has been crystallized with a part of the asparaginyl oligosaccharide $Man_6GlcNAc_2$-Asn. Only the first two residues close to the binding site are shown. MBP-C has been crystallized with the monosaccharide α--mannose. Types of representation: calcium (sphere), mannose and the residues at postion 189 and 207 (ball and stick), side chains of the lectins involved in the complex (thin stick).

vs. MBP-A causes a loss about 3 kcal/mol in interaction energy. This loss is compensated by a stronger interactions wih two amino acid side chains which also coordinate to the calcium ion. The substitution of valine for isoleucin at position 207 does not influence the interaction energy significantly. The existence of the two complexes showing opposite orientations of mannose can be well understood by comparison of the interaction energies.

The question if two possible orientations of the carbohydrate part must be considered is of great interest. Therefore the interaction energies of MBP-A and MBP-C complexes with several ligands have been investigated.

Tabel 3: Interaction energies for α mannose complexed with MBP-A and MBP-C.

MBP-A : (2MSB.pdb)	[kcal/mol]	MBP-C : (1RDL.pdb)	[kcal/mol]
GLU-185	-1.16		-1.71
ASN-187	-1.60		-3.19
HIS-189	-4.89	Val-198'	-1.96
GLU 193	-1.75		-2.83
ASN 205	-1.90		-1.26
ASP 206	-0.54		-0.24
ILE 207	-0.61	Val-207'	-0.41
Ca	-7.06		-7.19
total	**-19.49**		**-18.97**

TABLE 4: Interactions energies for two possible orientations of the glycosides in complexes with MBP-A

Ligand	*coord. site (normal)*	interactions energies [kcal/mol] *normal*	*inverted*
α-Me-Man	O3,O4	-19.5	-16.5
α-Me-GlcNAc	O3,O4	-26.8	-23.8
α-Me-Fuc	O3,O2	-21.9	-20.0
β-Me-Fuc	O3,O2	-18.9	-21.9
α-Gal	O1,O2	-21.9	-18.6

Except for the complex of MBP-A with β-Me-Fuc (see Table 4) the so-called *normal* orientation of the sugar, as found in the crystal structure, exhibits the stronger interaction (ca. 2-3 kcal/mol) than the *inverted* orientation. The reasons why a fucose ligand may populate the *inverted* orientation will be discussed in more detail below. In general, it can be concluded that the *normal* orientation will probably dominate for most of the MBP-A carbohydrate complexes.

For MBP-C carbohydrate complexes (see Table 5) the situation is quite different. The interaction energies do not reveal any strong preference for the *normal* or the *inverted* orientation. Except for α-Me-fucoside when the inverted complex would be preferred (1.5 kcal/mol difference). In all MBP-C carbohydrate complexes for which experimental data are availabel, with the exception of the α-Me-mannoside complex 1RDM (Fig. 6), the ligand exhibits the *normal* orientation (2MSB in Fig. 6).

TABLE 5: Interaction energies for two possible orientations of glycosides in complexes with MBP-C

ligand	interactions energies [kcal/mol] *normal*	*inverted*
α-Me-Man	-19.0	-19.0
α-Me-GlcNAc	-21.1	-20.3
α–Me-Fuc	-20.1	-21.6
β-Me-Fuc	-18.3	-18.1
α-Gal	-16.9	-17.6

Deeper insight into the factors influencing the orientation and the binding affinities of various carbohydrates can be obtained by analysing in detail specific interactions of individual residues. The residues at positions 198 and 207 are different for MBP-A and MBP-C (see Fig. 7). The crystal structures of MBP-A complexes reveal that His-189can potentially form two types of interaction with mannose. The His-C_β is within the van der Waals distance of the C4 postion of the sugar and the imidazole ring can ineract with the axial 2-OH group of mannose. The importance of these contacts has been studied by site- directed mutagenesis [26] in which the imidazole ring was removed (His-189 --> Ala) and in which both the ring and the β-carbon are absent (His-189 --> Gly).

TABLE 6: Binding constants determined by NMR [26] and calculated interaction energies for the complexation of α–Me-mannoside with wild-type MBP-A and two mutants

Mutant	K_D [mM]	relative interaction energy [kcal/mol] *normal*	*inverted*
His (WT)	2.9 ±1.0	0.0	3.0
His-189--> Ala	4.8	2.3	4.2
His189-->Gly	15.0	2.4	4.7

Table 6 reports the measured dissociation constants for the wild type and two mutant proteins and the calculated interaction energy for both possible orientations of the carbohydrate. These data indicate that the interactions of His-189 with α–Me-mannoside has a significant influence on the binding affinity. The calculated interaction energies suggest that the loss of the contact between the imidiazole ring and the sugar is more important than the loss on C_β for the decrease of the binding affinity. On the other hand the dissociation constants suggest that a loss of C_β (Ala-->Gly) has larger effect on the energetics of binding. However, the K_D values also more accurately reflect the effects of mutation on binding site geometry and global conformational energy, while interaction energies were calculated for a limited volume around the active site with a fixed backbone geometry and the exclusion of solvent effects.

The crystal structure and the interaction analysis of mannose complexed with MPB-A and MBP-C (see Table 3) reveal that there is an additional weak interaction between the carbohydrate and Ile-207. The importance of this contact was tested by constructing a series of mutants at position 207 in which Ile was replaced by residues with shorter side chains (Leu, Val, Ala). Table 7 lists binding constants [26] and calculated interaction energies for both possible orientations of the carbohydrate. Both the experimental and the theoretical data show only a minor decrease in the binding affinity for the analyzed mutants, suggesting that the net contribution of the van der Waals contacts between the bound carbohydrate and residue 207 of the lectin to the free energy of ligand binding is of minor importance.

TABLE 7: Measued binding constants [26] and calculated interaction energies for the complexation of Man-BSA with wild-type MBP-A and three mutants at position 207.

Mutant	$K_{DManBSA}$ [ng/ml]	relative interaction energy [kcal/mol]	
		normal	*inverted*
Ile207(WT)	4.0 ± 0.7	0.0	3.3
Ile207-->Leu	5.2 ± 0.9	-0.1	3.2
Ile207-->Val	6.1 ± 1.3	0.2	3.5
ile207-->Ala	6.2 ± 0.4	0.4	3.8

3.3. INVERTED ORIENTATION OF FUCOSE

It has been suggested [33] that fucose binds to the wild type MBP-A with its OH2 and OH3 in the positions occupied by the OH4 and OH3 respectively of mannose in the crystal structure (2MSB.pdb). In this orientation (see Fig. 8 left) the methoxy group at position 1 of methyl fucoside, like the 6 hydroxymethyl-group of mannose, is located near Ile-207. Modeling indicates (see Table 4) that α- and β- methyl fucoside can be accomodated in either orientation. In the *inverted* orientation (see Fig. 8 right) the methoxy group at position 1 has van der Waals contacts with His-189. According to the interaction energies these contacts should be stronger for β- than for α glycosides. In both cases the 1-O-methyl group can establish interactions with the imidazole ring of His-189.

The possible orientation of the anomeric methyl group is not only an interesting academic question. The crystal structure of E-selectin has been solved without any ligand. There is strong experimental evidence [25,53-56] that the interaction between the Ca^{2+} of selectins and the hydroxyl groups at position 2 and 3 of fucose in Lewis^{x} form the nucleus of the sugar attachment process.

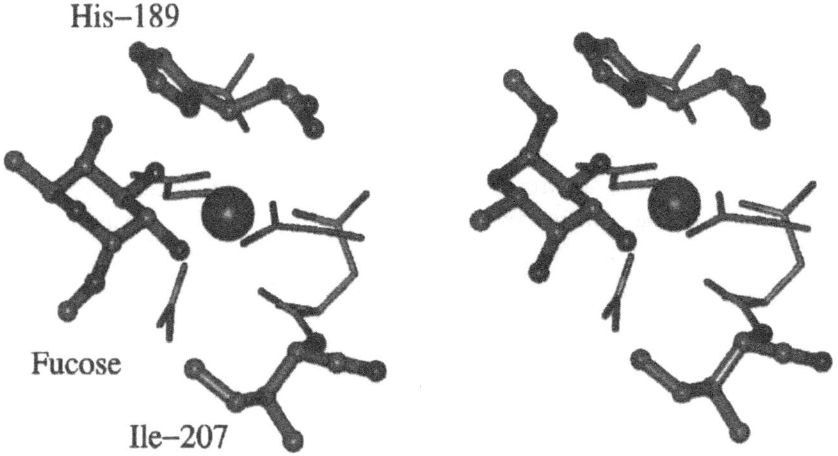

Fig. 8: Normal (left) *and* inverted (right) *orientation in MBP-A complexes. Types of representations: calcium (sphere), fucose and the residues 189 and 207 (ball and stick), other side chains of the lectins involved in the complex (thin sticks).*

TABLE 8: Relative interaction energies and measured inhibition binding constants [26] for methyl fucoside binding to MBP-A mutants. The energies given in parentheses are for the inverted orien-tation.

	relative interaction energy [kcal/mol]		K_I [ng/ml]	
Mutant	αMeFuc	βMeFuc	αMeFuc	βMeFuc
Wild Type	0.0 (1.9)	3.0 (0.0)	2.1 ± 0.2	9.5 ± 0.2
His-189--> Ala	3.1 (2.0)	4.3 (1.7)	5.5 ± 0.3	9.3 ± 1.4
Ile-207-->Ala	0.6 (-2.1)	2.6 (-2.1)	1.7 ± 0.4	6.3 ± 0.5

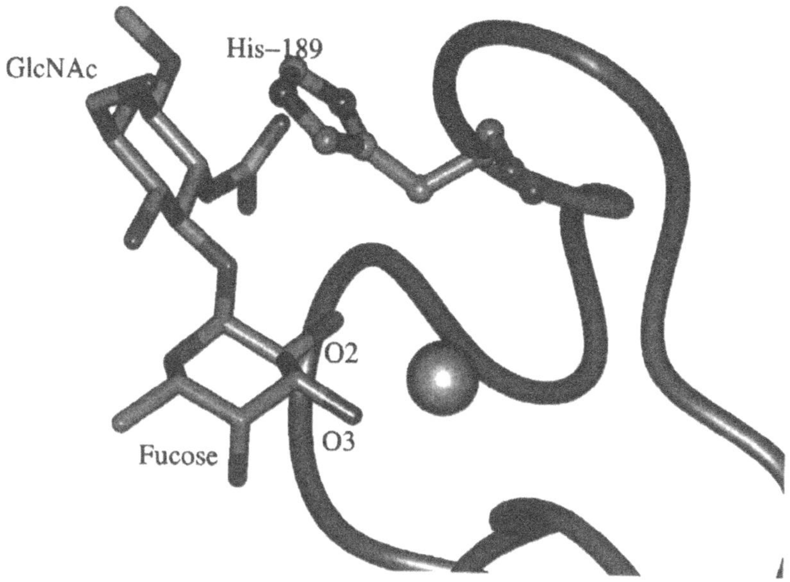

Fig. 9: Binding of Fuc β 1-3GlcNAc, a fragment of Lewis^X, with a selectin-like mutant of MBP-A. (PDB entry 2KBM.pdb) [34]. Types of representation: calcium (sphere), carbohydrates (stick) His-189 (ball and stick), backbone of the lectin (ribbon).

Since both orientations of the fucose seem to be possible it is of interest to evaluate the structural features which favor either orientation. Molecular modeling studies [57] have assumed that the O2 and O3 of fucose adopt the positions corresponding to O4 and O3 of mannose in the MBP-A complex. The calculated interaction energies as well as the reported inhibition binding constants (Table 8) indicate that the binding affinities are reduced by a mutation of His-189 and favored by the mutation at Ile-207. These results suggest that the O-methyl group of fucose forms an important interaction with His-189 and it may be concluded that the *inverted* orientation should also be populated at least some of the time. Recently 3D structures [34] of a selectin-like mutant of MBP-A complexed with Lewis^X oligosaccharides have been reported. The 2- and the 3-OH groups of fucose were shown to be coordinated to Ca^{2+} in the *inverted* orientation (compare Fig. 9 with Fig. 8).

The 2 and the 3-OH of fucose coordinate the calcium ion and form a network of cooperative hydrogen bonds with amino acid side chains that also coordinate the Ca^{2+}. However, the interaction of fucose with His-189 which was found to be significant for the binding in the case of monosaccharide ligands has nearly completely disappeared in the case of Lewis^X. It is the glucosamine which exhibits several strong interactions with His-189. Since fucose and glucosamine are 1-3 linked, the methoxy group at position 1 of methyl fucoside occupies the postion taken by C3 of GlcNAc in the binding site. Thus, it is not surprising that Lewis^X complexes adopt the *inverted* orientation, postulated (see Table 8) by site directed mutagenesis studies and interaction analysis.

4. Summary and Conclusions

Conventional molecular mechanics and molecular dynamics calculations are appropriate for closed shell ions such as Ca^{2+}, Zn^{2+}, Mg^{2+} and Cd^{2+} which do not have d-orbital based electronic effects. In this study these methods have been successfully used to study the binding affinities of various carbohydrates to Ca^{2+}-dependent lectins. A simple *nonbonded* model for the calcium ion was applied. The calcium ion is treated as a charged sphere with nondirectional Coulomb and van der Waals interactions with the rest of the protein and the ligand. Long-range electrostatic interactions up to 15 Å were taken into account. A reduced charge of +0.8 for the calcium ion produced the best agreement with experimental results. The distances between the calcium ion at the binding site and the coordinating oxygen atoms were used to judge how well different force field parameterization could reproduce the experimental results.

The same basic binding characteristics exist for all complexes where 3D coordinates are available. Two vicinal, equatorial hydroxyl groups of the carbohydrate act as direct coordination ligands for the Ca^{2+} and form the nucleus for the sugar binding site. Although this fundamental interaction is very similar for all complexes described the binding affinities vary considerably. The crystal structures do not directly provide information concerning for the specific interactions involved in ligand binding selectivity and affinity. A 3D model of the binding site has been derived which mimics the essential interactions between various lectin mutants and the carbohydrates. The calculated interaction energies for a variety of complexes provides a parameter which allows us to interpret the crystal structures and the results from site-directed mutagenesis studies on an energetic level.

However, the agreement of the calculated interaction energies and the measured K_D/K_I values was not always satisfying. The experimental values more accurately reflect the effects of mutation on binding site geometry and global conformational energy, while interaction energies were calculated for a limited volume around the active site, with a fixed backbone geometry and the exclusion of solvent effects. More comprehensive calculations including explicit water molecules are required to overcome this weakness of the chosen simplified model.

The relative binding affinities for both possible orientations of carbohydrates could be estimated. In MBP-A complexes (see Table 4) the *normal* orientation dominates. On the other hand it could be shown that fucose has a probability for complexing in the *inverted* orientation, as has previously been assumed. A recent determination of 3D structures of a selectin-like mutant of MBP-A complexed with Lewis^X oligosaccharides revealed that the 2- and the 3-OH of fucose do, in fact, adopt the *inverted* orientation.

From these findings it is obvious that the process of binding is very sensitive to slight structural changes. It has not been yet possible to predict precisely or to determine experimentally the structures of the ligands bound to selectin. A detailed evaluation of all possible energetic interactions including the effects of solvent molecules seems to be a promising way to gain deeper insights in the overall process of binding.

References

1. Lis, H. and Sharon, N. (1986) Lectins as molecules and as tools, *Annu.Rev.Biochem.* **55**, 35-67.
2. Drickamer, K. and E, T. M. (1993) Biology of Animal Lectins, *Ann. Rev. Cell. Biol.* **9**, 237-64.
3. Gabius, H. J. and Gabius, S. *Glycosciences - Status and Perspectives*; Chapman&Hall: D-69469 Weinheim, Germany, 1997.
4. Sharon, N. and Lis, H. (1990) Lectins as cell-recognition molecules, *Science* **246**, 227-234.
5. Feizi, T. (1993) Oligosaccharides that mediate mammalian cell-cell adhesion, *Curr.Opin. Struct.Biol.* **3**, 701-710.
6. Zanetta, J.-P. (197) Lectins and Carbohydrates in Animal Cell Adhesion and Control oProliferation, in ; Gabius, H.-J. and Gabius, S.(eds.), *Glycosciences: Status and Perspectives* , Chapman&Hall, Weinheim, pp 439-458.
7. Goldstein, I. J.,Hughes, R. C. and Monsigny, M. (1980) What should be called a lectin ?, *Nature* **285**, 66.
8. Kocourek, J. and Horejsi, V. (1981) Defining a lectin, *Nature* **290**, 188.
9. Drickamer, K. (1995) Increasing diversity of animal lectin structures, *Curr.Opin.Struct. Biol.* **5**, 612-616.
10. Drickamer, K. (1993) Ca^{2+}-dependant carbohydrate-recognition domains in animal proteins, *Curr. Opin. Struct. Biol.* **3**, 392-400.
11. Lasky, L. A. (1992) Selectins: interpreters of cell-specific carbohydrate information during inflammation, *Science* **258**, 964-69.
12. Lasky, L. A. (1995) Selectin-carbohydrate interactions and the initiation of the inflammatory response, *Annu. Rev. Biochem.* **64**, 113-139.
13. Lee, R. T.,Ichikawa, Y.,Fay, M.,Drickamer, K.,Shao, M. S. and Lee, Y. C. (1991) Ligand-binding Characteristics of Rat Serum-type Mannose-binding Protein (MBP-A), *J.Biol.Chem.* **226**, 4810-4815.
14. Lee, Y. C. and Lee, R. T. (1995) Carbohydrate-Protein Interactions: Basis of Glycobiology, *Acc. Chem. Res.* **28**, 321-327.
15. Toone, E. J. (1994) Structure and energetics of protein-carbohydrate complexes, *Curr. Opin. Struct. Biol.* **4**, 719-728.
16. Crocker, P. R. and Feizi, T. (1996) Carbohydrate recognition systems: functional triads in cell-cell interactions, *Curr. Opin. Struc. Biol.* **6**, 679-691.
17. Weis, W. I. and Drickamer, K. (1996) Structural Basis of Lectin-Carbohydrate Recognition, *Annu. Rev. Biochem.*, 441-473.
18. Rini, J. M. (1995) X-Ray crystal structures of animal lectins, *Curr. Opin. Struc. Biol.* **5**, 617-621.

19. Rini, J. M. (1995) Lectin Structure, *Ann. Rev. Biophys. Biomol. Struct.* **24**, 551-577.
20. Blanck, O.,Iobst, S. T.,Gabel, C. and Drickamer, K. (1996) Introduction of Selectin-like Binding Specifcity into a Homologous Mannose-binding Protein., *J.Biol.Chem.* **271**, 7289-7292.
21. Drickamer, K. (1992) Engineering galactose binding activity into a C-type mannose-binding protein, *Nature* **360**, 183-186.
22. Revelle, B. M.,Scott, D.,Kogan, T. P.,Zheng, J. and Beck, P. J. (1996) Structure-Function Analysis of P-selectin-Sialyl LewisX Binding Interaction: Mutagenic Alternation of Ligand Binding Specificity, *J.Biol.Chem.* **271**, 4289-4297.
23. Bajorath, J. and Arnuffo, A. (1994) Molecular Model of the Extracellular Lectin-like Domain in CD69, *J.Biol.Chem.* **269**, 32457-32463.
24. Bajorath, J. and Arnuffo, A. (1995) A Template for Generation and Comparison of Three-dimensional Selectin Models, *Biochem.Biophys. Res. Commun.* **216**.
25. Cooke, R. M.,Hale, R. S.,Lister, S. G.,Shah, G. and Weir, M. P. (1994) The Conformation of Sialyl LewisX Ligand Changes upon Binding to E-Selectin, *Biochemistry* **33**, 10591-10596.
26. Iobst, S. T.,Wormald, M. R.,Weis, W. I.,Dwek, R. A. and Drickamer, K. (1994) Binding of Sugar Ligands to Ca^{2+}-dependant Animal Lectins: I Analysis of Mannose Binding by Site-directed Mutagenesis and NMR, *J. Biol. Chem.* **269**, 15505-15511.
27. Iobst, S. T. and Drickamer, K. (1996) Selective Sugar Binding to the Carbohydrate Recognition Domains of the Rat Hepatic and Macrophage Asialoglycoprotein Receptors, *J. Biol. Chem.* **271**, 6686-6693.
28. Ng, K. K.-S.,Drickamer, K. and W, W. I. (1996) Structural Analysis of Monosaccharide Recognition of Rat Liver Mannose-binding Protein, *J.Biol. Chem.* **271**, 663-674.
29. Kolatkar, A. R. and Weis, W. I. (1996) Structural Basis of Galactose Recognition by C-type Animal Lectins, *J. Biol. Chem* **271**, 6679-6685.
30. Siebert, H.-C.,von der Lieth, C.-W.,Gilleron, M.,Reuter, G.,Wittmann, J.,Vliegenthart, J. F. G. and Gabius, H.-J. (1997) Carbohydrate-Protein Interaction, in H.-J Gabius and S. Gabius (eds.), *Glycosciences; Status and Perspectives*, Chapman&Hall, Weinheim, pp 291-310
31. Kollman, P. A. and Merz, K. M. (1990) Computer modeling of the interactions of complex molecules, *Acc. Chem. Res.* **23**, 246-252.
32. Weis, W. I.,Kahn, R.,Fourme, R.,Drickamer, K. and Hedrickson, W. A. (1991) Structure of the Calcium-Dependent Lectin Domain from a Rat Mannose-Binding Protein Determined by MAD Phasing, *Science* **254**, 1608-1615.
33. Weis, W. I.,Drickamer, K. and Hendrickson, W. A. (1992) Structure of a C-type mannose-binding protein complexed with an oligosaccharide, *Nature* **360**, 127-134.
34. Ng, K. K.-S. and Weis, W. I. (1997) Structure of a Selectin-like Mutant of Mannose-Binding Protein Complexed with Sialylated and Sulfated $Lewis^X$ Oligosaccharides, .
35. Graves, B. J.,Crowther, R. L.,Chandran, C.,Rumberger, J. M.,Li, S.,Huang, K.-S.,Presky, D. H.,Familetti, P. C.,Wolitzky, B. A. and Burns, D. K. (1994) Insight into E-selectin/ligand interactions from the crystal structure and mutagenesis of the lec/EGF domains, *Nature* **367**, 532-538.
36. Bajorath, J.,Hollenbaugh, D.,King, G.,Harte, W.,Eustice, D. C.,Darveau, R. P. and Aruffo, A. (1994) CD62/P-Selectin Binding Site for Myeloid Cells and Sulfatides Are Overlapping, *Biochemistry* **33**, 1332-1339.
37. Weis, W. I. and Drickamer, K. (1994) Trimeric structure of a C-type mannose-binding protein, *Structure* **2**, 1227-1240.
38. Bourne, Y.,van Tilbeurgh, H. and Cambillau, C. (1993) Protein-carbohydrate interactions, *Curr. Opinion Struct. Biol.* **3**, 681-686.
39. Weiner, S. R.,Kollman, P. A.,Nguyen, D. T. and Case, D. A. (1986) An all atom forcefield for simulations of proteins and nucleic acid, *J.Comput. Chem.* **7**, 230-252.
40. Weiner, S. R.,Kollman, P. A.,Case, D. Singh, U. C.,Ghio, C.,Alagona, G.,Profeta, S. and Weiner, P. (1984) A new force field for molecular mechanical simulation of nucleic acids and proteins, *J.Am.Chem.Soc.* **106**, 765-784.
41. Hagler, A. T.,Dauber, P. and Lifson, S. (1979) Consistent force field studies of intramolecular forces in hydrogen bonded crystals. II A benchmark for the objective comparison of alternative force fields, *J.Am.Chem.Soc*, 5122-5130.
42. Hagler, A. T.,Dauber, P. and Lifson, S. (1979) Consistent force field studies of intermolecular forces in hydrogen bonded crystals. III.The C=O...H-O hydrogen bond and the analysis of the energetics and packing of carboxylic acids, *J.Am.Chem.Soc.*, 5131-5141.
43. Hoops, S. C.,Anderson, K. W. and Merz, K. M. J. (1991) Force Field design for metalloproteins, *J.Am.Chem.Soc.* **113**, 4484-4490.

44. Banci, L.,Scgroder, S. and Kollman, P. A. (1992) Molecular dynamics characterization of the active cavity of carboxypeptidase A and some of its inhibitor adducts., *Proteins* **13**, 288-305.
45. Merz, K. M. J.,Murcko, M. A. and Kollman, P. A. (1991) Inhibition of carbonic anhydrase, *J.Am.Chem.Soc.* **113**, 4484-4490.
46. Vedani, A. and Huhta, D. W. (1990) A new force field for modeling metalloproteins, *J.Am.Chem.Soc.* **112**, 4759-4767.
47. Liang, J. H. and Lipscomb, W. N. (1990) Binding of CO_2 to the active site of human carbonic anhydrase II: A molec ular dynamics study, *Proc.Natl.Acad.Sci.U.S.A.* **87**, 3675 -3679.
48. Makinen, M. W.,Troyer, J. N.,van der Werff, H.,Berendsen, J. C. and van Gunsteren, W. F. (1989) Dynamical structure of carboxypeptidase A, *J.Mol.Biol.* **207**, 210-216.
49. Stote, R.H. and Karplus, M. (1995) Zinc Binding in Proteins and Solution: A Simple but Accurate Non-bonded Representation, *Proteins* **23**, 12-31.
50. Stote, R. H.,State, D. J. and Karplus, M (1991) On the treatment of electrostatic interactions in biomolecular simulation, *J.Chim.Phys.* **88**, 2419-2433.
51. Beseler, B. H.,Merz, K. M. and Kollmann, P. A. (1990) Atomic Charges Derived from Semiemperical Methods, *J.Comp.Chem.* **11**, 431-439.
52. Bohne, A.,Lang, E. and von der Lieth, C. W. (1996) SWEET - a quick way to generate reliable 3D-structures of Carbohydrates from sequence information alone., in ; Hofestädt, R.(eds.), *Computer Science and Biology* , Universität Leipzig, Leipzig, , pp 176-178.
53. Varki, A. (1994) Selectin ligands, *Proc. Natl..Acad. Sci. USA* **91**, 7390-7397.
54. Erbe, D. V.,Wolitzky, B. A.,Presta, L. G.,Norton, C. R.,Ramis, R. J.,Burns, D. K.,Rumberger, J. M.,Narasinga Rao, B. N.,Foxall, C. and Branddly, B. K. (1992) Indentification of an E-selectin Region Critical for Carbohydrate Recognition and Cell Adhesion, *J. Cell. Biol.* **119**, 215-227.
55. Erbe, D. V.,Watson, S. R.,Presta, L. G.,Wolitzky, B. A.,Foxall, C.,Brandly, B. K. and Lasky, L. A. (1993) P-and S-Selectin Use Common Sites for Carbohydrates Ligand Recognition and Cell Adhesion, *J: Cell. Biol.* **120**, 1227-1235.
56. Poppe, L.,Brown, G. S.,Philo, J. S.,Nikrad, P. V. and Shah, B. H. (1997) Conformation od sLe^X Tetrasaccharide, Free in Solution and Bound to E-, P-, and L-Selectin, *J.Am.Chem.Soc.* **119**, 1727-1736.
57. Kogan, T. P.,Revelle, B. M.,Tapp, S.,Scott, D. and Beck, P. J. (1995) A Single Amino Acid Residue Can Determine the Ligand Specificity of E-selectin, *J.Biol.Chem* **270**, 14047-14055.
58. Burling, F.T.,Weis, W.I.,Flaherty, K.M. and Brünger, A.T. (1996) Direct Observation of Protein Solvation and Discrete Disorder with Experimental Crystallographic Phases, *Science* **271**, 72-76

Appendix: Calcium oxygen distances [Å] for various experimentally determined structures of MBP-A and MBP-C. For each monomer the values for both calcium sites are listed. The nomenclature of the oxygen atoms is given for the MBP-C com plexes. In MBP-A complexes ASP-192 is replaced by ASN. The numbering of the residues is given MBP-C like.

resid.	SUG	SUG	GLN 190	ASP 192	GLU 198	ASN 210	ASP 211	ASP 211	ASP 166	ASP 166P	GLU 170	GLU 170	ASN 193	ASN 199	GLU 198
atom	O2	O3	OE1	OD1	OE1	OD1	O	OD1	OD1	OD2	OE1	OE2	OD1	OD1	O
PDB entry															
rdi_1	2.47	2.59	2.53	2.34	2.49	2.45	2.61	2.33	2.68	2.50	2.53	2.70	2.52	2.47	2.45
rdi_2	2.47	2.57	2.51	2.32	2.45	2.47	2.68	2.37	2.66	2.48	2.46	2.61	2.52	2.40	2.49
rdj_1	2.55	2.52	2.46	2.36	2.44	2.41	2.62	2.38	2.72	2.44	2.50	2.68	2.49	2.46	2.47
rdj_2	2.47	2.56	2.56	2.35	2.45	2.44	2.57	2.37	2.70	2.44	2.45	2.53	2.51	2.40	2.49
rdk_1	2.58	2.52	2.49	2.48	2.42	2.53	2.59	2.40	2.72	2.58	2.57	2.75	2.53	2.39	2.41
rdk_2	2.63	2.50	2.52	2.32	2.42	2.56	2.64	2.43	2.67	2.52	2.51	2.72	2.54	2.28	2.42
rdl_1	2.55	2.57	2.56	2.44	2.39	2.45	2.58	2.36	2.70	2.46	2.50	2.69	2.51	2.41	2.43
rdl_2	2.60	2.59	2.62	2.40	2.43	2.43	2.55	2.33	2.68	2.44	2.48	2.54	2.49	2.36	2.45
rdm_1	2.58	2.48	2.55	2.36	2.44	2.49	2.55	2.38	2.67	2.48	2.58	2.66	2.53	2.50	2.51
rdm_2	2.58	2.49	2.52	2.29	2.44	2.48	2.65	2.43	2.64	2.43	2.51	2.59	2.61	2.50	2.40
rdn_1	2.53	2.55	2.50	2.40	2.41	2.42	2.62	2.38	2.64	2.45	2.54	2.71	2.54	2.42	2.53
rdn_2	2.56	2.52	2.56	2.34	2.42	2.45	2.66	2.35	2.60	2.46	2.46	2.63	2.53	2.32	2.52
afa_1	2.59	2.76	2.51	2.43	2.43	2.44	2.55	2.23	2.69	2.30	2.54	2.56	2.51	2.47	2.49
afat_2	2.93	2.55	2.10	2.66	2.54	2.73	3.02	2.19	2.74	2.28	2.54	2.77	2.79	2.56	2.40
afa_3	2.42	2.65	2.50	2.45	2.33	2.48	2.51	2.31	2.62	2.31	2.61	2.67	2.51	2.53	2.42
afa_1	2.59	2.76	2.51	2.43	2.43	2.44	2.55	2.23	2.69	2.30	2.54	2.56	2.51	2.47	2.49
afa_2	2.93	2.55	2.10	2.66	2.54	2.73	3.02	2.19	2.74	2.28	2.54	2.77	2.79	2.56	2.40
afat_3	2.42	2.65	2.50	2.45	2.33	2.48	2.51	2.31	2.62	2.31	2.61	2.67	2.51	2.44	2.42
afb_1	2.60	2.66	2.57	2.45	2.32	2.43	2.55	2.20	2.61	2.34	2.57	2.65	2.67	2.42	2.43
afb_2	2.79	2.75	2.63	2.50	2.30	2.73	2.62	2.22	2.73	2.29	2.73	2.58	2.75	2.53	2.48
afb_3	2.51	2.56	2.52	2.44	2.41	2.46	2.50	2.27	2.64	2.27	2.55	2.63	2.60	2.38	2.47
kmb21	2.60	2.78	2.51	2.50	2.48	2.39	2.56	2.22	2.55	2.36	2.68	2.48	2.48	2.42	2.50
kmb22	2.69	2.64	2.52	2.51	2.33	2.58	2.56	2.50	2.80	2.42	2.61	2.57	2.51	2.39	2.50
kmb23	2.87	2.48	2.62	2.48	2.29	2.50	2.59	2.37	2.76	2.51	2.63	2.53	2.43	2.41	2.42
kmb31	2.52	2.68	2.52	2.35	2.35	2.55	2.57	2.30	2.57	2.33	2.62	2.60	2.64	2.39	2.44
kmb32	2.88	2.88	2.61	2.46	2.22	2.55	2.56	2.55	2.60	2.28	2.35	2.60	2.70	2.55	2.36
kmb33	2.73	2.42	2.65	2.47	2.50	2.61	2.62	2.27	2.74	2.47	2.58	2.48	2.47	2.32	2.44
kmb41	2.62	2.78	2.49	2.36	2.57	2.46	2.57	2.34	2.59	2.30	2.60	2.53	2.54	2.34	2.52
kmb42	2.78	2.84	2.58	2.51	2.18	2.57	2.66	2.56	2.67	2.42	2.42	2.65	2.52	2.58	2.37
kmb43	2.55	2.48	2.51	2.48	2.44	2.54	2.60	2.29	2.71	2.40	2.56	2.54	2.38	2.31	2.38
msb_1	2.45	2.54	2.67	2.47	2.31	2.41	2.45	2.38	2.69	2.47	2.54	2.55	2.54	2.35	2.38
mean	**2.66**	**2.65**	**2.51**	**2.48**	**2.38**	**2.53**	**2.61**	**2.31**	**2.67**	**2.35**	**2.57**	**2.60**	**2.57**	**2.44**	**2.44**
min	2.42	2.42	2.10	2.35	2.18	2.39	2.45	2.19	2.55	2.27	2.35	2.48	2.38	2.31	2.36
max	2.93	2.88	2.67	2.66	2.57	2.73	3.02	2.56	2.80	2.51	2.73	2.77	2.79	2.58	2.52
st.dev.	**0.17**	**0.13**	**0.15**	**0.08**	**0.11**	**0.11**	**0.15**	**0.11**	**0.07**	**0.07**	**0.08**	**0.08**	**0.12**	**0.09**	**0.05**

MOLECULAR DYNAMICS CALCULATIONS ON METALLOPROTEINS

LUCIA BANCI and GIOVANNI GORI SAVELLINI
Department of Chemistry
University of Florence
Via Gino Capponi, 7
50121 Florence, Italy

1. Introduction

Molecular Dynamics (MD) simulations have been and are extensively applied to biological molecules, and they are very useful and precious for the determination of structural properties when experimental data are not available. They can also, in a fast and efficient way, predict structural changes induced by mutations, inhibitor binding or interaction with a substrate. Furthermore, they can provide information on the internal motions of macromolecules. With the algorithms presently available for MD calculations, only motions on fast time scales (i.e. up to nanoseconds) can be calculated. Still, these motions can be relevant for the comprehension of the biological function. They can be important for the determination of the molecular recognition for the substrate or the partner in biological reactions. Also the dynamic behavior of residue-residue interactions, which can affect the biological process, can be successfully addressed by MD

L. Banci and P. Comba (eds.), Molecular Modeling and Dynamics of Bioinorganic Systems, 191–216.

calculations. Finally, MD can be useful for the characterization of solvation properties, specifically the structure of internal water molecules which cannot easily be addressed experimentally.

When MD calculations are applied to molecules containing metal ions, the parameterization of the force field for the metal ion and its ligands is needed. This is a central problem and a key on the way to meaningful results. We report here some examples of MD calculations on metalloproteins, which demonstrate how successful they can be in reproducing and predicting structural and dynamic properties of the investigated systems.

2. Superoxide dismutase

The Cu/Zn superoxide dismutase (SOD) is a dimeric enzyme which catalyzes the dismutation of superoxide radicals which are produced during the oxygen metabolic cycle and which are extremely reactive towards the cells [1,2]. Superoxide radicals are dismutated to molecular oxygen and hydrogen peroxide. Each of the two identical subunits contains a copper and a zinc ion [3,4]. The copper ion is coordinated to four histidines, one of which is also coordinated to the zinc ion, forming a bridge between the two metal centers. The coordination of the zinc ion is completed by two other histidine ligands and a monodentate aspartate. A water molecule is found close to the copper ion with a Cu-O distance of 2.8 Å in the X-ray structure [4]. This water molecule is conserved in all the SOD isoenzymes studied so far [3,5,6]. Furthermore, its presence close to the copper ion has been confirmed by relaxation measurements of the water protons as a function of the magnetic field, and the corresponding Cu-O distance in aqueous solution is 2.6 Å [7,8]. It is of importance to note here, that based on NMR studies of Cu2 Co2 SOD derivative [9] it was found that in solution the structure of the active site of this enzyme is essentially the same as in the crystal. In particular, the coordination of the catalytically essential copper ion is maintained.

Copper is located at the end of a wide channel with several charged residues which are relevant for the enzymatic reaction [3,9-11]. The catalytic reaction is a two-step process in which the copper ion is alternately reduced, with the formation of molecular dioxygen, and then reoxidized with the production of hydrogen peroxide. The enzyme is very efficient and the rates approach the diffusion limit [12,13]. This efficiency has been proposed to be the result of suitable electrostatic fields produced by the residues present in the active channel, which steer the substrate to the reaction site, i.e. the copper ion [9,14-17]. On one side of the channel there is Arg 143, which is completely invariant in all structures [18]. Its guanidinium group is close to the copper ion and at a distance that may enforce the orientation of the superoxide anion through an H-bond towards the copper ion. The other side of the channel is formed by loop VII, the electrostatic loop, which extends over residues 121-144. The most relevant among these residues are Glu 133, Glu 132, Lys 136, and Thr 137. Their mutation produces changes in the catalytic rates and in the coordination geometry of the copper site [11,16,17,19].

The fundamental contribution by the electrostatic interactions to the catalytic efficiency of SOD has been the subject of many experimental and theoretical studies [20-26]. The latter have predicted and rationalized the role of several residues in the active channel, and this has been supported by experimental data. MD calculations can provide significant information on the structural and dynamic properties of a protein and their mutants, which can then be related to the steps of the physiological process. For this reason we have undertaken MD studies on the WT and some mutants of SOD.

For modeling the coordination site of this protein, we used the charges obtained by ab-initio calculations by Mc Cammon [27]. The equilibrium distances for the metal-donor bonds were set to values obtained from X-ray structures, the force constants to those developed by Merz for zinc in zinc-enzymes [28,29]. These parameters were derived with a model that includes electrostatics and interactions with water molecules. No bond or distance constraint between the copper ion and the closest water molecule was placed in our calculations. Angular overlaps model calculations indicated, that, in agreement with the larger Cu-O distance (2.6-2.8 Å, see above), the electronic properties of the copper center are not effected by this ordered water molecule[30].

The MD calculations were performed in the presence of explicit solvent molecules, with a shell (10 Å), of TIP3P water molecules [31]. This is of importance for these simulations as solvation of the enzyme may affect the structure and dynamics of the hydrophilic parts of the molecule, as well as H-bonding interactions in these protein regions. This is particularly true for SOD, where the active channel contains many charged, hydrophilic residues whose conformation determines the actual electrostatic field and consequently the catalytic rates, as concluded from SDM studies (see above).

With this setup we first run MD simulations on the WT protein in order to check the validity of the model [32]. The calculations have shown that the structure is quite stable and that the entire structure and specifically that of the active site is identical within the error to that in the crystal [32].

Consistent with the angular overlap model analysis (see above) the water molecule, located close to copper center was not constrained. Significantly, this water molecule is completely stable in the entire trajectory, with the oxygen atom located at the same distance from the copper center as observed in solution [7,8]. This behavior was reproduced for a series of mutants on residues present in the

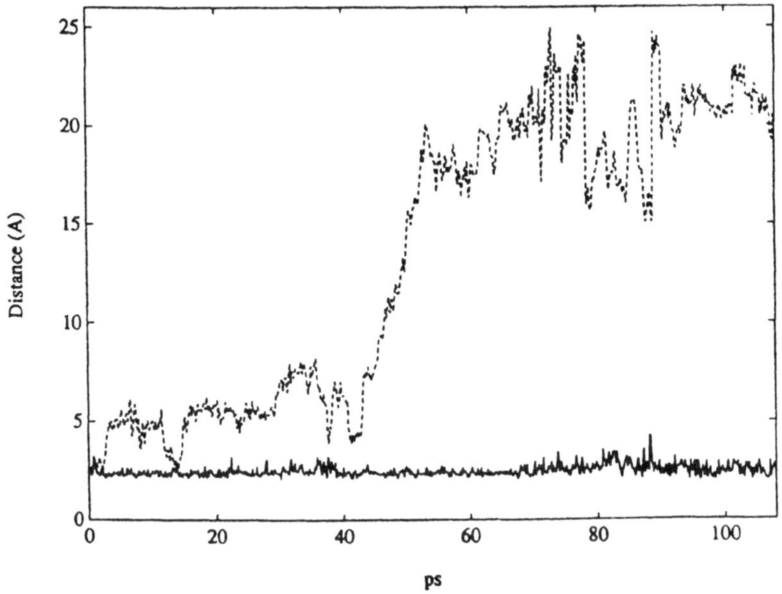

Fig. 1. Distance between Cu and water oxygen in WT SOD (solid line) and in Thr 137-Ile SOD (dashed line) as a function of the simulation time (taken with permission from [33]).

active channel, which have also been characterized experimentally [8]. In all but one of these mutants, a water molecule close to the copper center was found by water proton relaxation measurements. When Thr 137 is replaced by Ile, this water molecule is absent, and a small rearrangement of the copper coordination geometry is observed[33]. Our MD calculations have been able to reproduce even this subtle difference. Fig. 1 shows the copper-water oxygen distance during the simulation: the water molecule close to copper disappears in Thr 137-Ile mutant, and it is not replaced by any other water molecule. This is a remarkable

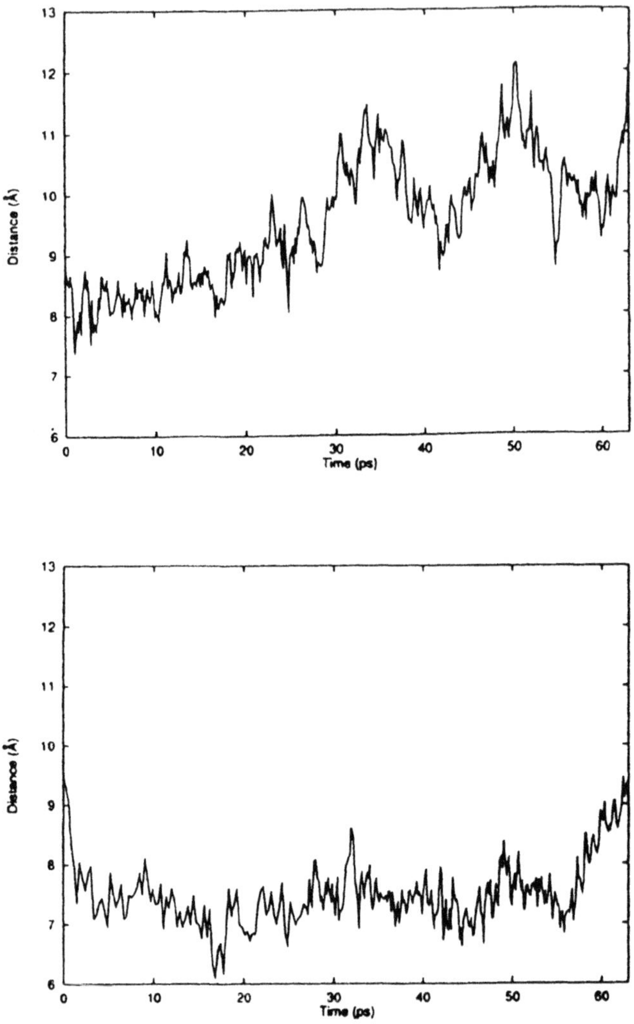

Fig. 2. Fluctuations of the distance between Cζ Arg 143 and Cβ Thr 137 in WT SOD (a) and Cδ Glu 143 and Cβ Thr 137 in Arg 143-Glu SOD (b) (taken with permission from [32]).

result of these calculations and a strong indication for the validity of our force field parameters. The activity of the Thr 137-Ile mutant is essentially maintained, indicating that this water molecule has no specific role in superoxide dismutation process. Indeed, in order to provide efficient electron transfer, superoxide should coordinate directly to the copper center. It is likely that the hydrophilic Thr residue stabilizes this water molecule, and it is displaced when a bulky hydrophobic residue is inserted.

The reaction rates are strongly determined by the electrostatic field produced by the charged residues of the channel, in particular by Arg 143. When this residue is substituted by a neutral one (Ile) the activity drops to 10% of that of WT, when it is replaced by a negatively charged residue (Glu) the activity is only 4 % [9,14,34]. The main effect of replacing residue 143 is a modification of the local electrostatic field, and this was determined experimentally by affinity studies for anions [9]. However, these changes have also structural and dynamic consequences [32]. The distance between Arg 143 and Thr 137 determines the narrowest section of the active site channel. In WT this distance increases and fluctuates during the simulation (Fig. 2a). In the Thr 137-Glu mutant it becomes much smaller and fluctuates less (Fig. 2b), despite the smaller size of the side chain of Glu with respect to Thr. This width and its fluctuations correlate strongly with the enzymatic efficiency. Since the rates are determined by the diffusion of the substrate through the channel, this indicates that its width and its mobility are important factors in the determination of the catalytic rates.

3. Heme proteins

3.1. DESCRIPTION OF THE INVESTIGATED SYSTEMS

One class of heme proteins which we have studied by MD calculations is that of peroxidases. These are oxidative heme enzymes with a five-coordinate high spin iron(III) which is bound to the four pyrrole nitrogens of the heme and to the Nε2 nitrogen of an axial histidine (proximal histidine). Hydrogen peroxide reacts with peroxidases as a two electron oxidant [35,36]. The resulting protein intermediate (Compound I) is able to oxidize a large variety of substrates. Peroxidases are highly selective with respect to the nature of these substrates and different reaction mechanisms may be involved. For example, both lignin peroxidase (LiP) and cytochrome *c* peroxidase (CcP) oxidize large substrate molecules, the aromatic polymer lignin and the protein cytochrome *c* , respectively. Due to their size, these substrates cannot approach the active site of the enzyme. Thus, a long-range electron transfer process is probably responsible for their oxidation. Redox mediators have been claimed to be of importance for this catalyzed reaction [37,38]. In contrast, horseradish peroxidase (HRP) oxidizes small aromatic molecules which may approach the heme moiety. However, for peroxidases the interaction with substrates is still an open question.

X-ray structures of several peroxidases are available (CcP [39,40], LiP [41-43], Manganese peroxidase (MnP) [44], *Coprinus cinereus* peroxidase (CiP) [45], *Arthromices ramosus* peroxidase [46], pea ascorbate peroxidase [47]) but the crystallographic structure of HRP is still lacking. From the analysis of the available X-ray structures and sequence alignment [48], it emerges that the iron chromophore and most of the residues in the active site are conserved in all known peroxidases [48]. The fifth ligand at the iron center is an axial histidine. This proximal histidine is H-bonded to an invariant Asp residue. The strong H-bond imparts anion character to the proximal histidine, which therefore stabilizes the high oxidation state. This facilitates the formation of Compound I [40,43,49-51]. Two residues (a histidine and an arginine) are invariant at the distal site of the peroxidases sequenced up to now [48]. These residues have also an important role in the formation and stabilization of Compound I [52-55].

In all the available crystallographic structures, a well ordered network with at least four water molecules was found in the heme surroundings. Three water molecules link through H-bonds the distal histidine with the distal arginine and a propionate. The fourth water molecule is located in the proximal site and is invariantly H-bonded to the conserved proximal aspartate residue. The solvation properties are relevant for the biological function, as H-bonds in the distal side stabilize the correct conformation of residues involved in the formation of Compound I [40].

All the fungal and plant peroxidases have two structural Ca^{2+} ions [56,57], and their binding sites are conserved in all the sequenced proteins [58]. The removal of the calcium ions affects the enzymatic activity; in the case of the HRP the decrease is 20-40% [59]. It was also shown that the removal of the calcium ions reduces the rate limiting step of the overall reaction, i.e. the rate of reduction of Compound II, an intermediate in the catalytic reaction [60]. Based on NMR studies it has been proposed that the calcium ions affect the structure of the active site [61,62].

Cytochromes *c* are heme electron transfer proteins and contain a *c*-type heme group, covalently bound to the protein frame [63,64]. The iron ion is coordinated by the four nitrogens of the heme pyrroles and two axial ligands. In eukaryotic cytochromes *c* the two axial ligands are a methionine and histidine. The iron ion switches the oxidation state between iron (III) and iron (II). Several X-ray and solution structures are available for various proteins at different oxidation states [65-72]. In particular, yeast iso-I cytochrome *c* has been well characterized, both in the solid state [68,69] and in solution [73,74]. Oxidation state-dependent properties of cytochrome *c* have not been completely clarified and they are the subject of extensive scientific debate [75]. Some structural changes have been observed which involved, among the others, groups around the heme as well as the H-bond network at the distal site. Some differences have also been observed between the crystal and the solution structure, particularly on some groups involved in the interaction with the substrate.

Water molecules in cytochromes *c* have been proposed to experience structural changes in dependence of the oxidation state [76]. In particular, the position of the catalitically relevant water molecule (WAT 166), observed by X-ray and NMR studies, is matter of debate in the literature [76,77]. The analysis of the X-ray data of oxidized and reduced yeast iso-I cytochrome *c* suggest that WAT 166 moves from 6.6 Å to 5.1 Å from the iron upon oxidation. This has been proposed to be a characteristic feature for all eukaryotic cytochromes *c*, and to provide a mechanism for modulating the electron-withdrawing power of the axial methionine ligand in dependence of the oxidation state of the heme group [76]. However, from NMR studies on horse heart cytochrome *c* it emerges that this water molecule essentially maintains its position, independent of the oxidation state [77].

3.2. MD SIMULATIONS AND SOLVATION PROPERTIES

We have carried out several MD calculations [78,79], starting with coordinates from the X-ray structures of CcP and LiP, with the aim of testing the force field parameters of the heme environment. Long MD trajectories (up to 160 picoseconds) were simulated both for CcP and LiP including a 10 Å shell of explicit solvent (TIP3P water molecules) around the protein structure. In addition, all the crystallographic water molecules were included in the calculations. The overall protein charge was neutralized by the addition of counter ions (Na^+ or Cl^-). The parameters for bonding interactions were taken from AMBER [80] and from data reported in the literature, obtained from some porphyrin model complexes [81,82]. The parameters for nonbonded interactions were taken from AMBER, with the exception of the charge on iron(III) which was set to 1.25, according to recent data obtained from INDO calculations [83]. No distance constraint was applied to the active site. The parameters used for these preliminary calculations were able to maintain the structure of the heme active site very close to that observed in the crystallographic structure. This is consistent with NMR solution studies which have shown that the active site has essentially the same conformation in the crystal and in solution [84-87].

The MD calculations have also been able to reproduce for both CcP and LiP, the structure of the conserved water molecules and the network of H-bonds. In CcP (Fig. 3) five crystallographic water molecules, located close to the heme

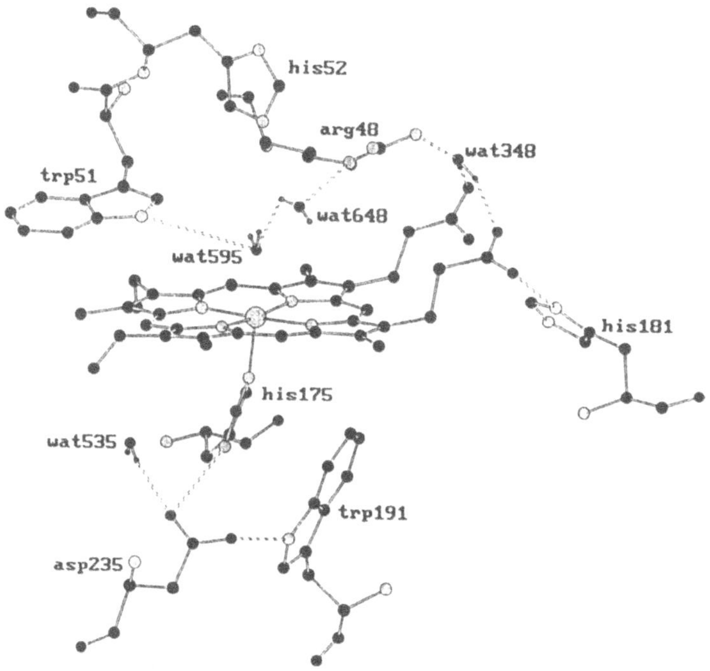

Fig. 3. Network of the ordered water molecules in the CcP active site. H-bonds are indicated by dashed lines (taken with permission from [78]).

moiety, maintain their original position during the entire simulation. The presence of these five water molecules leads to an extensive H-bond network which stabilizes Trp 51 and Arg 48 at the distal site, and link them to the propionate groups of the heme. One of the conserved water molecules is simultaneously H-bonded to the Arg 48 and the two propionate groups. The fifth conserved water molecule is at the proximal site and is H-bonded to Asp 235. For LiP (Fig. 4) a similar network of well ordered water molecules was observed, in the entire MD trajectory. As in CcP there are several well ordered conserved water molecules in

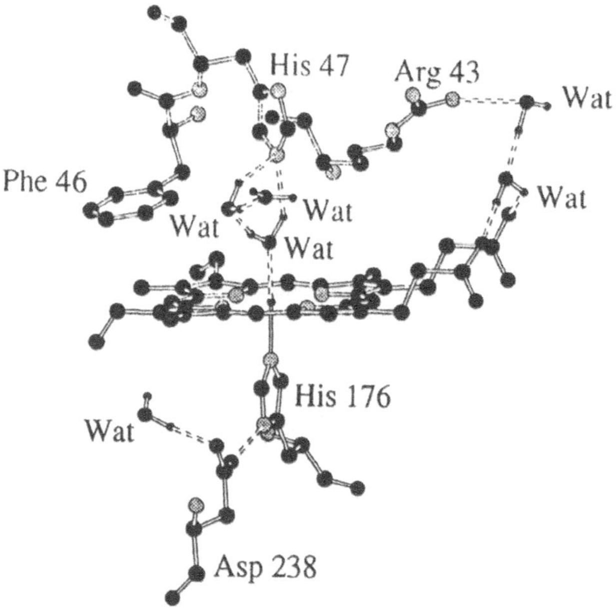

Fig. 4. Network of the ordered water molecules in LiP active site. H-bonds are indicated by dashed lines (taken with permission from [79]).

the distal cavity [79], and these are linked through H-bonds to the distal hydrophilic residues (His 47, Arg 43) and the heme propionate chains. Another conserved water molecule is H-bonded to the proximal Asp 238, in a position very similar to that in CcP. The hydrophilic character of the distal site is believed to be important for the catalytic activity of peroxidases. The network of H-bonds, which stabilizes the conformation of the heme environment, is probably also responsible for the five-coordination of the iron center [88], which leads to a free coordination site for the interaction with hydrogen peroxide. Our MD calculations indicate that, the water molecule close to the metal ion is not directly coordinated. It is fixed by the surrounding atoms and no bond constraint is needed to maintain the position.

We have extended the characterization of the solvation properties to the electron transfer heme protein cytochrome *c*. Long MD trajectories (600 ps) have been calculated on the reduced and oxidized forms of yeast iso-I cytochrome *c* [89] in order to characterize the mobility and the structural changes induced by the change of the oxidation state and to analyze the differences between the crystal and the solution structure. These calculations included both crystallographic and solvation water molecules. In both systems counter ions were added to the starting crystal structure to neutralize the overall protein charge (7 and 6 Cl^- ions to the oxidized and reduced forms, respectively). All force field parameters, except those of the heme and the axial ligands, were taken from the new AMBER force field database [90] except for the heme and its axial ligands. For the iron donors the parameters for the bonding interactions were derived from relevant X-ray structures or from data reported in the literature [78]. The nonbonded interaction parameters for iron and its ligands were taken from AMBER. For the oxidized form of iso-I cytochrome *c* the atomic point charges for the metal and its donors were obtained from *ab initio* calculations using GAUSSIAN 92 [91]. The model system considered was an Fe(III) porphyrin with axial methylimidazole and dimethylthioether donors. A series of pseudo-potentials was used for the iron center and the 6-31G* basis set for the rest of the system. This is consistent with the atomic point charges in AMBER which are also calculated with the 6-31G* basis set [90]. For the reduced yeast iso-I cytochrome *c*, the extra negative charge was distributed empirically on the iron center and its donor atoms. A 600 ps MD trajectory was calculated for both the oxidized and reduced forms of the protein, and the protein coordinates were saved every 3 ps.

After equilibration of the protein, the WAT 166 - iron distance remains stable around 7.8 Å in the oxidized protein. In the reduced form this water-iron distance reaches a mean value of 5.5 Å for about 130 ps, and then the water molecule diffuses to a new, approximately identical position to WAT 166 in the oxidized protein, where it remains for the rest of the trajectory. The displacement of WAT 166 in the reduced form is simultaneous to an observed shift in position of the Tyr 67 ring, which creates a solvent accessible channel in the distal site of the heme and accounts for an increased solvent accessibility. Several water molecules approach the distal site during the rest of the trajectory. One of these is always present at 5.5 Å from the iron center. In the case of the oxidized protein no water molecule moves closer than 7.8 Å to the iron center. These results are in agreement with solution NMR data on horse heart cytochrome *c* and point to a different behavior in solution and the solid state.

3.3. INTERACTIONS WITH METAL IONS

MD calculation can also be useful for the prediction of the structural role of ions present in a biological molecule. In the case of peroxidases, MD calculations have been important to understand the structural role of calcium ions [79]. Both Ca^{2+} ions (see above) could be relevant for maintaining the conformation of the active site. Indeed, one ligand of each Ca^{2+} is a residue which immediately follows either the distal or the proximal histidine. We have performed an MD simulation on LiP, starting from the crystallographic structure but not including the two calcium ions, to evaluate the structural effects of the two calcium ions.

In Fig. 5 The active site of LiP without the two Ca^{2+} ions in the X-ray and the final structure of our MD simulation are compared in Fig. 5. This clearly shows a drastic perturbation of the distal histidine (His 47) when the Ca^{2+} ions are removed. The rearrangement of this histidine already starts after a few picoseconds of MD and is maintained after 100 ps. At the end of the MD trajectory, His 47 has moved far away from the iron center, with the distance between the Nε2 nitrogen of His 47 and the iron center changing from 5.0 Å to a

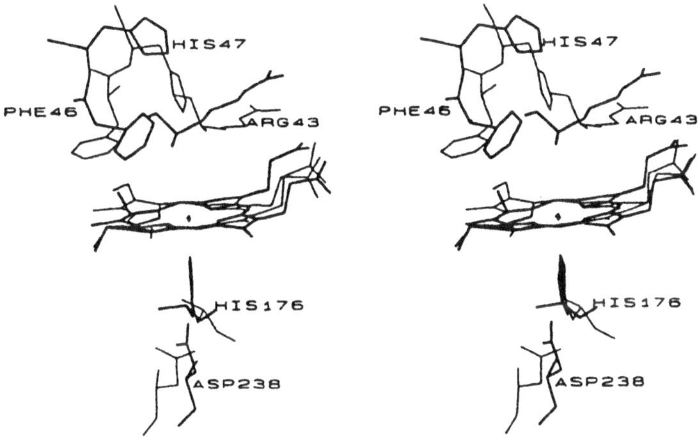

Fig. 5. Comparison of the active site of LiP in the X-ray (thin) and MD (bold) structure obtained without the Ca^{2+} ions (taken with permission from [79])

mean value of 8.4 Å. This rearrangement of the distal site was not observed during the entire MD trajectory of LiP in the presence of the two Ca^{2+} ions. Significantly, in CcP, where no calcium ion is present, the structure of the active and of the distal site in particular, are completely stable, even after long simulations performed without any constraint. These observations support the hypothesis that in fungal and plant peroxidases the conformation of the protein region around the heme is strongly determined by the calcium ions. In their absence the capability of His 47 to interact with hydrogen peroxide is drastically reduced, and this affects the catalytic efficiency of the protein.

4. Concluding remarks

The reported studies show how useful MD calculations are for the characterization of the mobility and the solvation properties of metalloproteins. In the case of metalloenzymes we have shown that the presence of one ordered water molecule close to the metal ion is can be analyzed by the potential generated by the surrounding atoms and that no bond is necessary to keep it in the position close to the metal ion. This behavior makes the water molecule easily displaceable upon interaction with substrates or small structural rearrangements. These are requirements for high efficiency in catalyzed reactions when they involve electron transfer processes. According to the Marcus theory [92], the rate of electron transfer through a coordinated water molecule is reduced by two or three orders of magnitudes [93,94].

In the case of cytochromes, the solvation properties of the heme moiety are different in the two oxidation states with the reduced form being more solvated. Due to the large dielectric constant of water, this stabilizes the reduced form where the iron-heme moiety has an higher charge.

In conclusion, these studies have shown how useful the application of MD simulations of metalloproteins is for the investigation of their structural and dynamic properties, and thus for their biological function.

Acknowledgments

We are grateful to Professor Ivano Bertini for helpful discussion.

References

1. Fridovich, I. (1974) Superoxide dismutase, *Adv. Enzymol.* **41**, 35-97.
2. Fridovich, I. (1986) Superoxide dismutase, *Adv. Enzymol. Relat. Areas Mol. Biol.* **58**, 61-97.
3. Tainer, J.A., Getzoff, E.D., Beem, K.M., Richardson, J.S. and Richardson, D.C. (1982) Determination and Analysis of 2 A Structure of Copper Zinc Superoxide Dismutase, *J. Mol. Biol.* **160**, 181-217.
4. Tainer, J.A., Getzoff, E.D., Richardson, J.S. and Richardson, D.C. (1983) Structure and Mechanism of Copper, Zinc Superoxide Dismutase, *Nature* **306**, 284-287.
5. Parge, H.E., Hallewell, R.A. and Tainer, J. (1992) Atomic structures of wild-type and thermostable mutant recombinant human Cu,Zn superoxide dismutase, *Proc. Natl. Acad. Sci. USA* **89**, 6109-6114.
6. Djinovic, K., Gatti, G., Coda, A., Antolini, L., Pelosi, G., Desideri, A., Falconi, M., Marmocchi, F., Rotilio, G. and Bolognesi, M. (1992) Crystal Structure of Yeast Cu,Zn, Superoxide Dismutase. Crystallographic refinement at 2.5 A Resolution, *J. Mol. Biol.* **225**, 791-809.
7. Bertini, I., Banci, L., Brown III, R.D., Koenig, S.H. and Luchinat, C. (1988) Electronic relaxation of a copper(II) dimer in a macromolecular complex as evaluated from solvent proton relaxation, *Inorg. Chem.* **27**, 951-953.

8. Banci, L., Bertini, I., Hallewell, R.A., Luchinat, C. and Viezzoli, M.S. (1989) Water in the active cavity of copper/zinc superoxide dismutase. A water ^{1}H-nuclear-magnetic-relaxation-dispersion study, *Eur. J. Biochem.* **184**, 125-129.

9. Banci, L., Bertini, I., Luchinat, C. and Hallewell, R.A. (1988) An investigation of superoxide dismutase Lys-143, Ile-143, and Glu-143 mutants: Cu_2Co_2SOD derivatives, *J. Am. Chem. Soc.* **110**, 3629-3633.

10. Banci, L., Bertini, I., Cabelli, D., Hallewell, R.A., Luchinat, C. and Viezzoli, M.S. (1990) Investigation of copper-zinc superoxide dismutase Ser-137 and Ala-137 mutants, *Inorg. Chem.* **29**, 2398-2403.

11. Banci, L., Cabelli, D.E., Getzoff, E.D., Hallewell, R.A. and Viezzoli, M.S. (1993) An Essential Role for the Conserved Glu-133 in the Anion Interaction with Superoxide Dismutase, *J. Inorg. Biochem.* **50**, 89-100.

12. Fee, J.A. and Bull, C. (1986) Steady-state kinetic studies of superoxide dismutases, *J. Biol. Chem.* **261**, 13000-13005.

13. Cabelli, D.E. and Bielski, B.H. (1983) Kinetic and mechanism for the oxidation of ascorbic acid/ascorbate by HO_2/O_2^- radicals. A pulse radiolysis and stopped-flow photolysis study, *J. Phys. Chem.* **87**, 1809-1812.

14. Beyer, W.F., Jr., Fridovich, I., Mullenbach, G.T. and Hallewell, R.A. (1987) Examination of the role of Arginine-143 in the human Copper and Zinc superoxide dismutase by site-specific mutagenesis, *J. Biol. Chem.* **262**, 11182-11187.

15. Bertini, I., Banci, L., Luchinat, C. and Hallewell, R.A. (1988) The exploration of the active-site cavity of copper-zinc superoxide dismutase, in Blanch, H.W. and Klibanov, A.M. (eds) *Annals of the New York academy of sciences*, New York Academy of Science Book, New York, pp.37-52.

16. Bertini, I., Banci, L., Luchinat, C., Bielski, B.H.J., Cabelli, D., Mullenbach, G.T. and Hallewell, R.A. (1989) An investigation of a human erythrocyte SOD modified at the position 137, *J. Am. Chem. Soc.* **111**, 714-719.

17. Getzoff, E.D., Cabelli, D.E., Fisher, C.L., Parge, H.E., Viezzoli, M.S., Banci, L. and Hallewell, R.A. (1992) Faster Superoxide Dismutase Mutants designed by Enhancing Electrostatic Guidance, *Nature* **358**,, 347-351

18. Getzoff, E.D., Tainer, J.A., Stempien, M.M., Bell, G.I. and Hallewell, R.A. (1989) Evolution of CuZn superoxide dismutase and the greek key b-barrel structural motif, *Proteins: Structure,Function,and Genetics* **5**, 322-336.

19. Banci, L., Bertini, I., Luchinat, C. and Viezzoli, M.S. (1993) pH dependent properties of SOD studied through mutants on Lys 136, *Inorg. Chem.* **32**, 1403-1406.

20. Cudd, A. and Fridovich, I. (1982) Electrostatic Interactions in the Mechanism of Bovine Erythrocyte Superoxide Dismutase, *J. Biol. Chem.* **257**, 11443-11447.

21. Sharp, K., Fine, R. and Honig, B. (1987) Computer simulations of the diffusion of a substrate to an active site of an Enzyme, *Science* **236**, 1460-1463.

22. Argese, E., Viglino, P., Rotilio, G., Scarpa, M. and Rigo, A. (1987) Electrostatic Control of the Rate-Detrmining Step of the Copper,Zinc Superoxide Dismutase Catalytic Reaction, *Biochemistry* **26**, 3224-3228.

23. Getzoff, E.D., Tainer, J.A., Weiner, P.K., Kollman, P.A., Richardson, J.S. and Richardson, D.C. (1983) Elecrostatic Recognition between Superoxide and Copper, Zinc Superoxide Dismutase, *Nature* **306**, 287-290.

24. Klapper, I., Hagstrom, R., Fine, R., Sharp, K. and Honig, B. (1986) Focusing of electric fields in the active site of Cu-Zn superoxide dismutase: effects of ionic strength and amino-acid modification, *Proteins: Structure,Function,and Genetics* **1**, 47-59.

25. Sines, J.J., Allison, S.A. and McCammon, J.A. (1990) Point Charge Distribution and Electrostatic Steering in Enzyme/Substrate Encounter: Brownian Dynamics of Modified Copper/Zinc Superoxide Dismutase, *Biochemistry* **29**, 9403-9412.

26. Bacquet, R.J., McCammon, J.A. and Allison, S.A. (1988) Ionic Strength Dependence of Enzyme-Substrate Interactions. Monte Carlo and Poisson-Boltzmann Results for Superoxide Dismutase. *J. Phys. Chem.* **92**, 7134-7141.

27. Shen, J., Wong, C.F., Subrasmaniam, S., Albright, T.A. and McCammon, J.A. (1990) Partial electrostatic charges for the active center of copper, zinc superoxide dismutase, *J. Comp. Chem.* **11**, 346-350.

28. Merz, K.M., Murcko, M.A. and Kollman, P.A. (1991) Inhibition of carbonic anhydrase, *J. Am. Chem. Soc.* **113**, 4484-4490.

29. Merz, K.M. (1991) CO_2 binding to human carbonic anhydrase II, *J. Am. Chem. Soc.* **113**, 406-411.

30. Banci, L., Bencini, A., Bertini, I., Luchinat, C. and Piccioli, M. (1990) ^{1}H NOE and ligand field studies of copper-cobalt superoxide dismutase with anions, *Inorg. Chem.* **29**, 4867-4873.

31. Jorgensen, W.L., Chandrasekhar, J., Madura, J., Impey, R.W. and Klein, M.L. (1983) Comparison of Simple Potential Functions for Simulating Liquid Water, *J. Chem. Phys.* **79**, 926-935.

32. Banci, L., Carloni, P., La Penna, G. and Orioli, P.L. (1992) Molecular dynamics studies on superoxide dismutase and its mutants: the structural and functional role of Arg 143, *J. Am. Chem. Soc.* **114**, 6994-7001.

33. Banci, L., Carloni, P. and Orioli, P.L. (1994) Molecular Dynamics Studies on Mutants of Cu,Zn Superoxide Dismutase: The Functional Role of Charged Residues in the Electrostatic Loop VII, *Proteins: Structure,Function,and Genetics* **18**, 216-230.

34. Fisher, C.L., Cabelli, D.E., Tainer, J.A., Hallewell, R.A. and Getzoff, E.D. (1994) The role of Arginine 143 in the Electrostatic and Mechanism of Cu,Zn Superoxide Dismutase: Computational and Experimental Evaluation of Site-directed Mutants, *Proteins: Structure,Function,and Genetics* **19**, 24-34.

35. Bosshard, H.R., Anni, H. and Yonetani, T. (1991) Everse, J., Everse, K.E. and Grisham, M.B. (eds) *Peroxidases in Chemistry and Biology*, CRC Press, Boca Raton,FL , pp.84

36. Dawson, J.H. (1988) *Science* **240**, 433-439.

37. Goodwin, D.C., Aust, S.D. and Grover, T.A. (1995) Evidence for veratryl alcohol as a redox mediator in lignin peroxidase-catalyzed oxidation, *Biochemistry* **34**, 5060-5065.

38. Harvey, P.J. and Candeias, L.P. (1995) Radical cation cofactors in lignin peroxidase catalysis, *Biochem. Soc. Trans.* **23**, 262-267.

39. Poulos, T.L., Freer, S.T., Alden, R.A., Edward, S.L., Skogland, U., Takio, K., Eriksson, B., Xuong, N., Yonetani, T. and Kraut, J. (1980) The crystal structure of Cytochrome *c* peroxidase, *J. Biol. Chem.* **255**, 575-580.

40. Finzel, B.C., Poulos, T.L. and Kraut, J. (1984) Crystal structure of yeast cytochrome *c* peroxidase at 1.7 Å resolution, *J. Biol. Chem.* **259**, 13027-13036.

41. Piontek, K., Glumoff, T. and Winterhalter, K. (1993) Low pH crystal structure of glycosylated lignin peroxidase from *Phanerochaete chrysosporium* at 2.5 Å resolution. *FEBS Lett.* **315**, 119-124.

42. Edwards, S.L., Raag, R., Wariishi, H., Gold, M.H. and Poulos, T.L. (1993) Cristal Structure of Lignin Peroxidase, *Proc. Natl. Acad. Sci. USA* **90**, 750-754.

43. Poulos, T.L., Edwards, S.L., Wariishi, H. and Gold, M.H. (1993) Crystallographic Refinement of Lignin Peroxidase at 2 Å, *J. Biol. Chem.* **268**, 4429-4440.

44. Sundaramoorthy, M., Kishi, K., Gold, M.H. and Poulos, T.L. (1994) The crystal structure of manganese peroxidase from *Phanerochaete chrysosporium* at 2.06 Å resolution, *J. Biol. Chem.* **269**, 32759-32767.

45. Petersen, J.F.W., Kadziola, A. and Larsen, S. (1994) Three-dimensional structure of a recombinant peroxidase from *Coprinus cinereus* at 2.6 Å resolution, *FEBS Lett.* **339**, 291-296.

46. Kunishima, N., Fukuyama, K., Matsubara, H., Hatanaka, H., Shibano, Y. and Amachi, T. (1994) Crystal structure of the fungal peroxidase from *Arthromyces ramosus* at 1.9 Å resolution. Structural comparisons with the lignin and cytochrome c peroxidase, *J. Mol. Biol.* **235**, 331-344.

47. Patterson, W.R. and Poulos, T.L. (1995) Crystal structure of recombinant pea cytosolic ascorbate peroxidase, *Biochemistry* **34**, 4331-4341.

48. Welinder, K.G. and Gajhede, M. (1993) Structure and Evolution of Peroxidases, in Greppin, H., Rasmussen, S.K., Welinder, K.G. and Penel, C. (eds) *Plant Peroxidases Biochemistry and Physiology*, University of Copenhagen & University of Geneva, Geneva, pp.35-42.

49. Poulos, T.L. and Kraut, J. (1980) The stereochemistry of Peroxidase catalysis, *J. Biol. Chem.* **255**, 8199-8205.

50. Goodin, D.B. and McRee, D.E. (1993) The Asp-His-Fe triad of cytochrome *c* peroxidase controls the reduction potentials, electronic structure, and coupling of the tryptophan free radical to the heme, *Biochemistry* **32**, 3313-3324.

51. Banci, L., Bertini, I., Turano, P., Tien, M. and Kirk, T.K. (1991) Proton NMR investigation into the basis for the relatively high redox potential of lignin peroxidase, *Proc. Natl. Acad. Sci. USA* **88**, 6956-6960.

52. Erman, J.E., Vitello, L.B., Miller, M.A. and Kraut, J. (1992) Active-Site Mutations in Cytochrome c Peroxidase: A Critical Role for Histidine 52 in the Rate of Formation of Compound I, *J. Am. Chem. Soc.* **114**, 6592-6593.

53. Erman, J.E., Vitello, L.B., Miller, M.A., Shaw, A., Broen, K.A. and Kraut, J. (1993) Histidine 52 Is a Critical Residue for Rapid Formation of Cytochrome c Peroxidase Compound I, *Biochemistry* **32**, 9798-9806.

54. Vitello, L.B., Erman, J.E., Miller, M.A., Wang, J. and Kraut, J. (1993) Effect of arginine-48 replacement on the reaction between cytochrome c peroxidase and hydrogen peroxide, *Biochemistry* **32**, 9807-9818.

55. Bujons, J., Dikiy, A., Ferrer, J.C., Banci, L. and Mauk, A.G. (1997) Charge reversal of a critical active-site residue in cytochrome *c* peroxidase. Characterization of the Arg48-->Glu variant, *Eur. J. Biochem.* **243**, 72-84.

56. Ogawa, S., Shiro, Y. and Morishima, I. (1979) Calcium binding by horseradish peroxidase c and heme environmental structure, *Biochem. Biophys. Res. Commun.* **90**, 674-678.

57. Welinder, K.G. (1985) Plant Peroxidases, *Eur. J. Biochem.* **151**, 497-450.

58. Welinder, K.G. (1992) Superfamily of plant, fungal and bacterial peroxidases, *Curr. Op. Struct. Biol.* **2**, 388-393.

59. Haschke, R.H. and Friedhoff, J.M. (1978) Calcium-related properties of horseradish peroxidase, *Biochem. Biophys. Res. Commun.* **80**, 1039-1042.

60. Shiro, Y., Kurono, M. and Morishima, I. (1986) Presence of Endogenous Calcium Ion and Its Functional and Structural Regulation in Horseradish Peroxidase, *J. Biol. Chem.* **261**, 9382-9390.

61. Morishima, I., Kurono, M. and Shiro, Y. (1986) Presence of endogenous calcium ion in horseradish peroxidase. Elucidation and metal-binding site by substitution of divalent and lanthanide ions for calcium and use of metal-induced NMR (proton and cadmium-113) resonances. *J. Biol. Chem.* **261**, 9391-9399.

62. Barber, K.R., Rodriguez Maranon, M.J., Shaw, G.S. and Van Huystee, R.B. (1995) Structural influence of calcium on the heme cavity of cationic peanut peroxidase as determined by ^{1}H-NMR spectroscopy, *Eur. J. Biochem.* **232**, 825-833.

63. Pettigrew, G.W. and Moore, G.R. (1987) *Cytochromes c; Biological Aspects*, Springer-Verlag, Berlin.

64. Moore, G.R. and Pettigrew, G.W. (1990) *Cytochromes c; Evolutionary, Structural and Physicochemical Aspects*, Springer-Verlag, Berlin.

65. Takano, T. and Dickerson, R.E. (1981) Conformation change in cytochrome c:II. Ferricytochrome c refinement at 1.8 A and comparison with ferrocytochrome c structure, *J. Mol. Biol.* **153**, 95-155.

66. Takano, T. and Dickerson, R.E. (1981) Conformation change of cytochrome c: I. Ferrocytochrome c structure refined at 1.5 A resolution, *J. Mol. Biol.* **153**, 79-94.

67. Leung, C.J., Nall, B.T. and Brayer, G.D. (1989) Crystallization of yeast iso-2-cytochrome c using a novel hair seeding technique, *J. Mol. Biol.* **206**, 783-785.

68. Bushnell, G.W., Louie, G.V. and Brayer, G.D. (1990) High-resolution three-dimensional structure of horse hearth cytochrome c, *J. Mol. Biol.* **214**, 585-595.

69. Berghuis, A.M. and Brayer, G.D. (1992) Oxidation state-dependent conformational changes in cytochrome c, *J. Mol. Biol.* **223**, 959-976.

70. Louie, G.V., Hutcheon, W.L.B. and Brayer, G.D. (1988) Yeast iso-1-cytochrome c: a 2.8 A resolution three-dimensional structure determination. *J. Mol. Biol.* **199**, 295-314.

71. Murphy, M.E.P., Nall, B.T. and Brayer, G.D. (1992) Structure determination and analysis of yeast iso-2-cytochrome c and a composite mutant protein, *J. Mol. Biol.* **227**, 160-176.

72. Ochi, H., Hata, Y., Tanaka, N., Kakudo, M., Sakuri, T., Achara, S. and Morita, Y. (1983) Structure of rice ferricytochrome c at 2.0 A resolution, *J. Mol. Biol.* **166**, 407-418.

73. Baistrocchi, P., Banci, L., Bertini, I., Turano, P., Bren, K.L. and Gray, H.B. (1996) Three-dimensional solution structure of *saccharomyces cerevisiae* reduced iso-1-cytochrome *c*, *Biochemistry* **35**, 13788-13796.

74. Banci, L., Bertini, I., Bren, K.L., Gray, H.B., Sompornpisut, P. and Turano, P. (1997) The solution structure of oxidized cytochrome *c*: hints to understand its function and folding, *Biochemistry* in press

75. Scott, R.A. and Mauk, A.G. (1996) *Cytochrome c. A multidisciplinary approach*, University Science Books, Sausalito, California.

76. Berghuis, A.M., Guillemette, J.G., McLendon, G., Sherman, F., Smith, M. and Brayer, G.D. (1994) The role of conserved internal water molecule and its associated hydrogen bond network in cytochrome c, *J. Mol. Biol.* **236**, 786-799.

77. Qi, P.X., Urbauer, J.L., Fuentes, E.J., Leopold, M.F. and Wand, A.J. (1994) Structural Water in Oxidized and Reduced horse heart cytochrome c, *Nature struct. Biol.* **1**, 378-382.

78. Banci, L., Carloni, P. and Gori Savellini, G. (1994) Molecular Dynamics Studies on peroxidases: A Structural Model for Horse Radish Peroxidase and a Substrate Adduct, *Biochemistry* **33**, 12356-12366.

79. Banci, L., Carloni, P., Diaz, A. and Gori Savellini, G. (1996) Molecular dynamics calculations on peroxidases: the effect of calcium ions on protein structure, *JBIC* **1**, 264-272.

80. Weiner, S.J., Kollman, P.A., Case, D.A., Singh, U.C., Ghio, C., Alagona, G. and Profeta, S., Jr. (1984) A new force field for molecular mechanical simulation of nucleic acids and proteins, *J. Am. Chem. Soc.* **106**, 765-784.

81. Angelucci, L., De Gioia, L. and Fantucci, P. (1993) The mol. mechanics approach to the study of inorganic systems. The conformational properties of the iron(II)-porphyrin-polypeptide complexes. *Gaz. Chim. Ital.* **123**, 111-117.

82. Ruf, H.H., Wende, P. and Ullrich, V. (1979) Models for ferric cytochrome P450. Characterization of hemin mercaptide complexes by electronic and ESR spectra, *J. Inorg. Biochem.* **11**, 189-204.

83. Collins, J.R., Du, P. and Loew, G.H. (1992) Molecular dynamics calculations on CcP, *Biochemistry* **31**, 11166

84. Satterlee, J.D., Russell, D.J. and Erman, J.E. (1991) Proton Homonuclear Correlated Spectroscopy as an Assignment Tool for Hyperfine-Shifted Resonances in Medium-Sized Paramagnetic Proteins: Cyanide-Ligated Yeast Cytochrome c Peroxidase as an Example, *Biochemistry* **30**, 9072-9077.

85. de Ropp, J.S., La Mar, G.N., Wariishi, H. and Gold, M.H. (1991) NMR study of the active site of resting state and cynide-inhibited Lignin Peroxidase from *Phanerochaete chrysosporium*. Comparison with Horseradish Peroxidase, *J. Biol. Chem.* **266**, 15001-15008.

86. Banci, L., Bertini, I., Pierattelli, R., Tien, M. and Vila, A.J. (1995) Factoring of the hyperfine shifts in the cyanide adduct of lignin peroxidase from *P. chrysosporium*, *J. Am. Chem. Soc.* **117**, 8659-8667.

87. Satterlee, J.D. and Erman, J.E. (1991) Proton NMR assignments of heme contacts and catalytically implicated amino acids in cynade-ligated cytochrome c peroxidase determined from one- and two-dimensional nuclear Overhauser effects, *Biochemistry* **30**, 4398-4405.

88. Smulevich, G. (1993) Structure-Function relationships in peroxidases via resonance raman spectroscopy and site-directed mutagenesis: cytochrome c peroxidase, in Clark, R.J.H. and Hester, R.E. (eds) *Biomolecular Specrtoscopy, Part A, Vol 20, Advances in Spectroscopy*, J. Wiley and Sons, London, pp.163-193

89. Banci, L., Gori Savellini, G. and Turano, P. (1996) *Comparison between the reduced and oxidized forms of yeast iso-1-cytochrome c: a MD study in explicit water*, submitted.

90. Cornell, W.D., Cieplak, P., Bayly, C.J., Gould, I.R., Merz, K.M., Jr., Ferguson, D.M., Spellmeyer, D.C., Fox, T., Caldwell, J.W. and Kollman, P.A. (1995) A Second Generation Force Field for the Simulation of Proteins, Nucleic Acid, and Organic Molecules, *J. Am. Chem. Soc.* **117**, 5179-5197.

91. Frisch, M.J., Trucks, G.W., Head-Gordon, M., Gill, P.M.W., Wong, M.W., Foresman, J.B., Johnson, B.G., Schlegel, H.B., Robb, M.A., Replogle, E.S., Gomperts, R., Andres, J.L., Raghavachari, K., Binkley, J.S., Gonzalez, C., Martin, R.L., Fox, D.J., Defrees, D.J., Baker, J., Stewart, J.J.P. and Pople, J.A. (1992) *GAUSSIAN 92*, Gaussian inc. Pittsburgh PA.

92. Marcus, R.A. and Sutin, N. (1985) Electron transfer in chemistry and biology, *Biochim. Biophys. Acta* **811**, 265-275.

93. Beratan, D.A., Onuchic, J.N., Betts, J.N., Bowler, B.E. and Gray, H.B. (1990) Electron-Tunneling Pathways in Ruthenated Proteins, *J. Am. Chem. Soc.* **112**, 7915-7921.

94. Casimiro, D.R., Richards, J.H., Winkler, J.R. and Gray, H.B. (1993) Electron Transfer in Ruthenium-Modified Cytochromes c. *J. Phys. Chem.* **97**, 13073-13077.

THE EFFECTIVE CRYSTAL FIELD METHODOLOGY AS USED TO INCORPORATE TRANSITION METALS INTO MOLECULAR MECHANICS

ANDREI L. TCHOUGRÉEFF
Karpov Institute of Physical Chemistry,
Vorontsovo pole 10, Moscow 103064, Russia
and
Center for Computational Chemistry,
Keldysh Institute for Applied Mathematics,
Miusskaya pl. 4, Moscow 125047, Russia

1. Introduction

Transition metal complexes (TMC) of biological significance are too large for a complete quantum chemical treatment. Thus incorporation of transition metal atoms (TMA) into the general molecular mechanics (MM) formalism [1] would be highly desirable [2, 3]. However, the problem of molecular modeling TMA must not be limited to obtaining appropriate MM potentials. The ability of the MM approach (when extended to TMA) to describe potential energy surfaces for different low energy electronic states of TMCs is of equal importance. Such states are accessible, for example, in experiments on spin transitions [4]. Multiple spin states are also involved in oxygen binding by the heme containing proteins [5]. The most important feature of such a combined method is the necessity to implement different levels of the description of electronic structure in the different parts of the system. These parts are (i) the isolated group of strongly correlated d-electrons on the metal atom and (ii) electrons in the closed shell of the ligands. The physical reason for a distinction between electrons in the ligands and those in the d-shell is that in an organic molecule one can never distinguish the multiplet terms of the component atoms. In contrast, the d-shell of a transition metal ion in a complex largely retains the system of electronic terms that it had in the free state. Due to specific properties of electrons in the d-shell, different levels of electron correlation must be achieved to describe electrons in the d-shell and those in the ligand orbitals

L. Banci and P. Comba (eds.), Molecular Modeling and Dynamics of Bioinorganic Systems, 217–232.

properly. This requirement has been successfully implemented in the effective crystal field (ECF) method [6]. In Sect. 2 of the present paper we discuss the underlying principles and give an account of the ECF method. Sect. 3 is a brief review of the current implementation of the ECF approach and of some recent results. In Sect. 4 new developments and extentions of the ECF methodology, including its combination with MM are reviewed.

2. Theoretical Background

Both semi-empirical and *ab-initio* methods based upon the self consistent field (SCF) approximation [7] have been used to calculate the ground and low energy excited states of TMCs [8]. Although *ab-initio* calculations involving large CI expansions or a special set of configurations in the MCSCF wave function can lead to a reasonable agreement of the calculated and observed characteristics of TMC, these calculations are restricted to relatively small systems and can hardly be used for screening large portions of the configurational space which is necessary for molecular dynamics modeling of metal containing enzymes.

Despite the reasonable agreement between the observed properties of TMCs and those calculated by SCF based semi-empirical methods some fundamental problems remain unsolved. These include (for more detailed discussion see [6, 9] and references therein):

(i) the Koopmans Theorem is not valid for the states with large contributions from the atomic *d*-states;

(ii) the *Aufbau*-principle frequently breaks down for orbitals with significant contributions from atomic *d*-states;

(iii) the iteration procedures implied by the SCF scheme oscillate or converge very slowly.

The Koopmans Theorem relates the SCF orbital energies to ionization potentials. For organic molecules the difference between the minus orbital energy (*i.e.* the ionization potential according to Koopmans Theorem) and the experimental ionization potential is usually small. For TMCs the disagreement with experiment can amount to $10 \div 20$ eV for levels with significant contribution from *d*-orbitals (see [9] and references therein). This suggests that the behavior of *d*-electrons in TMCs is not consistent with the SCF picture, where independent electrons move in the mean field induced by the nuclei and other electrons. In contrast, *d*-electrons are strongly correlated.

Another rarely recognized problem is the break down of the *Aufbau*-principle (*i.e.* the rule that the MO's are filled with pairs of electrons starting from the lowest energy orbital). This occurs frequently in the semi-empirical calculations of TMCs, where empty or singly occupied orbitals are calculated to have lower orbital energies than some doubly occupied orbitals. This generally happens to orbitals derived from *d*-atomic orbitals. It can be checked [6] that the Slater determinants which break the *Aufbau*-principle do not represent minima of the SCF energy functionals but saddle points. Namely this causes the well known but rarely reported problems with the convergence of the SCF iterations. More distressing than the convergence problems is the fact that the success of SCF based semi-empirical methods is achieved by the loss of the confidence that the obtained Slater determinants represent the minima of the SCF energy functional. The parameterization of a semi-empirical Fock operator for TMAs, that uses a trial wave function which is not a stable solution of the SCF problem is a doubtful procedure.

It follows that there are some fundamental problems in the quantum chemical description of the electronic structure of TMCs. They are due to the important role of electron correlations in the *d*-shell. The description of electrons in the ligands with SCF methods seems to cause less problems.

The state of the quantum chemical description of TMCs described above is in sharp contrast to the situation with understanding of their electronic structure. The experimental data on *d*-*d* excitation spectra can be successfully interpreted with models equivalent to the crystal field theory (CFT) [10, 11]. This is because the specific properties of TMCs are controlled by the *d*-electrons. Their low energy excitations are responsible for the characteristic absorption bands in the optical spectra and for the magnetic properties. The ground state spin depends on the balance between Coulomb repulsion of *d*-electrons and their interaction with the ligands which induce an external field. These features apparently correspond to the situation described by the naïve CFT where all interesting events happen in the *d*-shell of TMA's whose ligand environment remains inert [10]. Thus the qualitative physical picture provided by the CFT is correct to a large extent. This is due to the correct form of the electronic wave function which is used implicitly. The wave function of a complex in the CFT is a product of the multiplet (full CI) state for a fixed number of *d*-electrons and of an unspecified closed shell state of the remaining electrons in the ligands. The ligand electrons are not considered explicitly, and that is why crystal fields can not be satisfactorily calculated within the CFT's own framework. This has been achieved in the ECF method [6].

The formal derivation of the Effective Crystal Field method proceeds as described in the literature [6]. The whole set of the valence atomic orbitals (AO) of a TMC (including the 4*s*-, 4*p*-, and 3*d*-AO's of the metal center and the valence AO's of all ligand atoms) is separated into two parts. The first contains only 3*d*-orbitals of the TMA (*d*-system). The second part contains 4*s*- and 4*p*-AO's of the TMA and the valence orbitals of the ligand atoms (ligand system or l-system). The total Hamiltonian for a TMC can be written as the sum:

$$H = H_d + H_l + H_c + H_r \tag{1}$$

where H_d is the Hamiltonian for the d-electrons in the field of atomic cores of the TMC, H_l is the Hamiltonian for the electrons of the l-system, H_c and H_r are operators for the Coulomb and resonance interactions between the two systems.

The exact wave function $\Phi(N)$ for any N-electron state of a TMC can be presented in the form [12]:

$$\Phi(N) = \sum_{nkm} C_{km}(n; N-n)\Phi_d^k(n) \wedge \Phi_l^m(N-n) \tag{2}$$

where $\Phi_d^k(n)$ are the n-electron wave functions for the metal d-orbitals; $\Phi_l^m(N-n)$ are the $(N-n)$-electron wave functions for the l-orbitals; $C_{km}(n; N-n)$ are variation parameters. For most of the Werner-type TMCs $\Phi_d^k(n)$ functions with a specific value n give a major contribution to eq. (2). This particular value of n depends on the oxidation state of the transition metal atom in the TMC. The terms in eq. (2) with other values of n provide some smaller corrections. The wave function Φ_n for the n'th electronic state of a TMC is then assumed to take the form:

$$\Phi_n = \Phi_d^n(n_d) \wedge \Phi_l(N-n_d) \tag{3}$$

where $\Phi_d^n(n_d)$ is the spin and symmetry adapted n_d-electron wave function of the metal d-orbitals, and Φ_l is the $(N-n_d)$-electron wave function of the l-orbitals. For TMCs with closed-shell ligands, the ground state of the l-system can also be described by a closed-shell Slater determinant $\Phi_l({}^1A_1)$ with zero total spin. Thus, both the spin multiplicity and the point symmetry of the function eq. (3) coincide with those of the functions Φ_d^n of the d-system.

The two function multipliers in eq. (3) must be determined from the energy minimum condition. A serious problem arises due to the fact that wave functions of the type used in eq. (3) correspond to a fixed integer number of electrons in the d-shell. Therefore, all matrix elements of the resonance (electron hopping) operator H_r for the functions of that type vanish. In order to include the effects of the resonance interaction between the subsystems, we consider the *effective* Hamiltonian H^{eff} [6, 7] which operates in the subspace spanned by the functions of eq. (3). Its eigenvalues coincide with those of the exact Hamiltonian of eq. (1):

$$
\begin{aligned}
H^{eff} &= PH_oP + H_{RR} \\
H_o &= H_d + H_l + H_c \\
H_{RR} &= PH_rQ(EQ - QH_oQ)^{-1}QH_rP
\end{aligned}
\tag{4}
$$

Here, P is the operator projecting on the subspace of functions with the fixed number n_d of d-electrons; $Q = 1 - P$.

The variational problem for the effective Hamiltonian H^{eff} of eq. (4) in the P-subspace transforms into a system of equations for the functions Φ_d^n and Φ_l (see *e.g.* [6, 7]):

$$
\begin{aligned}
H_d^{eff}\Phi_d^n &= E_d^n\Phi_d^n \\
H_l^{eff}\Phi_l &= E_l\Phi_l \\
H_d^{eff} &= H_d + < \Phi_l \mid H_c + H_{RR} \mid \Phi_l > \\
H_l^{eff} &= H_l + < \Phi_d^n \mid H_c + H_{RR} \mid \Phi_d^n >
\end{aligned}
\tag{5}
$$

Since the l-system is described by a single Slater determinant Φ_l, the latter must be found from the SCF procedure applied to the Fockian F_l^{eff}, derived from the Hamiltonian H_l^{eff} by the standard method [7, 13]. Proceeding semi-empirically we use the CNDO parameterization [13] for all the l-AO's. The solution of the SCF problem for the l-system with the approximate CNDO Fock operator gives the one electron density matrix P_{kl}, the energies of the molecular orbitals (MO) ϵ_i, and MO-LCAO coefficients c_{ik}. These quantities completely describe the electronic structure of the l-system. They are used to construct the effective Hamiltonian H_d^{eff} of eq. (5) by averaging the operators H_c and H_{RR} over the ground state Φ_l of the l-system. This yields the effective Hamiltonian H_d^{eff} in the form [6]:

$$
H_d^{eff} = \sum_{\mu\nu\sigma} U_{\mu\nu}^{eff} d_{\mu\sigma}^+ d_{\nu\sigma} + \frac{1}{2} \sum_{\mu\nu\rho\eta} \sum_{\sigma\tau} (\mu\nu \mid \rho\eta) d_{\mu\sigma}^+ d_{\nu\sigma} d_{\rho\tau}^+ d_{\eta\tau} \tag{6}
$$

where the d-electron Coulomb interaction term is that of the free ion and the effective core attraction parameters $U_{\mu\nu}^{eff}$ contain contributions from the Coulomb and the resonance interaction of the d- and l-systems:

$$
U_{\mu\nu}^{eff} = \delta_{\mu\nu}U_{dd} + W_{\mu\nu}^{atom} + W_{\mu\nu}^{field} + W_{\mu\nu}^{cov} \tag{7}
$$

where

$$
W_{\mu\nu}^{atom} = \delta_{\mu\nu}\left(\sum_{i\in s,p} g_{\mu i}P_{ii}\right) \tag{8}
$$

is the repulsion of electrons in the d-shell from those in the 4s- and 4p-AO's of the metal;

$$W_{\mu\nu}^{field} = \sum_{L}(P_{LL} - Z_L)V_{\mu\nu}^{L} \tag{9}$$

is the Coulomb interaction of d-electrons with the net charges on the ligand atoms, in the standard CFT form; and

$$W_{\mu\nu}^{cov} = -\sum_{i}\beta_{\mu i}\beta_{\nu i}\left(\frac{1-n_i}{\Delta E_{di}} - \frac{n_i}{\Delta E_{id}}\right) \tag{10}$$

takes into account the resonance. In eqs. (8)-(10) P_{ii} is the diagonal matrix element of the one-electron density matrix of the l-system; $P_{LL} = \sum_{l\in L} P_{ll}$ is the electron population of the ligand atom L; Z_L is the core charge of the ligand atom L; $V_{\mu\nu}^{L}$ is the matrix element of the potential energy operator describing the interaction between a d-electron and an electron placed on the ligand atom $L; n_i$ is the occupation number of the i'th l-MO ($n_i = 0$ or 1); ΔE_{di} (ΔE_{id}) is the energy necessary to transfer an electron from the d-shell (from the i'th MO) to the i'th MO (to the d-shell):

$$\begin{aligned} \Delta E_{di} &= I_i - A_d - g_{id} \\ \Delta E_{id} &= I_d - A_i - g_{id} \end{aligned} \tag{11}$$

$\beta_{\mu i}$ is the resonance integral between the μ'th d-orbital and the i'th l-MO, described by the resonance integrals $\beta_{\mu k}$ between the μ'th d-AO and the k'th l-AO:

$$\beta_{\mu k} = -(I_d + I_k)S_{\mu k}\beta^{ML} \tag{12}$$

where I_d and I_k are the valence state ionization potentials, $S_{\mu k}$ is the overlap integral between the μ'th d-AO and the k'th l-AO, and β^{ML} is the only adjustable parameter (see below), specific for each ML pair, where M stands for a transition metal atom and L stands for a ligand donor atom (nitrogen, oxygen *etc.*).

After establishing the form of the effective Hamiltonian for d-electrons, the states of n_d electrons in the d-shell are calculated by diagonalization of the matrix associated with H_d^{eff} in the n_d-electron wave function basis set. The ground state of the complex is given by the external product of the ligand Slater determinant and the n_d-electron wave function, obtained by diagonalization of H_d^{eff}, which leads to the lowest eigenvalue. The d-d excited states are products of other n_d-electron functions with the same Slater determinant. The excitation energies are equal to differences between the eigenvalues of H_d^{eff}.

3. Standard Implementation and Recent Results

The ECF method has been developed originally with a specific aim to calculate *d-d* spectra of TMCs. The only interest in the electronic structure of the ligands was to estimate the ECF. Therefore, the simplest semi-empirical CNDO method [13] was used to calculate the characteristics of the *l*-system. From eqs. (7)-(12) it follows that all quantities that are involved in the determination of the ECF (*i.e.* ionization potentials, partial atomic charges) may be obtained with reasonable accuracy from CNDO calculations on organic molecules. Calculations for a series of octahedral ML_6^{2+} and ML_6^{4-} complexes, where M = Ti, V, Cr, Mn, Fe, Co, Ni, Cu; L = H_2O, NH_3, F^-, Cl^-, CN^- were reported [6, 14, 15] . From these, the parameters β^{ML} were obtained by comparision to experimental values of $10Dq$ [16]. In all cases we observe excellent agreement between calculated and experimental spin state and symmetry of the ground state. *d-d* Spectra of the excitations were reproduced with the accuracy of 2000 cm^{-1}. We also applied the ECF/CNDO method to ML_4Z_2 compounds with the same ligands L as above and Z = H_2O,Py, NCS^- [17], and to metalloporphyrins [9]. The spin multiplicities and the spatial symmetry of the ground states of the mixed ligand complexes are correctly reproduced by the ECF/CNDO method. The same is true for the energies and symmetries of the lower excited states.

Based on the successful prediction ot the ground state spin and symmetry for a variety of TMCs we studied the spin transition of *cis*-[Fe(bipy)$_2$-(NCS)$_2$] [18] which is well documented [4] and where the experimental molecular geometries are known for both spin isomers [19]. The ECF/CNDO calculations reproduce the change of the total spin of this quasi-octahedral d^6 complex from S = 0 (low-spin) to S=2 (high-spin) which accompanies the increase of the Fe-N distances by $\sim 0.15 - 0.20$ Å. It was also possible to predict low-energy *d-d* excited states for both spin forms in acceptable agreement with the experiment.

The successfull description of the Fe(II) spin isomenrs motivated us to address the problem of the spin states of myoglobine (Mb) [20]. The interpretation of the kinetics of the binding of small molecules like CO, O_2 or NO to Mb is controversial [5, 21, 22]. It is known that the reaction rates are controlled by the electronic states of the reactants and products. The ground state of Mb is a quintet, CO and O_2 have singlet ($^1\Sigma^+$) and triplet ($^3\Sigma_g$) ground states, respectively, and MbCO as well as MbO_2 both have singlet ground states. According to the spin coupling rules [23], the binding of both CO and O_2 is a spin forbidden process for Mb in the S=2 ground state since a change of the total spin of either systems (Mb + CO or Mb + O_2) is required. It emerges that an interpretation of the low-energy electronic spectum of Mb might be useful for the analysis of the binding kinetics. We note that SCF *ab-initio* calculations of iron porphyrins (FeP) may give erroneous results with a bias towards high-spin ground states. A high-quality *ab-initio* study on FeP the ground state has erroneously been found to be a quintet [24]. Moreover, *ab-initio* approaches are too expensive to use for screening of large portions of the nuclear configuration space. We have studied the ground state spin and symmetry of the iron(II) porphyrine imidazole complex in dependence of the coordination sphere geometry [20]. Following earlier suggestions [5] we computed its electronic ground and several low-energy excited states for a two dimensional area in the coordinate space defined by two internal coordinates: namely, by that of the shift of the Fe atom perpendicular to the porphyrine plane (r_{Fe-Ctr}), and by the distance between the Fe atom and the imidazole nitrogen atom (r_{Fe-N_ϵ}). The results are depicted on Fig. 1. Three types of ground states are possible in a rectangle with $r_{Fe-Ctr} = 0.0 - 0.5$ Å and $r_{Fe-N_\epsilon} = 2.0 - 2.3$ Å. The $^5B_{2g}$ state is the ground state in a small region with a large shift of Fe from the center of the porphytin ring. The experimental geometry of Mb falls in this area, and Mb has a quintet ground state [21]. With a short Fe $-$ N$_\epsilon$ distance ($r_{Fe-N_\epsilon} = 2.0$ Å) the ground state is a singlet ($^1A_{1g}$) provided the r_{Fe-Ctr} distance is smaller than 0.45 Å. A relatively small elongation of the axial Fe-N$_\epsilon$ bond results in a triplet (^3E_g) ground state. This is not surprising, since the ECF/CNDO calculation [9] of the four coordinated iron(II) porphyrine yields this ground state, in agreement with *ab-initio* calculations and experiment [25].

The analysis of the electronic states found in the the ECF/CNDO study [20] may have consequences for the interpretation of the binding mechanism. This follows from the accepted idea [5] that slow motions of the protein backbone are transferred to the iron center and can change its spin state and consequently its reactivity towards dioxygen binding. The motions of the axial imidazole ligand may bring the heme to areas in the coordinate space where it has either singlet or triplet electronic state. The binding of CO (singlet) or O_2 (triplet) are then spin allowed processes. The binding kinetics of these small molecules is therefore affected by the rate with which the heme enters the areas in the coordinate space where it has the ground state required for either reaction.

4. Further Developments

Up to this point we have described the ECF method and analyzed the results of its application to the calculation of the *d*-level splitting and *d*-*d* spectra in a series of TMCs. The ECF method allowed us to perform systematic calculations of the crystal field for various ligand environments. The results of these calculations are in fair agreement with the experimental

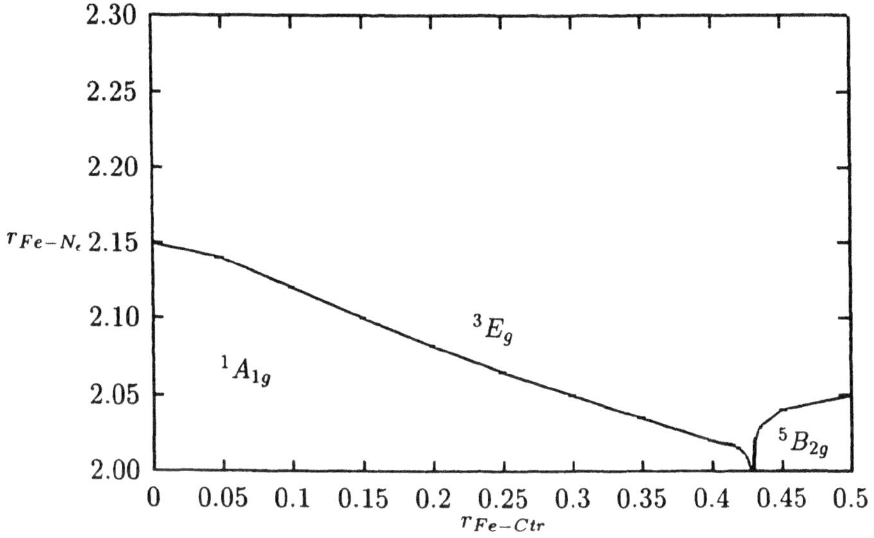

Figure 1. Map of the ground state spin of heme as a function of the Fe-N_ϵ and Fe-Ctr distances.

data, particularly with respect to the spin multiplicity of the ground states of the complexes. Let us turn now to possible extensions of the method.

In the simple version of the ECF/CNDO method the electronic structure of the ligands is treated within the CNDO approximation. A more detailed description of the ligand electronic structure can be obtained with the INDO parameterization [13]. This has been done for a series of octahedral complexes, metalloporphyrins, and chlorocuprates [26]. In all cases the ground states and the excitation spectra of the compounds are in agreement with available experimental data.

A further application of the ECF methodology is to develop a method for the calculation of potential energy surfaces (PESs) of TMCs. This is possible with the representation of the total energy of the TMC in the form [7, 29]:

$$E_n = E_d^{eff}(n) + E_L \tag{13}$$

where $E_d^{eff}(n)$ is the energy of the n'th state of the effective Hamiltonian of eq. (6) for d-electrons, E_L is the energy of the ligands. This expression is general [7] and represents the total electronic energy of the n'th state of a TMC if it is described by the electronic wave function of the form of eq. (3). The CNDO or INDO parameterizations for the ligands are probably of high enough accyracy when we only need the charge distribution in the ligands and the orbital energies at fixed experimental geometries. However, these methods are not accurate enough for geometry optimizations (or more generally for computations of PESs) of TMCs since an accurate core-core repulsion will be required. Therefore in a recent study [27] the SINDO/1 method [28] was used for the calculation of the electronic structure of the ligand sphere. So far the ECF/SINDO1 method has only been tested for a few molecules at their experimental geometries. These results are promising and they provide an accurate description of the ground state spins, symmetries and excitation energies. In cases where the SINDO/1 method can be used for the analysis of the geometry of the ligands the combined ECF/SINDO1 method must also lead to accepatble results in the analysis of complex geometries and PESs. In fact, any semi-empirical method that gives a reasonable description of the PES of organic molecules can be used in combination with the ECF methodology for the *d*-shell of TMAs. Currently, we are testing the possibility to combine the ECF approach with the MINDO/3 parameterization [30] for the ligands.

An alternative to the step by step improvement of semi-empirical descriptions of the ligand's electronic structure for the computation of the PES of TMCs is to use the MM approximation to calculate the ligand's energy E_L. Here the ECF/CNDO or ECF/INDO method is only retained for the *d*-shell electrons [29]. This simply replaces the ligand's energy E_L by E_{mm}, estimated by an MM calculation. The total energy of the n-th electronic state of the complex then has the form:

$$E_n = E_d^{eff}(n) + E_{mm} \tag{14}$$

The energies $E_d^{eff}(n)$ of the *d*-shells are calculated with the ECF/CNDO (or ECF/INDO) method. Two aspects are important here: First, the ECF/MM approach based on eq. (14) allows the use of a single MM potential for the ligands for all electronic states of the metal center. This is because the electronic energy of the *d*-shell is calculated explicitly. This approach does not require different parameters of the MM potentials for different spin states of the complex. The "different ionic radii" for the ions of the different spin states are obsolete (see below). Second, the MM potentials for TMAs extracted from structural data on TMCs cannot be used directly in eq. (14) since they include implicitly effects of the *d*-shells.

The ECF/MM approach has been used to calculate energy profiles of *cis*-$[Fe(bipy)_2(NCS)_2]$ (see also Sect. 3) [29]. The ECF/CNDO method used before [18] must now be completed by the MM potential representing ligand's energy. We used the standard MM2 parameterization for all atoms except iron, for which no bending or torsional terms were used and the bond stretching potential was modeled by a Morse function:

$$E_b^{Fe-N} = D_{Fe-N}(1 - e^{-\alpha_{Fe-N}(r - r^0_{Fe-N})})^2, \tag{15}$$

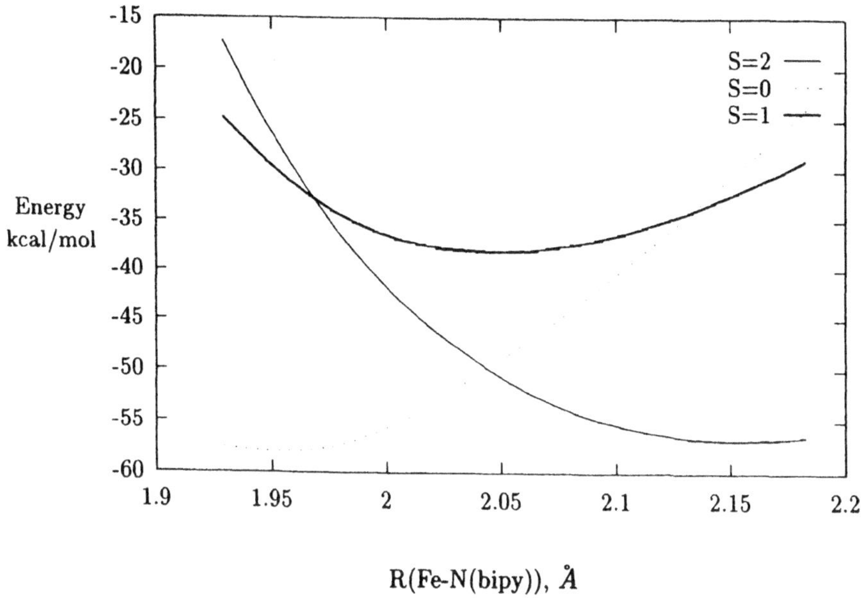

Figure 2. Calculated energy profiles for the spin states of *cis*-[Fe(bipy)$_2$(NCS)$_2$].

The following parameter values were used to reproduce the experimentally determined positions of the PES's minima along the Fe-N bond-length coordinates for the low-spin and high-spin states:

$$r^0_{Fe-N_{bipy}} = 2.035 \text{ Å}, D_{Fe-N_{bipy}} = 70.0 \text{ kcal/mol}, \alpha_{Fe-N_{bipy}} = 2.0 \text{ Å}^{-1};$$
$$r^0_{Fe-N_{NCS}} = 1.975 \text{ Å}, D_{Fe-N_{NCS}} = 40.0 \text{ kcal/mol}, \alpha_{Fe-N_{NCS}} = 3.0 \text{ Å}^{-1}. \quad (16)$$

Results of our calculation are shown in Fig. 2. It is indeed possible to reproduce the whole qualitative picture of the lowest electronic terms derived from experiment [4, 31] with a single MM potential for the ligands and the ECF/CNDO *d*-electron energy. Different Fe-N distances for the hs and ls forms of the complex result from the well balanced contributions of a single ligand potential for each donor (eq. (16)) and the geometry sensitive *d*-shell energy obtained by the ECF/CNDO procedure. The *d*-electron energy is specific for each spin state since it is obtained by diagonalization of the n_d-electron effective Hamiltonian of eq. (6) in each point along the path in the coordinate space. Thus energy profiles such as Fig. 2 may now be *calculated* [29] rather than drawn by hand [4, 31].

We can now return to the discussion of the Mb ground states. With eqs. (13), (14) the PES of Mb can be obtained with an appropriate MM potential. In the vicinity of the boundaries (solid lines in Fig.1) the energy differences between the states are comparable to the spin-orbit interaction.

Thus along these boundaries the ground states contain significant contributions from all states with total spins of 0, 1, and 2. The energy profile along the internal coordinate of a simultaneous shift of the iron atom and the imidazole ligand (with a constant $r_{Fe-N_\epsilon} = 2.0$ Å) is therefore likely to be adiabatic rather than nonadiabatic. That means that for the corresponding vibration anharmonic corrections may be significant. These must be taken into account, for example, in the analysis of protein dynamics by a combination of normal-mode and Mössbauer techniques [32].

5. General Discussion and Conclusions

In the present paper we have demonstrated how the ECF methodology can be applied to the calculation of the *d*-level splitting, *d*-*d* excitation spectra, and PES of TMCs. The ECF method allows the performance of systematic calculations of the crystal field for various ligand environments. The results of these calculations are in fair agreement with experimental data. The basis for this success is a physically justifiable form of the trial wave function of TMCs (eq.(3)).

We will now consider differences between the ECF methodology and other approaches that have been used together with MM of TMA. The angular overlap model (AOM) [33, 34] has to be mentioned here. The effective Hamiltonian for the *d*-shell (eq. (6)) has the form of the standard crystal field Hamiltonian. The covalence term of eq. (10) dominates the *d*-level splitting appears analogous and its form is similar to the AOM parameters e_σ and e_π. However, the differences between the two approaches are very important. The AOM is obtained by applying second order perturbation theory (PT) in H_r (eq. (1)) to obtain one-electron *d*-states of the TMC. This involves the matrix equation for the one-electron MO's of the whole complex and treats the part of the secular equation which corresponds to *d*-*l* mixing. Therefore the estimates for the complex MO's and for the orbital energies are of somewhat lower quality. This also applies to the formalism based on Green's function, reviewed in [35]. Thus the AOM approach is based on the SCF approximation but the MO's are replaced by estimates based on PT. The consequences of the SCF approximation with respect to the numerical procedures (see Sect. 2) are not known since corresponding calculations of the ligand field have not been reported [33, 35].

The ECF method applies second order PT in H_r to the many-electron states of the form of eq. (3). This formal difference leads to apparent differences in the final formulae for the crystal field matrices and to different numerical results. In the AOM, there is no difference between the influence of the occupied and empty *l*-MO's on the position of the *d*-levels and it does not distinguish between electron transfer from and to the *d*-shell. These differences are rather important. The ECF approach considers the energetics of the localized and strongly interacting electrons in the *d*-shell explicitly. This allows the *calculation* of the crystal field as a function of the ligand environment, both in terms of the ligand geometry and the chemical nature. In the AOM the crystal field parameters e_σ and e_π are fitted to experimental spectra. They are specific for each ligand and not transferable, even for constant donor atoms. In the ECF approach the electronic structure of the ligands is taken into account explicitly. Only the magnitude of the resonance (eq. (12)) between orbitals of a given donor atom and *d*-orbitals of a given transition metal is parameterized. The same value is used for all the ligands containing a given donor atom, since the details of the ligand electronic structure are taken into account by the SCF calculation of the ligands.

A Cellular Ligand Field Stabilization Energy (CLFSE) term has been added to the MM energy of TMCs [36]. This term is by definition the sum of one-electron energies of the *d*-levels, obtained by diagonalization of the 5×5 crystal field matrix taken as a sum of cellular contributions from different ligands [33, 34]. The electronic states of the *d*-shell are then simulated by ascribing different occupation numbers to the one-electron *d*-levels obtained by the crystal field matrix diagonalization. This is an approximate procedure in contrast to the ECF approach, where the *d*-electron energy is obtained by exact diagonalization of the effective Hamiltonian H_d^{eff} of eq. (6) in the many-electron basis. However, to the extent to which the electronic structure of TMCs can be described by a single determinant with some definite *d*-level filling the CLSFE/MM method [36] is capable to reproduce characteristic features of the PES of TMCs. For example, the structures of Cu(II) complexes which are dominated by the Jahn-Teller distorsion can be described within the CLFSE/MM approach, due to the fact that the d^9 complexes of Cu(II) can be exactly described on the single determinant level. Nevertheless, some problems may arise with complexes where electron correlations are more important. This is the case with complexes like *cis*-$[Fe(bipy)_2(NCS)_2]$, where spin transitions are possible. In this example it has been shown that the electronic wave function of the low spin form is sufficiently multiconfigurational [37]. It is known that configuration mixing is an important contribution to the stabilization of the low spin configuration of the *d*-shell. The spin isomerism itself is a consequence of a balance between the crystal field strength, electron correlation and the exchange energy. The high spin states can be represented by a single determinant and thus, they are not stabilized by configuration mixing. The crystal field and the CI push into the same direction – they both stabilize the low spin states. For this reason, in any single determinant based approach the crystal field required to achieve the low spin state (when going from the high spin free ion) is stronger than it is if the full CI description of the *d*-shell is used. This may lead to certain misbalances in the parameterization of one-electron schemes.

Acknowledgments

This work has been supported by the Russian Foundation for Basic Research (RFBR) through the grants 96-03-32781, 96-03-32824, and 97-03-42536 and by INTAS through the grant 94-4089. The support from NATO and the Organizing Committee of the Workshop for the visit of the author to San-Miniato is gratefully ackowledged as are the valuable discussions with Prof. G. Náray-Szabó, Prof. E.-W. Knapp, Prof. W. Nowak, Prof. C.R. Landis, Prof. R.J. Deeth, Prof. A. Warshel, and Prof. J. Kozelka.

References

1. Burkert, U., Allinger, N.L. (1982) *Molecular Mechanics*. ACS, Washington.
2. Comba P. (1993) The relation between ligand structures, coordination stereochemistry, and electronic and thermodynamical preperties. *Coord. Chem. Rev.* **123**, 1 - 48; Comba, P., Hambley, T.W. (1995) *Molecular Modeling*. VCH, Weinheim.
3. Hay P.B. (1993) Methods for molecular mechanics modeling of coordination compounds. *Coord. Chem. Rev.* **126**, 177 - 236.
4. König, E. (1991) Nature and dynamics of the spin-state interconversion in metal complexes. *Struct. & Bonding.* **76**, 53 - 152.
5. Perutz, M.F. (1970) Stereochemistry of cooperative effects in haemoglobin. *Nature.* **228**, 726 - 739; (1972) Nature of haem-haem interactions. *ibid.* **237**, 495 - 499; Perutz, M.F., Fermi, G., Luisi, B., Shaanan, B., and Liddington, R.C. (1987) Stereochemistry of Cooperative Mechanisms in Hemoglobin. *Acc. Chem. Res.* **20**, 309-321.
6. Soudackov, A.V., Tchougréeff, A.L., Misurkin, I.A. (1991) Electronic structure and optical spectra of transition metal complexes via the effective hamiltonian method. in *Electron-Electron Correlation Effects in Low-Dimensional Conductors and Superconductors*. Ovchinnikov, A.A. and Ukrainskii, I.I. (eds.) Springer, Berlin; A.V. Soudackov. (1991) Электроная структура комплексов переходных металлов. PhD Thesis. Karpov Institute of Physical Chemistry, Moscow, [in Russian]; Soudackov, A.V., Tchougréeff, A.L., Misurkin, I.A.(1992) Electronic structure and optical spectra of transition metal complexes by the effective Hamiltonian method. *Theor. Chim. Acta.* **83**, 389 - 416.
7. McWeeny, R. (1992) *Methods of molecular quantum mechanics*. AP, London.
8. Russo, N., Salahub, D.R. (eds.) (1996) *Proceedings of the NATO Advanced Study Institute on Metal-Ligand Interactions: Structure and reactivity, Cetraro, Italy, September 4-16, 1994*, Kluwer, Dordrecht.
9. Soudackov, A.V., Tchougréeff, A.L., Misurkin, I.A. (1996) Ground State Multiplicities and d-d Excitations of Transition Metal Complexes by Effective Hamiltonian Method. *Int. J. Quant. Chem.* **58**, 161 - 173.
10. Bethe, H. (1929) Termaufspaltung in Kristallen. *Annalen der Physik.* **3**, 133-215.
11. Jørgensen, C.K. (1971) *Modern aspects of ligand field theory*. North-Holland, Amsterdam.
12. Constanciel, R. (1975) Element analysis of wave functions by group density method. in *Localization and Delocalization in Quantum Chemistry*. Chalvet, O., Daudel, R., Diner, S., and Malrieu, J.-P. (eds.) D. Reidel Publ., Dordrecht.
13. Pople, J.A., Beveridge, D.L. (1970) *Approximate molecular orbital theory*. McGraw-Hill, NY.

14. Soudackov, A.V., Tchougréeff, A.L., Misurkin, I.A. (1994) Расчеты электронной структуры октаэдрических гексааква- и гесааминокомплексов металлов первого переходного ряда методом эффективного гамильтониана. *Журн. Физ. Хим.* **68**, 1256 - 1264 [in Russian]; Calculations on electronic structure of octahedral hexaqua- and hexamino complexes of first row transition metals by the effective Hamiltonian method.*Russ. J. Phys. Chem.* **68**, 1135 [in English].
15. Soudackov, A.V., Tchougréeff, A.L., Misurkin, I.A. (1994) Расчеты электронной структуры хлоридных комплексов металлов первого переходного ряда методом эффективного гамильтониана. *Журн. Физ. Хим.* **68**, 1265 - 1270 [in Russian]; Calculations on electronic structure of chlorocomplexes of first row transition metals by the effective Hamiltonian method. *Russ. J. Phys. Chem.* **68**, 1142 [in English].
16. Lever, A.B.P. (1986) *Inorganic Electronic Spectroscopy*. Elsevier, Amsterdam.
17. Soudackov, A.V., Tchougréeff, A.L., Misurkin, I.A. (1996) Crystal Field Splittings and Optical Spectra of Transition Metal - Mixed Ligand Complexes by Effective Hamiltoninan Method. *Int. J. Quant. Chem.* **57**, 663 - 671.
18. Tchougréeff, A.L., Soudackov, A.V., Misurkin, I.A., Bolvin, H., and Kahn, O. (1995) High-Spin - Low-Spin Transitions in Fe(II) Complexes by an Effective Hamiltonian Method. *Chem. Phys.* **193**, 19 - 26.
19. Gallois, B., Real, J.-A., Hauw, C., and Zarembowitch, J. (1990) Structural changes associated with the spin transition in $Fe(phen)_2(NCS)_2$: A single crystal X-ray study. *Inorg. Chem.* **29**, 1152 - 1157.
20. Tchougréeff, A.L. (1997) Спиновые переходы в геме и кинетика связывания CO и O_2 миоглобином. *Хим. Физика.* Accepted. [in Russian]; Spin states of the heme and kinetics of CO and O_2 binding by myoglobin. *Chem. Phys. Reps.* Accepted. [in English].
21. Frauenfelder, H., Wolynes, P.G. (1985) Rate theories and puzzles of hemeprotein kinetics. *Science.* **229**, 337 - 345.
22. Springer, B.A., Sligar, S.G., Olson, J.S., and Phillips, G.N. (1994) Mechanisms of ligand recognition in Myoglobin. *Chem. Rev.* **94** 699 - 714.
23. Landau, L.D., Lifshitz, E.M. (1989) *Квантовая механика.* Наука, M. [in Russian]; (1959) *Quantum Mechanics.* Pergamon Press, London [in English].
24. Rawlings, D.C., M. Gouterman, M., Davidson, E.R., and Feller, D. (1985) Theoretical investigations of the electronic states of porphyrins. III. Low-lying electronic states of porphinatoiron(II). *Int. J. Quant. Chem.* **28**, 773 - 796.
25. Dedieu, A., Rohmer, M.-M., Veillard, A. (1982) *Ab-initio* calculations of metalloporphyrins. *Adv. Quant. Chem.* **16**, 43 - 95.
26. Tokmachev, A.M., Tchougréeff, A.L. (1997) Эффективное кристаллическое поле для расчета возбуждений d-электронов в комплексах переходных металлов. *Хим. Физика.* Accepted. [in Russian]; Effective crystal field for calculation of d-electron excitation in transition metal complexes. *Chem. Phys. Reps.* Accepted. [in English].
27. Soudackov, A.V., Jug, K. (1997) Effective Hamiltonian – crystal field on the INDO level: Calculations of d-d spectra of some iron(II) compounds. *Int. J. Quant. Chem.* **62**, 403 - 418.
28. Nanda, D.N., Jug, K. (1980) SINDO1. A semiempirical SCF MO method for molecular binding energy and geometry. Approximations and Parametrization. *Theoret. Chim. Acta.* **57**, 95 - 106.
29. Darkhovskii, M.B., Tchougréeff, A.L. (1997) Об объединении квантовой химии и молекулярной механики. ППЭ комплексов переходных металлов на примере спинового перехода в *cis*-$[Fe(bipy)_2(NCS)_2]$. *Хим. Физика.* Accepted. [in Russian]; On unification of quantum chemistry and molecular mechanics. PES of transition metal complexes as exemplified by spin transition in *cis*-$[Fe(bipy)_2(NCS)_2]$. *Chem. Phys. Reps.* Accepted. [in English].

30. Bingham, R.C., Dewar, M.J.S., Lo, O.H. (1975) Ground states of molecules. XXV. MINDO/3. An improved version of the MINDO semiempirical SCF MO method. *J. Amer. Chem. Soc.* **97**, 1285 - 1294.
31. Gütlich, P. (1981) Spin crossover in iron(II)-complexes. *Struct. & Bonding.* **44**, 83 - 195.
32. Melchers, B., Knapp, E.-W., Parak, F., Cordone, L., Cupane, A., and Leone, M. (1996) Structural fluctuations of myoglobin from normal-modes, Mössbauer, Raman, and absorption spectroscopy. *Biophys. J.* **70**, 2092 - 2099.
33. Gerloch, M., Wooley, R.G. (1984) The Functional Group in Ligand-Field Studies: The Empirical and Theoretical Status of the Angular Overlap Model. *Progr. Inorg. Chem.* **31**, 371-446.
34. Schäffer, C.E. (1973) Two Symmetry Parameterizations of the Angular-Overlap Model of the Ligand-Field. *Struct. & Bonding.* **14**, 69-110.
35. Kikoin, K.A., Fleurov, V.N. (1994) *Transition Metal Impurities in Semiconductors.* World Scientific, Singapore.
36. Burton, V.J., Deeth, R.J., Kemp, C.M., and Gilbert, P.J. (1995) Molecular mechanics for coordination complexes: the impact of adding d-electron stabilization energies. *J. Amer. Chem. Soc.* **117**, 8407-8415.
37. Tchougréeff, A.L. unpublished result.

QUANTUM CHEMICAL STUDIES OF TRANSITION METAL CATALYZED ENZYME REACTIONS

PER E.M. SIEGBAHN
Department of Physics
Stockholm University, Box 6730
S-113 85 Stockholm, Sweden

1. Introduction

The treatment of an enzymatic reaction containing transition metals by high accuracy quantum chemical methods was until very recently considered an almost impossible task. Therefore, the appearance of such calculations last year [1,2] was quite unexpected and can be regarded as a major breakthrough in the area of quantum chemistry. New important areas for quantum chemical applications have suddenly been opened and it can be expected that a very large number of biochemical applications on transition metal catalyzed enzymatic reactions will appear in the coming years. These applications will not only change the area of applied quantum chemistry but are also likely to have a large impact in biochemistry, leading to a new understanding of many enzymatic reactions. In this presentation an overview of the recent work done in Stockholm will be made, with examples mainly taken from the chemistry of Methane Monooxygenase (MMO), of Photosystem II (PSII) and of Ribonucleotide Reductase (RNR).

There are several reasons for the sudden breakthrough of quantum chemistry in the area of transition metal enzyme chemistry, but the major one is certainly the development of parameterized hybrid Density Functional Theory (DFT) methods. DFT methods of reasonable accuracy have been around for almost twenty years but they were usually considered as inferior in accuracy and less reliable than standard quantum chemical *ab initio* methods. The main advantage with the DFT methods was instead that they were very fast. The first major improvement of the accuracy of the DFT methods came with the in-

L. Banci and P. Comba (eds.), Molecular Modeling and Dynamics of Bioinorganic Systems, 233–253.

troduction of gradient terms of the density [3-6] but the main and final step was taken when Hartree -Fock exchange was also included, and when a few semi-empirical parameters were introduced [7]. The decision to parameterize the DFT methods was taken at a time when it was already clear that even *ab initio* methods need to be parameterized to achieve high accuracy for most problems of chemical interest [8]. The hesitation to take the step towards a parameterized theory should be seen in light of the rather bad experience from the heavily parameterized semi-empirical schemes developed and used previously. It should therefore be emphasized that the present parameterized schemes are very different in nature compared to the old schemes and only use a quite limited number of parameters. Also, these parameters only affect the results to a small, but quite essential, extent. The method used in the present applications is the B3LYP method of Becke [7] which contains three semi-empirical parameters. It has been very carefully tested against experiment for standard benchmark tests for molecules containing first and second row atoms [7,9] and has been shown to be highly competitive in accuracy compared to the most accurate, parameterized *ab initio* methods. A large number of comparisons have also been made recently for transition metal containing systems, where the accuracy is also found to be high, although slightly lower than for molecules with only lighter atoms [10].

The development of parameterized DFT methods is not the only reason why successful applications on transition metal containing enzyme reactions have been made the past year. This development has become possible also because the quantum chemical understanding of transition metal complexes has grown sufficiently high that such applications are meaningful. Even though the DFT methods are fast, the enzyme centers have to be modeled by simpler systems to be computationally tractable. As will be shown in this presentation the models used are quite different from what would have been selected based on previous chemical knowledge. Instead, they are built on the experience gained over the past decade on computational modeling of transition metal complexes [11]. Modeling aspects concerned are, for example, the choice of ligands and the spin states. In fact, if the total spin states determined experimentally would have been used directly in the present study, progress in understanding the mechanisms of these reactions would have been very difficult. Also, if the ligands determined experimentally would have been used, the calculations would have taken so long that little progress would have been achieved either. The present model studies are characterized by a very large number of models and calculations. Typically, up to a hundred different structures have been fully optimized in each of the projects discussed below.

The final aspect that will be emphasized here in connection with the recent development is the new and quite different understanding of the chemistry that has evolved during recent years of quantum chemical modeling. In this presentation this aspect is most obvious when the subject of electron transfer is discussed. For example, for Ribonucleotide Reductase (RNR) an entirely new mechanism is suggested where, instead of an electron transfer, a sequence of hydrogen transfers is suggested [12]. With this mechanism the need for long range electron transfer is eliminated in many biochemical situations. This new type of understanding is not connected to the above mentioned development of DFT methods but is based on a large number of model studies performed in the last ten years. This is perhaps the most important aspect of the development taking place and it is predicted here that this type of model thinking will have a strong influence on the understanding of biochemical reactions in the future.

2. Computational details

The calculations discussed in this presentation were performed in two steps. First, an optimization of the geometry was performed using the B3LYP method with rather small, double zeta, basis sets. In the second step the energy was evaluated for the optimized geometry using very large basis sets including diffuse functions and with two polarization functions on each atom. The final energy evaluation was also performed at the B3LYP level. All calculations were performed using the GAUSSIAN-94 program [13].

The B3LYP functional can be written as [7,14],

$$F^{B3LYP} = (1-A)*F_x^{Slater} + A*F_x^{HF} + B*F_x^{Becke} + C*F_c^{LYP} + (1-C)F_c^{VWN}$$

where F_x^{Slater} is the Slater exchange, F_x^{HF} is the Hartree-Fock exchange, F_x^{Becke} is the gradient part of the exchange functional of Becke [3,7],

F_c^{LYP} is the correlation functional of Lee, Yang and Parr [4] and F_c^{VWN} is the correlation functional of Vosko, Wilk and Nusair [5]. A, B and C are the coefficients determined by Becke [7] using a fit to experimental heats of formation. However, it should be noted that Becke did not use F_c^{VWN} and F_c^{LYP} in the expression above when the coefficients were determined, but used the correlation functionals of Perdew and Wang instead [6].

The B3LYP energy calculations were made using the large 6-311+G(2d,2p) basis sets in the GAUSSIAN-94 program. This basis set has two sets of polarization functions on all atoms including two f-sets on the metals, and also additional diffuse functions. In the B3LYP geometry optimizations a much smaller basis set, the LANL2DZ set of the Gaussian-94 program, was used. For the transition metal atoms this means that a non-relativistic ECP according to Hay and Wadt [15] was used. The valence basis set used in connection with this ECP is essentially of double zeta quality. The rest of the atoms are described by standard double zeta basis sets.

3. Results and discussion

The present discussion will be divided into three subsections, each one discussing a separate biochemical problem studied recently. The first example is from a study on Methane Monooxygenase (MMO) [1], where an iron dimer center is the active site. In the second example, the chemistry of Photosystem II (PSII) is discussed both concerning hydrogen abstraction [2] and oxygen evolution. Manganese dimer complexes are used to model the active Mn_4 site. Finally, in the third example a hydrogen transfer model is suggested [12] rationalizing the chemistry occurring in Ribonucleotide Reductase (RNR).

3.1. METHANE MONOOXYGENASE

Methane monooxygenases are a group of enzymes which convert methane to methanol via a monooxygenase pathway in which the dioxygen molecule is activated via reduction :

$$CH_4 + O_2 + NADH + H^+ \longrightarrow CH_3OH + H_2O + NAD^+ \qquad (1)$$

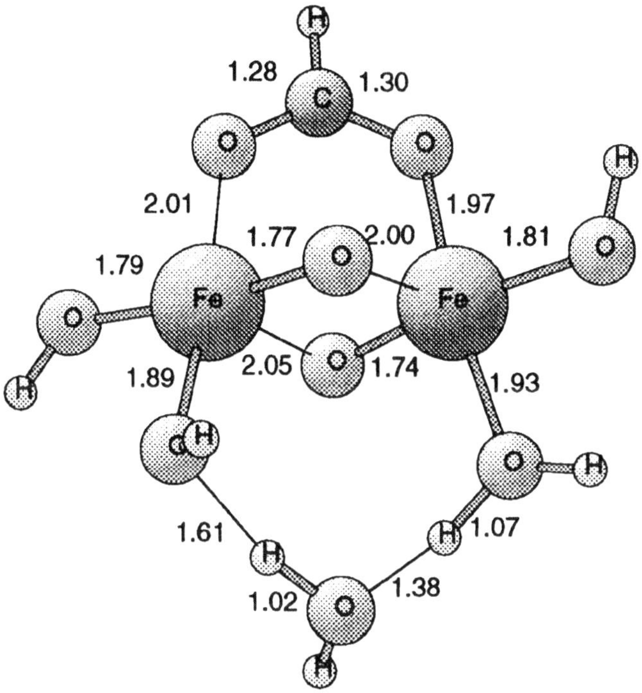

Figure 1: The optimized [9]A Fe(IV)-(μ-O)$_2$-Fe(IV) structure, modeling compound **Q** of Methane Monooxygenase.

The X-ray structure of the 251 kD hydroxylase from *M. capsulatus* [16], which has recently become available has a dinuclear iron center bridged by a hydroxide, a glutamate, and an acetate from the buffer, where waters are expected to be bound in vivo. The terminal ligands are mostly oxygen derived but there is a histidine ligand on each iron center. Experiments have indicated that the active species in the reaction with methane, commonly denoted compound **Q**, probably has bridging μ-oxo groups and with both irons having oxidation state IV [17,18]. The complex is diamagnetic. To model this compound, a complex shown in Fig.1 was adopted with only oxygen derived ligands for simplicity.

The geometry obtained after a B3LYP optimization, shown in Fig.1, is quite surprising. First, both iron centers are only five coordinated. In fact, models with 6-coordinated starting structures for the optimization, quickly rearrange to adopt 5-coordination. The second, and perhaps even more surprising feature, is that the two oxo groups are asymmetrically placed between the Fe atoms so that the two M-O (oxo) bond lengths for each metal are substantially unequal. This is very different from what is known from similar structures, most notably Mn(III) dimeric complexes.

The selection of spin state for the complex in Fig.1 is an extremely important modeling aspect. In fact, if the experimental result that the complex is diamagnetic is followed, leading to a spin state of zero, it is unlikely that any progress had been obtained in explaining the chemistry of MMO. The reason is that there are many possible singlet states and only one of them is the correct one. Convergence to this particular state is very difficult. If, on the other hand, the spin-coupling is modeled as *ferromagnetic* rather than *anti-ferromagnetic*, there is only one possible state and convergence is much easier. Model calculations show that the energy loss is very small using the former coupling. The second important model assumption is that the simple ligands chosen actually are reasonable models of the true, mostly amino acid, ligands. This rather drastic type of modeling has been shown to be reasonable by a very large number of model studies performed during the last ten years [11]. In the present case of MMO, both these critical model assumptions are shown to be correct as demonstrated by the results obtained (see below).

In the Fe_2(IV,IV) species of Fig.1, the structure can be regarded as being derived from an octahedron, but showing strong Jahn Teller (JT) distortions. The JT axis, along which the M-L bonds are elongated, is not the axis normal to the M-(μ-O)$_2$-M plane, as in the known Mn(III) cases, but is the axis defined by the long M-oxo bond and the vacant site trans to it; this vacancy can be interpreted as an extreme case of JT distortion leading to the departure of the H_2O ligand, assisted by the relatively strong second sphere hydrogen bonding that it can engage in when no longer bound to iron. Also seen in the figure are the expected four short M-L bonds in the plane normal to the JT axis. Of these four short bonds, one is formed to a bridging oxo. The quantum chemical studies indicate that only the structure shown in Fig.1 has any significant stability–no other Jahn-Teller distortional isomers were found.

An alternative way to view the structure of the diiron cluster is that the JT axis along which M-L bonds are elongated is automatically defined by the position of the ligand lost from the cluster. In this way the enzyme can control the orientation of the JT axis, because bound waters may be lost easily, but the ligating groups derived from the polypeptide are much less likely to be lost. The bridging carboxylate present both in the model in Fig.1 and in the actual complex of MMO, forces the ligand lost to be in the Fe-(μ-O)$_2$-Fe plane and therefore also the JT axis to be in this plane. This causes the large asymmetry in the bridging Fe-oxo distances, which in turn has important consequences for the reactivity of the complex. This aspect of the structure/function relationship could not have been postulated from a knowledge of the behavior of other d^4 ions such as Mn(III) but, in the absence of structural data, could only come from quantum chemical studies.

After the work described above was finished a striking confirmation of the structure in Fig.1 appeared [19]. New EXAFS data on intermediate **Q** of MMO was interpreted in terms of an Fe(IV)-(μ-O)$_2$-Fe(IV) diamond core structure with asymmetric bridging oxo groups, essentially identical with the intermediate proposed here. Fe-O (oxo) distances of 1.77 Å and 2.05 Å were reported. Even though ferromagnetic coupling and simple water and hydroxyl ligands are used, a structure is thus obtained from the calculations which in the most critical region for the chemistry is almost identical to the true structure.

The adequacy of using the structure in Fig.1 as a model for compound **Q** is further demonstrated by the resulting chemistry as obtained from the calculations. A rather low barrier of 6.3 kcal/mol for the initial reaction with methane is obtained in line with experimental findings. The transition state structure indicates a hydrogen abstraction type reaction in line with isotopic labeling experiments. An iron-methyl intermediate is then quickly formed explaining the short life-time of the methyl radicals in radical clock experiments. A very similar energy was finally obtained for the Fe(III)-O_2-Fe(III) peroxo structure compared to compound **Q** in line with the easy rearrangement between these structures as required by the reaction. Further details of this work can be found in Ref.1.

3.2. PHOTOSYSTEM II

Photosystem II (PSII) is a large enzyme complex located to the thylakoid membranes in green plants, algae and cyanobacteria. PSII is a key enzyme in the biosphere in that it catalyzes the light-driven electron transfer from water to the membrane bound electron carrier plastoquinone. In the process, two molecules of water are oxidized to molecular oxygen,

$$2H_2O \longrightarrow 4H^+ + 4e^- + O_2 \tag{2}$$

This process occurs in a complex five step reaction, known as the Kok or S-state cycle. Thus, the water oxidizing complex in PSII provides the photosynthetic organisms with an unlimited source of electrons and protons. The formation of O_2 is catalyzed by an Mn-cluster comprised of 4 Mn atoms that are held together by di-μ-oxobridges and carboxylate bridges, and a tetrameric structure was recently proposed based mainly on EXAFS and EPR data [20].

After the absorption of a light quantum, an electron is rapidly ejected from the primary electron donor P680 (a dimer or multimer of chlorophylls) to the electron acceptor system. The oxidized donor $P680^+$ is rapidly reduced (in nanoseconds) by a nearby (10-15 Å) tyrosyl residue denoted Tyr_Z, which functions as an interface to the Mn-cluster where the water is ultimately oxidized. The mechanism for oxidation of the manganese cluster in PSII has earlier been assumed to occur by simple electron-transfer steps to the nearby oxidized Tyr_Z^{ox} radical, but recent results obtained from EPR studies of the Tyr_Z radical in PSII are reported to be inconsistent with a pure electron transfer function of this radical. It was therefore suggested that Tyr_Z acts as a hydrogen abstractor, mediating electron transfer between the manganese cluster and the photooxidized chlorophyll in Photosystem II. The mode of action was suggested to be oxidation of the manganese cluster by abstraction of hydrogen atoms from coordinated water or hydroxide [21,22].

To test the above hypothesis, that hydrogen atoms are abstracted by the tyrosyl radical from the manganese cluster complex, a set of calculations were performed for different monomeric and dimeric manganese models. The key question is whether the O-H bond energy for a coordinated water or hydroxyl group is weak enough for a hydrogen to be abstracted by the Tyr_Z radical. The experimental O-H bond strength of phenol is 86.5 $\pm$ 2 kcal/mol to be compared to the calculated value of 83.1 kcal/mol. This error of about 3 kcal/mol is systematic and is the same for water. The experimental O-H bond strength in water is 118.0 kcal/mol and the calculated value 114.4 kcal/mol. The calculated value for tyrosine, 82.2 kcal/mol, is quite similar to the value in phenol, as expected. For the hydrogen abstraction model to be valid the coordination of water to manganese should thus reduce the O-H bond strength by about 30 kcal/mol, which has been questioned by many workers in the field. However, the calculations are very much in line with the possibility for hydrogen abstraction. Starting with a monomeric Mn(III)-(H_2O) model and with only water and hydroxyl ligands, the first O-H bond strength is only 1.1 kcal/mol (O-H bond strength 83.3 kcal/mol) stronger than the one in tyrosine and the second O-H bond strength only 1.3 kcal/mol (O-H=83.5 kcal/mol) stronger. Going to a more realistic dimer model does not change these results significantly. The corresponding O-H bond strength differences are 2.1 kcal/mol (O-H=84.3 kcal/mol) and 2.8 kcal/mol (O-H=85.0 kcal/mol), for a Mn(IV)-(μ-O)$_2$-Mn(III)-(H_2O) model. The dimer model is expected to be quite realistic since the actual Mn_4 cluster can be regarded as two rather loosely coupled dimers. It can be concluded that the calculations predict almost thermoneutral hydrogen abstractions, which is actually a requirement for the reaction since the overall reaction energy of the catalytic cycle is quite small. It is obvious that a high endothermicity would have violated the hydrogen abstraction model but a too high exothermicity would also have violated the model since this would have required a large endothermicity in some other step, such as in the O_2 formation step. Further details of this work can be found in Ref.2.

To obtain the mechanism for O_2 formation in PSII turns out to be a substantially more difficult task than to obtain a reasonable picture for the hydrogen abstraction. It is also much harder than to obtain the reaction mechanism of MMO, discussed above. The main reason for this is that not much is known about the structure and oxidation states of the Mn_4 cluster complex. The EXAFS data suggest a model with two fairly parallel Mn-dimers connected in one of the ends (the inner end) to each other by carboxylate bridges and rather far away from each other in the other end (the outer end). The hypothesis [22] used in the present work is that water molecules are initially bound to the outer manganese atoms. The Tyr_Z radical then abstracts hydrogen atoms, one by one in each S-state of the Kok cycle, from these water molecules. At some stage towards the end of the Kok cycle the outer ends meet and the O-O bond formation can take place between the oxygens of the initial water molecules. An important additional piece of information is that the S_1 state lacks an EPR signal. This leaves essentially only two possibilities for the oxidation states of the Mn atoms in the cluster in the starting S_0 state. The first, and most likely possibility is that one of the dimers is Mn_2(III,III) and the other dimer Mn_2(IV,III). The other possibility is Mn_2(IV,II) for one dimer and Mn_2(IV,III) for the other one.

To test the above reaction hypothesis several sets of calculations have been performed for the O-O bond formation, so far not leading to any definite conclusion on the mechanism. In these test calculations monomer models have been used for each of the dimers. In the first set of calculations the oxygens of two Mn(V)=O oxo-monomers form the O-O bond. In the second set, one oxygen from an Mn(IV)=O and one from an Mn(V)=O oxo-monomer react. In the third set, one Mn(V)=O and one Mn(IV)-(OH) meet to form an O_2H peroxide group. A point close to the transition state for this bond formation is shown in Fig.2. In the fourth set finally, one Mn(IV)=O and one Mn(IV)-(OH) react to form a peroxide group. In all cases, too high barriers for O-O bond formation have been found. The experimental barrier height is about 10 kcal/mol and barriers in the range 17-28 kcal/mol were obtained. There is one reason in common for the high barriers in three of these cases and this is the difficulty to reach a point where an oxyl group (oxygen radical) is formed. Very early in the reactions the Mn=O oxo-bond needs to be promoted to the reactive Mn-O· oxyl-bond and this costs too much energy in these cases. The only case where the promotion to the oxyl-state is not exceedingly high is for the

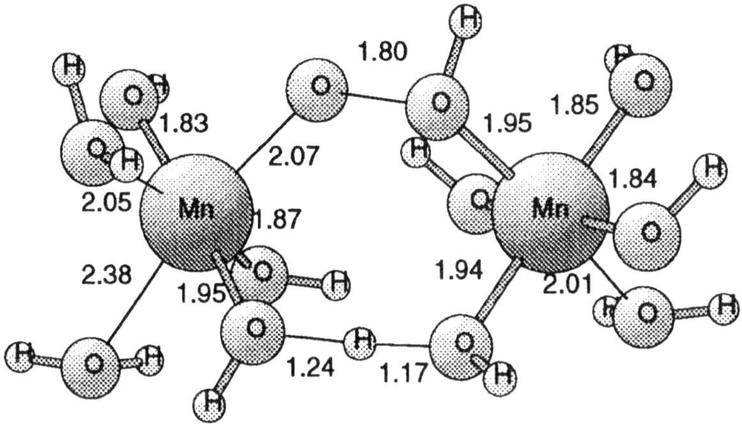

Figure 2: The transition state for O-O bond formation for an Mn(V)=O... OH-Mn(IV) dimer model.

Mn(V)=O...(OH)-Mn(IV) case shown in Fig.2, where the excitation energy is only 7 kcal/mol, but in this case there is a substantial barrier at a later stage leading to a total barrier height of 28 kcal/mol.

In order to attack the problem of reaching the reactive oxyl state, another set of calculations were performed based on the results from the MMO reaction. In that case, a Jahn-Teller effect is of key importance, as described above. In order to obtain a similar JT effect for an Mn_2 dimer a bridging carboxylate was introduced, forcing the weak JT axis to be in the Mn-(μ-O)$_2$-Mn plane. At least one of the manganese atoms has to be in the Mn(III) oxidation state, since this is the JT unstable state. A large number of different dimer complexes were optimized, the most interesting one shown in Fig.3. As seen in the figure, this structure is JT-distorted in an even more dramatic way than compound **Q** in Fig.1. In fact, one of the bridging Mn(V)-oxo bonds is now so short, 1.66 Å, that it is almost equal to the double

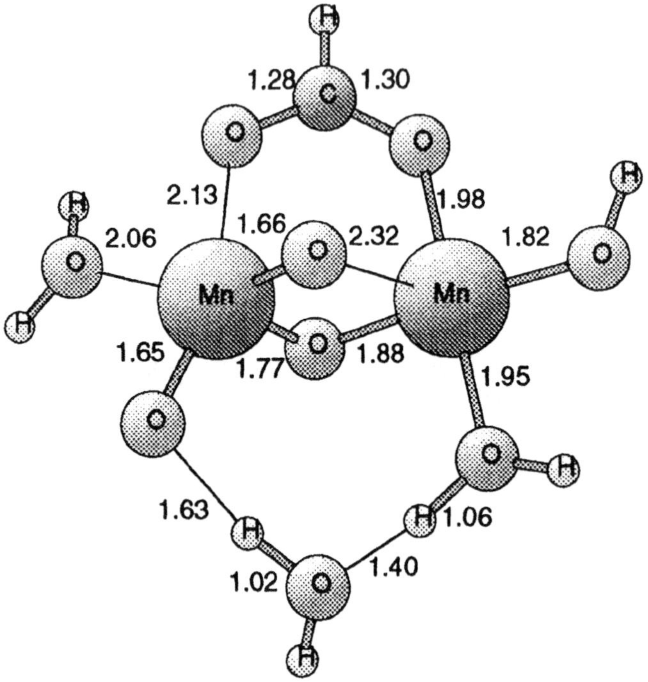

Figure 3: The Jahn-Teller distorted O=Mn(V)-(μ-O)$_2$-Mn(III) dimer model.

bonded Mn(V)=O oxo-bond of 1.65 Å. This bridging oxygen has a very long distance to the JT unstable Mn(III) center of 2.32 Å. The site on the other side of the Mn(III) atom is empty in line with the compound **Q** structure of MMO. The most interesting consequence of this Mn(V)=O oxo structure is that the corresponding Mn(IV)-O· oxyl structure is only 8 kcal/mol higher in energy. The reason for this low excitation energy is that the radical spin is spread out on two oxygens, the oxygens with the shortest Mn-O bond distances in Fig.3. This result becomes even more interesting as the leftmost Mn atom is made 6-coordinated, which could be forced by the protein environment. In this case the oxyl state is actually more stable than the oxo state, since there is a very strong trans effect making the site opposite to the Mn=O oxo bond quite unfavorable.

However, the O-O bond formation problem in PSII is not solved yet, even after finding the interesting structure in Fig.3. The major remaining problem is how to reach the reactive oxyl state by hydrogen abstraction. The Mn(V)=O state has to be reached by hydrogen abstraction from an Mn(IV)-(OH) state. Since the Mn(IV) state is the most stable Mn-state this will lead to a high energy for hydrogen abstraction. If, on the other hand, the Mn(IV)=O oxo state is reached it is expected that the excitation energy to reach the Mn(III)-O· oxyl state should be high since it is JT unstable. Further work on this reaction is in progress where, for example, a possible role of chlorine is investigated. Chlorine is known to be necessary for O_2 formation.

3.3. HYDROGEN TRANSFER

The final subject discussed in this brief review of recent work, concerns the general phenomenon normally referred to as Electron Transfer (ET). In particular, the occurence of long range (> 5 Å) electron transfer, quite often postulated in biochemical reaction sequences, is critically examined. If an electron transfer model is assumed in the enzyme RNR, for example, ET is implied to occur over a distance as long as 35 Å [23]. Very long range ET is also normally advocated in enzymes like nitrogenase and hydrogenase. In a recent computational study a different, and much simpler, mechanism is suggested in which hydrogen atoms instead are effectively transferred in a sequence of steps, each one requiring only very short range electron transfer on the order of one Å [12].

The enzyme RNR can serve as a typical example of where the present model study is applicable. The following experimental observations give the essential background. A neutral tyrosyl radical has been observed in this enzyme in its resting state [24]. This radical is created by an iron dimer complex in a process very similar to the one described above for MMO. The tyrosyl radical is involved in a hydrogen bonding chain of amino acids with a cysteine residue at the other end of the chain, 34 Å away. The other amino acids in this chain are three additional tyrosines, a histidine, an aspartic acid, a glutamic acid and a tryptophane. The iron dimer complex is also part of the chain. In the active state of the enzyme the radical is transferred from tyrosine to cysteine and this is the process of interest here. Conventionally, this

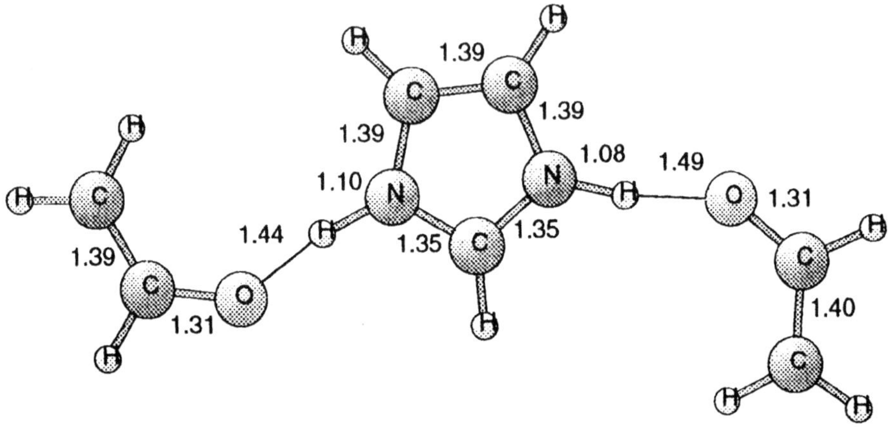

Figure 4: A typical transition state for Proton Governed Hydrogen Transfer (PGHT).

is thought to occur by a long range ET. In the computational model study of the radical transfer, simpler models were used for the amino acids. For example, for tyrosine a vinyl alcohol was used and for histidine an imidazole was used. In order to be as general as possible, the hydrogen transfers between pairs and triples of amino acids were identified by types. Models considered include, for example, cases where the amino acid radical is a Σ or a Π state (i.e. has the unpaired electron in a σ or a π orbital), and cases where the intermediate residue is an acid or a base. Water is also considered as a possible intermediate group. The model reactions studied are all such that the net result of the reaction is a transfer of one neutral hydrogen atom.

When the results of the model calculations are summarized, mainly two different types of hydrogen transfers can be identified. A key feature of these mechanisms is that charge separation is always kept small in contrast to the previous electron transfer mechanisms. A typical example of the first type of mechanism is shown in Fig.4. This figure shows a hydrogen transfer from the tyrosine to the left over to the tyrosyl to the right, with a histidine in between. A typical feature of this mechanism is that the system is planar. The hydrogen transfer occurs in the following way. First, a proton is transferred from tyrosine over to histidine leading to the transition state in the figure. In this transition state the odd π-electron is symmetrically delocalized over the system. From this point another proton moves from the histidine over to the other tyrosine and the odd π electron becomes localized on this tyrosine, to complete the hydrogen transfer and thereby the radical transfer. The barrier height for hydrogen transfer is 8.3 kcal/mol and the reaction is obviously thermoneutral. Since the first key step is a proton transfer, this mechanism is termed Proton Governed Hydrogen Transfer (PGHT).

The second main mechanism for hydrogen transfer is shown in Fig.5. This again shows two tyrosines, but with a water as an intermediate group in this case. The main chemical differences between the systems in Fig.4 and Fig.5 in this context is that histidine, in contrast to water, has a π system to match the tyrosine π systems and that histidine is more basic than water. These differences lead to entirely different types of transition states in the two cases. Rather than being a planar system, the tyrosines in Fig.5 are arranged such that the π systems are on top of each other so that there is some π overlap. For real tyrosines this means that the planes of the rings are parallel. The distance between the tyrosines is 2.56 Å. The hydrogen transfer mechanism in this case is more of a one step reaction with the proton passing over the water and the electron going directly between the π systems. The barrier height for this hydrogen transfer is 5.4 kcal/mol. Since the transition state is arranged to allow for overlap between the π systems this mechanism is termed Overlap Governed Hydrogen Transfer (OGHT).

The minimization of charge separation is a key feature of the above mechanisms. The assistance of protons in the electron transfer has been suggested previously and is termed Proton Coupled Electron Transfer (PCET) [25], and this is an essential part also of the present mechanism. The main, and chemically very significant, difference between PCET on the one hand and PGHT and OGHT on the other, occurs

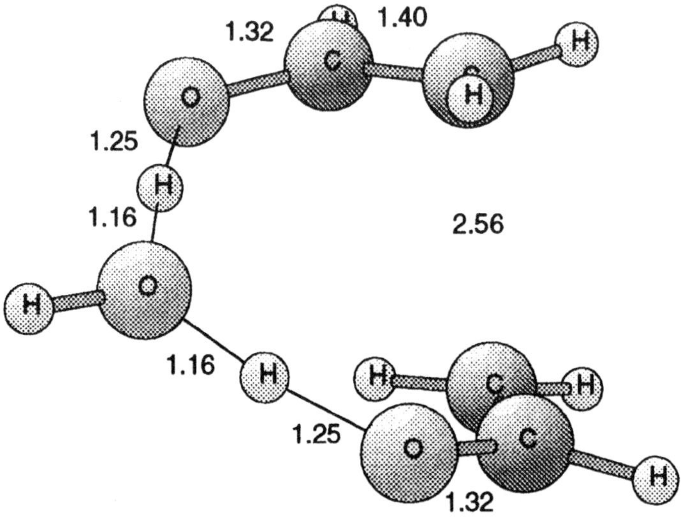

Figure 5: A typical transition state for Overlap Governed Hydrogen Transfer (OGHT).

after the first proton transfer. From the reactants and to the transition state in Fig.4, the mechanisms are thus basically the same. After this point the PCET mechanism implies that the electron is transferred from one tyrosine to the other, and most importantly, the proton must *go back* to the original position. Otherwise this would not be an electron transfer. If the proton is allowed *to follow* the electron, as in the PGHT mechanism, this leads to a hydrogen atom transfer instead. Inspection of the transition state in Fig.4 makes this a rather obvious suggestion, with charge neutralization as the driving force. It is clear that the energy for the neutral system is much lower than for the system with charge separation. An important consequence of this trivial suggestion is that for the following steps of the long range transfer, the situation is essentially identical to the first step for PGHT and OGHT and these steps can therefore occur in the same way. This is not so for PCET where proton assitance is not as easy for the following steps, since there will be one more hydrogen atom present when a situation like the one in Fig.4 is encountered again. This means that for the present mechanisms of hydrogen transfer, long range radical transfer is easily completed. From the cysteine to the tyrosyl radical in RNR, this means about six separate steps in which the radical moves altogether 34 Å. It should be pointed out that each hydrogen atom only moves a small distance of about one Å, and that the maximum charge separation is of a similar small magnitude.

The work to find support for the above mechanisms has only just started but several protein structures and experiments are in line with these findings. For example, it has been shown that electron transfer between cysteine and tyrosine does not occur in RNR if the hydrogen bonded network is destroyed in some way. If the tyrosines next to each other in the RNR chain are exchanged by phenylalanines by mutation, the electron transfer is stopped [26]. This means that even though the π ring systems are still there, which might be considered enough for electron transfer, there is still no electron transfer because the hydrogen bonding proton is missing in phenylalanine. Another interesting aspect of the RNR structure is that two neighboring tyrosines in the RNR chain are arranged with their rings parallel to each other, very similar to the optimal arrangement for the OGHT transition state in Fig.5. When the present study was performed [12] it was initially considered as quite surprising that two tyrosines are placed next to each other in the RNR chain without any intermediate group in between them compared to Fig.5. This was considered surprising since the calculations clearly show that this leads to a very high barrier for hydrogen transfer. The barrier is lowered considerably by the presence of a bridging water, because in this way the costly bending of the O-H bond out of the tyrosine ring plane is avoided. However, more detailed experimental structures including water, which were not published and therefore not available at the time of the theoretical study, have shown that there is indeed a water molecule present at the desired position between the tyrosines [27]. A similar arrangement as in Fig.5 is also present in Cytochrome C Peroxidase (CCP) where a tryptophane radical has been found, with a parallel arrangement between a histidine and a tryptophan, and with a bridging aspartic acid in between [28], suggesting an OGHT mechanism for radical transfer.

4. Conclusions

In this presentation three different projects performed during the last year have been briefly described. These projects serve to illustrate the very rapid development taking place in this area of high accuracy quantum chemical applications in biochemistry. In particular, it is now possible to perform reliable calculations on reactions involving enzymes containing transition metals. The availability of carefully parameterized DFT methods with a very small number of parameters is one key to the possibility to perform these studies. Another important factor for this progress is the increased understanding of quantum chemical models for transition metal complexes obtained during recent years. Important simplifications of the actual active complexes are necessary in order to allow for an accurate treatment of a sufficiently large number of different models. Typically, hundreds of structures have been optimized in each one of the projects described above. Another key for the progress has been the understanding of the spin-coupling in the dimeric complexes, showing that it is possible to model these by ferromagnetic coupling. A recent verification of the computed structure for compound **Q** of MMO is perhaps the best illustration of the adequacy of the methods and models used. EXAFS measurements confirm the highly JT distorted diamond structure with substantially unequal iron-oxo bond lengths [19]. The rapid development in this area is expected to continue the coming years with a large number of mechanistic studies of transition metal catalyzed enzyme reactions. At present, studies are almost finished also for nitrogenase, tyrosinase and hydrogenase.

References

1. Siegbahn, P.E.M., and Crabtree, R.H. (1997) The mechanism of C-H activation by di-iron methane monooxygenases: quantum chemical studies, *J. Am. Chem. Soc.* in press.

2. Blomberg, M.R.A., Siegbahn, P.E.M., Styring, S., Babcock, G.T., Åkermark, B., and Korall, P. (1997) A quantum chemical study of hydrogen abstraction from manganese coordinated water by a tyrosyl radical: a model for water oxidation in photosystem II, submitted to *J. Am. Chem. Soc.*

3. Becke, A.D. (1988) Density-functional exchange-energy approximation with correct asymptotic behaviour, *Phys. Rev.* **A38**, 3098;

4. Lee, C., Yang, W. and Parr, R.G. (1988) Development of the Colle-Salvetti correlation-energy formula into a functional of the electron density, *Phys. Rev.* **B37**, 785.

5. Vosko, S.H., Wilk, L., and Nusair, M. (1980) Accurate spin-dependent electron liquid correlation energies for local spin density calculations: a critical analysis, *Can. J. Phys.* **58**, 1200.

6. Perdew, J.P., and Wang, Y. (1992) Accurate and simple analytic representation of the electron-gas correlation energy, *Phys. Rev.* **B45**, 13244; Perdew, J.P., Chevary, J.A., Vosko, S.H., Jackson, K.A., Pederson, M.R., Singh, D.J., and Fiolhais C. (1992) Atoms, molecules, solids and surfaces: Applications of the generalized gradient approximation of exchange and correlation, *Phys. Rev.* **B46**, 6671.

7. Becke, A.D. (1993) A new mixing of Hartree-Fock and local density-functional theories, *J. Chem. Phys.* **98**, 1372; Becke, A.D. (1993) Density-functional thermochemistry. III. The role of exact exchange, *J. Chem. Phys.* **98**, 5648.

8. Pople, J.A., Head-Gordon, M., Fox, D.J., Raghavachari, K., and Curtiss, L.A. (1989) Gaussian-1 theory: A general procedure for prediction of molecular energies, *J. Chem. Phys.* **90**, 5622.

9. Bauschlicher, Jr., and Partridge, H. (1995) The sensitivity of B3LYP atomization energies to the basis set and a comparison of basis set requirements for CCSD(T) and B3LYP, *Chem. Phys. Lett* **240**, 533.

10. Blomberg, M.R.A., Siegbahn, P.E.M., and Svensson, M. (1996) Comparisons of results from parametrized configuration interaction (PCI-80) and from hybrid density functional theory with experiments for first row transition metal compounds, *J. Chem. Phys.* **104**, 9546.

11. Siegbahn, P.E.M. (1996) Electronic structure calculations for molecules containing transition metals, in I. Prigogine and S.A. Rice (eds.), *Adv. Chem. Phys. Vol. XCIII*, Wiley, New York, pp.333.

12. Siegbahn, P.E.M., Blomberg, M.R.A., and Crabtree, R.H. (1997) Hydrogen transfer in the presence of amino acid radicals, *Theor. Chem. Acc.* in press.

13. GAUSSIAN 94 (Revision A.1), Frisch, M.J., Trucks, G.W., Schlegel, H.B., Gill, P.M.W, Johnson, B.G., Robb, M.A., Cheeseman, J.R., Keith, T.A., Petersson, G.A., Montgomery, J.A., Raghavachari, K., Al-Laham, M.A., Zakrzewski, V.G., Ortiz, J.V., Foresman, J.B., Cioslowski, J., Stefanov, B.R., Nanayakkara, A., Challacombe, M., Peng, C.Y., Ayala, P.Y., Chen, W., Wong, M.W., Andres, J.L., Replogle, E.S., Gomperts, R., Martin, R.L., Fox, D.J., Binkley, J.S., Defrees, D.J., Baker, J., Stewart, J.P., Head-Gordon, M., Gonzales C., and Pople, J.A. (1995) Gaussian, Inc., Pittsburgh PA.

14. Stevens, P.J., Devlin, F.J., Chablowski, C.F., and Frisch, M.J. (1994) Ab initio calculation of vibrational absorption and circular dichroism spectra using density functional force fields, *J. Phys Chem.* **98**, 11623.

15. Hay, P.J., and Wadt, W.R. (1985) Ab initio effective core potentials for molecular calculations. Potentials for K to Au including the outermost core orbitals, *J. Chem. Phys.* **82**, 299.

16. Rosenwieg, A.C., Frederick, C.A., Lippard, S.J., and Nordlund, P. (1993) Crystal structure of a bacterial non-haem iron hydroxylase that catalyses the biological oxidation of methane *Nature* **366**, 537.

17. Feig, A.L., and Lippard, S.J. (1994) Reactions of non-heme iron(II) centers with dioxygen in biology and chemistry, *Chem. Rev.* **94**, 759.

18. Wallar, B.J., and Lipscomb, J.D. (1996) Dioxygen activation by enzymes containing binuclear non-heme iron clusters, *Chem. Rev.* **96**, 2625.

19. Shu, L., Nesheim, J.C., Kauffmann, K., Munck, E., Lipscomb, J.D., and Que, Jr., L. (1997) An $Fe_2^{IV}O_2$ diamond core structure for the key intermediate Q of methane monooxygenase, *Science* **275**, 515.

20. Yachandra, V.K., Sauer, K., and Klein, M. P. (1996) Manganese cluster in photosynthesis: Where plants oxidize water to dioxygen, *Chem. Rev.* **96**, 2927.

21. Hoganson, C.W., and Babcock, G.T. (1988) Electron transfer events near the reaction center in O_2-evolving photosystem II preparations, *Biochemistry* **27**, 5848.

22. Hoganson, C.W., Lydakis-Simantiris, N., Tang, X.-S., Tommos, C., Warncke, K., Babcock, G. T., Diner, B. A., McCracken, J., and

Styring, S. (1995) A hydrogen-atom abstraction model for the function of Y_Z in photosynthetic oxygen evolution, *Photosynth. Res.* **46**, 177.

23. Sjöberg, B.M. (1994) The ribonucleotide reductase jigsaw puzzle: a large piece falls into place, *Structure* **2**, 793.

24. Ehrenberg, A., and Reichard, P. (1972) Electron spin resonance of the iron-containing protein B2 from ribonucleotide reductase, *J. Biol. Chem.* **247**, 3485.

25. Cukier, R.I. (1994) Mechanism for proton-coupled electron transfer reactions, *J. Phys. Chem.* **98**, 2377.

26. Ekberg, M., Sahlin, M., Eriksson, M., and Sjöberg, B.M. (1996) Two conserved tyrosine residues in protein R1 participate in an intermolecular electron transfer in ribonucleotide reductase, *J. Biol. Chem.* **271**, 20655.

27. Uhlin, U., and Eklund, H. (1994) Structure of ribonucleotide reductase protein R1, *Nature* **370**, 533.

28. Goodin, D.B., and McRee, D.E. (1993) The Asp-His-Fe triad of cytochrome c peroxidase controls the reduction potential, electronic structure and coupling of the tryptophan free radical to the heme, *Biochemistry* **32**, 3313.

AB INITIO AND DENSITY FUNCTIONAL THEORY APPLIED TO MODELS FOR THE OXO-TRANSFER REACTION OF DIOXOMOLYBDENUM ENZYMES

SNEŽANA ZARIĆ
Department of Chemistry
University of Belgrade
Studentski trg 16
11001 Belgrade
Yugoslavia

MICHAEL B. HALL
Department of Chemistry
Texas A&M University
College Station
Texas 77843-3244

1. Introduction

Molybdenum oxotransferases catalyze the transfer of an oxygen atom to and from a substrate. In the active site of these enzymes molybdenum is surrounded with one or two oxo groups and two or three sulfur groups [1]. Additional oxygen or nitrogen ligands may be present [1]. Based upon the known physical and chemical properties of the oxotransferases a minimal catalytic cycle is shown in Scheme 1 [2].

In the catalytic cycle the molybdenum center cycles among the VI, IV and V oxidation states. With the addition of a substrate, X, and H_2O, $[Mo(VI)O_2]^{2+}$ is reduced to $[Mo(IV)O(OH_2)]^{2+}$ and a substrate oxide, XO, is formed. By one-electron processes, $[Mo(IV)O(OH_2)]^{2+}$ is first oxidized to $[Mo(V)O(OH)]^{2+}$, which is then oxidized back to $[Mo(VI)O_2]^{2+}$.

L. Banci and P. Comba (eds.), Molecular Modeling and Dynamics of Bioinorganic Systems, 255–277.

To make an appropriate synthetic model system for this catalytic cycle one has to prevent reaction between Mo(IV)O and Mo(VI)O_2 forming the biologically irrelevant μ-oxo Mo(V) dimer. In the enzyme this reaction is prevented by bulky substituents and rigidity in the active site. In the model systems it has been prevented by sterically bulky ligands [2-8], or by the electronic properties of ligands [9].

Scheme 1

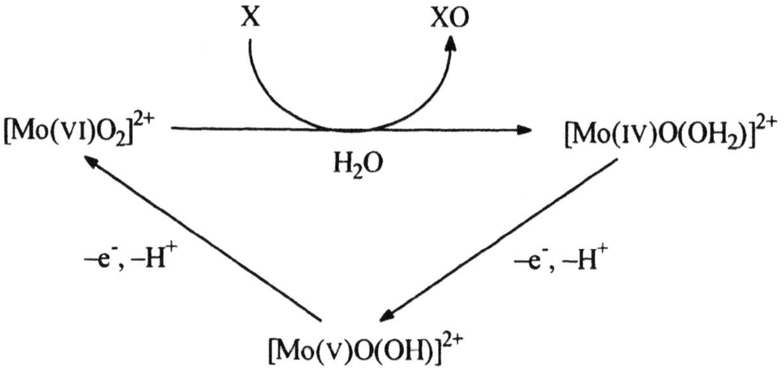

A number of Mo(VI) and Mo(IV) complexes have been synthesized to model the active site of molybdenum oxotransferases [2,10-21]. The first step of the catalytic cycle, known as the oxo-transfer reaction, has been studied by many experimental groups [3-7,9]. A comprehensive study of the reverse oxo-transfer reaction was done for the MoO_2(*t*-BuL-NS)$_2$/MoO(*t*-BuL-NS)$_2$ system (*t*-BuL-NS is (bis(4-*tert*-butylphenyl)-2-pyridylmethanethiolate(1-)) [6]. In this system the steric properties of the ligand prevent the formation of the binuclear μ-oxo Mo(V) complex. Reactions with several XO substrates exhibited well-behaved second-order kinetics. Relatively large activation entropies indicated an associative transition state. The authors proposed a reaction scheme with Mo(IV)O(OX)-(*t*-BuL-NS)$_2$ as a transition state or intermediate. Only a few groups have synthesized monomeric Mo(V) complexes, postulated in the reoxidation of the catalyst. These complexes have bulky ligands or coordinated solvents [2,22,23], which prevent formation of the Mo(V) dimer.

A model system which acts in a complete catalytic cycle was synthesized by Xiao et al. [2]. In this system Mo is surrounded by the bulky hydrotris(3,5-

dimethyl-1-pyrazolyl)-borate anion (L), SPh^-, and oxygens. The catalytic cycle with this model system is shown in Scheme 2. The species $L(SPh)Mo(VI)O_2$ reacts with the substrate PPh_3 and forms the coordinatively unsaturated or weakly solvated "L(SPh)Mo(IV)O". This species can be oxidized back to the Mo(VI) complex with Me_2SO. In dry toluene the Mo(IV) species reacts with the Mo(VI) species to form the binuclear μ-oxo Mo(V) complex. However, in the presence of H_2O, the Mo(IV) species reacts with the Mo(VI) species to form L(SPh)Mo(V)O(OH). It was proposed that the reaction involves the intermediate $L(SPh)Mo(IV)O(OH_2)$. The Mo(V) species is oxidized by O_2 back to the original Mo(VI) species, and in this way the catalytic cycle is completed. By isotopic labeling it was shown that the oxygen in the OH group of the Mo(V) species and one oxygen on the re-formed Mo(VI) species originate from H_2O and not from O_2.

Scheme 2

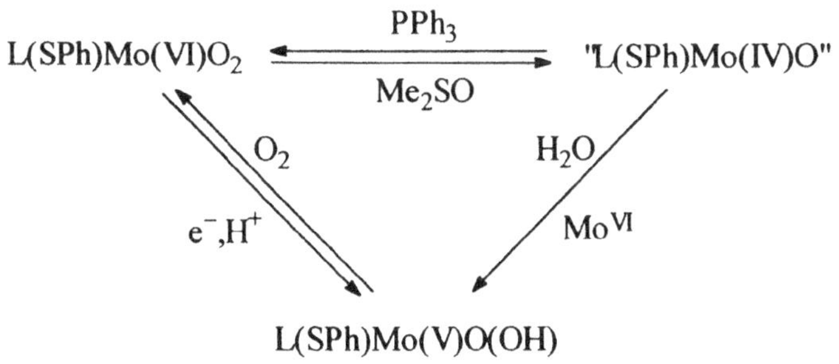

There are some theoretical studies on oxo-molybdenum systems. Rappé and Goddard [24] suggest that in *cis*-dioxo complexes the second oxo group is important in the reaction sequence. When both oxygens are oxo groups, the formation of a formal triple bond with the metal center is prevented. However, upon addition of substrate and formation of the substrate-O bond, the spectator oxygen (the one not involved in the bond) forms a triple bond with the metal center, which stabilizes the resulting species.

The results of semiempirical and *ab initio* studies [25] on several dioxo Mo(V) and Mo(VI) complexes with tetradentate N_2S_2 and N_2O_2 ligands indicate that the ligand has a profound effect on the geometry and electronic structure of the complex. These studies predict that the HOMO is mainly Mo d in character, a result which is consistent with the experimental EPR spectra.

In this work, we describe new results on a theoretical model system for a catalytic cycle, with the pattern that primarily follows the synthetic model system of Xiao et al. [2]. The reaction mechanism has been studied by identifying the structures and energies of possible intermediates and transition states. The transfer of the oxygen atom from the molybdenum center to the substrate and subsequent displacement of the substrate oxide by water has been studied in detail. The aim of this work is also to compare different *ab initio* and DFT methods.

It is well known that transition metal complexes, especially those with multiple M-L bonds are difficult to describe accurately by theory. These systems usually need high level correlation methods. On the other hand, these systems are usually rather large, so these calculations are often too demanding. Recently DFT has been shown to be very successful for calculations on transition metal complexes and DFT methods have recently become very popular for these systems.

2. Theoretical details

The geometries were optimized with Hartree-Fock and two different DFT methods (B3LYP and B3P86). In some cases the geometries were completely optimized whereas in other cases some constraints were imposed, these constraints are described in the particular cases below. Zero point-energy corrections, which might amount to several kilocarolies per mole, have not been made.

The results shown in this work have been obtained using two basis sets. In basis set 1 (bs1) for each atom, except for hydrogen, the core and valence electrons were divided, and ECP were used for molybdenum (1s2s2p3s3p3d) [26], oxygen (1s), nitrogen (1s), carbon (1s), sulfur (1s2s2p), and phosphorous (1s2s2p) [27], while the valence electrons were described by a double-zeta basis set [26,27]. In addition, a polarization function of 0.340 was added to phosphorus to properly describe the hypervalent molecules containing phosphorous [28]. This polarization function is more important for P than for S because P is directly involved in the reaction. This basis set was used for calculations in which different levels of *ab initio* methods were compared (Section 3) [29]. In basis set 2 (bs2) standard lanl2dz basis sets were used for all atoms except for Mo. For Mo the lanl2dz basis set was modified to include the recently proposed 5p functions which have been used successfully by this group [30-33]. Again, a polarization function of 0.340 was added to phosphorus. Bs2 was used in all geometry optimizations and energy calculations for the complete catalytic cycle (Sections 4-7).

Calculations were performed with GAMESS-UK [34], Gausian92 [35] and Gausian94 [36] and with a separate code for GMO2 [37].

3. Comparison of different levels of correlation and different model systems

Three model systems were developed based on the experimental model systems [2,6] and used to determine the necessary chemical properties of an accurate model and the level of correlation necessary to understand the changes in the electronic structure for the oxo-transfer reaction [29]. These systems were compared for the first step of the catalytic cycle, the oxo-transfer from the molybdenum center to the substrate

$$\mathrm{Mo(VI)O + PR_3 \longrightarrow Mo(IV) + OPR_3.}$$

For this reaction the intermediate $\mathrm{Mo(IV)OPR_3}$, was also optimized. The first model system,

$$\mathrm{[H_4Mo(VI)O] + PH_3 \longrightarrow [H_4Mo(IV)OPH_3] \longrightarrow [H_4Mo(IV)] + OPH_3}$$

is the smallest system that reproduces the correct oxidation state of the molybdenum center for both the oxotransferases and the experimental models. The second model system is

$$\mathrm{[H_2Mo(VI)O_2] + PH_3 \longrightarrow [H_2Mo(IV)O(OPH_3)] \longrightarrow [H_2Mo(IV)O] + OPH_3}$$

Here, the transition state from $\mathrm{[H_2Mo(VI)O_2] + PH_3}$ to $\mathrm{[H_2Mo(IV)O(OPH_3)]}$ is studied. In this model system the reactant has two oxygen atoms *cis* to each other (as the oxotransferases and the experimental model systems) and two hydrogen atoms *trans* to each other. For the third model system only the first part of the oxo-transfer reaction,

$$\mathrm{[(NH_3)_2(SH)_2Mo(VI)O_2] + PMe_3 \longrightarrow [(NH_3)_2(SH)_2Mo(IV)O(OPMe_3)]}$$

along with the transition state has been studied in order to compare the result with the other two model systems. The third model system is very similar to the experimental model systems.

The geometries for these three model systems were optimized at the HF level with constraints in the geometry [29]. In the first model system the four hydrogens are fixed in the same plane and are at 90° angles with respect to the single oxygen atom. Constraints in the second model system were applied to keep the model in a quasi-octahedral environment, similar to that of the experimental systems. An approximate transition state was obtained by following an approximate reaction coordinate involving the difference between the Mo-O and O-P distances. The electron correlation energy was evaluated by single point Moller-Plesset perturbation theory (up to fourth order) and configuration interaction type calculations (CASSCF and GMO2) for each optimized geometry of the model systems.

TABLE 1. Relative energies (kcal/mol) at different levels of *ab initio* theory for the first model system [29]

level	Mo(VI)O+PR_3	Mo(IV)OPR_3	Mo(IV)+OPR_3
HF	0.0	-37.4	19.1
MP2	0.0	67.2	109.7
MP3	0.0	11.8	73.2
MP4	0.0	86.5	141.6
GMO2(FCI)	0.0	16.6	67.0
GMO2(SDTQ)	0.0	15.4	65.8
GMO2(SDTQ/SD)	0.0	19.3	70.5
CASSCF	0.0	16.9	67.2

Relative energies along the reaction path for the first model system (Table 1) show that there is a large difference between the HF and CASSCF levels of theory. This difference implies that electron correlation is very important in calculating the correct trend and relative energies for oxo-transfer reactions. The difference in energy between Mo(IV)OPR_3 and Mo(IV) + OPR_3 is similar at various levels of theory because both are geometrically and electronically very similar.

The comparison of relative energies for MP2, MP3, and MP4 levels with CASSCF results shows that MP2 overestimate the relative stability of Mo(IV) + OPR_3, because there are low-lying unoccupied orbitals in Mo(IV)O + PR_3, that are not in Mo(IV) + OPR_3. Although the error at the MP4 level is even larger than that at the MP2 level, the MP3 energies reproduce the trend in the CASSCF results. GMO2(FCI) and GMO2(SDTQ) relative energies are in good agreement with CASSCF energies. Although the GMO2(FCI) energies are better than those of GMO(SDTQ), only GMO2(SDTQ) calculations have been used for larger systems.

TABLE 2. Relative energies (kcal/mol) at different levels of *ab initio* theory for the second model system [29]

level	Mo(VI)O+PR_3	transition state	Mo(IV)OPR_3	Mo(IV)+OPR_3
HF	0.0	25.2	-57.5	-18.7
MP2	0.0	60.1	39.7	74.6
MP3	0.0	32.1	-12.5	26.8
MP4	0.0	53.8	48.6	85.0
GMO2(SDTQ)	0.0	26.7	-19.0	10.5

For the second model system the MP2, MP3, and MP4 energies were compared with the GMO2(SDTQ) energies (Table 2). The comparison of HF and GMO2(SDTQ) energies shows that correlation energy is very important. The MP2 and MP4 methods again overestimate the correlation energy while the MP3 results are the closest to the GMO2(SDTQ) results.

The calculated MP2, MP3, and GMO2(SDTQ) energies for the third model system are reported in Table 3. The energies at all levels of theory have the same trend. However, the MP2 method overestimates the reaction barrier by 30.7 kcal/mol, and underestimate the relative stability of Mo(IV)OPR_3 by 60.6 kcal/mol compared to the GMO2(SDTQ) results. Again, for this system, a model very close to the experimental one, MP3 results are in closer agreement with the GMO2(SDTQ) results. It has been shown for some other transition metal systems that the MP3 method can be in better agreement with higher correlation methods compared to other MP orders [32] .

The comparison of the GMO2(SDTQ) energies (Table 1-3) for these three model systems shows that only the third model system has overall energetics in

TABLE 3. Relative energies (kcal/mol) at different levels of *ab initio* theory for the third model system [29]

level	Mo(VI)O+PR_3	transition state	Mo(IV)OPR_3
HF	0.0	3.2	-113.7
MP2	0.0	39.1	-10.6
MP3	0.0	4.9	-68.9
GMO2(SDTQ)	0.0	8.4	-71.2

agreement with the experimental data. For the first one, the reaction is too endothermic, because this system is too simple, and does not have chemical features comparable to the real system. The replacement of two hydrogens from the first model with one oxygen in the second model stabilizes the Mo(IV)OPR_3 by 34 kcal/mol. This improvement illustrates the importance of the spectator oxygen, a result that is in agreement with previous results for molybdenum dioxo complexes [24]. However, the reaction is still too endothermic. Our third model is very close to the experimental model system, and the reaction for this system is exothermic enough to accurately model the rest of the reaction. It is well known that ligands attached to a metal center can play an important role because of their electronic and steric effects. This molybdenum system is an especially clear example for this phenomena. The ligands attached to the molybdenum center, especially the spectator oxygen, are very important for the energetics of the oxo-transfer reaction. We used the smaller model systems to compare different *ab initio* methods, and the third model system to study a complete catalytic cycle.

4. Modeling the complete catalytic cycle

The third model system is closely related to the experimental model of Xiao et al. [2] and Holm and co-workers [6]. Two amine groups and SH^- group, fixed at 90° angles with respect to each other, replace the hydrotris(3,5-dimethyl-1-pyrazolyl)borate anion used by Xiao et al. Using two amine groups and SH^- group reproduces the charge, the π-donor characteristics, and the softness of the bulky ligand as well as the oxidation state of the metal. Fixing the ligands mimics the local steric effects associated with the bulky ligand used experimentally. Another SH^- group replaces the SPh^- group.

Scheme 3

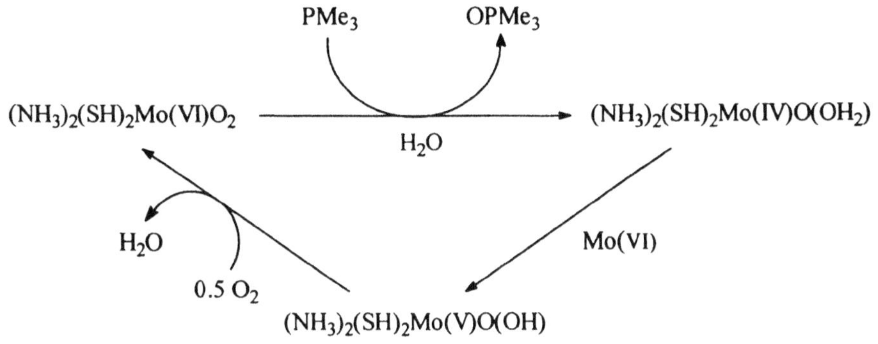

Our model catalytic cycle is shown in Scheme 3. In the first step trimethylphosphine approaches an oxygen atom of $(NH_3)_2(SH)_2Mo(VI)O_2$. Phosphine extracts the oxygen and forms trimethylphosphine oxide, and a water molecule enters the site vacated by the extracted oxygen atom. In this process $(NH_3)_2(SH)_2Mo(VI)O_2$ is reduced to $(NH_3)_2(SH)_2Mo(IV)O(OH_2)$. In the next step, $(NH_3)_2(SH)_2Mo(IV)O(OH_2)$ is oxidized by an one-electron process to $(NH_3)_2(SH)_2Mo(V)O(OH)$. Then, by another one-electron process, $(NH_3)_2(SH)_2Mo(V)O(OH)$ is oxidized back to $(NH_3)_2(SH)_2Mo(VI)O_2$. The driving force for the reaction is the production of trimethylphosphine oxide from trimethylphosphine and oxygen, and the cycle continues until all the phosphine is converted.

5. Driving force of the reaction

Experimentally, the driving force for the reaction is the production of triphenylphosphine oxide from triphenylphosphine and oxygen,

$$PPh_3 + 0.5\ O_2 \longrightarrow OPPh_3.$$

There are two values reported for the exothermicity of this reaction, -67 kcal/mol [11] and -73.8 kcal/mol [38]. The exothermicity for the reaction

$$PMe_3 + 0.5\ O_2 \longrightarrow OPMe_3$$

TABLE 4. Experimental [39,40] and calculated geometrical data for PMe_3 and PMe_3O molecules.

	exp.	calculated		
		B3LYP	B3P86	HF
PMe_3				
P-C	1.847	1.88	1.88	1.87
C-P-C		98.8	98.8	99.7
PMe_3O				
P-C	1.809	1.85	1.84	1.83
C-P-C		104.7	104.8	105.4
P-O	1.476	1.54	1.54	1.52
C-P-O	114.4	113.8	113.8	113.2

is close to these values, -79.7 kcal/mol [38]. Thus PMe_3 should be a suitable model since it has an O-P bond energy similar to the experimental one.

The geometry of trimethylphosphine and trimethylphosphine oxide were optimized at HF, B3LYP and B3P86 levels. The calculated and experimental bond lengths and angles are given in Table 4. For PMe_3 the calculated P-C bond lengths, 1.88 Å with DFT methods and 1.87 Å with the HF method, are slightly longer than the electron diffraction value of 1.847 Å [39]. For PMe_3O the experimental P-O and P-C distances are 1.476 and 1.809 Å, respectively [40]. The calculated bond lengths with all methods are longer than the experimental values; the bond lengths at the HF level (1.52 for the P-O and 1.83 Å for the P-C) are closest to the experimental ones. The optimized O-P-C angles are slightly smaller than those found experimentally.

TABLE 5. Experimental [38] and calculated ΔH for the $PMe_3 + 0.5\ O_2 \longrightarrow OPMe_3$ reaction

	ΔH (kcal/mol)
exp	-79.7
B3LYP	-58.7
B3P86	-61.4
MP3//B3LYP	-54.5
MP3//HF	-55.5
MP3//B3P86	-54.5

The experimental and calculated exothermicity for the

$$PMe_3 + 0.5\ O_2 \longrightarrow OPMe_3$$

reaction is presented in Table 5. The calculated exothermicity at the MP3 level is very similar, regardless of the chosen geometry. The values with DFT methods are slightly larger, but all of the calculated values are smaller than the experimental one.

6. Geometries of the model system

The geometries of the molecules involved in the theoretical model system were optimized with B3LYP, B3P86 and HF methods. During optimization two amine groups and a SH^- were fixed at 90° angles with respect to each other, to mimic constraints in experimental model system [2]. With the B3LYP method the geometries were also optimized without any constraints.

Some important bond lengths and angles for the optimized geometries of $(NH_3)_2(SH)_2Mo(VI)O_2$ are shown in Table 6. The Mo-O bonds formally have

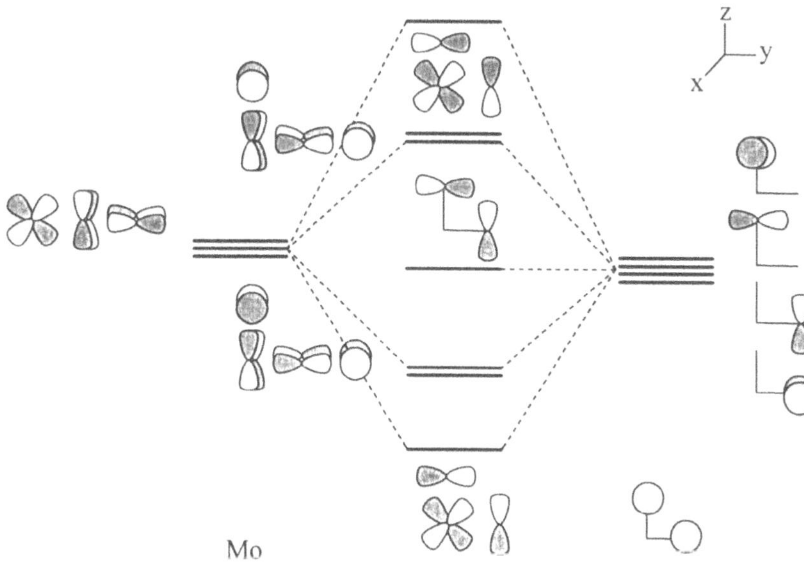

Fig. 1. Molecular orbital diagram of the interaction between the p orbitals of the oxygen atoms and the molybdenum d_{xy}, d_{xz}, and d_{yz} orbitals.

a bond order of 2.5 since both the p_y orbitals of one oxygen and the p_z orbital of the other oxygen compete for the same Mo d_{yz} orbital (Fig. 1).

The optimized Mo-O, and Mo-S bond distances are very similar in both DFT geometries, and are slightly shorter, and slightly longer, respectively, in the HF geometry. The Mo-N bond lengths are very similar for the B3P86 and the HF geometries, and are slightly longer for the B3LYP geometry. There are experimental bond lengths (Mo-O 1.696 Å, Mo-S 2.418 Å and Mo-N 2.411 Å) for MoO_2(*t*-BuL-NS)$_2$ [7]. Our calculated values are in reasonably good agreement with the experimental data. However, we cannot compare the experimental data

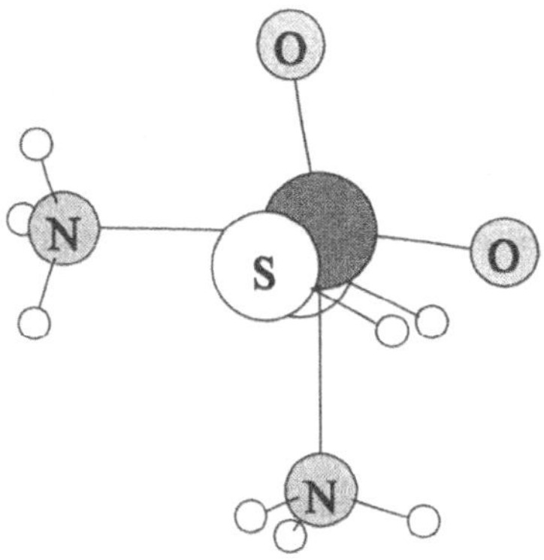

Fig. 2. The structure of $(NH_3)_2(SH)_2MoO_2$ optimized at B3LYP level with constraints (see text). Geometrical data for this structure are given in Table 6.

TABLE 6. Some geometrical data for $(NH_3)_2(SH)_2MoO_2$ molecule calculated with different methods

	B3LYP*	B3LYP	B3P86	HF
Mo-O	1.74	1.74	1.73	1.69
Mo-O	1.74	1.74	1.74	1.69
Mo-N	2.42	2.42	2.38	2.40
	2.42	2.43	2.38	2.40
Mo-S	2.53	2.54	2.53	2.58
	2.53	2.54	2.52	2.58
O-Mo-O	105.8	106.2	102.7	105.4
N-Mo-N	87.4	90.0	90.0	90.0

*completely optimized geometry

directly with the calculated parameters, because our molecule is only a model system for $MoO_2(t\text{-BuL-NS})_2$, and the ligand differences will cause small discrepancies in the bond lengths. The fully optimized B3LYP geometry is very similar to the geometry with constraints. The major changes that slightly stabilize the fully optimized structure are slightly shorter Mo-N and Mo-S bond lengths and a slightly smaller N-Mo-N angle.

For the reduced (d^2) species with one oxygen vacancy, $(NH_3)_2(SH)_2Mo(IV)O$, the lowest energy singlet state has both electrons in the Mo d_{xy} orbital. In order to determine whether the ground state is a singlet or a triplet,

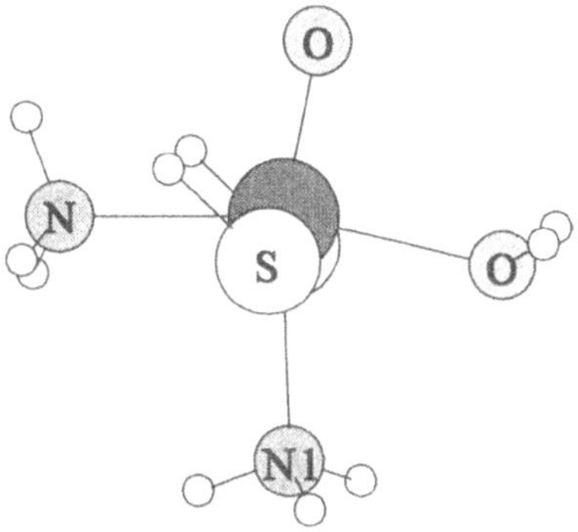

Fig. 3. The structure of $(NH_3)_2(SH)_2MoO(OH_2)$ optimized at B3LYP level with constraints (see text). Geometrical data for this structure are given in Table 7.

TABLE 7. Some geometrical data for $(NH_3)_2(SH)_2MoO(OH_2)$ molecule calculated with different methods.

	B3LYP*	B3LYP	B3P86	HF
Mo-O	1.73	1.71	1.72	1.69
Mo-OH_2	3.58	2.22	2.20	2.19
Mo-N	2.23	2.24	2.21	2.26
Mo-N1	2.25	2.44	2.40	2.46
Mo-S	2.45	2.56	2.52	2.62
	2.46	2.56	2.53	2.62
O-Mo-O	48.5	91.0	90.0	94.8
N-Mo-N	159.9	90.0	90.0	90.0

*completely optimized geometry

the two configurations were used in the partial geometry optimization (all ligand angles were fixed at a 90° angle) at the HF level with the basis set I [41]. In the singlet state the Mo-O bond length (1.65 Å) is shorter than that in the triplet state (1.71 Å). Configuration interaction calculations [41] with single and double excitations (CISD) show that the singlet is lower in energy than the triplet but only by about 1 kcal/mol. However, access to the triplet state surface is formally forbidden by symmetry and it can be assumed that the reaction remains on the singlet surface. $(NH_3)_2(SH)_2Mo(IV)O$ was optimized only at the B3LYP level and only in the singlet state because we needed this structure and energy only to determine whether if the second transition state is associative or dissociative. The Mo-O bond optimized to 1.71 Å, which is 0.03 Å shorter than the Mo-O bond in $(NH_3)_2(SH)_2Mo(IV)O_2$. The ligand rearrangement observed in Holm's model has not been considered in this study because Xiao et al.'s tridentate ligand system prevents such a rearrangement.

$(NH_3)_2(SH)_2Mo(IV)O(OH_2)$, where water replaces the vacant site, has two electrons in the Mo d_{xy} orbital, consistent with the $(NH_3)_2(SH)_2Mo(IV)O$ results. The most important geometrical data on the optimized structures at different levels of theory are given in the Table 7. The Mo-O bond in this molecule can have a formal bond order of 3, and in the DFT geometry this bond is shorter by 0.03 Å than that in $(NH_3)_2(SH)_2Mo(IV)O_2$ but there is almost no difference for the HF geometries. In the angularly constrained structures, the Mo-OH_2 bond is considerably longer (0.5 Å) for all methods than the Mo-O bond. The nitrogens show a strong *trans* influence as the Mo-N bond *trans* to the oxo group is longer by 0.2 Å (for all methods) than the Mo-N bond *trans* to the OH_2 group. The completely optimized structure with the B3LYP method is very different from the geometry with constraints. The N-Mo-N angle is 159.9°, and the Mo-OH_2 bond distance is 3.57 Å. Hence, for this molecule the constraints imposed by bulky ligands are very important.

In the optimized structures of $(NH_3)_2(SH)_2MoO(OH)$ the Mo-OH bond is longer for all the methods than the Mo-O bond, but it is shorter than the Mo-OH_2 bond in $(NH_3)_2(SH)_2MoO(OH_2)$ (Table 8). The *trans* influence is less pronounced than it is in $(NH_3)_2(SH)_2Mo(IV)O(OH_2)$. The Mo-N bond *trans* to the oxo group is longer by 0.13 Å (for both DFT methods) that the Mo-N bond *trans* to the OH group. The completely optimized structure for this molecule is similar to the optimized geometry with constraints. The N-Mo-N angle is 79.4°, and the Mo-O bonds are slightly shorter than those in the geometry with constraints.

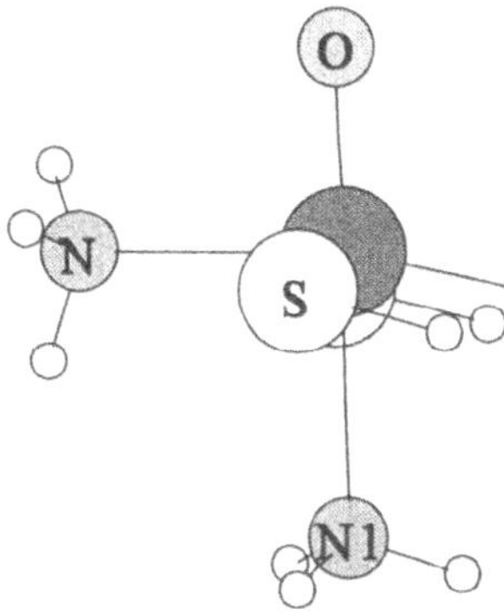

Fig. 4. The structure of $(NH_3)_2(SH)_2MoO(OH)$ optimized at B3LYP level with constraints (see text). Geometrical data for this structure are given in Table 8.

TABLE 8. Some geometrical data for the optimized structures of $(NH_3)_2(SH)_2MoO(OH)$ calculated with different methods.

	B3LYP*	B3LYP	B3P86	HF
Mo-O	1.72	1.72	1.71	1.87
Mo-OH	1.95	1.96	1.95	1.90
Mo-N	2.30	2.30	2.27	2.31
Mo-N1	2.44	2.43	2.40	2.33
Mo-S	2.53	2.54	2.51	2.55
	2.55	2.54	2.51	2.56
O-Mo-O	109.1	103.3	103.5	105.4
N-Mo-N	79.4	90.0	90.0	90.0

*completely optimized geometry

7. Energetics of the model system

Energetics of the model system have been calculated at the B3LYP, B3P86 and MP3 methods (Table 9). In previous work, MP3 method was shown to be the first reliable order of perturbation theory for the energies of the various oxo-molybdenum complexes (Section 3) [29]. The calculated relative energies for the steps of the catalytic cycle are very similar for the two DFT methods. The comparison of the energetics from the DFT and MP3 calculations shows that the DFT energies for intermediate, intermediate product 1 and 2 are less exothermic than the MP3 energies (Table 9).

TABLE 9. Relative energies (kcal/mol) of the model system for catalytic cycle with respect the reactants

	B3LYP*	B3LYP	MP3//B3LYP	B3P86
Reactants	0.0	0.0	0.0	0.0
Intermediate	-31.6	-26.6	-59.6	-29.5
Intermediate product 1	-35.5	-19.5	-43.8	-22.1
Intermediate product 2	-43.6	-31.1	-64.3	-34.3
Reactants'	-58.7	-58.7	-54.5	-61.4

Reactants: 2 $(NH_3)_2(SH)_2Mo(VI)O_2 + PMe_3 + H_2O + 0.5O_2$
Intermediate: $(NH_3)_2(SH)_2Mo(VI)O_2 + (NH_3)_2(SH)_2Mo(IV)O(OPMe_3) + H_2O + 0.5O_2$
Intermediate product 1: $(NH_3)_2(SH)_2Mo(VI)O_2 + (NH_3)_2(SH)_2Mo(IV)O(OH_2) + OPMe_3 + 0.5O_2$
Intermediate product 2: 2 $(NH_3)_2(SH)_2Mo(V)O(OH) + OPMe_3 + 0.5O_2$
Reactants': 2 $(NH_3)_2(SH)_2Mo(VI)O_2 + OPMe_3 + H_2O$

In the first step of the catalytic cycle, the substrate PMe_3 approaches an oxygen atom of $(NH_3)_2(SH)_2Mo(VI)O_2$ and forms the intermediate $(NH_3)_2(SH)_2Mo(VI)O(OPMe_3)$. This process is exothermic and the calculated exothermicity varies from -59.6 kcal/mol for the MP3//B3LYP method to -26.6 kcal/mol for the B3LYP method (Table 9). The next process, which substitutes $OPMe_3$ with H_2O and generates the $(NH_3)_2(SH)_2Mo(VI)O(OH_2)$ complex, is less exothermic than the intermediate by 7.1 kcal/mol with the B3LYP method, and by 7.4 kcal/mol with B3P86 method. The calculated energetics for this step with the completely optimized geometries (B3LYP*) will not be considered further, because the completely optimized geometry of $(NH_3)_2(SH)_2Mo(VI)O(OH_2)$ is very different from the geometry with constraints, and this new geometry is impossible for the experimental model system. The oxidation of $(NH_3)_2(SH)_2Mo(VI)O(OH_2)$ to $(NH_3)_2(SH)_2Mo(V)O(OH)$ is exothermic by -11.6 kcal/mol with the B3LYP method and -12.2 kcal/mol with the B3P86 method. The final step, oxidation of $(NH_3)_2(SH)_2Mo(V)O(OH)$ to $(NH_3)_2(SH)_2Mo(IV)O_2$, is exothermic by -27.6 kcal/mol at the B3LYP level, by -27.1 at the B3P86 level, and by -14.9 kcal/mol at the B3LYP* level. This step is, however, endothermic by 9.8 kcal/mol at the MP3//B3LYP level.

In this cycle the calculated energies are for the gas phase species. However, because of the large dipole moment of $OPMe_3$ compared to that of PMe_3, solvent effects will stabilize $OPMe_3$, and the energies for all products beyond the intermediate will be lower. This effect will generate a progressively more exothermic cycle. The overall reaction,

$$PMe_3 + 0.5\ O_2 \longrightarrow OPMe_3$$

is exothermic by -58.7 kcal/mol at B3LYP method.

8. The oxo-transfer step

The transfer of an oxygen atom from the molybdenum center to the substrate has been studied in more detail. The first transition state is the transition state for PMe_3 to approach an oxygen of $(NH_3)_2(SH)_2MoO_2$. Previous work has shown that PMe_3 approaches an oxygen with a Mo-O-P angle of 130°, and with a dihedral O-Mo-O-P angle of 90°, for the regular octahedral geometry of the complex [41]. The 90° dihedral angle is necessary because the appropriate antibonding molecular orbital has lower energy [41]. This approximate transition state was not fully optimized in previous work as it was determined by following an assumed reaction coordinate. Here, the geometry was optimized as a transition state with geometry constraints (Section 6) at the HF and B3LYP levels, and completely optimized at the HF level (Table 10). The geometries obtained by transition state optimizations are in agreement with the previously reported approximate transition state.

The optimized structures at the HF and B3LYP level are very similar. The completely optimized structure at the HF level is only slightly different from the structure with constraints. In the transition state structure the Mo-O bond that is being attacked by PMe_3 is slightly longer than that in the reactant (Table 6), and the O-P bond that is formed is still fairly long. The Mo-O-P angle is 129° and the O-Mo-O-P dihedral angle is 83° (HF) and 77° (B3LYP). When the O-P bond is formed, the $OPMe_3$ ligand rotates, decreasing the O-Mo-O-P dihedral angle. During this rotation the Mo-$OPMe_3$ bond lengthens, and the O-P bond shortens. When the dihedral angle reachs 0°, we get the intermediate species $(NH_3)_2(SH)_2MoO(OPMe_3)$.

Bond lengths in the transition state are intermediate between those in the reactant and those in the intermediate (Tables 6, 10, 11). With the B3LYP method the Mo-$OPMe_3$ bond (1.84 Å) is slightly longer than that in the reactant (1.74 Å, Table 6) and shorter than that in the intermediate (2.22 Å, Table 11). The O-P bond is beginning to form in the transition state and it is longer (2.06 Å) than it

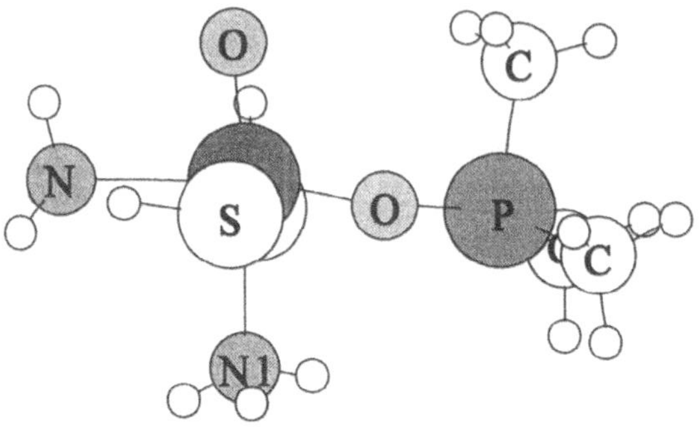

Fig. 5. The structure of first transition state $(NH_3)_2(SH)_2MoO\text{-}(OPMe_3)$ optimized at B3LYP level with constraints (see text). Geometrical data for this structure are given in Table 10.

TABLE 10. Some geometrical data for the optimized structures of the first transition state $(NH_3)_2(SH)_2MoO\text{-}(OPMe_3)$ calculated with different methods.

	HF*	HF	B3LYP
Mo-O	1.68	1.68	1.74
Mo-OPMe$_3$	1.78	1.78	1.84
Mo-N	2.33	2.36	2.36
Mo-N1	2.43	2.41	2.41
Mo-S	2.56	2.53	2.52
	2.56	2.57	2.56
O-Mo-O	105.7	104.1	105.2
N-Mo-N	81.7	90.0	90.0
O-P	2.37	2.38	2.06
P-C	1.85	1.85	1.86
Mo-O-P	129.0	128.7	128.8
O-Mo-O-P	83.8	83.2	77.5

*completely optimized geometry

is in the intermediate (1.57 Å). The calculated vibrational frequencies of the completely optimized geometry show one imaginary frequency with a motion appropriate for this transition state. The B3LYP barrier is 28.4 kcal/mol, while the MP3 barrier was 14.1 kcal/mol [41].

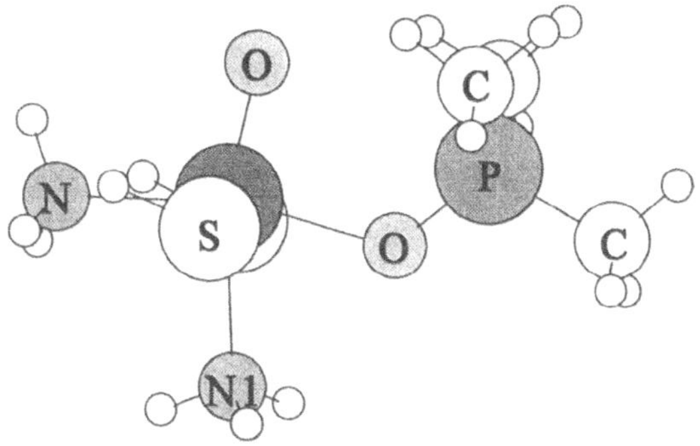

Fig. 6. The structure of intermediate $(NH_3)_2(SH)_2MoO(OPMe_3)$ optimized at B3LYP level with constraints (see text). Geometrical data for this structure are given in Table 11.

TABLE 11. Some geometrical data of the optimized structures of the intermediate $(NH_3)_2(SH)_2MoO(OPMe_3)$ calculated with different methods.

	B3LYP*	B3LYP	B3P86	HF
Mo-O	1.73	1.73	1.72	1.69
Mo-$OPMe_3$	2.30	2.22	2.19	2.19
Mo-N	2.23	2.24	2.22	2.29
Mo-N1	2.36	2.41	2.38	2.44
Mo-S	2.54	2.52	2.49	2.58
	2.54	2.60	2.55	2.64
O-Mo-O	84.9	94.2	94.3	94.9
N-Mo-N	105.9	90.0	90.0	90.0
O-P	1.63	1.57	1.57	1.55
P-C	1.86	1.83	1.83	1.81
Mo-O-P	122.8	125.3	123.4	129.2
O-Mo-O-P	1.4	-1.8	-3.2	-1.4

*completely optimized geometry

The intermediate $(NH_3)_2(SH)_2MoO(OPMe_3)$ was optimized with constraints at different levels and without any constraints at the B3LYP level (Table 11). In the optimized structures the O-Mo-O-P dihedral angle is close to 0°, and it varies from 1.4° for the completely optimized structure with the B3LYP method to -3.2° with the B3P86 method. The O-P bond is nearly fully formed, and the Mo-$OPMe_3$ bond distance with all methods is similar to the Mo-OH_2 bond distance (Table 7). In this structure the nitrogens again show a *trans* influence.

The completely optimized structure is very similar to the structures with constraints, the N-Mo-N angle increases to 105.9°.

The next step in the oxo-transfer reaction, after the formation of the intermediate, is the substitution of $OPMe_3$ with H_2O. There are two possible mechanisms: a dissociative mechanism where $OPMe_3$ leaves before OH_2 enters; and an associative mechanism where a concerted exchange occurs between $OPMe_3$ and OH_2. $(NH_3)_2(SH)_2MoO + OPMe_3$ separated at infinite distance is higher in energy by 35.4 kcal/mol than the intermediate at the B3LYP level and at the MP3//B3LYP by 60.6 kcal/mol. Thus, this step in the oxo-transfer reaction happens with an associative mechanism.

9. Conclusion

Our theoretical model system for the molybdenum oxotransferases and the energy profile for the reaction are in qualitative agreement with what is known about these systems. The calculated energies for the complete catalytic cycle show that the cycle is progressively exothermic. The optimized transition state geometry is in accord with the assumption that the direction of approach of the substrate is decisive. The transition state barrier is 28.4 kcal/mol. The mechanism for the release of the substrate oxide product is associative, a result which is consistent with experimental results.

The comparison of different methods in this work shows that the two DFT methods give very similar results for the optimized geometries and for the calculated energies. Optimized DFT geometries are slightly different than HF geometries. While the DFT and MP3 energies are in qualitative agreement, there is a difference for the relative stability of the intermediates. These results indicate that DFT methods can be successfully used for geometries and energies of systems with multiple M-L bonds.

Acknowledgments. The authors would like to thank the National Science Foundation (Grant No. CHE 94-23271), the Robert A. Welch Foundation (Grant No. A-648) and Serbian Research Council (Grant No. 02E09) for financial support. This research was conducted in part with use of the Cornell Theory Center, a resource for the Center for Theory and Simulation in Science and Engineering at Cornell University, which is funded in part by the National Science Foundation, New York State, and IBM Corporation.

References

1. Holm, R.H. (1990) The biologically relevant oxygen atom transfer chemistry of molybdenum: from synthetic analogue systems to enzymes, *Coord. Chem. Rev.* **110**, 183-221
2. Xiao, Z.; Young, C. G.; Enemark, J. H.; Wedd, A. G. (1992) A single model displaying all the important centers and processes involved in catalysis by molybdoenzymes containing $[Mo^{VI}O]^{2+}$active sites *J. Am. Chem. Soc.* **144**, 9194-9195.
3. Roberts, S. A.; Young, C. G. ; Cleland, Jr. W. E.; Ortega, R. B.; and Enemark, J. H. (1988) Synthesis, characterization, and oxygen atom transfer reaction of $\{HB(Me_2C_3N_2H)_3\}MoO\{S_2P(OR)_2\}$ and $\{HB(Me_2C_3N_2H)_3\}MoO_2\{\eta^1{-}S_2P(OEt)_2\}$, *Inorg. Chem.* **27**, 3044-3051.
4. Harlan, E. W.; Berg, J. M.; and Holm, R. H. (1986) Thermodynamic fitness of molybdenum(IV,VI) complexes for oxygen atom transfer reactions, including those with enzymatic substrates, *J. Am. Chem. Soc.* **108**, 6992-7000.
5. Gheller, S. F.; Schultz, B. E.; Scott, M. J.; Holm, R. H. (1992) A broad-substrate analogue reaction system of the molybdenum oxotransferases, *J. Am. Chem. Soc.* 114, 6934-6935.
6. Schultz, B. E. and Holm, R. H. (1993) Kinetics of oxygen atom transfer in an analogue reaction system of the molybdenum oxotransferases, *Inorg. Chem.* **32**, 4244-4248.
7. Schultz, B. E.; Gheller, S. F.; Muetterties, M. C.; Scott, M. J.; and Holm, R. H. (1993) Molybdenum-mediated oxygen atom transfer: an improved analogue reaction system of molybdenum oxotransferases, *J. Am. Chem. Soc.* **115**, 2714-2722.
8. Xiao, Z.; Bruck, M. A.; Doyle, C.; Enemark, J. H.; Grittini, C.; Gable, R. W.; Wedd, A. G.; and Young, C. G. (1995) Dioxomolybdenum(VI) complexes of tripodal nitrogen-donor ligands: syntheses and spectroscopic, structural, and electrochemical studies, including the generation of EPR-active molybdenum(V) species in solution, *Inorg. Chem.* **34**, 5950-5962.
9. Oku, H.; Ueyama, N.; Kondo, M.; and Nakamura, A. (1994) Oxygen atom transfer systems in which the (μ-oxo)dimolybdenum(V) complex formation does not occur: syntheses, structures, and reactivities of monooxomolybdenum(IV) benzenedithiolato complexes as models of molybdenum oxidoreductases, *Inorg. Chem.* **33**, 209-216.
10. Enemark, J. H. and Young, C. G. (1993) Bioinorganic chemistry of pterin-containing molybdenum and tungsten enzymes, *Adv. Inorg. Chem.* **40**, 1-88.
11. Holm, R. H. (1987) Metal-centered oxygen atom transfer reactions, *Chem. Rev.* **87**, 1401-1449.
12. Holm, R. H. and Berg, J. M. (1986) Toward functional models of metalloenzyme active sites: analogue reaction systems of the molybdenum oxo transferases, *Acc. Chem. Res.* **19**, 363-370.
13. Das, S. K.; Chaudahury, P. K.; Biswas, D.; Sarkar, S. (1994) Modeling for the active site of sulfite oxidase: synthesis, characterization, and reactivity of $[Mo^{VI}O_2(mnt)_2]^{2-}$ (mnt^{2-} = 1,2-dicyanoethylenedithiolate), *J. Am. Chem. Soc.* **116**, 9061-9070.
14. Barbaro, P.; Bianchini, C.; Scapacci, G.; Masi, D.; and Zanello, P. (1994) Dioxomolybdenum(VI) complexes stabilized by polydentate ligands with NO_3, N_2O_2, and NS_2 donor-atom sets, *Inorg. Chem.* **33**, 3180-3186.
15. Caradonna, J. P.; Reddy, P. R.; Holm, R. H. (1988) Kinetics, mechanisms, and catalysis of oxygen atom transfer reaction of S-oxide and pyridine N-oxide substrates with molybdenum(IV,VI) complexes: relevance to molybdoenzymes, *J. Am. Chem.Soc.* **110**, 2139-2144.
16. Arzoumanian, H.; Lopez, R.; and Agrifoglio, G. (1994) Synthesis and X-ray characterization of tetraphenylphosphonium tetrathiocyanatodioxomolybdate(VI): a remarkable oxo transfer agent, *Inorg. Chem.* **33**, 3177-3179.
17. Kusthardt, U.; Albach, R. W. and Kiprof, P. (1993) New insights into the electrochemistry of molybdenum(VI) dioxo complexes containing a tetradentate $S_2(NH)_2$-Type ligand, *Inorg. Chem.* **32**, 1838-1843.

18. Eagle, A. A.; Laughlin, L. J.; Young, C. G.; Tiekink, E. R. T. (1992) An oxothio-molybdenum(VI) complex stabilized by an intramolecular sulfur-sulfur interaction: implications for the active site of oxidized xanthine oxidase and related enzymes, *J. Am. Chem. Soc.* **114**, 9195-9197.
19. Oku, H.; Ueyama, N.; and Nakamura, A. (1995) Thiolato-activated oxo-metal bond features in molybdenum and tungsten oxidoreductase models as revealed by raman spectroscopy, *Inorg. Chem.* **34**, 3667-3676.
20. Craig, J. A.; Harlan, E. W.; Snyder, B. S.; Whitener, M. A.; and Holm, R. H. (1989) Oxomolybdenum(IV,V,VI) complexes: structures, reactivities, and criteria of detection of binuclear (μ-oxo)molybdenum(V) products in oxygen atom transfer systems, *Inorg. Chem.* **28**, 2082-2091.
21. Cervilla, A.; Corma, A.; Fornes, V.; Llopis, E.; Perez, F.; Rey, F.; and Ribera, A. (1995) Model reactions of molybdo-reductase. A novel and highly efficient reduction of nitrobenzene to aniline catalyzed by a molybdenum-mediated oxygen atom transfer reaction, *J. Am. Chem.Soc.* **117**, 6781-6782.
22. Laughlin, L. J. and Young, C. G. (1996) Oxygen atom transfer, coupled electron-proton transfer, and correlated electron-nucleophile transfer reactions of oxomolybdenum(IV) and dioxomolybdenum(VI) complexes, *Inorg. Chem.* **35,** 1050-1058.
23. Dhawan, I. K.; Pacheco, A.; and Enemark, H. (1994) Structural and Spectroscopic characterization of a monooxomono(dithiolene)molybdenum(V) compound and its implications for the low pH form of sulfite oxidase, *J. Am. Chem. Soc.* **116**, 7911-7912.
24. Rappé, A. K. and Goddard, W. A. III (1982) Hydrocarbon oxidation by high-valent group 6 oxides, *J. Am. Chem. Soc.* **104**, 3287-3294.
25. Peng G., Nichols, J., McCullough, E. A. Jr., and Spence, J. T. (1994) Models for the molybdenum(VI/V) centers of the molybdenum hydroxylases and related enzymes. Geometry, electronic structure, and EPR g-tenzor predictions from *ab initio* and semiempirical molecular orbital studies, *Inorg. Chem.* **33**, 2857-2864.
26. Hay, P. J. and Wadt, W. R. (1985) *Ab initio* effective core potentials for molecular calculations. Potentials for K to Au including the outermost core orbitals, *J. Chem. Phys.* **82**, 299-310.
27. Stevens, W. J.; Basch, H.; and Krauss (1984) Compact effective potentials and efficient shared-exponent basis sets for the first- and second-row atoms, *J. Chem. Phys.* **81**, 6026-6034.
28. Huzinaga, S.; Andzelm, J.; Klobukowski, M.; Radzio-Andzelm, E; Sakai, Y.; Tatewaki, H. (1984) *Gaussian Basis Sets for Molecular Calculations*, Elsevier, Amsterdam.
29. Pietsch, M. A.; Couty, M.; and Hall, M. B. (1995) Comparison of Moller-Plesset perturbation methods, complete active space self-consistent field theory, and a new generalized molecular orbital method for oxygen atom transfer from a molybdenum complex to a phosphine, *J. Phys. Chem.* **99**, 16315-16319.
30. Couty, M. and Hall, M. B. (1996) Basis sets for transition metals: optimized outer p functions *J. Comp. Chem.* **17**, 1359-1370.
31. Strout, D. L.; Zaric, S.; Niu, S.; and Hall, M. B. (1996) Methane metathesis at a cationic iridium center, *J. Am. Chem. Soc.* **118,** 6068-6069.
32. Zaric, S. and Hall, M. B. (1997) *Ab initio* calculations of the geometry and bonding energies of alkanes and fluoroalkanes complexes with tungsten pentacarbonyl, *J. Phys. Chem.*, submitted
33. Zaric, S.; Couty, M.; and Hall, M. B. (1997) *Ab initio* calculations of the geometry and vibrational frequencies of the triplet state of tungsten pentacarbonyl amine: a model for the unification of the preresonance raman and the time-resolved infrared experiments, *J. Am. Chem. Soc.* **119,** 2885-2888.
34. Dupuis, M.; Spangler, D.; Wendolowski, J. (1993) NRCC Software Catalog; Vol.1, Program No. QG01 (GAMESS). Guest, M.F.; Fantucci, P.; Harrison, R.J.; Kendrick, J.; van Lenthe, J. H.; Schoeffel, K.; Sherwood, P. GAMESS-UK; CSF Ltd.

35. Gaussian 92 Frisch, M. J.; Trucks, G. W.; Head-Gordon, M.; Gill, P. M. W.; Wong, M. W.; Foresman, J. B.; Johnson, B. G.; Schlegel, H. B.; Robb, M. A.; Replogle, E. S.; Gomperts, R.;Andres, J. L.; Raghavachari, K.; Binkley, J. S.; Gonzalez, C.; Martin, R. L.; Fox, D. G.; Defrees, D. J.; Baker, J.; Stewart, J. P.; and Pople, J. A. (1992) Gaussian, Inc., Pittsburgh, PA.
36. Gaussian 94 Frisch, M. J.; Trucks, G. W.; Schlegel, H. B., Gill, P. M. W.; Johnson, B. G.; Robb, M. A.; Cheeseman, J. R.; Keith, T. A.; Petersson, G. A.; Montgomery, J. A.; Raghavachari, K.; Al-Laham, M. A.; Zakrzewski, V. G.; Ortiz, J. V.; Foresman, J. B.; Cioslowski, J.; Stefanov, B. B.; Nanayakkara, A.; Challacombe, M.; Peng, C. Y.; Ayala, P. Y.; Chen, W.; Wong, M. W.; Andres, J. L.; Replogle, E. S.; Gomperts, R.; Martin, R. L.; Fox, D. G.; Binkley, J. S.; Defrees, D. J.; Bajer, J.; Stewart, J. P.; Head-Gordon, M.; Gonzalez, C.; and Pople, J. A. (1995) Gaussian, Inc., Pittsburgh, PA
37. Couty, M. and Hall, M. B. (1997) Generalized molecular orbital theory II (GMO2), *J. Phys. Chem.* in press.
38. Holm, R.H. and Donahue, J. P. (1993) A thermodynamic scale for oxygen atom transfer reaction, *Polyhedron* **12**, 571-589
39. Bartell, L. S. and Brockway, L. O. (1960) Electron diffraction study of the structure of trimethylphosphine, *J. Chem. Phys.* **32**, 512-515.
40. Wilkins, C. J.; Hagen, K.; Hedberg, L.; Shen, Q.; and Hedberg, K. (1975) An electron diffraction investigation of the molecular structures of gaseous trimethylphosphine oxide, trimethylphosphine sulfide, trimethylarsine oxide, and trimethylarsine sulfide, *J. Am. Chem.Soc.* **97**, 6352-6358.
41. Pietsch, M. A. and Hall, M. B. (1996) Theoretical studies on models for the oxo-transfer reaction of dioxomolybdenum enzymes, *Inorg. Chem.* **35**, 1273-1278.

QUANTUM MECHANICAL MODELING OF ACTIVE SITES IN METALLOPROTEINS. ELECTROSTATIC COUPLING TO THE PROTEIN/SOLVENT ENVIRONMENT

J. LI, P. BEROZA, L. NOODLEMAN and D.A. CASE
Department of Molecular Biology
The Scripps Research Institute
La Jolla, California 92037 USA

1. Introduction

Metalloproteins are quite common in biochemistry: more than 50% of the structures in the Protein Data Bank, and an estimated one-third of all purified enzymes, require a metal ion for their biological function [1]. The study of methods to treat such systems in simulations is very active. In this chapter, we will discuss some approaches to energetic issues involving transition metal clusters in proteins, drawing examples primarily from recent work in our lab. We will focus on two issues. The first involves the use of density functional theory, and the extent to which it can provide accurate and reliable results for the first row transition metals. A second area will discuss our experience, and that of others, in using continuum solvent models as a framework for constructing descriptions of the interaction of metalloprotein active sites with their protein and solvent environments. We will consider primarily energetic and thermodynamic issues, with an emphasis on energetic effects of charge changes, such as adding protons or electrons to a system. This is a fairly "static" picture that leaves out important kinetic and dynamic effects; nevertheless, establishing sound information about thermodynamic "end points" is a key (and often very difficult) step in constructing models for the function of metalloproteins.

2. Density functional theory of transition metal complexes

Electronic structure calculations on transition metal complexes face a number of difficulties beyond the obvious problem that most interesting systems contain

L. Banci and P. Comba (eds.), Molecular Modeling and Dynamics of Bioinorganic Systems, 279–306.

large numbers of atoms and electrons. One principal difficulty is that molecular orbital theory, which can supply a satisfying qualitative account of a great many features of transition metal systems, exhibits a variety of systematic deficiencies, so that quantitative results from Hartree-Fock calculations are often unreliable. It is useful to think in terms of "dynamical" and "non-dynamical" correlation effects that go beyond the Hartree-Fock molecular orbital approach [2]. Roughly speaking, dynamical correlation is a short-ranged effect that decreases electron-electron repulsion within a single configuration; additional "non-dynamical" effects are often long-ranged, and arise when there is a near-degeneracy of more than one configuration. Transition metal complexes turn out to be more difficult than typical organic molecules on both counts [2]. First, the small spatial extent of the *3d* shell leads to large dynamical correlation effects, and high angular momentum basis functions are needed to describe these in a CI or perturbation approach. Second, systems with unpaired spins frequently have a number of near-degenerate configurations for both ground and excited states, so that non-dynamical correlation effects are frequently important. Even at a qualitative level, the presence of unpaired electrons can lead to large spin-polarization effects that can complicate or invalidate simple molecular orbital arguments. Since there is no practical single approach that handles all of these problems, a variety of approaches have to be considered [3].

2.1 GENERAL FEATURES OF DFT APPLIED TO TRANSITION METALS

It is clear that no short review can possibly cover such a broad topic, but we will take this opportunity to lay out some general conclusions we have drawn from our experience about the performance of various types of density functional theory in transition metal systems.

(1) Spin polarization effects are often an important part of the energetics of complexes of the type seen in proteins, where ligand fields are often weak and high-spin metal configurations are quite common. An extensive discussion of the qualitative and quantitative effects of these changes is given elsewhere [4].

(2) Bond energies are generally overestimated in the simplest, local spin density DFT theory, but are considerably better with gradient corrections. Since density gradients tend to be smaller in molecules than in the isolated atoms or fragments, an approximation neglecting gradient information works better for molecules than for atoms: the "non-local" models favor density inhomogeneities more than local theories do. For many purposes, non-local DFT provides a useful level of approximation for considering geometries and chemical energetics. Organic systems have been well-studied, both for energetics and

structures. For small organic molecules, non-local density functional methods usually give bond energies accurate to 3 to 5 kcal/mol, bond lengths accurate to 0.02 Å, and bond angles accurate to a few degrees. In addition, accurate energies have been obtained for a variety of chemical reactions, excitation energies, electron affinities, proton affinities, and ionization potentials [3,5-7]. Transition metal complexes provide greater challenges for all quantum chemical methods. Geometries and vibrational frequencies have been calculated with good accuracy in recent work (average bond length error 0.03 Å) [8]. We have calculated bond energies for a variety of small spin polarized transition metal complexes ($MnCl_2$, $FeCl_2$, $FeCl_3$, ScO, TiO, VO, CuO), which gave bond energies with errors in the range 1-17% (0.8 to 11 kcal/mol.), in agreement with the findings of other groups for example, on simple metal hydrides [9].

(3) Correlation effects often tend to oppose exchange effects, particularly in open shell systems. In the absence of correlation, exchange provides the only direct pathway to reduce electron-electron repulsion by correlating electrons of the same spin, but this must compete with related changes in kinetic and nuclear attraction energies. Correlation can allow opposite spin electrons to avoid each other as well, reducing the "need" for large exchange effects. In practice, this means that exchange forces that favor localization of parallel-spin electrons (such as in a high-spin transition metal ion) will be reduced. DFT models are considerably better than the Hartree-Fock, which significantly overestimates the strength of local exchange forces, as we discuss below.

(4) When correlation is turned on, the hole around a given electron becomes deeper and more short-ranged, so that approximations based on local information become more useful. The distinction between exchange and correlation is not always a useful concept, especially in the presence of non-dynamical correlation effects; further, the so-called "exchange-correlation" energy of density functional theory also contains a kinetic energy component [10,11]. Many non-local models show some cancelation of errors in their treatment of exchange and correlation, so that the sum is more accurate than the individual components, and accounts of correlation are expected to be more trustworthy for antiparallel spins than for parallel spins [12], although the practical consequences of this for open-shell systems are not yet clear.

2.2 ENERGETICS OF METAL IONS AND HEXA-AQUA COMPLEXES

The energetics of redox processes for metal ions in solution can be partitioned into two contributions: the gas phase ionization potential of the reduced species and the solvation energy difference between the oxidized and reduced states. The importance of ionization and excitation processes is even broader, since these are fundamental to both bonding and net electron transfer processes. We begin with an assessment of the performance of density functional calculations for ionization and excitation energies of the first row transition metal ions. Following this, we examine the energetics of formation of the corresponding metal ion-hexaaqua complexes.

Fig. 1 shows the calculated vs. experimental ionization potentials for the $M^{2+,3+}$ (IP_{III}) and $M^{1+,2+}$ (IP_{II}) processes. Despite some quantitative

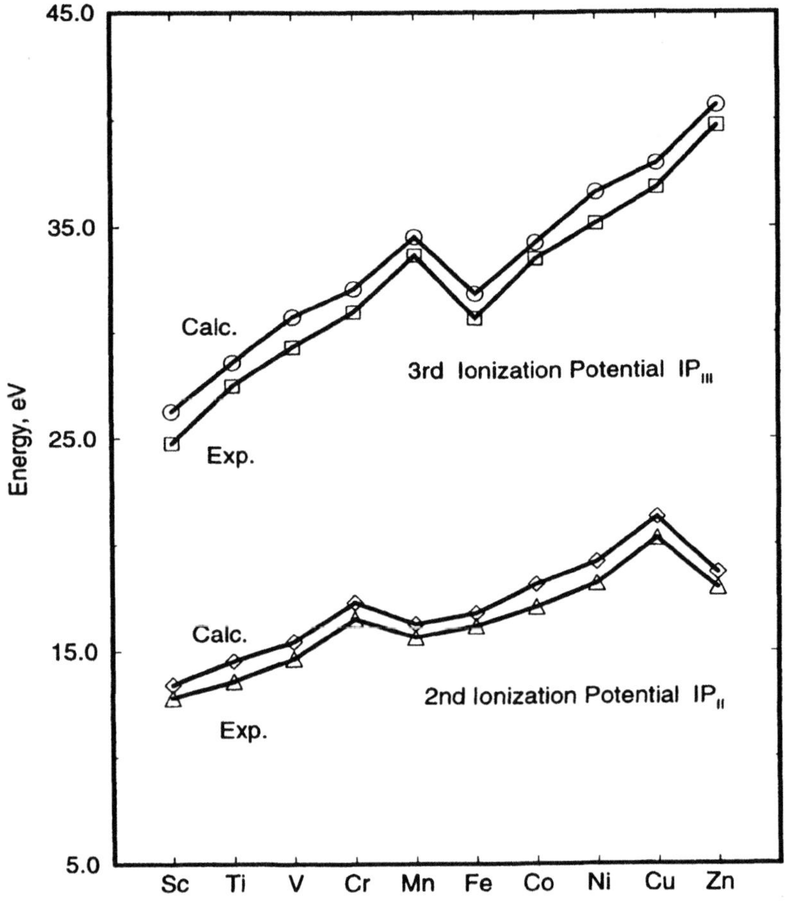

Fig. 1. DFT (Becke88 + Perdew86) calculations of transition-metal ionization potentials.

differences, the calculated IP's track the observed IP's quite well, with errors of about 0.5-1 eV. In these calculations, core orbitals were frozen through M(3s,3p) and relativistic effects were neglected. Relaxation of these restrictions will improve the overall agreement with experiment. The relativistic correction improves the IP_{III} of $Mn^{2+,3+}$ and $Fe^{2+,3+}$ by 0.2 eV [13], and the use of a smaller frozen core has a similar effect. Nevertheless, the method illustrated in Fig. 1 is not an expensive one, and can readily be applied to rather large molecules with complex ligands.

The IP_{III} plot shows a relative maximum at $Mn^{2+,3+}$, where removing a Mn(3d) electron corresponds to losing the enhanced stability of the half-filled 3d shell $d^5 \rightarrow d^4$, and the relative minimum at $Fe^{2+,3+}$ corresponds to having the half-filled shell for the oxidized state $d^6 \rightarrow d^5$. A similar argument can be

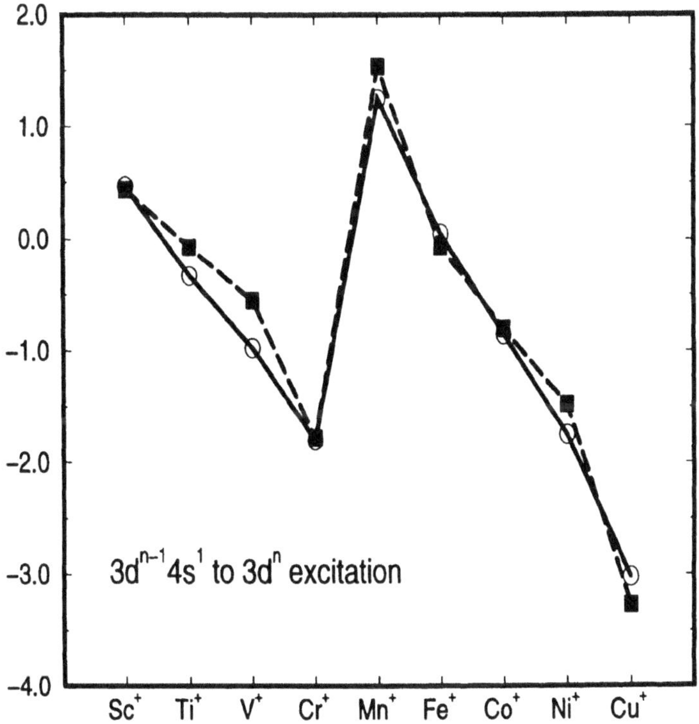

Fig. 2. DFT (Becke88 + Perdew86) calculations of transition metal excitation energies, in eV. Squares are experiment; circles are calculated.

made for the relative maxima of $Cr^{1+,2+}$ and $Cu^{1+,2+}$, losing a half-filled shell and filled shell for the oxidized species. The gradual increase shown across the series Sc → Zn is a consequence of the increasing nuclear charge across the series, and that electron screening of this charge is only partially effective. The other notable observation is that the IP's are very large, particularly for the $M^{2+,3+}$ couple. Comparing this with the experimental first IP of the water molecule, 12.6 eV, we see that the IP_{III} values are much larger for the entire series, and the very large electron relaxation effects observed below for the M(III)-hexaaqua ions becomes plausible.

Fig. 2 shows the calculated *vs.* experimental $3d^{(n-1)}4s^1 \rightarrow 3d^n$ promotion energy of the first row transition metal cations M^+, which is a model for the d → s contribution to bonding in transition metal complexes. Here the agreement of the calculations with experiment is closer than for the IP's, probably because core relaxation effects, which are omitted in the calculations, are less important when the charge on the atom is not changing.

Transition metal-hexaaqua complexes provide good examples of the comparative influences of ligand field stabilization and spin polarization energies in determining the ground electronic state. These energies are often in competition, depending on the metal d electron count. This is particularly evident when comparing the energies of high spin vs. low spin complexes.

In Fig. 3 we compare the calculated binding energies for $M(H_2O)_6^{2+,3+}$ formation from the ion and six separated H_2O molecules with the experimental hydration enthalpies for the first row transition metal series. The calculations used a triple-ζ basis, plus one set of polarization functions, and the Becke88 and Perdew86 gradient corrections for exchange and correlation. The experimental hydration energies are much more negative because of the presence of the second solvation shell and the extended solvent environment. We will examine such effects later when we discuss condensed phase results. Both the binding and hydration energies display very similar double bowl shapes for both oxidation states. All the ground state complexes are high spin both experimentally and from the calculations with the single exception of $Co(H_2O)_6{}^{3+}$, which is observed to be low spin by UV/visible spectroscopy [14]. The effects of spin polarization combined with the increasing nuclear charge across the series can be seen in the downward sloping dashed line which interpolates between the empty (d^0), half-filled ($d^5\alpha$) and filled ($d^5\alpha d^5\beta = d^{10}$) 3d shell. The double bowl shape (open circles for calculated high-spin states) is then associated with the additional ligand field stabilization energy, which is strongest at $d^3\alpha = t_{2g}^3\alpha$ and at $d^5\alpha d^3\beta = t_{2g}^3\alpha e_g^2\alpha t_{2g}^3\beta$. The low spin states are

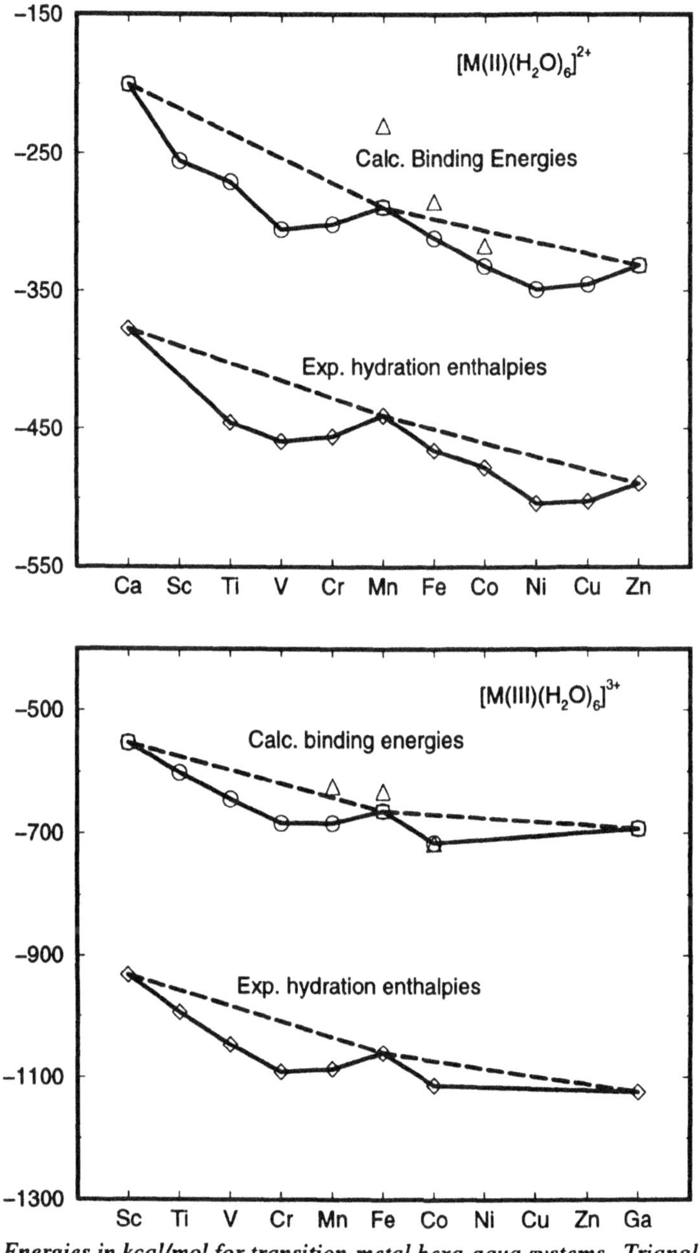

Fig. 3. Energies in kcal/mol for transition-metal hexa-aqua systems. Triangles show calculations for low-spin systems; all others are high spin.

usually higher in energy since the loss of spin polarization is not sufficiently compensated by LFSE energy. The low spin state energies (triangles) are plotted for the Mn, Fe and Co complexes. The oxidizing power of the M(III) ions is so large that the Ni(III), Cu(III), and Zn(III) ions are unstable in water, leading to rapid oxidation of H_2O and reduction of the metal to M(II). Co(III) is the last metal ion in the series for which the hydration energy, and aqueous redox potential are measurable [15-17]. Consequently, Ni(III), Cu(III), and Zn(III) hydration energies were omitted from the figure.

Only for Co(III) is the low spin complex the ground state. The predicted low spin complex is nearly degenerate with the high spin $Co(H_2O)_6^{3+}$. The low-spin state is only 0.02 eV higher in energy than high-spin; however, this ordering reverses (low-spin becoming 0.14 eV lower than high-spin) when a smaller frozen core is used [freezing only Co(1s,2s,2p) instead of Co(1s,2s,2p,3s,3p)]. Even rather large ab initio multi-reference CI (MRCI) calculations do not yield the experimental HS vs. LS state ordering, with the predicted E(LS)-E(HS) energy difference being 10 kcal/mol [17]. This illustrates a principal advantage of DFT in providing a good intial description of spin-state energetics in transition metal complexes.

2.3 ELECTRONIC STRUCTURES OF MULTIPLE-METAL COMPLEXES

The Kohn-Sham procedure uses orbitals from a single-determinant picture to construct the densities, and builds in correlation effects through an approximate density functional. This procedure works well for systems whose wavefunction is dominated by a single configuration, and whose correlation is therefore primarily a dynamic one; the description of short-range, dynamical correlation in terms of a transferable function of the density and its gradients is accurate enough to be quantitatively useful in a wide range of systems. For systems with substantial mixing of configurations, however, where non-dynamical correlation is important, currently available Kohn-Sham procedures run into both conceptual and practical difficulties. Basically, these non-dynamical effects are longer ranged [11] and can have a wide variety of consequences, since the nature of the near-degenerate configurations that mix can be completely different from one system to another. No simple approximate density functional, especially if based on local information, can be expected to describe such diverse behavior.

One approach breaks the underlying symmetry of the spin density distribution [18]. Unlike the Pauli principle, which is a precise symmetry requirement for fermions, the Coulomb hole is essentially a means by which a wave function can lower its electron-electron (e-e) repulsion, and also its total energy.

One way to achieve this is to remove whatever constraints are keeping the e-e repulsions too high. In a standard spin-restricted framework, the densities satisfy $\rho^{\alpha}(r) = \rho^{\beta}(r) = \rho/2$, and the electron repulsion between α and β electron densities is high. The broken symmetry model changes the form of ρ^{α}, ρ^{β}, allowing the electron densities for α and β spin to differ in space; how they differ precisely will be determined by the self-consistent field of the system in the density functional calculation. For the individual orbitals, this means removing the constraint that the orbitals be members of the irreducible representations of the molecular point group defined by the nuclear geometry. The orbitals are instead calculated in a subgroup compatible with the symmetries of ρ^{α} and ρ^{β}, although it is the localization property more than the point group symmetry that is essential. By using unrestricted spin-orbitals, all of the energy terms (kinetic energy, nuclear electron attraction, and e-e repulsion) will change in a variationally optimal way. Both the classical electron-electron repulsion and the exchange energy terms are altered when passing from the spin restricted to the spin unrestricted case. In general, the α and β densities become more compact, and occupy different regions in space. These effects are quite strong when there are spin polarized transition metal atoms, because the energetic cost on the kinetic energy of having a compact α and β densities is outweighed by the advantageous lowering of some combination of the "classical" electron-electron repulsion, the exchange part of the spin polarization, and the nuclear-electron attraction energies. Further, the broken symmetry state and its energy will differ significantly from the high spin state and its energy, even though both will experience spin polarization, because the two states have different covalency and delocalization, affecting all the energy terms.

The broken symmetry approach for spin-polarized and spin-coupled systems, in combination with spin projection methods, provides a intuitive picture of energetic interactions that contains most of the physics involved. The basic idea is to construct "broken-symmetry" single determinant wavefunctions, which are not pure-spin states, and therefore are not expected to represent the true wavefunctions of the system. Rather, spin-projection techniques are used to estimate the energies and properties of the "correct" pure-spin states.

We can represent an antiferromagnetic spin-coupled state (broken symmetry state) within density functional theory as analogous to a spin-unrestricted determinant in which spin-up electrons are predominantly located on some metal sites, and spin-down electrons on others [18,19]. The broken symmetry method has been applied to a variety of transition metal complcxcs [20,21] Although the spin localization provides important stabilization for the system, and leads to single-determinant models that more faithfully represent many aspects of the underlying (multi-configurational) electronic wavefunction, there are important and difficult questions about how the energies and properties of the broken symmetry states are to be interpreted, *e.g.* whether and how the energies of multiconfigurational states can be estimated from such density functional studies.

The theory for a system with only two unpaired electrons is especially simple [22,23]. Consider an open shell composed of two orthogonal space orbitals a,b with opposite spins

$$\Psi_B = A(ab\alpha\beta) \equiv |ab\alpha\beta| \tag{1}$$

Following common practice, the coordinate labels can be omitted, with the convention that the labels increase from left to right. The function Ψ_B is not an eigenfunction of S. Proper spin eigenfunctions $\Psi(S, M_S)$ are given by [22]

$$\Psi(1,1) = |ab\alpha\alpha|, \quad \Psi(1,-1) = |ab\beta\beta|,$$

$$\Psi(1,0) = 1/\sqrt{2}\,(|ab\alpha\beta| + |ab\beta\alpha|),$$

$$\Psi(0,0) = 1/\sqrt{2}\,(|ab\alpha\beta| - |ab\beta\alpha|) \tag{2}$$

from elementary spin algebra. Therefore,

$$\Psi_B = 1/\sqrt{2}\,[\Psi(1,0) + \Psi(0,0)] \tag{3}$$

The state Ψ_B has $M_S = 0$, but it is not an eigenstate of spin. Rather, it is a state of mixed spin (with both S=1 and S=0 contributing,) and will be of broken spatial symmetry if $a(r) \neq b(r)$.

In general, a mixed spin, broken symmetry state takes the form [23,24]

$$\Psi_B = \sum_S C(S)\,\Psi_S \tag{4}$$

Since the "true" electron Hamiltonian of a many electron system cannot mix eigenfunctions of different spin (in the absence of spin-orbit coupling), and since the total spin operator squared $\mathbf{S}^2$ commutes with this Hamiltonian, we have the energy equation for the broken symmetry state [23,24]:

$$E_B = \langle \Psi_B|H|\Psi_B \rangle = \sum_S [C(S)]^2 \langle \Psi_S|H|\Psi_S \rangle \tag{5}$$

There are no cross terms here between states of different S.

To evaluate the energy of the singlet (S=0) and triplet (S=1) states, we need also a separate equation for S=1. Keeping always with energy evaluations involving only single configuration wave functions:

$$E(S=1) = <\Psi(1,1)|H|\Psi(1,1)> = <\Psi(1,-1)|H|\Psi(1,-1)> \tag{6}$$

Note that we have chosen not to make use of $\Psi(1,0)$ since this has more than one configuration. There is nothing "wrong" with multiconfigurational wave functions, but local density functional theory is most easily given a concrete and well defined form in terms of single configuration wave functions [23,24].

The energy of the broken symmetry wave function in the present case is now easily found

$$E_B = (1/2)[E(S=0) + E(S=1)] \tag{7}$$

Of more interest is the singlet-triplet splitting:

$$E(S=1) - E(S=0) = 2[E(S=1) - E_B] \equiv J \tag{8}$$

where J is the Heisenberg parameter in a Hamiltonian of the form $H = J\mathbf{S}_1 \cdot \mathbf{S}_2$. Eq. (8) has been used for many years to estimate singlet-triplet splittings in molecular excited states using $X\alpha$ or local density functional theory [25]. This approach can be generalized to the situation with more than two unpaired electrons, as we discuss elsewhere [18].

3. Electronic structure in condensed phases

Although the application of quantum chemistry to structural and energetic problems in gas-phase molecules has reached a high level of sophistication and maturity, many interesting problems remain for molecules in more complex condensed-phase environments, such as might be encountered at the active site of a protein. The electrostatic aspects of such surroundings can often be usefully described with continuum dielectric theory; for example, the protein interior might be described with a low local dielectric that contains embedded charges (to represent charged side chains and polar groups within the protein) surrounded by a high dielectric solvent. Even for complex geometries, the resulting electrostatic interactions can be determined by numerical solutions to the underlying differential equations, and such theories have come into widespread use in the past few years [26,27].

The idea of modeling the electrostatic portion of solvation by placing the solute in a cavity immersed in a dielectric continuum dates back to the Born model for ionic solvation [28] and to work by Kirkwood [29] and Onsager [30] on solvation effects in polar molecules, and the subsequent history of continuum-model calculations has recently been reviewed [31,32]. The simplest models use spherical cavities and consider just the coupling of the molecular dipole to the reaction field arising from polarization of the solvent dielectric [33]. This model involves straightforward computations, adding a term proportional to the dipole-moment operator to the gas-phase Hamiltonian. It provides reasonable results for simple solutes, but has obvious limitations for systems that are not very spherical, and it is known that both the shape of the cavity and higher-order moments of the charge distribution have important effects on the results [34,35]. Analytic expressions based on multipole expansions in ellipsoidal cavities are also feasible [34,36], but these are still restricted to relatively compact systems, and the multipole expansions involved often converge very slowly.

A more general approach has been developed by Tomasi and co-workers [37] that permits the boundary between the solute and the solvent to take on an arbitrary shape. In this approach the reaction potential is considered to arise from surface charge densities at the boundaries between different dielectric regions. A boundary-element method is used to discretize this charge density, turning the electrostatic integral equations into linear systems that can be solved by iterative or matrix-inversion methods [37,38]. Because the reaction potential arises from discrete charges (at each boundary element), integrals of this potential over the quantum-mechanical basis functions can be accomplished in the same manner as nuclear attraction integrals, and no multipole expansion of the charge distribution is required. This model, when combined with careful quantum calculations, has been shown to give an excellent account of many aspects of aqueous solvation of small organic molecules [31,39,40].

An alternative approach computes the reaction potential from a finite-difference solution to the Poisson-Boltzmann equation. Although the underlying physical models for the reaction potential are similar in the finite-difference and boundary-element methods, the finite-difference approach has two principal advantages: First, the effects of mobile counter-ions (salt effects) can easily be incorporated at the Debye-Hückel level; this can be done with some difficulty in boundary-element methods [41-43], but important elaborations like the inclusion of a counter-ion exclusion region are not readily incorporated. Second, the incorporation of non-homogeneous environments, such as a protein-solvent environment consisting of multiple dielectric regions with embedded charges, is straightforward and imposes little additional computational cost in the finite-difference approach but adds significantly to the number of boundary elements and the complexity of the relations between them in the boundary-element method.

3.1. BASIC EQUATIONS FOR CONTINUUM SOLVATION MODELS

Within the density functional approximation, the electronic energy of a molecule can be written as,

$$E = \sum_i < \psi_i | h(1) | \psi_i > + \frac{1}{2} \int \rho_{tot}(\mathbf{r})\, G(\mathbf{r},\mathbf{r}')\rho_{tot}(\mathbf{r}')\, d\mathbf{r}d\mathbf{r}' + E_{xc}[\rho_{el}(\mathbf{r})] \quad (9)$$

where the one-electron Hamiltonian, $h(1)$, contains the kinetic energy integrals, the ψ_i are the Kohn-Sham orbitals, ρ_{tot} is the total charge density due to both electrons and nuclei, and $E_{xc}[\rho_{el}(\mathbf{r})]$ represents the exchange and correlation energy for the electron density $\rho_{el}(\mathbf{r})$. $G(\mathbf{r},\mathbf{r}')$ is the Coulomb interaction operator, which will just be $1/|\mathbf{r}-\mathbf{r}'|$ in the gas-phase, but will have a more complex form in solution, as we discuss below.

Applying the variational principle, $\delta E/\delta \rho_{el}(\mathbf{r}) = 0$ to Eq. (9) yields the Kohn-Sham equations for the one-electron orbitals:

$$\left[h(1) + \int G(\mathbf{r},\mathbf{r}')\, \rho_{tot}(\mathbf{r}')\, d\mathbf{r}' + v_{xc}[\rho_{el}(\mathbf{r})]\right]\psi_i(\mathbf{r}) = \varepsilon_i \psi_i(\mathbf{r}) \quad (10)$$

where

$$v_{xc} = \frac{\delta E_{xc}[\rho_{el}(\mathbf{r})]}{\delta \rho_{el}(\mathbf{r})} \quad (11)$$

and

$$\rho_{el}(\mathbf{r}) = \sum_i |\psi_i(\mathbf{r})|^2 \quad (12)$$

The crux of this approach to solvation is to take the effective electron-electron interaction, $G(\mathbf{r},\mathbf{r}')$ to be the Green function of the Poisson equation,

$$\nabla\, \varepsilon(\mathbf{r})\, \nabla\, G(\mathbf{r},\mathbf{r}') = -4\pi\, \delta(\mathbf{r},\mathbf{r}') \quad (13)$$

where $\varepsilon(\mathbf{r})$ is the (non-uniform) dielectric constant, and $\delta(\mathbf{r},\mathbf{r}')$ is a Dirac's delta function representing a unit charge at $\mathbf{r}'$. For the molecular "interior" that we treat by quantum mechanics, we set $\varepsilon(\mathbf{r}) = 1$. This makes it convenient to divide the Green function into a gas-phase-like Coulomb part and a remainder which we define as the "reaction potential" contribution, G^*, representing the effect of solvent:

$$G(\mathbf{r},\mathbf{r}') = \frac{1}{|\mathbf{r}-\mathbf{r}'|} + G^*(\mathbf{r},\mathbf{r}') \quad (14)$$

As long as $\mathbf{r}$ and $\mathbf{r}'$ are both in the interior, $G^*(\mathbf{r},\mathbf{r}')$ can be thought of as the potential due to the surface charge induced on the dielectric interface by a unit charge at $\mathbf{r}'$. In the exterior, $G^*(\mathbf{r},\mathbf{r}')$ can be thought of as the amount by which the Coulomb potential is reduced due to dielectric screening. With this separation, Eq. (9) can be rewritten:

$$E = \sum_i <\psi_i|h(1)|\psi_i> + \frac{1}{2}\int \frac{\rho_{tot}(\mathbf{r})\,\rho_{tot}(\mathbf{r}')}{|\mathbf{r}-\mathbf{r}'|}\,d\mathbf{r}d\mathbf{r}'$$

$$+ E_{xc}[\rho_{el}(\mathbf{r})] + \frac{1}{2}\int \rho_{tot}(\mathbf{r})\,G^*(\mathbf{r},\mathbf{r}')\rho_{tot}(\mathbf{r}')\,d\mathbf{r}d\mathbf{r}' \qquad (15)$$

where the solvent dependence enters only in the last term, which describes the interaction of the solute and solvent, and includes the work of polarizing the solvent. The other terms correspond to the gas-phase problem and are handled in usual way in the quantum mechanics program while the solvent reaction term is calculated by a finite difference method. The gas-phase and solvent parts of the calculation are brought to self-consistency by an iterative procedure. This division makes good numerical sense because the Coulomb integrals of the gas-phase part can be performed analytically (or semi-analytically, depending upon the code), and much work has gone into ensuring high accuracy for the estimation of these interactions. The reaction potential contribution, on the other hand, is much smaller and is a relatively smooth function of position, and more straightforward numerical integration techniques can be used for it. Details of our approach are given elsewhere [44].

There has been a revival of interest in these sorts of methods in quantum organic chemistry over the past few years, and there is not space to review these advances here, and other overviews are available [45,46]. Most approaches that allow an arbitrary cavity shape use boundary element methods to compute the solvent response, rather than the finite-difference approach outlined above. Among the many trends that might be mentioned, we would point out the increasing availability of gradients with respect to nuclear position (and hence, of automatic geometry optimization) [47,48], and the use of a simplified model called COSMO, which starts from the infinite dielectric (conductor) limit for the solvent, allowing simplified equations and derivatives [49]. Much of the work in the organic chemistry area has concentrated on neutral species, which are certainly of interest but are often easier to treat than the highly charged moieties characteristic of many transition metal complexes.

3.2. ABSOLUTE pK_as FOR ORGANIC IONS

As an illustration of work on organic ions, we present in Fig. 4 results of "absolute" pK_as for models of ionizable groups in proteins [50]. This is just the free energy difference (in solution) between the acid and its conjugate base, plus a constant that represents the solvation free energy of the proton. (The latter is related to the absolute energy of the standard hydrogen electrode, for which we use an experimental value.) Calculations of absolute pK_as thus require accurate estimates of both gas-phase proton affinities and of solvation free energies. The former are part of the traditional province of quantum chemistry, and it is known that DFT methods generally perform rather well [51].

Solvation energies for neutral species can often be measured, and a good many theories have been proposed and tested to compute and rationalize trends in solvation energies [52]. Many such studies have employed continuum solvent models, and generally, the agreement with experiment has been favorable, especially for aqueous solvation [45,53,54]. It is important to note that the Poisson-Boltzmann and similar treatments of solvation free energies provide only the "electrostatic" part of solvation, and that there are expected to be additional contributions to solvation free energies arising from the work required to insert (hypothetical) completely uncharged molecules into aqueous solution. This work involves the formation of a cavity of the appropriate size (usually a positive contribution to solvation free energy), plus any free energy arising from van der Waals and dispersion interactions between solute and solvent (usually a negative contribution to free energy). The sum of these non-electrostatic effects

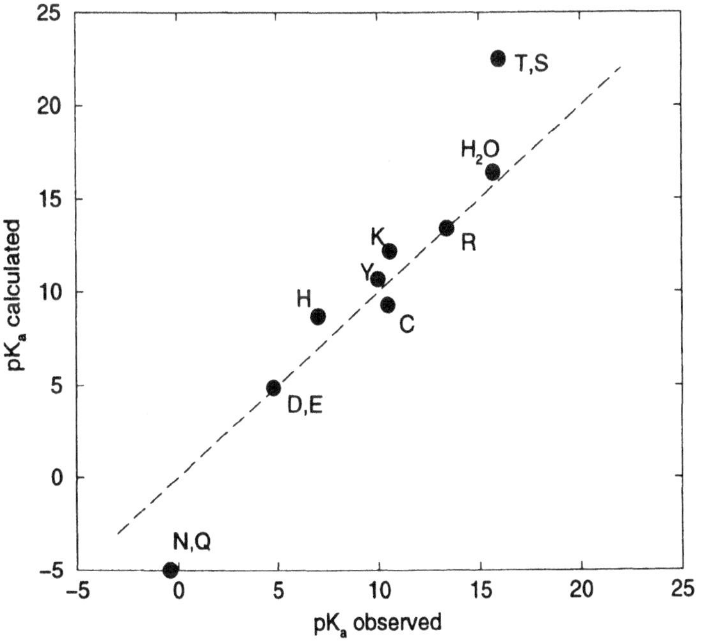

Fig. 4. Absolute pK_a values for models of protein sidechains.

tends to provide a positive contribution to the free energy of solvation. In our current model, non-electrostatic contributions for the deprotonated species are taken to be the same as their protonated counterparts, so that these contributions have no effect on the computed pK_a values.

Overall results are shown in Fig. 4. Although very strong and very weak acids are not predicted reliably, all groups titrating in the normal pH range show excellent agreement with observed values. An analysis of the individual contributions to these values [50] suggests that the gas-phase proton affinities are fairly accurate, with most of the error arising from estimates of the solvation free energies for the charged species; nevertheless, the general level of agreement is impressive, and provides a useful foundation for studies of general acid-base reaction mechanisms.

4. Redox potentials in transition-metal systems.

4.1. REDOX POTENTIALS OF AQUEOUS TRANSITION METAL IONS

Of greater interest to the construction of models for metalloproteins are similar calculations on oxidation/reduction and deprotonation of hydrated transition-metal cations. In a recent study [13], models for Mn^{2+}, Mn^{3+}, Fe^{2+} and Fe^{3+} were constructed with either one or two shells of water molecules (six or eighteen molecules) treated explicitly in the density functional calculation, and the effects of the remaining water approximated by a continuum solvent. These are extremely challenging calculations: in the gas phase, the third ionization potentials of iron and manganese are about 750 kcal/mol, and hydration enthalpies are around 450 kcal/mol for the divalent cations, and over 1000 kcal/mol for the

Table 1. Properties of iron and manganese in water.

		Calc.	*Exp.*
Hydration enthalpies	Fe^{2+}	431	465
kcal/mol	Fe^{3+}	1046	1060
	Mn^{2+}	409	440
	Mn^{3+}	1073	1087
Redox potentials	$Fe^{3+} \rightarrow Fe^{2+}$	1.06	0.77
eV	$Mn^{3+} \rightarrow Mn^{2+}$	1.59	1.56

trivalent species. Table 1 shows results for hydration enthalpies and redox potentials. The overall results are in impressive agreement with experiment, and suggest that there are no major deficiencies in either the density functional or the continuum solvent models being used.

4.2. REDOX POTENTIALS IN BLUE COPPER PROTEINS

We recently carried out an extension of these ideas to blue copper proteins, comparing plastocyanin (Pc) and rusticyanin (Rc) [55]. These are both single-copper electron transfer proteins, with very similar active site geometries of distorted tetrahedra about copper, with two histidine, one cysteine and one methionine ligand. A key goal of the modeling was to explain the 300 mV higher redox potential for rusticyanin compared to plastocyanin. We used a continuum electrostatic model to investigate what structural features might be responsible for this difference. In this molecular-mechanics type model, the protein is represented as a set of partial charges (representing the dipoles and charged groups of the protein) contained in a continuum of low dielectric solute, which itself is embedded in a high dielectric ionic solvent.

The calculations were based on atomic coordinates from NMR-refined structures of rusticyanin [55] and French bean plastocyanin [56]. For rusticyanin at pH 2, we assume acidic residues are neutral and basic residues (including histidine residues not coordinated to the metal) are charged. For plastocyanin at pH 5.5, we assume acidic and basic residues are ionized.

The total electrostatic energy is given by:

$$E = \frac{1}{2}\sum_{i,j} q_i q_j g(i,j) \tag{16}$$

where the sums are over all partial charges $\{q_i\}$ in the protein and $g(i, j)$ is the electrostatic Green's function between atom centers i and j. The Green's function can be calculated from a finite-difference solution to the linearized Poisson-Boltzmann equation [57], and need only be calculated for atoms that have different charge in the reduced and oxidized states of the protein. The "diagonal" terms where i and j refer to the same atom give a Born or desolvation term, representing the gain or loss of solvent polarization upon reduction when no other charges or dipoles are present. It is this term that effectively penalizes the (partial) burial of charge within the protein interior; this term is often partially offset by interactions between the active site charges and other charges and dipoles inside the protein [58].

We used two different models for the charge change upon reduction of the copper atom in blue copper proteins. In the first and simpler model, we simply assigned a charge of +2 to the copper atom in the oxidized state, and +1 in the reduced state, setting the ligand charges to those of the AMBER 1994 force field [59]. It is clear that this simple model concentrates the charge change far too much on the copper ion, whereas quantum mechanical calculations of transition metal redox events consistently show that most of the charge change occurs at the ligands, in spite of the fact that their formal oxidation numbers do not change [18]. Hence we have used a second, "distributed" charge model based on $X\alpha$ scattered wave calculations of small clusters that mimic the active sites of blue copper proteins [60]. In this model, we took the net ligand charges from the $X\alpha$ calculations, and adjusted the copper charge so that the net active site charge (copper plus four ligands) was +1 for the oxidized state and zero for the reduced state. This gives a model where almost all of the charge change is on the ligand atoms. Although the details of the actual charge distribution in the protein are not known, results for these two models will give some idea of the sensitivity of the results to choice of active site charges, with the "distributed" charge results arguably a more realistic model for the charge changes that actually occur upon reduction.

Table 2 summarizes the differences in calculated redox potential (in 100 mM monovalent salt) between rusticyanin and French bean plastocyanin. The first column reflects the average over 15 rusticyanin structures obtained using the "distributed" charge model discussed above, and the second column contains the values calculated for the simpler model where only the copper charge changes. Both sets of numbers are shown as differences from the average redox energies calculated from 21 nmr refined coordinates for plastocyanin. The energetic contributions are separated into three parts: *active site*, which contains the difference in the solvation energy of the set of charges that change upon reduction and the interactions of those charges with the side chains of the ligands to the copper; *backbone*, which contains the contributions from the polypeptide backbone; and *side chains*, which summarizes the contributions of charged amino acid side chains. The overall difference in redox potential changes by only 35 mV between the two charge models, although it is clear that the partition between "active site" and "backbone" components is sensitive to this choice. As would be expected, interactions that reflect large distances, such as those with charged side chains, tend to be insensitive to the details of the active site charge model.

Table 2 shows that the polypeptide backbone makes the largest contribution to the difference in redox potential between plastocyanin and rusticyanin. The contribution to the Rc/Pc difference is dominated by the different orientation of peptide dipoles that are close to the active site, notably the peptide

Table 2. Contributions to $\Delta E^0(Rc - Pc)$

		distributed charges		*simple charges*	
Active site		-8		-125	
	desolvation		43		-179
	ligands		-51		54
Backbone		153		225	
	leu12/pro52		-38		-53
	ala33/lys81		33		21
	gly34/gly82		32		55
	pro36/gly84		53		129
	his37/his85		2		21
	asn38/ser86		-6		-4
	cys84/cys138		42		34
	his87/his143		9		23
	others		26		1
Side chains		84		93	
	lysines		26		26
	histidines		14		14
	glutamates		44		42
	aspartates		13		14
	others		-13		-3
Total		228		193	

Redox contributions for 100 mM salt are listed in meV. For the individual backbone contributions, the plastocyanin residue is listed to the left of the corresponding rusticyanin amino acid.

groups of the cysteine ligand, and those in the loop "above" the active site (in the orientation of Fig. 5), which contains the more buried histidine ligands and is comprised of residues 33-38 in plastocyanin and 81-86 in rusticyanin. For almost all of these residues, the negative (carbonyl) end of the peptide dipole is closer to the metal site in plastocyanin than in rusticyanin. For example, the distance from the carbonyl oxygen of pro 36 in plastocyanin to the Sγ atoms of the cysteine ligand (which has the biggest charge change upon reduction in the distributed charge model) is 4.0 Å; the corresponding distance in rusticyanin is 6.2 Å, and this shift increases the redox potential of rusticyanin by about 50

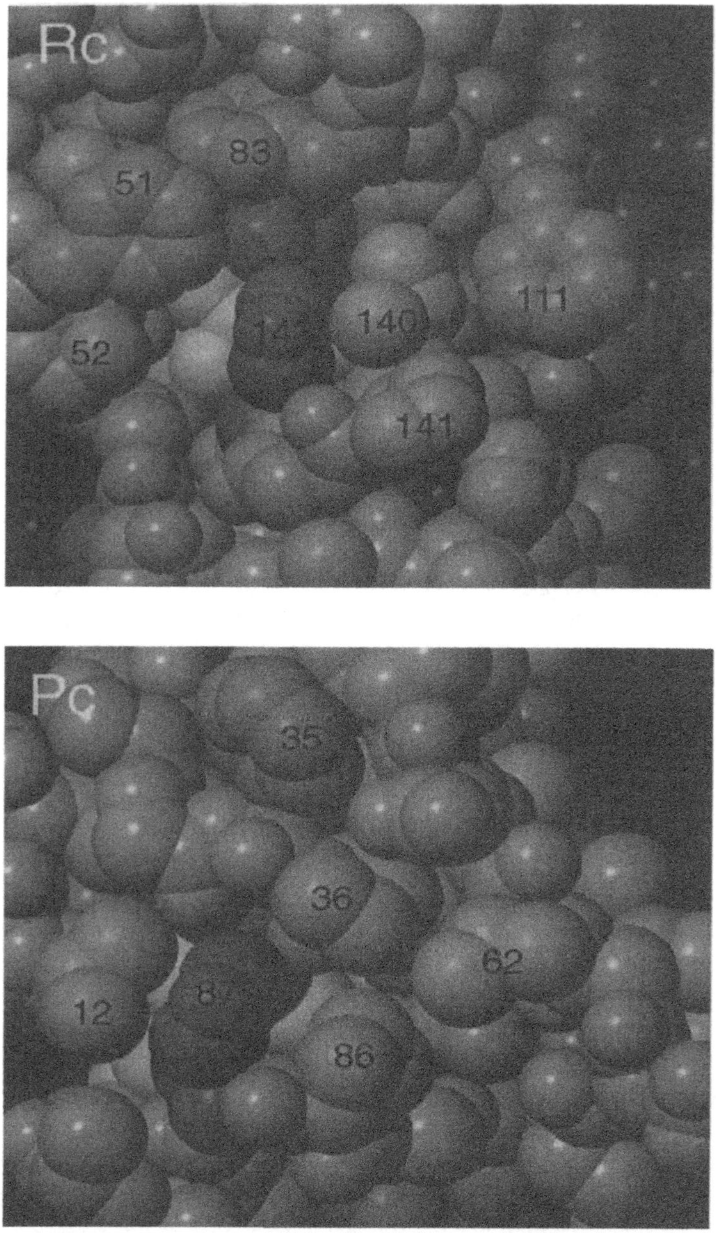

Fig. 5. Active site geometries of rusticyanin and plastocyanin.

meV relative to plastocyanin (distributed charge model). Similarly, the distance between the backbone carbonyl oxygen of the cysteine ligand and its own sidechain Sγ atom increases from about 3.3 Å in plastocyanin to about 3.7 Å in rusticyanin. The net effect of these and other rearrangements of the peptide backbone are predicted to make a significant contribution to the overall change in redox potentials.

There is overall a large charge difference between rusticyanin and plastocyanin, which can contribute to the ease with which the copper site can accept an electron. The nominal charge for rusticyanin at pH 2 is +18 (in the oxidized state) whereas it is -8 for plastocyanin at pH 5.5. However, most of the charged sidechains are distant from the active site, and their effect on the redox potential is screened by solvent and mobile counterions. Table 2 shows the contributions of various charged sidechains to the redox potential difference. The excess positive charge on rusticyanin relative to plastocyanin indeed stabilizes the reduced form of Rc, but only by an amount that is roughly 1/3 of the total redox potential difference.

Overall, these electrostatic calculations offer interesting insights into modulation of redox potential by the protein environment, but the detailed numbers should be treated with caution. Within our model of a rigid protein with a uniform internal dielectric constant, the model is fairly consistent, so that the standard deviation of predicted redox potentials over 15 rusticyanin structures and 21 plastocyanin structures is about 30 meV. But the fundamental structural assumptions (that the protein is rigid and is in the same conformation in oxidized and reduced forms) limit the overall realism of the model. Nevertheless, all of the calculations predict a significant increase in redox potential on going from a plant plastocyanin to the bacterial rusticyanin, in accord with experimental data, and three qualitative conclusions about the origins of this behavior are likely to be valid, even within the uncertainties of the overall method: first, the contribution of desolvation (in the distributed charge model) is about the same in the two proteins; second, the charged side chains contribute less than half of the overall redox shift; and third, the orientation of backbone dipoles relative to the active site makes the biggest contribution, with a large part of this coming from the residues adjacent to the more buried histidine ligand.

4.3. ABSOLUTE REDOX POTENTIALS FOR IRON-SULFUR CLUSTERS

Returning to quantum-mechanical descriptions, we have also carried out density functional calculations for 1Fe, 2Fe, and 4Fe iron-sulfur clusters [61]. From broken-symmetry and high spin state calculations, the energies of the spin ground states can be calculated, along with spin Hamiltonian parameters

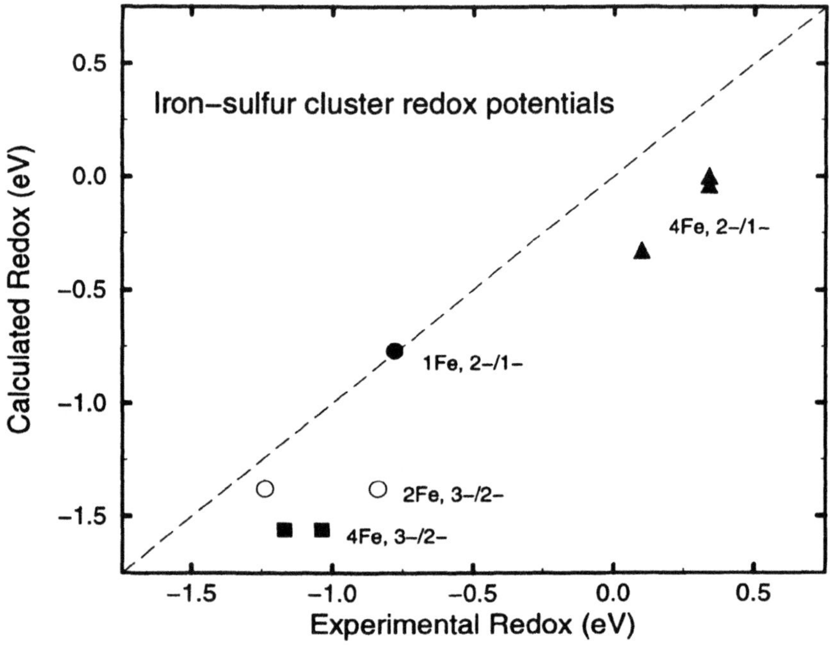

Fig. 6. Redox potentials for iron-sufur clusters.

J, B, B' for Heisenberg coupling, resonance delocalization within a mixed valence pair, and resonance between pairs, respectively. Environmental effects are modeled through a continuum dielectric, so that the solvent contribution to the redox potential can be calculated. There is a good correlation between predicted and measured redox potentials, and shown in Fig. 6. The Heisenberg spin coupling produces a negative redox potential shift for the 2Fe redox couple compared with one-iron clusters. Both spin coupling and resonance delocalization contribute to the much higher redox potential found for the 4Fe high potential $(2-/1-)$ couple compared to the reduced ferredoxin couple $(3-/2-)$ in synthetic analogues, and probably in proteins as well. The solvation contribution is large in all clusters and for all redox couples, but the calculations suggest that solvation mainly compensates for differences in electron-electron repulsion energies in vacuum.

A global comparison of our calculations (using $\varepsilon=37$) with experiment is given in Fig. 6. Overall the correlation between theory and experiment is acceptable, bearing in mind that these are "absolute" redox calculations. Experimentally, the redox potentials of these synthetic clusters span a total range of about 1.6 eV from about −1.24 to +0.34eV. The reduced 2Fe, 3−/2− and 4Fe, 3−/2− redox couples have the most negative redox potentials, covering overlapping ranges, followed by the 1Fe, 2−/1−, and finally with the oxidized HP couple 4Fe, 2−/1− at the most positive potential. The theoretical calculations (with $\varepsilon = 37$ as in DMF) give a similar total span of about 1.67 eV, with similar ordering, ranging from the reduced 4Fe, 3−/2− to 2Fe, 3−/2− at somewhat higher potential, followed by 1Fe, 2−/1−, and finally by the electronic states for the oxidized HP couple 4Fe, 2−/1− . The maximum calculation error is about −0.5eV. compared to experiment, with a range of error from 0.0 to −0.5eV., or generally less than 10% of the related solvation energy difference.

In the Born model for solvation, the solvation energy difference of the oxidized minus reduced state is given by $-[(Q_{ox}^2 - Q_{red}^2)/2R_{eff}](1-(1/\varepsilon))$, where the Q's are the total molecular charges, and R_{eff} is the effective molecular radius of the molecule, (R_{eff} can be approximated as a spherical cavity encompassing the Van der Waals envelope around the atoms.) The ratio of solvation energies for the 2−,3− couple versus the 2−,1− couple for 4Fe4S systems should then be approximately 5/3, since the molecular radius does not vary much among the three oxidation states. The calculated ratio obtained directly from solution of the Poisson-Boltzmann equation for our complicated molecular shape gives a ratio of $\Delta E_{PB}(solv)$ as 1.63 at both $\varepsilon = 37$ and $\varepsilon = 80$. This is entirely consistent with the simple Born model for these systems.

Another aspect of the Born equation above is the simple dependence of the solvation energy difference on the dielectric constant ε, which can be checked both against the more complex Poisson-Boltzmann result, and against the experimentally observed dependence of cluster redox potentials on solvent. For the 4Fe, 2−/1− couple, the ratios from the Born equation for the solvation energy difference at $\varepsilon = 9$, 80 to that at $\varepsilon = 37$ are 0.91,1.02, respectively. The Born equation prediction for ΔE_{PB} with $\varepsilon = 9$ or 80, starting from the calculated ΔE_{PB} for $\varepsilon = 37$, almost exactly corresponds to the values directly calculated from the simple formulas. While the Poisson-Boltzmann energies behave precisely as expected theoretically as a function of ε, the experimental behavior is more complicated. The predicted positive shift in redox potential with increasing ε of the solvent is not found experimentally with any uniformity. Experimental redox shifts of 4Fe4S clusters in solvent are not at all monotonic with dielectric constant, and can differ by 0.2 eV for solvents with nearly equal dielectric constants like acetonitrile and N,N dimethylformamide, both with $\varepsilon = 37$. Further, the positive shift observed in going from methanol ($\varepsilon = 32.6$), DMF, or DMSO ($\varepsilon = 47.1$) to water ($\varepsilon = 78.5$) can be as large as 0.5 eV, far more positive than the 0.1 eV shift predicted theoretically. (These calculations did not use different probe radii for different solvents, which might be appropriate.) It is likely, however, that the detailed molecular structure of the solvent is important for the finer scale aspects of redox potentials. Whether this is a result of bonding between the cluster and solvent, specific hydrogen-bonding interactions, electron donation, solvent size, or boundary layer effects, among other possibilities, is unknown, but there are clear indications of limitations to the continuum dielectric method.

Extension of these ideas to protein environments is expected to be a straightforward but challenging problem. The dielectric response of proteins is certainly more complex than that of simple solvents, but results like those discussed above for plastocyanin suggest that useful information is likely to be obtained. The relative redox potentials of iron-sulfur clusters in various protein environments have been studied by others [62-64], and we are currently working on extensions of these ideas to absolute redox potential calculations for protein systems.

Acknowledgments.

This research was supported by NIH Grants GM39914 and GM50154..

References

1. Holm, R.H., Kennepohl, P., and Solomon, E.I. (1996) Structural and functional aspects of metal sites in biology. *Chem. Rev.* **96,** 2239-2314.

2. Siegbahn, P.E.M. (1996) Electronic structure calculations for molecules containing transition metals. *Adv. Chem. Phys.* **93,** 333-387.

3. Ziegler, T. (1991) Approximate density functional theory as a practical tool in molecular energetics and dynamics. *Chem. Rev.* **91,** 651-667.

4. Li, J., Noodleman, L., and Case, D.A. (1997) Electronic structure calculations: Density functional methods with applications to transition metal complexes, in E.I. Solomon and A.B.P. Lever, Ed., *Inorganic Electronic Structure and Spectroscopy,* John Wiley, New York. To appear.

5. Baker, J., Muir, M., and Andzelm, J. (1995) A study of some organic reactions using density functional theory. *J. Chem. Phys.* **102,** 2063-2079.

6. Baker, J., Muir, M., Andzelm, J., and Scheiner, A. (1996) Hybrid Hartree-Fock density-functional theory functionals: The adiabatic connection method, in B.B. Laird, R.B. Ross, and T. Ziegler, Ed., *Chemical Applications of Density Functional Theory,* American Chemical Society, Washington, DC. pp. 342-367.

7. Curtiss, L.A., Raghavachari, K., Redfern, P.C., and Pople, J.A. (1997) Assessment of Gaussian-2 and density functional theories for the computation of enthalpies of formation. *J. Chem. Phys.* **106,** 1063-1079.

8. Sosa, C., Andzelm, J., Elkin, B.C., Wimmer, E., Dobbs, K.D., and Dixon, D.A. (1992) A local density functional study of the structure and vibrational frequencies of molecular transition-metal complexes. *J. Phys. Chem.* **96,** 6630-6636.

9. Ziegler, T. and Li, J. (1994) Bond Energies for Cationic Bare Metal Hydrides of the first Transition Series: a Challenge to Density Functional Theory. *Can. J. Chem.* **72,** 783-789.

10. Parr, R.G. and Yang, W. (1989) "Density-Functional Theory of Atoms and Molecules." Oxford University Press, New York.

11. Ernzerof, M., Burke, K., and Perdew, J.P. (1996) Density functional theory, the exchange hole, and the molecular bond, in J.M. Seminario, Ed., *Recent Developments and Applications of Modern Density Functional Theory*, Elsevier, Amsterdam.

12. Burke, K. and Perdew, J.P. (1995) Real space analysis of the exchange-correlation energy. *Int. J. Quantum Chem.* **56,** 199.

13. Li, J., Fisher, C.L., Chen, J.L., Bashford, D., and Noodleman, L. (1996) Calculation of redox potentials and pK_a values of hydrated transition metal cations by a combined density functional and continuum dielectric theory. *Inorg. Chem.* **35,** 4694-4702.

14. Lever, A.B.P. (1984) "Inorganic Electronic Spectroscopy, 2nd Ed.." Elsevier, Amsterdam, Netherlands.

15. Johnson, D.A. (1982) "Some thermodynamic aspects of inorganic chemistry, 2nd edition." Cambridge University Press, Cambridge, England.

16. Shriver, D.F., Atkins, P.W., and Langford, C.H. (1990) "Inorganic Chemistry." W.H. Freeman, New York.

17. Åkesson, R., Pettersson, L.G.M., Sandström, M., and Wahlgren, U. (1994) Ligand field effects in the hydrated divalent and trivalent metal ions of the first and second transition periods. *J. Am. Chem. Soc.* **116,** 8691-8704.

18. Noodleman, L. and Case, D.A. (1992) Density-functional theory of spin polarization and spin coupling in iron-sulfur clusters. *Adv. Inorg. Chem.* **38,** 423-470.

19. Noodleman, L. (1981) Valence bond description of antiferromagnetic coupling in transition metal dimers. *J. Chem. Phys.* **74,** 5737-5743.

20. Solomon, E.I., Tuczek, F., Root, D.E., and Brown, C.A. (1994) Spectroscopy of binuclear dioxygen complexes. *Chem. Rev.* **94,** 827-856.

21. Bencini, A. and Gatteschi, D. (1986) $X\alpha$-SW calculations of the electronic structure and magnetic properties of weakly coupled transition-metal clusters. The $[Cu_2Cl_6]^{2-}$ dimers. *J. Am. Chem. Soc.* **108,** 5763-5771.

22. McWeeney, R. and Sutcliffe, B.T. (1976) *Methods of Molecular Quantum Mechanics* Academic Press,

23. Ziegler, T., Rauk, A., and Baerends, E.J. (1977) On the calculations of multiplet energies by the Hartree-Fock-Slater method. *Theor. Chim. Acta* **43,** 261.

24. Noodleman, L. and Norman, J.G., Jr. (1979) The $X\alpha$ valence bond theory of weak electronic coupling. Application to the low-lying states of $Mo_2Cl_8^{4-}$.. *J. Chem. Phys.* **70,** 4903-4906.

25. Case, D.A. (1982) Electronic structure calculations using the $X\alpha$ method. *Ann. Rev. Phys. Chem.* **33,** 151-171.

26. Honig, B., Sharp, K., and Yang, A.-S. (1993) Macroscopic models of aqueous solutions: Biological and chemical applications. *J. Phys. Chem.* **97,** 1101-1109.

27. Davis, M.E. and McCammon, J.A. (1993) Electrostatics in biomolecular structure and dynamics. *Chem. Rev.* 509-521.

28. Born, M. (1920) Volumes and heats of hydration of ions. *Z. Phys.* **1,** 45-48.

29. Kirkwood, J.G. (1939) *J. Chem. Phys.* **7,** 911.

30. Onsager, L. (1936) *J. Am. Chem. Soc.* **58,** 1486.

31. Tomasi, J., Bonaccorsi, R., Cammi, R., and Valle, F.J. Olivares del (1991) Theoretical chemistry in solution. Some results and persepctives of the continuum methods and in particular of the polarizable continuum model. *J. Mol. Struct. (Theochem)* **234,** 401-424.

32. Karelson, M.M. and Zerner, M.C. (1992) Theoretical treatment of solvent effects on electronic spectroscopy. *J. Phys. Chem.* **96,** 6949-6957.

33. Wong, M.W., Frisch, M.J., and Wiberg, K.B. (1991) Solvent effects. 1. The mediation of electrostatic effects by solvents. *J. Am. Chem. Soc.* **113,** 4776-4782.

34. Rivail, J.L., Terryn, B., Rinaldi, D., and Ruiz-Lopez, M.F. (1985) Liquid state quantum chemistry: A cavity model. *J. Mol. Struct. (Theochem)* **120,** 387-400.

35. Rinaldi, D., Rivail, J.-L., and Rguini, N. (1992) Fast geometry optimization in self-consistent reaction field computations on solvated molecules. *J. Computat. Chem.* **13,** 675-680.

36. Ford, G.P. and Wang, B. (1992) The optimized ellipsoidal cavity and its application to the self-consistent reaction field calculation and hydration energies of cations and neutral molecules. *J. Computat. Chem.* **13,** 229-239.

37. Miertus, S., Scrocco, E., and Tomasi, J. (1981) Electrostatic interaction of a solute with a continuum. A direct utilization of ab initio molecular potentials for the prevision of solvent effects. *Chem. Phys.* **55,** 117-129.

38. Grant, J.A., Williams, R.L., and Scheraga, H.A. (1990) Abinitio self-consistent field and potential-dependent partial equalization of orbital electronegativity calculations of hydration properties of N-acetyl-N'-methyl-alanineamide.. *Biopolymers* **30,** 929-949.

39. Tuñón, I., Silla, E., and Tomasi, J. (1992) Methylamines basicity calculations. In vacuo and in solution comparative analysis. *J. Phys. Chem.* **96,** 9043-9048.

40. Tuñón, I., Silla, E., and Pascual-Ahuir, J.-L. (1993) Theoretical study of the inversion of the alcohol acidity scale in aqueous solution. Toward and interpretation of the acid-base behavior of organic compounds in solution. *J. Am. Chem. Soc.* **115,** 2226-2230.

41. Yoon, B. J. and Lenhoff, A. M. (1990) A boundary element method for molecular electrostatics with electrolyte effects. *J. Comp. Chem.* **11,** 1080-1086.

42. Zauhar, R.J. (1991) The incorporation of hydration forces determined by continuum electrostatics into molecular mechanics simulations. *J. Computat. Chem.* **12,** 575-583.

43. Juffer, A.H., Botta, E.F.F., van Keulen, B.A.M., Ploeg, A.van der, and Berendsen, H.J.C. (1991) The electric potential of a macromolecule in a solvent: A fundamental approach. *J. Computat. Phys.* **97,** 144-171.

44. Chen, J.L., Noodleman, L., Case, D.A., and Bashford, D. (1994) Incorporating solvation effects into density functional electronic structure calculations. *J. Phys. Chem.* **98,** 11059-11068.

45. Tomasi, J. and Persico, M. (1994) Molecular interactions in solution: an overview of methods based on continuous distributions of solvent.. *Chem. Rev.* **94,** 2027-2094.

46. Tawa, G.J., Martin, R.L., Pratt, L.R., and Russo, T.V. (1996) Solvation free energy calculations using a continuum dielectric model for the solvent and gradient-corrected density functional theory for the solute. *J. Phys. Chem.* **100,** 1515-1523.

47. Cortis, C.M., Langlois, J.-M., Beachy, M.D., and Friesner, R.A. (1996) Quantum mechanical geometry optimization in solution using a finite element continuum electrostatics method. *J. Chem. Phys.* **105,** 5472-5484.

48. Cossi, M., Tomasi, J., and Cammi, R. (1995) Analytical expressions for the free energy derivatives for molecules in solution. Application to the geometry optimization. *Int. J. Quant. Chem. Quant. Chem. Symp.* **29,** 695-702.

49. Andzelm, J., Kölmel, C., and Klammt, A. (1995) Incorporation of solvent effects into density functional calculations of molecular energies and geometries. *J. Chem. Phys.* **103,** 9312-9320.

50. Richardson, W.H., Peng, C., Bashford, D., Noodleman, L., and Case, D.A. (1997) Incorporating solvation effects into density functional theory: Calculation of absolute acidities. *Int. J. Quantum Chem.* **61,** 207-217.

51. Chandra, A.K. and Goursot, A. (1996) Calculation of proton affinities using density functional procedures: A critical study. *J. Phys. Chem.* **100,** 11596-11599.

52. Cramer, C.J. and Truhlar, D.G. (1994) Develpment and biological applications of quantum mechanical continuum solvation models, in P. Politzer and J.S. Murray, Ed., *Quantitative Treatments of Solute/Solvent Interactions*, Elsevier, Amsterdam. pp. 9-54.

53. Tannor, D.J., Marten, B., Murphy, R., Friesner, R.A., Sitkoff, D., Nicholls, A., Ringnalda, M., Goddard, W.A., III, and Honig, B. (1994) Accurate first principles calculation of molecular charge distributions and solvation energies from ab initio quantum mechanics and continuum dielectric theory. *J. Am. Chem. Soc.* **116,** 11875-11882.

54. Sitkoff, D., Sharp, K.A., and Honig, B. (1994) Accurate calculation of hydration free energies using macroscopic solvent models. *J. Phys. Chem.* **98,** 1978-1988.

55. Botuyan, M.V., Toy-Palmer, A., Chung, J., Blake, R.C., II, Beroza, P., Case, D.A., and Dyson, H.J. (1996) NMR solution structure of Cu(I) rusticyanin from *Thiobacillus ferrooxidans:* Structural basis for the extreme acid stability and redox potential. *J. Mol. Biol.* **263,** 752-767.

56. Moore, J.M., Lepre, C., Gippert, G.P., Chazin, W.J., Case, D.A., and Wright, P.E. (1991) High-resolution solution structure of reduced French bean plastocyanin and comparison with the crystal structure of poplar plastocyanin. *J. Mol. Biol.* **221,** 533-555.

57. Beroza, P. and Fredkin, D.R. (1996) Calculation of amino acid pK_as in a protein from a continuum electrostatic model: method and sensitivity analysis.. *J. Computat. Chem.* **17,** 1229-1244.

58. Bashford, D. and Karplus, M. (1990) pK_a's of ionizable groups in proteins: Atomic detail from a continuum electrostatic model. *Biochemistry* **29,** 10219-10225.

59. Cornell, W.D., Cieplak, P., Bayly, C.I., Gould, I.R., Merz, K.M., Jr., Ferguson, D.M., Spellmeyer, D.C., Fox, T., Caldwell, J.W., and Kollman, P.A. (1995) A second generation force field for the simulation of proteins, nucleic acids, and organic molecules. *J. Am. Chem. Soc.* **117,** 5179-5197.

60. Guckert, J.A., Lowerty, M.D., and Solomon, E.I. (1995) Electronic structure of the reduced blue copper active site: Contributions to reduction potentials and geometry. *J. Am. Chem. Soc.* **117,** 2817-2844.

61. Mouesca, J.-M., Chen, J.L., Noodleman, L., Bashford, D., and Case, D.A. (1994) Density functional/Poisson-Boltzmann calculations of redox potentials for iron-sulfur clusters. *J. Am. Chem. Soc.* **116,** 11898-11914.

62. Stephens, P.J., Jollie, D.R., and Warshel, A. (1996) Protein control of redox potentials of iron-sulfur proteins. *Chem. Rev.* **96,** 2491-2513.

63. Correll, C.C., Ludwig, M.L., Bruns, C.M., and Karplus, P.A. (1993) Structural prototypes for an extended family of flavoprotein reductases: Comparison of phthalate dioxygenase reductase with ferredoxin reductase and ferredoxin. *Prot. Sci.* **2,** 2112-2133.

64. Banci, L., Bertini, I., Savellini, G.G., and Luchinat, C. (1996) Individual reduction potentials of the iron ions in Fe_2S_2 and high-potential Fe_4S_4 ferredoxins. *Inorg. Chem.* **35,** 4248-4253.

SEMI-EMPIRICAL MO CALCULATIONS ON ENZYME REACTION MECHANISMS

TIMOTHY CLARK, PETER GEDECK, HARALD LANIG AND GUDRUN SCHÜRER
Computer-Chemie-Centrum
Institut für Organische Chemie
Friedrich-Alexander-Universität Erlangen-Nürnberg
Nägelsbachstraße 25
D-91054 Erlangen
Germany

1. Introduction

Historically, almost the entire range of possible types of structure and energy calculations (including dynamics) has been applied to the study of metalloenzymes. It is useful to consider the strengths and weaknesses of these types of calculation and their role in determining the characteristics of metalloenzymes and their reaction mechanisms. Table 1 shows a schematic view of the types of calculation and their role in the investigative process.

Ab initio and density functional (DFT) calculations have been used extensively for calculations on highly simplified model systems of enzyme reaction mechanisms. They serve both to calibrate semi-empirical calculational methods that will be used on larger active site models and to provide a baseline for the interpretation of the effects of enzyme catalysis. This latter function is particularly important. Good *ab initio* or DFT calculations are accurate enough to provide almost experimental quality thermodynamic and structural data for small reaction systems and serve to define the characteristics of the unperturbed gas phase reaction in order to be able to assess the effects of either solvent, the enzyme or both.

Classical mechanical molecular dynamics calculations, at the other end of the scale, provide enzyme-specific information, especially for systems where, for instance, access to the active site may be the rate-determining step in the enzymatic reaction. Molecular dynamics simulations may be used to choose between binding modes that cannot separate on the basis of enthalpy considerations alone [1]. These are naturally the calculations that are closest to the real experimental situa-

L. Banci and P. Comba (eds.), Molecular Modeling and Dynamics of Bioinorganic Systems, 307–317.

TABLE 1. Calculations used to investigate metalloenzyme reaction mechanisms.

Method	System	Purpose
Ab initio MO and density functional theory	Usually an extremly small model of the active site or of the uncatalyzed reaction	To determine the basic characteristics of the reaction and the role of the metal center in its enzymatically catalyzed counterpart. Often used to calibrate semi-empirical MO techniques for the system in question.
Semi-empirical MO	An extended active site model (typically 100–200 atoms)	Exact identification of the role of the metal and the residues around the active site in the mechanism of the enzyme-catalyzed reaction. Identification of a probable reaction mechanism by calculation of the activation energies. Possibly calculation of isotope effects for comparison with values measured for the enzymatic reaction.
Molecular Dynamics (based on classical dynamics)	Often the entire enzyme plus surrounding waters	Useful for determining aspects of the reaction mechanism not given by active site models (such as access to the active site, stereoselectivity, binding modes, etc.).

tion and should therefore provide the most relevant data. They do, however, have the disadvantage that they are comparably complicated and therefore often difficult to interpret. It is important to realize that the conceptually simplest models are also those that allow the clearest interpretation. Michael Dewar once observed that calculations that mimic experiment exactly are also as difficult to interpret as the experiment. Conceptual theoretical organic chemistry is, remarkably, almost entirely based on early Hückel calculations and, similarly, qualitative inorganic molecular orbital theory is based on extended Hückel theory. The combination of the methods outlined in Table 1 and the stepwise elaboration of the interpretation of the new effects that enter into each larger model system allow a uniquely detailed view of the exact nature of the reaction and a conceptual division of the final result into the contributions of various effects or chemical entities within the complete system.

Semi-empirical MO calculations on realistic active site models became possible about 10 years ago. Merz, Dewar and Hoffmann published a pioneering AM1-study of the mode of action of carbonic anhydrase in 1989 [2], after which the Erlangen group published similar studies on carboxypeptidase A [3] and liver alcohol dehydrogenase [4]. It is significant that these first three systems are all zinc-containing enzymes. The zinc parametrization of AM1 is particularly well suited to the predominantly ionic coordination bonding situation found in the active site of zinc enzymes, where the metal dication is complexed by one or more histidines, cysteines and water. Another important advantage of divalent metal ion-containing active sites is that they can often be optimized fully and still retain a structure very similar to that found in the enzyme itself (i.e. they are capable of existence as isolated entities in an enzyme-like structure). This means that an active site model can be calculated without geometrical constraints, so that, for instance, transition states can be located and characterized as such by calculation of the force constant (Hessian) matrix. If, for instance, some of the backbone atoms of the residues adjacent to the active site must be fixed in order to retain a structure similar to that found in the enzyme, transition state searches are often limited to reaction path calculations or two-dimensional grid searches, which for a system of 200 atoms are unlikely to reveal the true transition state. However, the highest energy structure found in such searches cannot be characterized as a minimum, true transition state or higher order stationary point because of the constraints used in the optimization. Thus, technically, the strongly bound active site models containing a central metal dication are preferable to the typically more loosely bound hydrogen-bond networks of active sites without a metal ion. It is not clear in many cases what the energetic effect of fixing atoms on the calculated thermodynamics and activation energy for a reaction will be. It is often the case that changing the position of fixed atoms slightly has a major effect on activation energies in particular, so that the investigator can never be sure that he is not biasing the results by the choice of fixed atoms and their positions.

An attractive alternative to fixing backbone atoms is to constrain to a given Cartesian position with a harmonic potential, the "hardness" of which can be adjusted to suit the environment. This technique has the advantages that it can be used effectively with the common transition state optimizers and that the stationary points can be characterized by calculation of the Hessian matrix because they represent real stationary points within the combination of quantum mechanical forces and applied harmonic potentials used for the calculation.

We have implemented both harmonic constraints to Cartesian positions and harmonic constraints of internal coordinates (useful for performing reaction coordinate searches using Cartesian coordinates) in our semi-empirical MO program VAMP. We will now describe the use of this technique to investigate the mode of action of phospholipase A_2.

2. Description of the biological system

2.1. OCCURENCE AND ACTIVITY OF THE PHOSPHOLIPASES

Phospholipases hydrolyze phospholipids, which are an important component of biological membranes. This class of enzymes is found in mammalian pancreas,

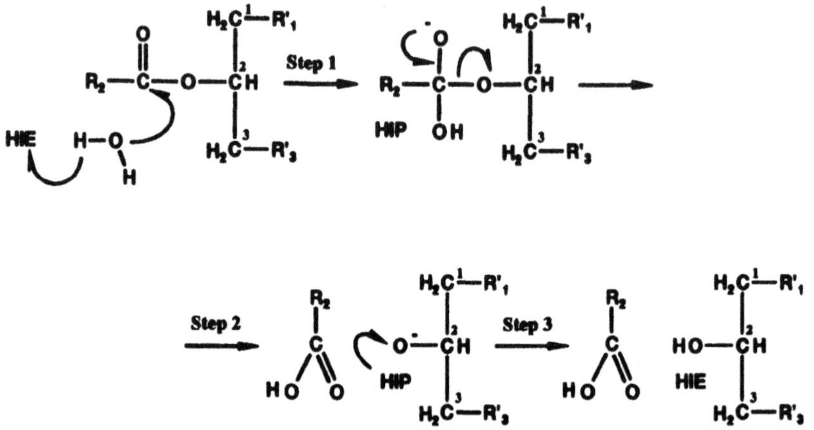

Fig. 1. Reaction mechanism of phospholipase A_2.

reptile and insect venoms and the synovial fluid.

Phospholipase A_2 (PLA_2) catalyses the hydrolytic cleavage of the *sn*-2-ester bond of L-glycerophospholipids. Often, this hydrolysis releases arachidonic acid, which is a precursor of the eicosanoides. These are mediators in inflammatory processes. Therefore, PLA_2 inhibitors are expected to have an anti-inflammatory effect [5]. PLA_2 is a small crystallizable enzyme. Its high degree of secondary structure and the large number of disulfide bridges make it very stable against denaturing conditions.

In this study, X-ray structures of the Chinese Cobra (*naja naja atra*) venom PLA_2, taken from the Brookhaven Protein Database, were used as starting point. The PDB entry 1poa contains the structure of the uninhibited enzyme, 1pob that of a complex with a phosphonate transition state analogue [6].

2.2. MECHANISM OF REACTION

The active site of PLA_2 contains a heptacoordinated Ca(II) ion. Substrate binding process takes place in this region. The diacylphosphatide is coordinated to the metal ion via an oxygen of the phosphate group and the carbonyl oxygen of the hydrolyzable ester group.

After coordination, the further reaction pathway can be divided into three steps. The interaction of the carbonyl group with the Ca(II) ion polarizes the carbonyl carbon, which can be attacked easily by a nucleophilic water molecule. This water molecule is present at the active site and is polarized and finally deprotonated by His48. Nucleophilic attack and proton transfer produce a tetrahedral intermediate (T.I.). The collapse of the T.I. is initiated by the cleavage of the bond between the carbonyl carbon and the ester oxygen. This results in a carbonic acid

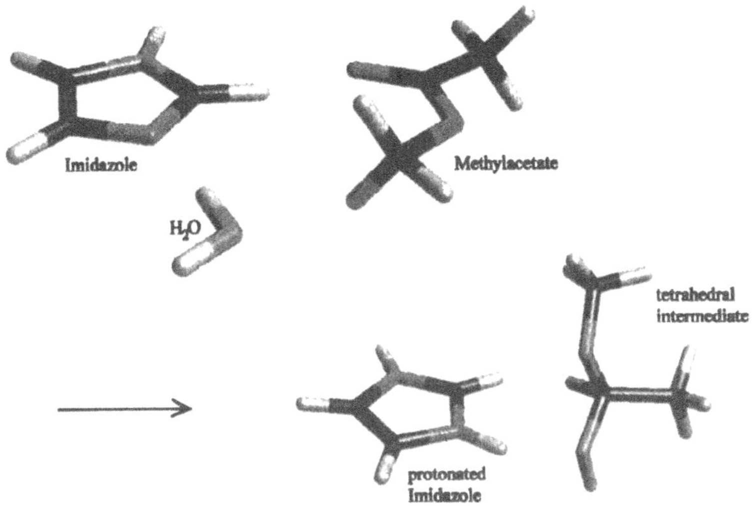

Fig. 2. Minimal system I - formation of the tetrahedral intermediate.

and an *sn*-2-alkoxide. Protonation of the alkoxide by His48 completes the reaction (Fig. 1).

3. Description of the method

For the investigation of the enzyme catalysis two different model systems have been used to describe the enzymatic system.

In the minimal system I (Fig. 2), the reactands are represented by model molecules. These are methylacetate as substrate, imidazole as His48 and a water molecule. The reaction mechanism of the system corresponds to the usually assumed general base catalysis process [7]. This model allows the investigation of the thermodynamics and the activation barrier of the base-catalyzed reaction, without the further influence of the enzyme.

Model system II (Fig. 3) represents the active site of PLA_2. This extended active site approximation contains the Ca(II) ion, the residues of its inner ligand sphere, the catalytic residues His48 and Asp99, a substrate molecule and the nucleophilic water. The coordination sphere of the metal ion consists of a calcium binding loop, Asp49 and the substrate. The residues of the calcium binding loop, which are not involved in calcium coordination, were replaced by glycine. 1-Phosphato-2-acetoxy-3-propanol is used as substrate. This approximation allows interactions of the reactands with an environment very similar to the native

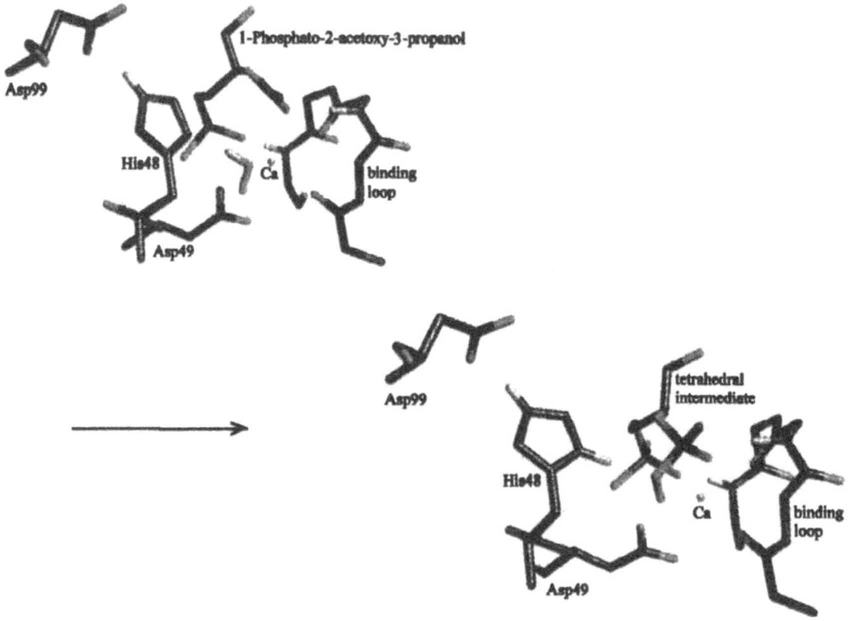

Fig. 3. Extended model system II - formation of the tetrahedral intermediate.

enzyme.

Comparing the extended active site approximation with the minimal system, the influence of the enzyme on the ester hydrolysis and its energetic course can be estimated. The calculations were carried out with the semi-empirical MO program VAMP [8] using the PM3 [9, 10] Hamiltonian. The extended active site approximation contains harmonic constraints fixing the backbone nitrogens and carbonyl carbon atoms of the amino acid segments to the positions given by the X-ray structure (1pob). The force constants chosen correspond to 1.6 mdyn/Å for the backbone atoms in the middle of the chains and 4.6 mdyn/Å for the atoms at the chain end. Even this slight influence causes a considerable improvement of the geometry conservation of the active site model during the optimization compared to the X-ray structure. The calculation of the reaction path was also carried out using harmonic constraints.

4. Results

4.1. STEP 1

In Fig. 4, the transition states of step 1, the formation of the tetrahedral intermediates (T.I.) are shown. This formation is the rate determining step. Comparing the energy profiles between the two model systems, the activation energy is lowered

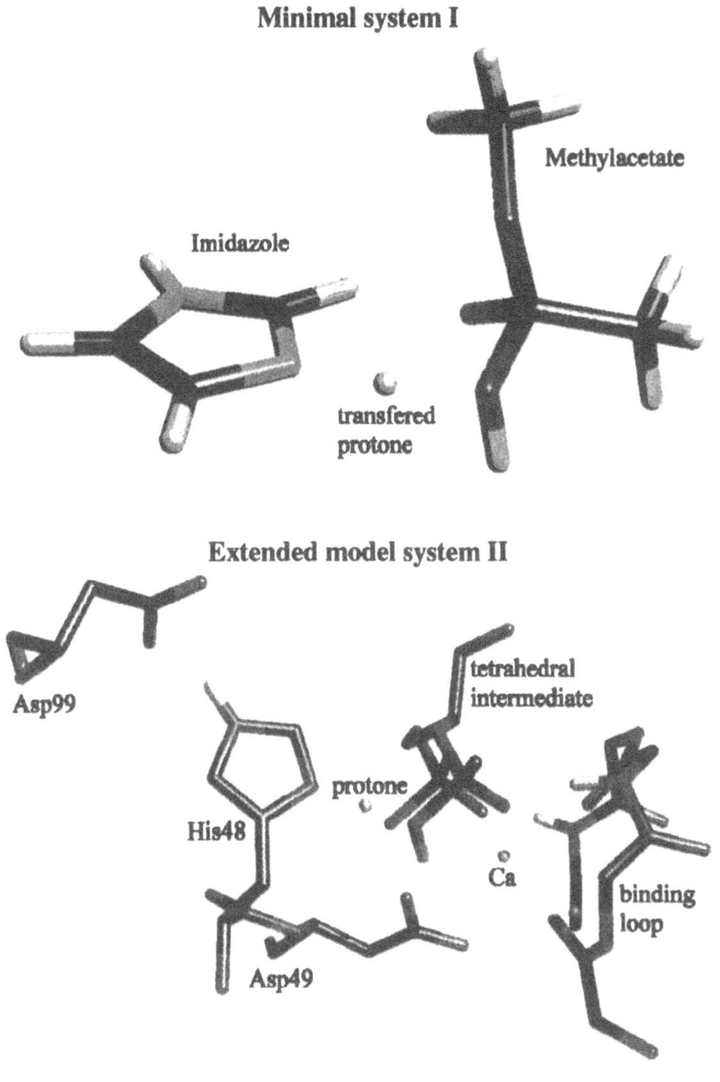

Fig. 4. Transition state of step 1, the formation of the tetrahedral intermediate.

from 45 kcal/mol in the minimal system to 22 kcal/mol in the extended active site approximation.

The reaction consists of a nucleophilic attack and a proton transfer from the water molecule to His48. Because of the resulting charge separation the proton transfer has an important contribution to the activation energy. The T.I. of the minimal system can only be stabilized by a favorable orientation of the charged molecules. However, the extended system shows a very similar geometry to the

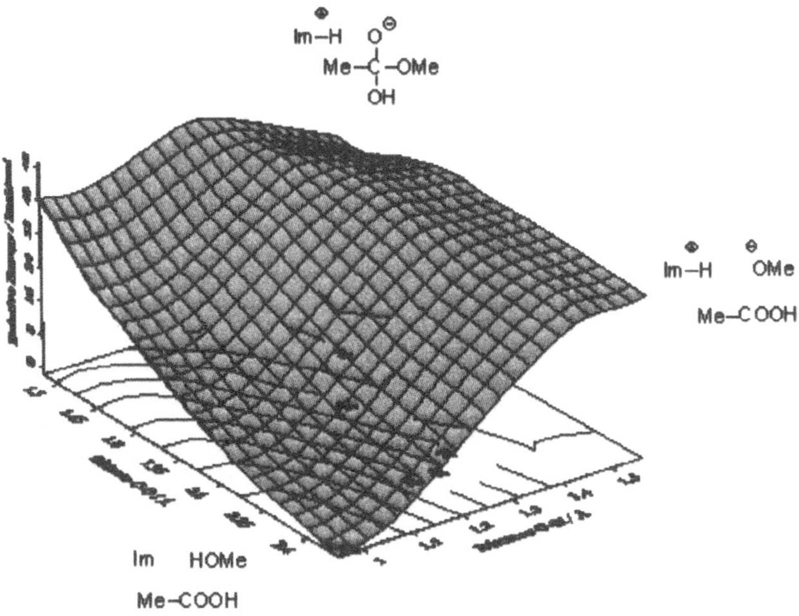

Fig. 5. Calculated energy surface for steps 2 and 3 in model system 1

optimized minimal system and further provides an environment stabilizing the charges. The negatively charged T.I. is able to interact with the metal ion and the amino acids in the coordination sphere. The positive charge on the protonated His48 can be stabilized by a hydrogen bond between the nitrogen Nε2 of His48 and the carboxy group of Asp99.

The geometries of the transition states are very similar in both model systems. They show that the bond formation between the carbonyl carbon and the oxygen of the water molecule is almost finished while the proton transfer is going on. We therefore conclude that both processes are concerted to a certain degree. The difference of the activation energy between the minimal and the extended system is 23 kcal/mol.

4.2. STEPS 2 AND 3

The formation of the products is initialized by the bond cleavage between the carbonyl carbon and the ester oxygen. The resulting *sn*-2-alkoxide is coordinated to the Ca(II) ion as an additional ligand. This structure is a second intermediate. Finally, a proton transfer from the protonated His48 to the alkoxide completes the reaction sequence.

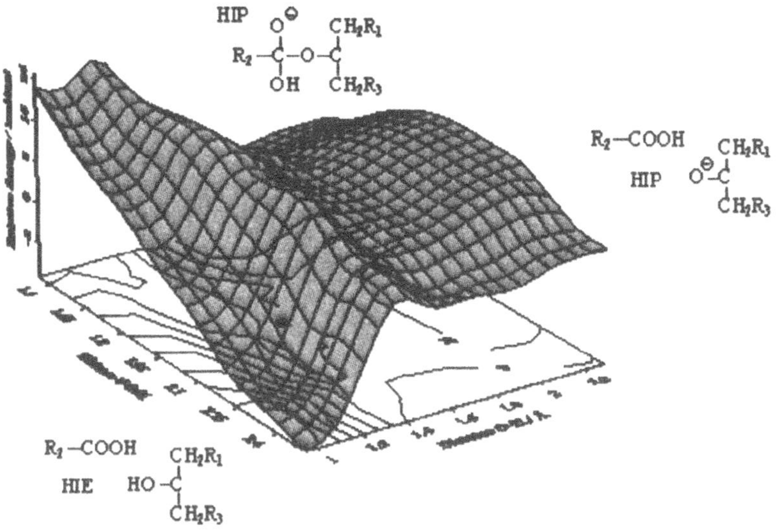

Fig. 6. Calculated energy surface for steps 2 and 3 in model system 2

Figs. 5 and 6 show the energy surfaces of the minimal and the extended models resulting from a grid calculation. The distance between the carbonyl carbon and the ester oxygen and the distance between the ester oxygen and the transfered proton were chosen as variables.

Comparison of the surfaces shows a similarity between the reaction course in the two model systems. The activation barrier for the proton transfer is much lower when the bond cleavage has already taken place. The influence of the enzyme environment on the activation energies is not very important for product formation from the tetrahedral intermediate. For the proton transfer, the activation barrier in the extended active site model (6 kcal/mol) is actually higher than in the minimal system (≈0.25 kcal/mol). However, this is not the rate determining step and therefore has no consequences for the kinetics of the catalytic process.

5. Conclusions

The rate determining formation of the T.I. is more favourable in the extended system, which better represents the active site environment. The considerable lowering of the activation barrier compared to the minimal system shows that the extended active site approximation describes the processes occuring in the native enzyme in better agreement to reality than the minimal system.

Considering that the extended active site approximation neglects the long-

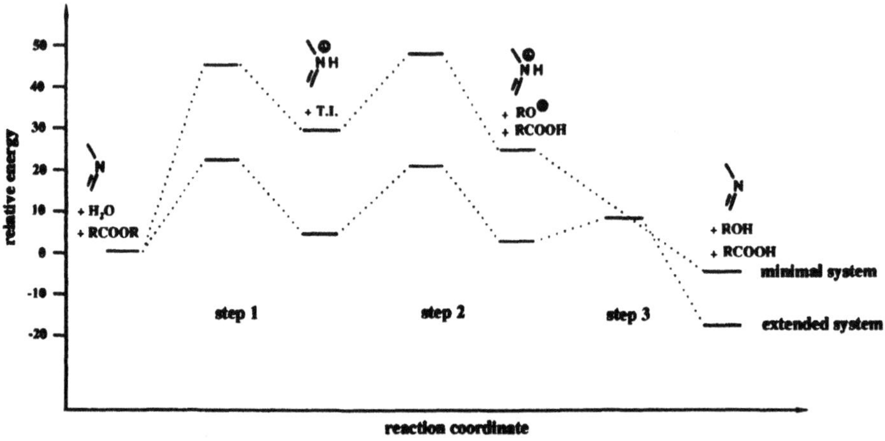

Fig. 7. Energy profile of the reaction mechanism of Phospholipase A_2.

range interactions of the reactands with the enzymatic environment, the calculated activation energy of 22 kcal/mol for the rate determining step is a good result. It demonstrates that semi-empirical methods combined with the use of harmonic constraints are a valuable approach for the investigation of enzymatic reaction mechanisms. Harmonic constraints provide a useful tool for simulating the active site. They conserve the geometry of the native enzyme and allow a complete geometry optimization at the same time.

The above study illustrates many of the points made in the introduction. Above all, it makes the role of the immediate environment of the reacting center clear, as shown by the schematic reaction profile given in Fig. 7. Although there can be no guarantee that the calculated mechanism really is the most favorable, the overall activation energy suggests that it is reasonable. As a next step, the effect of the more complete enzyme environment will be investigated using QM/MM techniques.

References

1. Merz, K.M. (1997) Lecture presented at the NATO Advanced Research Workshop on Molecular Modeling and Dynamics of Biological Molecules Containing Metal Ions, 15-21 March, San Miniato, Pisa, Italy.
2. Merz, K.M., Dewar, M.J.S., Hoffmann R. (1989) Mode of Action of Carbonic Anhydrase, *J. Am. Chem. Soc.* **111**, 5636–5649.
3. Alex, A., Clark, T. (1992) A Model Molecular Orbital Study of the Cleavage of Peptides by Carboxypeptidase A, *J. Comput. Chem.* **13**, 704.
4. Ritter von Onciul, A., Clark, T. (1993) Molecular Orbital Studies of Enzyme Mechanisms II: The Catalytic Oxidation of Alcohols by Liver Alcohol Dehydrogenase, *J. Comput. Chem.* **14**, 392.
5. Scott, D.L., White, St.P., Browning, J.L., Rosa, J.J., Gelb, M.H., Sigler P.B. (1991) Structures of Free and Inhibited Human Secretory Phospholipase A_2 from Inflammatory Exudate, *Science*

254, 1007–1010.
6. Scott, D.L., White, St.P., Otwinowski, Z., Yuan, W., Gelb, M.H., Sigler P.B. (1990) Interfacial Catalysis: The Mechanism of Phospholipase A_2, *Science* **250**, 1541–1546.
7. Dugas H. (1996) *Bioorganic Chemistry*, 3rd Edition, Springer-Verlag, New York, Berlin, Heidelberg, 165–166.
8. Rauhut, G., Alex, A., Chandrasekhar, J., Steinke, T., Sauer, W., Beck, B., Hutter, M., Gedeck, P., Clark, T. (1996) *Vamp 6.1*, Oxford Molecular Group PLC, The Medawar Centre, Oxford Science Park, Oxford, OX4 4GA, United Kingdom.
9. Stewart, J.J.P. (1989) Optimization of Parameters for Semiempirical Methods I. Method, *J. Comput. Chem.* **10**, 209–220.
10. Stewart, J.J.P. (1989) Optimization of Parameters for Semiempirical Methods II. Applications, *J. Comput. Chem.* **10**, 221–264.

NORMAL MODE ANALYSIS OF PROTEINS TO INTERPRET RESONANT AND INELASTIC SCATTERING OF γ QUANTA

ERNST WALTER KNAPP AND BERND MELCHERS
Freie Universität Berlin
Fachbereich Chemie
Institut für Kristallographie
Takustr. 6
D-14195 Berlin

FRITZ PARAK
Technische Universität München
Fakultät für Physik E17
James-Frank-Straße
D-85747 Garching

1. Introduction

Information on structural fluctuations in proteins can be obtained from the B-factors of X-ray structure analysis [1,2], NMR spectroscopy [3,4], inelastic neutron scattering [5,6], and Rayleigh scattering of Mössbauer radiation (RSMR) [7,8]. For iron containing proteins Mössbauer spectra [9-12] can also be used. Specific information on different local conformations in proteins can also be gained from measurements of IR [13,14] and UV-VIS [15,16] as well as resonance Raman spectra [17]. Recently, a new method namely inelastic X-ray scattering with high energy resolution became available [18,19]. The first application of this method to a biological sample was with myoglobin [20].

While Mössbauer spectra provide a most detailed energy resolution of the elastic peak, the inelastic X-ray scattering yields also the phonon side bands in grater detail. The B-factors from X-ray structure analysis of proteins provide the most detailed information on mean square fluctuations (MSF) for each atom but there is no

L. Banci and P. Comba (eds.), Molecular Modeling and Dynamics of Bioinorganic Systems, 319–342.

inherent time resolution. That is, it is not possible to discriminate between static and dynamic contributions to the MSF values. Incoherent neutron scattering as well as Mössbauer absorption spectroscopy yield time sensitive data of structural fluctuation of proteins but the structural information is limited. Inelastic neutron scattering provides an average over all hydrogen atoms, inelastic X-ray scattering data average over all protein atoms, whereas Mössbauer absorption spectroscopy probes the motion of the iron atom in iron containing proteins only. Mössbauer spectra with their high energy resolution display the central band of elastic transitions in all detail. However, the energy window is too narrow to exhibit also the inelastic sidebands involving phonon excitation and de-excitation. The opposite is the case for the inelastic X-ray scattering. Its energy resolution is instrumentally limited to 4.4 meV at present. This is not sufficient to discriminate the elastic from the quasi-elastic lines but the energy window is wide enough to display the inelastic phonon sidebands. Thus the two methods supplement themselves in an ideal way.

Theory may also contribute to the understanding of structural fluctuations in proteins. Conventional molecular dynamics (MD) simulations can in principle provide all information on protein structure and dynamics in atomic detail. In the first approaches to understand protein dynamics by MD simulation methods the protein molecules were considered without water. These simulation data obtained under vacuum conditions, showed a number of artifacts. Namely the mean square deviation (MSD) values with respect to crystal structures were too large and mean square fluctuations (MSF) did not agree well with the corresponding values of the B-factors from X-ray structure analysis. The inclusion of explicit water molecules and appropriate boundary conditions in the MD simulations of protein dynamics has removed most of these artifacts [21,22]

However, one problem remains unsolved so far and was pointed out by us some time ago [23]. Subnanosecond MD simulations of proteins provide MSF values for the individual atoms i which are generally as large or even larger then the MSF values

$$< \vec{r}_i^{\,2} >^x = \frac{3B_i}{8\pi^2} \tag{1}$$

derived from the B-factors of X-ray structure analysis. This agreement is astonishing since the B-factors contain also a considerable contribution from static intramolecular and lattice disorder. This becomes evident from the observation that the MSF of the iron atom in myoglobin is much larger if derived from X-ray structure analysis than it is by use of Mössbauer spectroscopy. Even at room temperature Mössbauer spectra of myoglobin [9-12] indicate that a considerable fraction of the MSF of the iron atom fluctuates with a characteristic time larger then nanoseconds. Contributions to the

MSF which are due to disorder can also be inferred from the fact that the MSF values derived from the crystal structures of myoglobin [24-26] do not vanish, if linearly extrapolated to zero temperature (see $<\vec{r}_{Fe}^{\,2}>^{x}$ in Fig. 2). Thus, it seems that a trajectory from MD simulation samples phase space faster than suggested by the value of the time parameter entering the time propagation algorithm. On the other hand, with conventional methods the time horizon of MD [27] simulations of proteins is limited to nanoseconds. Since many measurable dynamic processes in proteins are slower only in a few cases, a direct comparison between experimental and MD simulation data was possible. One example is an MD simulation of ribonuclease T1 [28] to evaluate the fluorescence depolarization rates of tryptophane, where the calculated rates were too short by about a factor of two.

Another issue is the temperature above which anharmonicities become significant in the dynamics of protein water systems. Mössbauer spectra of myoglobin exhibit a linear temperature dependence of the MSF of the iron atom up to a temperature of 180K. MD simulation data averaged over all backbone atoms of Mb [27] exhibit a significant anharmonic contribution to MSF values already below 100K (see Fig. 2).

Nevertheless, we recently obtained quantitative agreement of the low temperature MSF of the iron atom of deoxymyoglobin (Mb) $<\vec{r}_{Fe}^{\,2}>^{\gamma}$, measured by Mössbauer spectroscopy, with results from normal mode analysis of carboxymyoglobin (MbCO) including 170 crystal water molecules [29]. The spectra from inelastic X-ray scattering could well be interpreted up to room temperature using a single vibrational mode which is representative for the low frequency normal modes of myoglobin [20]. However, an explicit computation of the spectrum from inelastic X-ray scattering using normal mode data is still missing.

In this article we like to address some of the problems mentioned above. In section 2 the concept of normal modes is introduced. The normal modes results of Mb are summarized in section 3. The basis for a description of Mössbauer spectra and inelastic X-ray scattering data is the intermediate scattering function [30-32], as depicted in section 4. The intermediate scattering function is evaluated for a Brownian oscillator for the underdamped case in section 5 and for the overdamped case in section 6. In section 7 an underdamped low temperature Brownian oscillator is used for the interpretation of the inelastic X-ray scattering data of myoglobin [20]. In the Conclusions future avenues are discussed. In the Appendix it is shown that the recoil energy absorbed by a Mössbauer nucleus described by the low temperature Brownian oscillator is equal to the average recoil energy absorbed by a free Mössbauer nucleus.

2. The normal mode formalism for a molecular system

There is little to add to the seminal work from Wilson, Decius and Cross [33] on normal mode analysis of small molecules. As a result of the developments in computer technology computations of normal modes are now also possible for much larger molecular systems like proteins, comprising many thousands of atoms [34].

Normal modes of molecules can be used to describe small amplitude fluctuations of the atoms around their equilibrium positions. As opposed to conventional MD simulations, low temperature quantum effects are easily incorporated and a compact description of correlated motions of atoms in biomolecules is provided. However, the concept of normal modes may also be combined with MD simulation by characterizing structural fluctuations of proteins in a quasi-harmonic description [35,36], which includes anharmonic effects on the average.

The starting point for a derivation of the normal mode method is the equations of motion of all atoms of a molecule or a molecular system (protein and water)

$$m_i \frac{d^2}{dt^2} \vec{r}_i(t) = -\vec{\nabla}_{r_i} V(\vec{r}_1, \vec{r}_2, \ldots \vec{r}_n), \qquad i = 1,2,\ldots\ldots n, \tag{2}$$

where m_i is the mass and $\vec{r}_i$ the position vector of atom i, $\vec{\nabla}_{r_i}$ the gradient with respect to $\vec{r}_i$ and V the energy function. To simplify the equations, mass weighted coordinates according to

$$\vec{R}_i = \sqrt{m_i}\; \vec{r}_i, \qquad i = 1,2,\ldots\ldots n \tag{3}$$

are introduced. In these coordinates the gradient reads

$$\vec{\nabla}_{r_i} = \left(\frac{\partial}{\partial x_i}, \frac{\partial}{\partial y_i}, \frac{\partial}{\partial z_i}\right) = \sqrt{m_i}\left(\frac{\partial}{\partial X_i}, \frac{\partial}{\partial Y_i}, \frac{\partial}{\partial Z_i}\right) = \sqrt{m_i}\; \vec{\nabla}_{R_i},$$

yielding the equations of motion

$$\frac{d^2}{dt^2} \vec{R}_i(t) = -\vec{\nabla}_{R_i} U(\vec{R}) \tag{4}$$

where $U(\vec{R}) = V(\vec{r})$, with $\vec{R} = (\vec{R}_1, \vec{R}_2, \ldots\ldots \vec{R}_n) = (Q_1, Q_2, \ldots\ldots Q_{3n})$ and $Q_{3(m-1)+k} = (\vec{R}_m)_k$, which denotes the mass weighted k^{th} (k = 1,2,3) Cartesian components of the position vector of atom m.

The energy function U of the molecular system is expanded around an equilibrium conformation $\vec{R}^{(0)}$, where the 3n-component gradient operator $\vec{\nabla}_R$ applied to U vanishes, i.e. is set equal to a vector whose 3n components are zero.

$$\vec{\nabla}_R U\left(\vec{R}^{(0)}\right) = \vec{0}. \tag{5}$$

The expansion of the energy function up to second order with respect to the 3n dimensional displacement vector $\Delta\vec{R}(t) = \vec{R}(t) - \vec{R}^{(0)}$ reads

$$U\left(\vec{R}\right) = U\left(\vec{R}^{(0)}\right) + \frac{1}{2}\Delta\vec{R}\cdot\Omega^2\cdot\Delta\vec{R} . \tag{6}$$

It contains the (3n x 3n) dimensional Hesse matrix Ω^2 which is given by

$$(\Omega^2)_{ij} = \frac{\partial^2 U}{\partial Q_i \partial Q_j}. \tag{7}$$

With these definitions the equations of motion can be cast into a form linearized with respect to the displacement vector $\Delta\vec{R}(t)$

$$\frac{d^2}{dt^2}\Delta\vec{R}(t) = -\Omega^2\cdot\Delta\vec{R}(t) \tag{8}$$

These coupled equations of motion are valid for small $\Delta\vec{R}$. They can be decoupled by diagonalization of the Hesse matrix. This leads to the eigenvalue problem

$$[\Omega^2 - \omega_k^2\,\mathbf{1}]\cdot\vec{d}^k = \vec{0} \tag{9}$$

with eigenvalues ω_k^2 and eigenvectors $\vec{d}^k = \left(\vec{d}_1^k, \vec{d}_2^k, \ldots\ldots \vec{d}_n^k\right)$. The vector $\vec{d}_i^k$ describes the relative displacement of atom i with respect to the normal mode k in mass weighted form.

The solution of the eigenvalue problem of a molecular system of n atoms provides 3n different normal modes. The eigenvalues are the squares of the circular frequencies and should all have positive values. For an isolated molecular system, i.e. for a system which is allowed to translate and rotate freely, the six eigenvalues referring to these degrees of freedom should vanish. In practice it can happen that some eigenvalues become negative. This is indicative of a negative curvature of the energy function and happens if a sattle point is situated at the reference point $\vec{R}^{(0)}$, where the energy function $U(\vec{R})$ is expanded. But normally it is due to insufficient energy minimization, such that the gradient of the energy function, eq. (5), does not vanish. The positive definiteness of the eigenvalues and the occurrence of six vanishing eigenvalues is used as a quality criterion for sufficient energy minimization and appropriate diagonalization of the Hesse matrix.

Application of the normal mode method to large molecular systems is mainly limited by the requirements on memory rather than on CPU time. To diagonalize the Hesse matrix efficiently it must fit in the memory of an available computer. To save on memory for the diagonalization procedure a computer C program was written which uses the Hesse matrix in symmetric storage mode and represents the matrix elements in single precision (8 significant digits). With this program for n atoms about 4 n^2 Megabyte of memory are needed. To avoid the need to store a complete set of eigenvectors the eigenvectors are computed one after another by solving the corresponding linear equation system (9). The orthogonality of the eigenvectors was subsequently checked. For nearly degenerate eigenvalues the corresponding eigenvectors were evaluated simultaneously. With exception of the six vanishing eigenvalues corresponding to free rotation and translation so far no degeneracy was encountered in practical applications.

The general solution of the time dependent mass weighted displacement can be written as

$$\Delta\vec{R}(t) = \sqrt{2}\sum_{k=1}^{3n} \vec{d}^k \ \sin(\omega_k t + \varphi_k) \ a^k, \tag{10}$$

where the φ_k are arbitrary phase factors and a^k are amplitude factors. The $a^k(T)$ are temperature dependent, if they account for the average fluctuations in a canonical ensemble. For a molecular system obeying classical mechanics

$$\left(a_c^k(T)\right)^2 = \frac{k_B T}{\omega_k^2} \tag{11a}$$

and for a quantum mechanical system

$$\left(a_q^k(T)\right)^2 = \frac{\hbar}{2\omega_k} \coth\left(\frac{\hbar\omega_K}{2k_B T}\right) \tag{11b}$$

is valid. The free rotations and translations yield a diverging contribution which must be ignored if the considered molecular system is embedded in a solid phase, i.e. $a^k = 0$ for k = 1,2,....6. The average MSF of atom i can be computed from

$$< \vec{r}_i^{\,2} > = \frac{< \Delta\vec{R}_i^{\,2} >}{m_i} = \frac{1}{m_i}\sum_{k=7}^{3n}\left(\vec{d}_i^k\right)^2 \left(a^k(T)\right)^2 \tag{12}$$

3. Application of the normal mode method to myoglobin

In contrast to a recent contribution, where the normal modes of carboxymyoglobin (MbCO) where calculated [29], here deoxymyoglobin (Mb) is considered. Mb was described with all polar hydrogen atoms. The non-polar hydrogen atoms bound to carbon atoms were represented by an extend carbon atom type whose atomic partial charges and Lennard-Jones potential parameters where adapted. Together with 170 crystal waters, whose coordinates were taken from the crystal structure of MbCO [26], the molecular system contained n = 2075 atoms. The water molecules were modeled with flexible bond lengths and bond angles. For the non-bonded interactions of water, TIP3P [37] parameters were used. Starting from the X-ray structure of Mb [25] the CHARMM [38] energy function was carefully minimized. 2200 steps of steepest descent (SD) and 400 steps of conjugated gradient (CG) were applied while all non-hydrogen atoms were fixed. Then for 3000 steps of SD and 18000 steps of CG all atoms were free to move. From the energy minimized structure the (6225 x 6225) dimensional Hesse matrix of partial second order derivatives was evaluated and subsequently diagonalized by using the QL method of Eispack routines [39]. In units of cm^{-2} the seven lowest eigenvalues ω_k^2 adopt the values -1.237 10^{-7}, 1.006 10^{-9}, 1.269 10^{-9}, 9.339 10^{-6}, 4.333 10^{-5}, 1.190 10^{-4}, 50.60. The maximum frequency belonging to a hydrogen stretching vibration is at ω_{max} = 3646 cm^{-1}. Note, that the six smallest eigenvalues are close to zero and that the 7th eigenvalue significantly deviates from zero. This indicates that the energy minimization and the subsequent computation of the eigenvalues form the single precision Hesse matrix are properly performed.

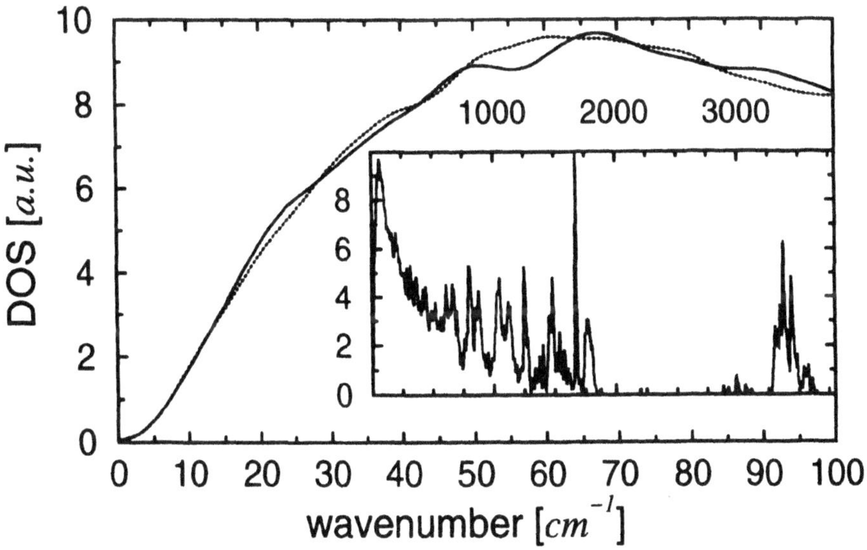

Figure 1. Density of states (DOS) spectrum form the normal modes of deoxymyoglobin (Mb) in arbitrary units. Insert: full spectrum, main part: spectrum between 0 - 100 cm^{-1}. The solid line refers to Mb the dotted line to MbCO. The spectrum is computed according to eq. (13). At each eigenvalue ω_k a Gaussian of unit intensity is placed. The width parameter of the Gaussians σ = 3.5 cm^{-1} is chosen such that the spectrum is smooth at low frequencies.

To obtain an overview of the results from the normal mode computation of Mb the density of states (DOS) spectrum evaluated according to

$$I_{DOS}(\omega) = \frac{1}{\sqrt{2\pi\sigma^2}} \sum_{k=7}^{3n} \exp\left[-\frac{(\omega - \omega_k)^2}{2\sigma^2}\right] \tag{13}$$

is displayed in the insert of Fig. 1. The line width parameter $\sigma = 3.5\,cm^{-1}$ is chosen to obtain a smooth spectrum. The DOS spectrum of Mb deviates only in some details from the DOS of MbCO in Ref. [29]. It shows the contributions from hydrogen stretching vibrations above 3000 cm^{-1} and the prominent sharp amide I band at 1700 cm^{-1} which is due to the CO stretching vibrations of the protein backbone. The broad band above 1750 cm^{-1} is from the bending vibrations of the 170 crystal water molecules.

The main part of Fig. 1 exhibits the low frequency DOS spectrum of Mb (solid line) and MbCO (dotted line) which are similar but differ in some details. Note, that the slope below 10cm^{-1} is not indicative for a Debye spectrum but is governed by the low frequency wings of the Gaussians corresponding to the smallest eigenfrequencies in expression (13). Hence, a Debye spectrum with its characteristic frequency dependence $I_{DOS}(\omega) \sim \omega^2$ which is typical for low frequency acoustic phonons cannot be observed. This is not surprising, since a molecular system consisting of a single protein molecule does not possess acoustic modes. Nevertheless, the six modes of free rotation and translation can be interpreted as degenerate acoustic modes. Even if one would consider the normal mode data from a large number of unit cells, say L of a protein crystal with n atoms per protein, only a small fraction of modes, namely 6L from a total of 3nL modes, would correspond to acoustic phonon modes. Under these conditions a Debye type phonon spectrum is hardly visible.

In Fig. 2 the temperature dependence of the MSF value of the iron atom of Mb is displayed, as obtained from the B-factor of the crystal structure [25] (diamonds), the elastic line of the Mössbauer spectrum [9] (open circles), and the normal mode analysis [29]. MD simulation data of the MSF of the iron atom are not available for all temperatures. Therefore, the MSF values, averaged over all protein backbone atoms of Mb derived from normal modes (upper solid line) and of MbCO derived from MD simulation [27] (crosses) are compared. The MSF values averaged over all protein backbone atoms as derived from normal mode analysis are for Mb and MbCO almost identical (not shown in the figure). This is in contrast to the MSF values of the iron atom which are markedly larger for Mb (lower solid line) than for MbCO (dotted line). This is due to the CO ligand whose presence in the heme pocket reduces the mobility of the iron atom. The same was observed for the vibrational modes above 500 cm^{-1} which are probed by the line broadening of the Soret band [29]. Furthermore, nearly all contributions (97.3%) to the MSF of the iron atom come from normal modes with frequencies below 200 cm^{-1} (dashed line). Hence, the intra heme vibrational modes of the iron atom, which have frequencies above 200 cm^{-1}, provide a negligible contribution to the MSF's observed by Mössbauer absorption spectroscopy. Furthermore, the MSF values of the iron atom relative to the carbon atom skeleton of the heme observed by the Soret band broadening are about 15% of the MSF values which the iron atom possesses with respect to a spatially fixed reference point [29].

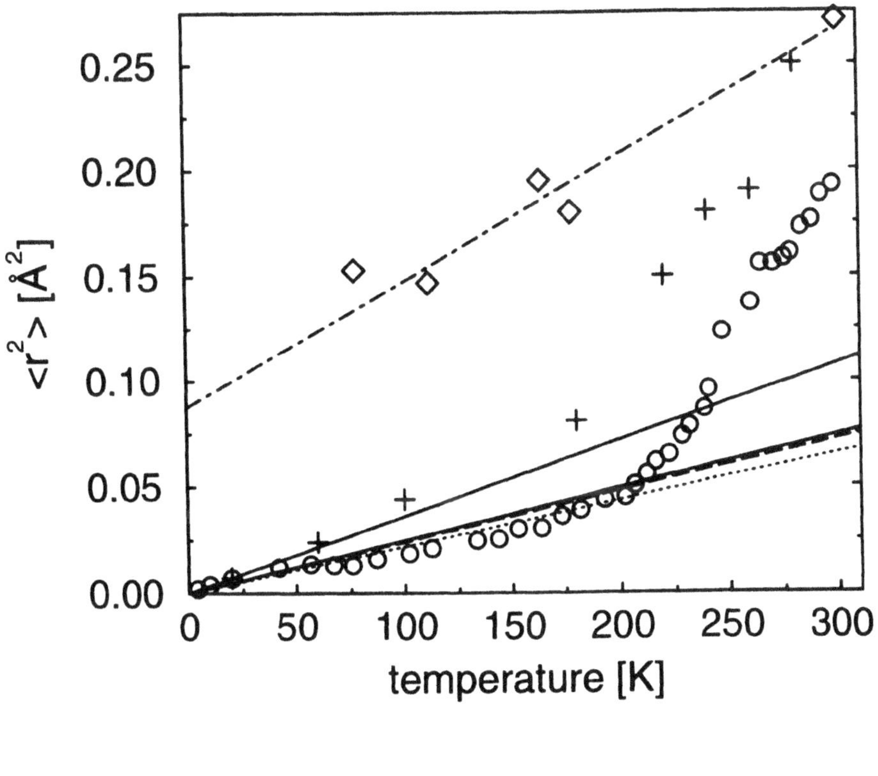

Figure 2. Temperature dependence of the MSF values of myoglobin. The MSF values of the iron atom derived from the B-factors of MbCO [26] (diamonds) and from the elastic line of Mössbauer spectra of Mb [9] (circles) are displayed. From normal mode analysis one obtains for the MSF of the iron atom in Mb the lower solid line, in MbCO the dotted line [29], and considering only normal modes of Mb with frequencies below 200cm^{-1} the dashed line. The upper solid line displays the MSF values from normal modes of Mb, averaged over all protein backbone atoms. The corresponding MD simulation data of MbCO are given by crosses [27].

4. Computation of Mössbauer absorption spectra from the intermediate scattering function

The intermediate scattering function is the basis for a description of incoherent neutron scattering and resonant-γ-ray absorption [30,31]. For a Mössbauer nucleus at position $\vec{r}(t)$ it can be expressed as

$$I(\vec{k},t) = < \exp(i\vec{k}\cdot\vec{r}(t)) \;\; \exp(-i\vec{k}\cdot\vec{r}(0)) > \tag{14}$$

and relates the position of the iron nucleus $\vec{r}$ at times t and 0. The angular brackets denote an ensemble average, $\vec{k}$ is the wave vector of the γ quantum absorbed from the Fe^{57} Mössbauer nucleus. The absorption spectrum can be calculated as real part ($\Re$) of the Fourier-Laplace transform of the intermediate scattering function $I(\vec{k},t)$ as follows

$$\sigma_a(\omega) \approx S_a(\omega) = \Re\left\{\frac{1}{\pi}\int_0^\infty dt \exp(-i\omega t) I(\vec{k},t)\right\}, \tag{15}$$

where σ_a, the cross section for Mössbauer absorption, is proportional to the scattering function $S_a(\omega)$. For a rigorous derivation see Ref. [32].

If the motion of the Mössbauer nucleus can be described by harmonic motions the intermediate scattering function, eq. (14), is cast in the form

$$I(\vec{k},t) = \exp[- < (\vec{k}\cdot\vec{r})^2 >]\cdot\exp[< \vec{k}\cdot\vec{r}(t) \;\; \vec{k}\cdot\vec{r}(0) >], \tag{16}$$

where

$$< \vec{r}^2(0) > \; = \; < \vec{r}^2(t) > \; = \; < r^2 >. \tag{16a}$$

Now the ensemble average has moved to the exponent. The first of the two exponential functions is the Lamb Mössbauer factor. It accounts for the intensity of the spectrum. The second term contains the position time correlation function $< \vec{r}(t)\cdot\vec{r}(0) >$ and accounts for the spectral line shape. Expression (16) is also valid if the cumulant expansion up to second order is a suitable approximation of eq. (14). This expansion scheme provides generally very efficient approximations of statistical averages involving exponential functions [40]. For an isotropic probe the scalar product $\vec{k}\cdot\vec{r}$ can be approximated by 1/3 kr, where $k = |\vec{k}|$ and $r = |\vec{r}|$. Now, the intermediate scattering function reads

$$I(k,t) = f(< r^2 >)\cdot\exp\left[-\tfrac{k^2}{3} < \vec{r}(t)\cdot\vec{r}(0) >\right], \tag{17}$$

where

$$f(< r^2 >) = \exp\left[-\tfrac{k^2}{3} < r^2 >\right] \tag{17a}$$

is the Lamb Mössbauer factor. The last term of the intermediate scattering function is expanded to yield

$$I(k,t) = f(<r^2>) \cdot \sum_{N=0}^{\infty} \frac{\left[k^2 <\vec{r}(t)\cdot\vec{r}(0)>\right]^N}{3^N N!}. \qquad (18)$$

The N = 0 term in the expansion (18) describes elastic processes where no phonons are excited or de-excited. The intensity of the corresponding elastic line is given by $f(<r^2>)$, the Lamb Mössbauer factor. The inelastic and quasi-elastic processes involving a change of the phonon excitation pattern in the medium are represented by the higher order expansion terms (N>0).

5. Absorption spectra from underdamped Brownian oscillators

For complex molecular systems like proteins [2] and glasses [41] Mössbauer absorption spectra contain broad lines above a characteristic temperature. To interpret the broadening of a Mössbauer absorption spectrum by quasi elastic lines, several authors proposed to use an overdamped Brownian oscillator model [9,42,43]. Also in the underdamped case, a Brownian oscillator is a useful generalization of undamped vibrational phonon modes, used for a conventional description with normal modes. The time correlation function of the position of the Mössbauer nucleus which enters the intermediate scattering function can be expressed by the general complex valued solution of a Brownian oscillator which reads

$$<\vec{r}(t)\cdot\vec{r}(0)> = \frac{<r^2>}{a+b}\left\{a\cdot\exp[t(+i\omega_1-\beta_0)] + b\cdot\exp[t(-i\omega_1-\beta_0)]\right\}, \qquad (19)$$

with

$$\omega_1^2 = \omega_0^2 - \beta_0^2. \qquad (19a)$$

The MSF of the Brownian oscillator is $<r^2>$, β_0 is the friction constant of the oscillator, ω_0 the bare and ω_1 the dressed frequency. For $\beta_0 < \omega_0$ the dressed frequency ω_1 is real valued, i.e. the Brownian oscillator is underdamped. Note, that the time correlation function, eq. (19), is normalized such that at t=0 it yields $<\vec{r}(t=0)\cdot\vec{r}(0)> \; = <\vec{r}^{\,2}>$. as expected. A derivation of the time correlation function of a Brownian oscillator can be found in Ref. [10]. The conventional real valued correlation function of a Brownian oscillator [44,45] is obtained from eq. (19), by using

$$a = \frac{1}{2}\left(1 - \frac{i\beta_0}{\omega_1}\right) \quad \text{and} \quad b = \frac{1}{2}\left(1 + \frac{i\beta_0}{\omega_1}\right), \qquad (20)$$

yielding

$$< \vec{r}(t)\cdot\vec{r}(0) >^{conv} = < r^2 > \exp(-t\beta_0)\left[\cos(\omega_1 t) + \frac{\beta_0}{\omega_1}\sin(\omega_1 t)\right] \tag{21}$$

Note, that for a conventional underdamped Brownian oscillator the expansion coefficients, in the correlation function (19), are complex valued. As a consequence the spectrum of the oscillator, i.e. the real part of the Fourier-Laplace transform of the correlation function, eq. (21), contains not only absorptive but also dispersive contributions. The latter give rise to terms of the type

$$\frac{\omega \pm \omega_1}{\beta_0^2 + (\omega \pm \omega_1)^2}.$$

To avoid these dispersive terms in expressions of absorption spectra derived from eq. (19), the coefficients a and b must be real valued. With the choice of a = ½ = b the spectrum of the correlation function is symmetric with respect to $\omega = 0$ and contains only the absorptive parts from the conventional Brownian oscillator. A suitable choice for the coefficients a and b in eq. (19) can be obtained by making the following considerations: The first term in the time correlation function, eq. (19), inserted in the intermediate scattering function accounts for resonant absorption at frequency $\omega = +\omega_1$, where a phonon in the medium surrounding the Mössbauer nucleus is excited. The second term at frequency $\omega = -\omega_1$ corresponds to a resonant absorption which goes along with the removal of phonon excitation with frequency ω_1 from the medium. Phonon excitations in the medium are available only according to a Boltzmann weight factor. At low temperatures the contributions from processes of phonon de-excitation must be reduced correspondingly. On the other hand the process of phonon excitation does not depend on temperature. Consequently, the appropriate weight factors a and b for the complex valued low temperature underdamped Brownian oscillator in expression (19) are

$$b = \exp\left[-\frac{\hbar\omega_1}{k_B T}\right] \quad \text{and} \quad a = 1. \tag{22}$$

With this choice, one obtains a complex valued low temperature Brownian oscillator which possesses an asymmetric spectrum. Inserting the time correlation function, eq. (19), of this low temperature underdamped Brownian oscillator with the weight factors eq. (22) in the intermediate scattering function eq. (18), and taking the real part from the Fourier-Laplace transform eq. (15), one obtains the absorption spectrum

$$S_a^{under}(\omega) = f(< r^2 >)\cdot\sum_{N=0}^{\infty}\frac{1}{N!}\left(\frac{k^2 < r^2 >}{3(1+b)}\right)^N \cdot S_N^{under}(\omega) \tag{23}$$

with

$$S_N^{\text{under}}(\omega) = \sum_{n=0}^{N} \binom{N}{n} b^n L_N^{\text{under}}\left[\omega + (2n - N)\,\omega_1\right]. \tag{23a}$$

The Lorentzian lines

$$L_N^{\text{under}}(\omega') = \frac{1}{\pi}\frac{\gamma_0 + N\beta_0}{(\gamma_0 + N\beta_0)^2 + \omega'^2} \tag{24}$$

are of unit intensity. Their width contains contributions from the Brownian oscillator (β_0), depending on the order N of the phonon excitation as well as on life time broadening and inhomogeneities of the Mössbauer transition in source and absorber. The latter are accounted for by a constant width parameter γ_0. The Lorentzians in expressions (23) and (24) can also be replaced by Gaussians and Voigtians of corresponding width. Note, that with this complex valued low temperature Brownian oscillator the average recoil energy absorbed by the medium during the resonant absorption of γ quanta is equal to the recoil energy taken by a free Mössbauer nucleus [46], according to

$$< \hbar\omega > = \int_{-\infty}^{+\infty} \hbar\omega \; S_a^{\text{under}}(\omega) = \frac{\hbar^2 k^2}{2m}. \tag{25}$$

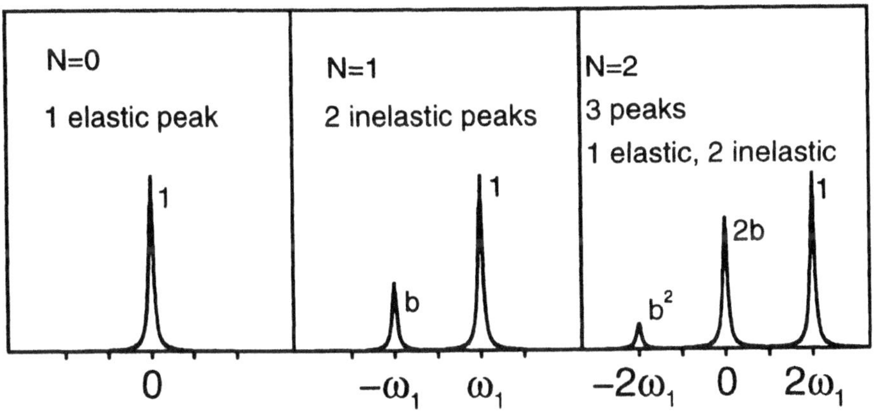

Figure 3. Scheme of phonon excitation processes described by the intermediate scattering function, eq. (23), of a single underdamped low temperature Brownian oscillator with the time correlation function eqs. (19) and (22). If no phonon excitation is involved (N=0) the absorption occurs at the resonance frequency ω=0, with unit intensity. In order N=1 of expansion (23), there is phonon excitation at frequency $+\omega_1$ with unit intensity and phonon de-excitation at $-\omega_1$ with an intensity weighted by the Boltzmann factor. In second order (N=2), there are three peaks: a twofold excitation at frequency $+2\omega_1$ with intensity 1, a combined excitation and de-excitation (quasi-elastic component) at frequency 0 with intensity 2b, and twofold de-excitation at frequency $-2\omega_1$ with intensity b^2.

The complex valued underdamped low temperature Brownian oscillator gives rise to a rather complicated scheme of phonon excitations and de-excitations, depicted in Fig. 3. In zero order (N=0) of the expansion (23), there is only one elastic line. In first order (N=1), there are two inelastic contributions, one for phonon excitation, the other for de-excitation. In second order (N=2), there are two inelastic contributions, one with twofold phonon excitation, the other with twofold de-excitation and one quasi-elastic contribution involving the combination of one phonon excitation and one phonon de-excitation. The lines corresponding to phonon excitations have unit intensity. For each event of de-excitation, the corresponding line is weighted with the Boltzmann factor b, eq. (22). More generally the Nth order expansion term corresponds to a skewed N^{th} order binomial distribution of lines.

6. Absorption spectra from overdamped Brownian oscillators

For a Brownian oscillator with a friction constant larger than the frequency ($\beta_0 > \omega_0$) the dressed frequency ω_1 adopts imaginary values. Now, instead of a dressed frequency a dressed friction constant is used

$$\beta_1^2 = \beta_0^2 - \omega_0^2 \,. \tag{26}$$

For large friction ($\beta_o >> \omega_o$) the square root can be expanded to yield approximately

$$\beta_1 = \beta_0 - \alpha_0 \,, \tag{26a}$$

where

$$\alpha_0 = \frac{\omega_0^2}{2\beta_0} \,. \tag{26b}$$

By inserting the above expression of the dressed friction constant in eq. (19) one obtains the time correlation function for a strongly overdamped Brownian oscillator

$$< \vec{r}(t) \cdot \vec{r}(0) >^{over} = \frac{< r^2 >}{a+b} \left[a \cdot \exp(-t\alpha_0) + b \cdot \exp(-t(2\beta_0 - \alpha_0)) \right]. \tag{27}$$

The second term decays very fast, whereas the first term decays even slower if the friction constant β_0 becomes larger. The fast decaying term is related to small amplitude rattling motions of the Brownian particle. It is due to fast fluctuations of small amplitude in narrow but deep valleys of the energy landscape in which the Brownian particle moves, and due to instant changes in the direction of movement by collisions with neighbor particles. Applied to proteins this decay term may be interpreted as the motion of the protein in a conformational substate [2]. The first term in eq. (27) describes the much slower motion of a Brownian particle which is traversing different valleys of the energy landscape. Transferred to proteins these motions of larger amplitude may correspond to transitions between different conformational substates. If one neglects the fluctuations of the fast decaying term which are only of small amplitude the time correlation function of a strongly overdamped Brownian oscillator simplifies to the familiar expression [10]

$$< \vec{r}(t) \cdot \vec{r}(0) >^{over} = < r^2 > \cdot \exp(-t\alpha_0). \tag{28}$$

After inserting this correlation function into the intermediate scattering function, eq. (18), one obtains for the absorption spectrum, eq. (15), the following expression [10]

$$S_a^{over}(\omega) = f(< r^2 >) \cdot \sum_{N=0}^{\infty} \frac{1}{N!} \left(\frac{k^2 < r^2 >}{3} \right)^N \cdot L_N^{over}(\omega), \tag{29}$$

where

$$L_N^{over}(\omega) = \frac{1}{\pi} \frac{\gamma_0 + N\alpha_0}{(\gamma_0 + N\alpha_0)^2 + \omega^2}. \tag{29a}$$

Here, the bare resonance frequency ω_r is also arbitrarily set to zero. Besides the narrow line at order N = 0 there are quasi-elastic broad lines at the same frequency which were used to interpret the line broadening of Mössbauer absorption spectra of myoglobin above T = 200K [9-11].

7. Inelastic X-ray scattering interpreted with an underdamped Brownian oscillator

Monochromatic γ quanta from a synchroton source near to the energy of 14.4 keV of the Mössbauer transition in Fe^{57} are scattered by a probe and energy analyzed by the Mössbauer effect in a Fe^{57} foil. A detailed explanation of the experiment of inelastic X-ray scattering and its set-up can be found in Refs. [18,19]. The conditions of X-ray scattering on Mb are given in Ref. [20].

The inelastic X-ray scattering data can be interpreted by the Fourier Laplace transform of the intermediate scattering function, eq. (15). Here, the wave vector $\vec{k}$ is given by the change of momentum $\Delta\vec{k} = \vec{k}_i - \vec{k}_f$ from the initial $(\vec{k}_i)$ to the final $(\vec{k}_f)$ wave vector. Since the energy change in the inelastic scattering event is negligibly small as compared to the energy of 14.4. keV for the incoming γ quanta, one can use the approximation $|\vec{k}_i| = |\vec{k}_f| = k_0 = 7.3 Å$, such that

$$\Delta k^2 = 2k_0^2(1-\cos\theta). \tag{30}$$

In the above expression θ is the scattering angle between $\vec{k}_i$ and $\vec{k}_f$. The change of momentum obeys a broad distribution which has a maximum towards small scattering angles at $\theta = 17°$. Hence, the most probable change of momentum is $\Delta k^2 = 4.66\,\text{Å}^{-2}$. The energy resolution of 4.4 meV is determined by the nested monochromator used for the incoming synchrotron radiation. Accordingly, the line width parameter γ_0 in eq. (24), corresponding to half width at half height, is given by 2 meV. Note, that the energy of 1 meV = 8.1 cm^{-1}.

Instead of using a complete set of normal modes to compute the spectrum of inelastic X-ray scattering only a simplified model is used here, where the vibrational modes of Mb are represented by a single Einstein mode of frequency ω_1. To avoid exaggerated structures of vibrational lines in the inelastic wings of the X-ray scattering spectrum the line width parameter γ_0 is enlarged for the inelastically scattered lines. Hence, γ_0 = 7 meV if $n \neq N/2 > 0$, and γ_0 = 2 meV if $n = N/2$ in the expression (23a), providing the absorption spectrum of an underdamped Brownian oscillator. Besides the width parameter γ_0 which is independent of the order (N) of the phonon process the N-dependent width parameter corresponding to the friction of the Brownian oscillator is given by β_0 = 0.6 meV. It describes vibrational dephasing and life time broadening effects [47] with a characteristic time of $1/\beta_0$ = 1ps. The resulting width parameter of the lowest order (N = 2) quasi-elastic line is given by $\gamma = \gamma_0 + 2\,\beta_0$ = 3.2 meV.

The intensity of the individual elastic and inelastic lines in the expression (23) are governed by the quantity $k^2 < r^2 >$, where k^2 is defined by Δk^2, eq. (30). Since the spectrum is calculated using a single vibrational mode, only the MSF value $<r^2>$ must be representative for all atoms and all vibrational modes. For that purpose, one can use MSF values of individual atoms $< r_i^2 >$ which are evaluated according to eq. (12) and contain MSF contributions accumulated from all normal modes.

The normal modes of Mb were evaluated with only 170 crystal waters. The Mb sample used for the inelastic X-ray scattering experiment had, however, a water content corresponding to N_w = 400 water molecules per Mb molecule. To account for this difference in water content the MSF averaged over all 170 water molecules was calculated from the normal modes, yielding at T = 300K the value $<r^2>_{water} = 0.383\text{Å}^2$. Similarly, the MSF averaged over all n_p = 1568 protein atoms at T = 300K provides the value $<r^2>_{protein} = 0.168\text{Å}^2$. From these data one can approximately calculate the MSF value averaged over N_w water molecules and n_p protein atoms according to

$$\overline{<r^2>} = \frac{3 \cdot N_w <r^2>_{water} + n_p <r^2>_{proteins}}{3 \cdot N_w + n_p}, \tag{31}$$

yielding $\overline{<r^2>} = 0.261 Å^2$.

In Fig. 4 the scattering intensity of the inelastic X-ray scattering on Mb at the temperature T = 300K is displayed as a function of the energy of the incoming γ quanta. The reference energy of 14.4 keV for resonant γ absorption in Fe^{57} is subtracted in the figure. A constant background counting rate of 0.018 s^{-1} was subtracted from the measured counting rate, whose maximum value is 1 s^{-1}. An adjustment of the scattering geometry of the measurement at 300K increased the counting rates as compared to the data at lower temperatures (62K, 124K, 189K). This requires to increase the common scale factor used for the data of the other three temperatures by 11% to fit the spectra of the Brownian oscillator model to the experimental data at 300K [20].

For energies of the incoming synchrotron radiation below 14.4 keV a phonon excitation must be taken up from the probe material to reach the energy needed for resonant γ quanta absorption in the analyzer Fe^{57} foil. At low temperatures the corresponding inelastic wing carries no significant intensity [20]. Also at room temperature the intensity of this inelastic phonon wing is smaller than the phonon wing at frequencies above 14.4 keV, where the excess energy is used to create phonon excitations in the probe material (see Fig. 4). The solid and dashed lines in Fig. 4 depict the spectra calculated with a underdamped Brownian oscillator according to eqs. (23) and (24). The line width parameters and the MSF value were chosen as explained above. For the frequency of the Brownian oscillator the values ω_1 = 100 cm^{-1} (solid line) and ω_1 = 80 cm^{-1} (dashed line) were used. The model parameters used with the ω_1 = 100 cm^{-1} are practically identical to the model fit used recently for the inelastic X-ray spectra of Mb at four different temperatures [20]. Though a better fit quality would be possible the line width parameters were not readjusted for the smaller frequency value ω_1 = 80 cm^{-1} (dashed line).

Note, that here the MSF value from normal mode analysis of Mb, averaged over all atoms and linearly extrapolated to room temperature, could be used to explain the experimental data. In a recent paper, the phonon order (N) dependent line width parameter was absent (β_0 = 0 in eq. (24)). Then, the lack of additional line broadening for the quasi-elastic lines with n = N/2 > 0 starting with N = 2 required to increase the MSF value at 300K by 23% as compared to the value derived from normal mode analysis, eq. (31).

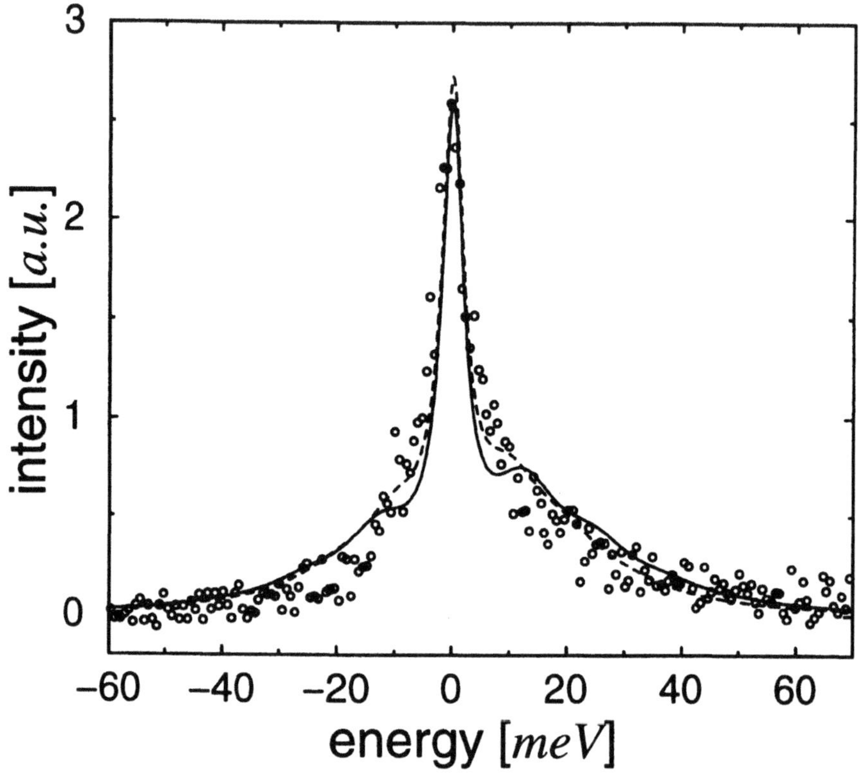

Figure 4. Inelastic X-ray scattering of Mb at 300K [20]. The inelastic X-ray scattering signal of Mb is depicted as a function of the incoming energy relative to the reference energy of 14.4 keV, where the analyzer foil absorbs. The solid and dashed lines are fits with spectra from underdamped low temperature Brownian oscillator according to eq. (23). The momentum transfer is $k^2 = 4.66Å^{-2}$, the average MSF value is 0.261 $Å^2$. Width parameters are $\beta_0 = 0.6$ meV, $\gamma_0 = 2$ meV for n = N/2 and $\gamma_0 = 7$ meV for n ≠ N/2 > 0. The frequency value of the oscillator is $\omega_1 = 9.9$ meV (12.3 meV) dashed (solid) line.

8. Conclusion

Normal mode analysis of proteins in conjunction with Brownian oscillator models are useful tools to obtain a more consistent interpretation of different experimental data, characterizing structural fluctuations in proteins. Interestingly, the mean square fluctuations (MSF) in myoglobin derived from the harmonic approximation of normal modes and linearly extrapolated to room temperature agree with the inelastic X-ray scattering experiments. This contrasts with MD simulation data which exhibit a marked non-linear temperature dependence of the MSF values already below 100K. On the other hand the MSF values derived from Mössbauer absorption spectra exhibit a non-linear temperature dependence above 200K.

While inelastic X-ray scattering data are interpreted with underdamped Brownian oscillators, the Mössbauer absorption spectra so far were interpreted with an overdamped Brownian oscillator. However, this oscillator carries only a 20 % fraction of the total MSF value of the iron atom [9]. Hence, in agreement with Mössbauer spectra on Mb the major part of the MSF value can still be due to underdamped vibrational modes. Nevertheless, it can also not be excluded, that the quasi-elastic line broadening observed in Mössbauer spectra above 200K is due to broad quasi-elastic lines which come from underdamped Brownian oscillators.

The energy resolution is not sufficient to resolve the quasi-elastic lines in the inelastic X-ray spectra of proteins. The fact that no anharmonicity is needed to explain these spectra, suggests that also in Mössbauer spectra of proteins no intensity would be missing if the energy window could be made wide enough to catch all contributions from quasi-elastic lines. This would resolve the conflict of the room temperature data from Mössbauer spectroscopy and inelastic X-ray scattering. However, a careful analysis of Mössbauer spectra and inelastic X-ray scattering data with an ensemble of Brownian oscillators whose parameters are determined by normal mode analysis will be required to further clarify these points.

Acknowledgments

The authors are grateful to Klaus Achterhold for valuable discussions. The CHARMM source code was provided by Martin Karplus and MSI Inc. The contribution was financially supported by the Deutsche Forschungsgemeinschaft SFB 312 D7, EU-net „The dynamics of protein structure“, and the BMBF Project 05 643 WOC3.

Appendix

For a quantum mechanical description of a Mössbauer nucleus embedded in a solid, it was shown that the average recoil energy absorbed by the solid during the absorption of a resonant γ quantum is equal to the recoil energy of the freely moving Mössbauer nucleus [46]. This relation can be expressed as

$$\sum_{G'} E_{G'}\, p(G',G) - E_G = \frac{\hbar^2 k^2}{2m}, \tag{A1}$$

where k is the wave number and m the mass of the Mössbauer nucleus. Before (after) absorption of the γ quantum the lattice state is denoted by G (G'). The lattice energy of the solid in state G' is $E_{G'}$. The transition probability to change from lattice state G to G' is given by the square of a transition matrix element

$$p(G',G) = \left| < G' | \exp(-i\vec{k}\cdot\vec{r}) | G > \right|^2. \tag{A2}$$

The recoil energy is absorbed from the solid in form of phonon excitations and gives rise to the asymmetric shape of the Mössbauer spectrum. The asymmetry of the spectrum can directly be used to compute the recoil energy which must be equal to the average energy absorbed by the medium, according to

$$< \hbar\omega > = \int_{-\infty}^{+\infty} \hbar\omega \; S_a(\omega)\, d\omega. \tag{A3}$$

To simplify the evaluation of the average (A3) the individual Lorentzian lines in eqs. (23) and (24) are replaced by appropriate Dirac delta-functions, yielding

$$S_a(\omega) = f(< r^2 >) \sum_{N=0}^{\infty} \frac{1}{N!} \left(\frac{k^2 < r^2 >}{3(1+b)} \right)^N \sum_{n=0}^{N} \binom{N}{n} b^n \delta(\omega - (N-2n)\omega_1), \tag{A4}$$

where the coefficient

$$b = \exp\left[-\frac{\hbar\omega_1}{k_B T} \right]. \tag{A5}$$

Inserting expression (A4) in (A3) yields

$$< \hbar\omega > = \hbar\omega_1 \cdot f(< r^2 >) \cdot \sum_{N=0}^{\infty} \frac{1}{N!} \left(\frac{k^2 < r^2 >}{3(1+b)} \right)^N \cdot \sum_{n=0}^{N} \binom{N}{n} b^n (N-2n). \tag{A6}$$

The last sum can be rearranged to yield

$$\left(N - 2b\frac{d}{db} \right) (1+b)^N = N(1+b)^N \frac{1-b}{1+b}. \tag{A7}$$

By using the definition (A5) for the coefficient b one obtains

$$\frac{1-b}{1+b} = \coth^{-1}\left(\frac{\hbar\omega_1}{2k_B T} \right). \tag{A8}$$

For a Brownian oscillator of mass m, frequency ω_0 and vanishing friction, one has

$$\frac{<r^2>}{3} = \frac{\hbar}{2m\omega_0} \coth\left(\frac{\hbar\omega_1}{2k_B T}\right). \tag{A9}$$

By using expressions (A7) to (A9) the last term in eq. (A6) can be rewritten to yield

$$\sum_{n=0}^{N} \binom{N}{n} b^n (N-2n) = N(1+b)^N \frac{3}{<r^2>} \frac{\hbar}{2m\omega_1}. \tag{A10}$$

Finally, with the above relation the expression (A6) for the recoil energy can be written as

$$<\hbar\omega> = \hbar\omega_1 f(<r^2>) \frac{k^2\hbar}{2m\omega_1} \sum_{N=1}^{\infty} \frac{1}{(N-1)!} \left(\frac{k^2 <r^2>}{3}\right)^{N-1} = \frac{k^2\hbar^2}{2m}. \tag{A11}$$

Thus, the asymmetry of the spectrum provides the recoil energy of the free Mössbauer nucleus. The same scheme of argument holds for several harmonic modes i.e. for normal modes.

Note, that an overdamped Brownian oscillator behaves like a classical system such that the line of arguments from a quantum description [46] does not hold. The overdamped oscillator contains only quasi-elastic lines, which are centered at the elastic line such that $<\hbar\omega> = 0$, corresponding to vanishing recoil energy.

References

1. Frauenfelder H., Petsko, G.A., Tsernoglou, D. (1979) Temperature-dependent X-ray diffraction as a probe of protein structural dynamics, *Nature* **280**, 558-563.

2. Frauenfelder H., Parak, F., Young, R.D. (1988) Conformational substates in proteins, *Ann. Rev. Biophys. Chem.* **17**, 451-479.

3. Wüthrich, K. (1989) Protein structure determination in solution by nuclear magnetic resonance spectroscopy, *Science* **243**, 45-50.

4. Palmer III., A.G., Rance, M., Wright, P.E. (1991) Intramolecular motions of a zinc finger DNA-binding domain from Xfin characterized by proton-detected natural abundance ^{13}C heteronuclear NMR spectroscopy, *J. Am. Chem. Soc.* **113**, 4371-4380.

5. Doster, W., Cusack, S., Petry, W. (1989) Dynamical transition of myoglobin revealed by inelastic neutron scattering, *Nature* **337**, 754-756.

6. Bellissent-Funel, M.-C., Lal, J., Bradley, K.F., Chen, S.H. (1993) Neutron structure factors of in-vivo deuterated amorphous protein C-phycocyanin, *Biophys. J.* **64**, 1542-1549.

7. Albanese, G., Deriu, A. (1979) High energy resolution X-ray spectroscopy, ***Rivista del nuovo cimento*** **2,** 1-40.

8. Nienhaus, G.U., Heinzl, J., Huenges, E., Parak, F. (1989) Protein crystal dynamics studied by time-resolved analysis of X-ray diffuse scattering, *Nature* **338**, 665-666.

9. Parak, F., Knapp, E.W., Kucheida, D. (1982) Protein dynamics: Mössbauer spectroscopy on deoxymyoglobin crystals, *J. Mol. Biol.* **16**, 177-194.

10. Knapp, E.W., Parak, F., Fischer, S.F. (1982) The influence of protein dynamics on Mössbauer spectra, *J. Chem. Phys.* **78**, 4701-7411.

11. Knapp, E.W., Fischer, S.F., Parak, F. (1982) Protein dynamics from Mössbauer spectra. The temperature dependence, *J. Phys. Chem.* **86**, 5042-5047.

12. Parak, F., Heidemeier, J., Nienhaus, G.U. (1988) Protein structural dynamics as determined by Mössbauer spectroscopy, *Hyperfine Interactions* **40**, 147-158.

13. Krimm, S. Bandekar, J. (1986) Vibrational spectroscopy and conformation of peptides, polypeptides and proteins, *Advan. Protein Chem.* **38**, 181-364.

14. Gerwert, K. (1993) Molecular reaction mechanisms of proteins as monitored by time-resolved FTIR spectroscopy, *Current Opinion in Struct. Biol.* **3**, 769-773.

15. Srajer, V., Champion, P.M. (1991) Investigations of optical line shapes and kinetic hole burning in myoglobin, *Biochem.* **30**, 7390-7402.

16. Leone, M., Cupane, A., Militello, V., Cordone, L (1994) Thermal broadening of Soret band in heme complexes and in heme-proteins: role of the iron dynamics, *Eur. Biophys. J.* **23**, 349-352.

17. Chen, X.G., Asher, S.A., Schweitzer-Stenner, R., Mirkin, N.G., Krimm, S. (1995) UV Raman determination of the $\pi\pi^*$ excited state geometry of N-methylacetamide: vibrational enhancement pattern, *J. Am. Chem. Soc,* **117**, 2884-2895.

18. Baron, A.Q.R., (1995) Report on the X-ray efficiency and time response of a 1 cm^2 reach through avalanche diode, *Nucl. Instr. Meth. A* **352**, 665-667.

19. Chumakov, A., Baron, A.Q.R., Rüffer, R., Grünsteudel, H., Grünsteudel, H.F., Meyer, A. (1996) Nuclear resonance energy analysis of inelastic X-ray scattering, *Phys. Rev. Lett.* **76**, 4257-4260.

20. Achterhold, K. Keppler, C., van Bürck, U., Potzel, W., Schindelmann, P., Knapp, E.W., Melchers, B., Chumakov, A.I., Baron, A.Q.R., Rüffer, R., Parak, F. (1996) Temperature dependent inelastic X-ray scattering of synchrotron radiation on myoglobin analyzed by the Mössbauer effect, *Eur. Biophys. J.* **25**, 43-46.

21. Daggett, V., Levitt, M. (1993) Realistic simulations of native-protein dynamics in solution and beyond, *Annu. Rev. Biophys. Biomol. Struct.* **22**, 353-380.

22. York, D.M., Wlodawer, A., Pedersen, L.G., Darden, T.A. (1994) Atomic-level accuracy in simulations of large protein crystals, *Proc. Natl. Acad. Sci. USA* **91**, 8715-8718.

23. Parak, F., Knapp, E.W. (1984) A consistent picture of protein dynamics, *Proc. Natl. Acad. Sci. USA* **81**, 7088-7092.

24. Hartmann, H. Parak, F., Steigemann W., Petsko, G.A., Ringe Ponzi, D. Frauenfelder, H. (1982) Conformational substates in a protein; structure and dynamics of metmyoglobin at 80 K., *Proc Natl. Acad. Sci. USA* **79**, 4967-4971.

25. Parak, F., Hartmann, H., Aumann, K.D., Reusher, H. Rennekamp, G., Bartunik, H., Steigemann, W. (1987) Low temperature X-ray investigation of structural distributions in myoglobin, *Eur. Biophys. J.* **15**, 237-249.

26. Hartmann, H., Zinser, S., Komninos, P., Schneider, R.T., Nienhaus, G.U., Parak, F. (1996) X-ray structure determination of a metastable state of carbonmonoxy myoglobin after photodissociation, *Proc. Natl. Acad. Sci. USA*, **93**, 7013-7016.

27. Lonarich, R.J., Brooks, B.R., (1990) Temperature dependence of dynamics of hydrated myoglobin *J. Mol. Biol.* **215**, 439-455.

28. MacKerell, Jr. A.D., Nilsson, L., Rigler, R., Heinemann, U. Saenger, W. (1989) Molecular dynamics simulations of ribonuclease T_1: comparison of the free enzyme and the 2' GMP-enzyme complex, *Proteins: Structure, Function and Genetics* **6**, 20-31.

29. Melchers, B., Knapp, E.W., Parak, F., Cordone, L., Cupane, A., Leone, M. (1996) Structural fluctuations of myoglobin from normal modes, Mössbauer, Raman, and absorption spectroscopy, *Biophys. J.* **70**, 2092-2099.

30. Singwi, K.S., Sjölander, A. (1960) Resonance absorption of nuclear Y-rays and the dynamics of atomic motions, *Phys. Review* **120**, 1093-1102.

31. Mössbauer, R.L. (1976) Debye Waller factors and Lamb factors in the scattering of X-rays, Y-rays and thermal neutrons, *Journal de Physique* **37**, 11-13.

32. Mössbauer, R.L. (1959) Kernresonanzabsorption von γ-Strahlung in Ir^{191}, *Z. Naturforsch.* **14a**, 211-116.

33. Wilson, Jr. E.B., Decius, J.C., Cross, P.C. (1955) *Molecular vibrations. The theory of infrared and Raman vibrational spectra*, Dover Publ., Inc., N.Y.

34. Case, D.A. (1994) Normal mode analysis of protein dynamics, *Current Opinion in Struct. Biol.* **4**, 285-290

35. Levy, R.M. Karplus M., Kushick, J., Perahia, D. (1984) Evaluation of the configurational entropy for proteins: Application to molecular dynamics simulations of an α-helix, *Macromolecules* **17**, 1370-1374.

36. Teeter, M.M. Case, D.A. (1990) Harmonic and quasiharmonic dynamics of crambin, *J. Phys. Chem.* **94**, 8091-8097.

37. Jorgensen, W.L. Chandrasekhar, J., Madura, J.D., Impey, R.W., Klein, M.L. (1983) Comparison of simple potential functions for simulating liquid water, *J. Chem. Phys.* **79**, 926-935.

38. Brooks, B.R., Bruccoleri, R.E., Olafson, B.D., States, D.J. Swaminathan, S., and Karplus, M. (1983) CHARMM: A program for macromolecular energy, minimization, and dynamics calculations, *J. Comp. Chem.* **4**, 187-217.

39. Smith, B.T., Boyle, J.M., Dongarra, J.J., Garbow, B.S., Ikebe, Y., Klema, V.C., Moler, C.B. (1976) in Goos, G., Hartmanis, J. (eds.) *Matrix Eigensystem Routines - EISPACK guide*, Springer Verlag.

40. Kubo R. (1961) A stochastic theory of line-shape and relaxation, in Haar D. Ter (ed.), *Fluctuation, relaxation and resonance in magnetic systems,* Plenum Press, N.Y., pp. 23-68.

41. Nienhaus, G.U., Frauenfelder, H., Parak, F. (1991) Structural fluctuations in glass-forming liquids: Mössbauer spectroscopy on iron in glycerol, *Phys. Rev. B* **43,** 3345-3350.

42. Shaitan, K.V. Rubin A.B. (1981) Theory of Mössbauer effects in proteins, *Sov. Biophys.,* **25**, 809ff.

43. Nowik, I., Cohen, S.G., Bauminger, E.R., Ofer, S. (1983) Mössbauer absorption in over-damped harmonically bound particles in Brownian motion, *Phys. Rev. Lett.* **50**, 1528-1531.

44. Uhlenbeck, G.E., Ornstein, L.S. (1930) On the theory of the Brownian motion, *Phys. Rev.* **36**, 823-841.

45. Chandrasekhar, S., (1943) Stochastic problems in physics and astronomy, *Rev. Mod. Phys.* **15**, 1-87.

46. Lipkin, H.J. (1962) Some simple features of the Mössbauer effect, *Ann. Phys.* **18,** 182-197.

47. Knapp, E.W., Fischer, S.F. (1981) On the theory of homogeneous and inhomogeneous line broadening. An exactly solvable model, *J: Chem. Phys.***74**, 89-95.

COMPUTER SIMULATIONS OF THE ACTION OF METALLOENZYMES

ARIEH WARSHEL
Department of Chemistry
University of Southern California
Los Angeles, California 90089-1062

1. Introduction

Metalloenzymes play a major role in many biological processes, including the action of G-proteins (e.g. refs. [1-4]), DNA-polymerases [5-9] and many other biologically important proteins [10]. Elucidating the origin of the catalytic activity of such systems, as well as understanding their detailed mechanism, is of major fundamental and practical importance. Thus, it is not surprising that many proposals have been put forward to explain the role of metals in metalloenzymes and the role of enzymes in general (e.g. ref. [11]). Yet, despite the enormous progress in structural and biochemical studies, it is hard to assess the importance of different catalytic factors based on current experimental information. For example, although mutation experiments are very useful, they only provide circumstantial evidence since they can be interpreted in different ways. In view of these difficulties, it seems clear that any progress in quantitative understanding of enzyme action requires the use of computer aided structure function correlation. This work will address the current state of the art in computer modeling of the action of metalloenzymes and consider some future options.

2. Modeling Metalloenzymes

Several possible options are available for quantum mechanical calculations of enzymatic reactions. In particular, one can use molecular orbital (MO)

L. Banci and P. Comba (eds.), Molecular Modeling and Dynamics of Bioinorganic Systems, 343–359.

and valence bond (VB) approaches. In modern (gas-phase) computational quantum chemistry, the former method has become the favored choice because of its more convenient algorithmic implementation for computing. However, as much as evaluation of activation free energies in enzymes are concerned, the VB approach is at present more convenient and more reliable. Here, we will briefly discuss both approaches.

The effect of the environment (which is denoted by s) surrounding the reacting system (r, referred to as solute) can be included in MO calculations in a fairly straightforward way if the overlap integrals between the solute and the solvent are neglected [12, 13]. The effect of the solvent polarization on the solute electronic states can be estimated within the zero differential overlap approximation by rewriting the diagonal matrix elements of r as

$$F_{vv} = F_{vv}^{0} + \sum_k V_{v,k} \tag{1}$$

where F^0 is the Fock matrix of the system r in a vacuum and $V_{v,k}$ is the potential at the site of the vth orbital due to the permanent charge distribution and electronic polarization of the atoms (k) in s. The eigenvectors of the modified Fock matrix give the solute molecular orbital vectors, $\boldsymbol{C}$, and the corresponding charges, $Q_i = Z_i - \Sigma_{v\varepsilon i}\Sigma_j n_j C_{jv}^2$ (where Z is the nuclear charge and v and j enumerate the atomic and molecular orbitals, respectively). The total potential energy of the system ($r + s$) can then be written as

$$E = E^0 + \sum_i \sum_k Q_i q_k / r_{ik} - \tfrac{1}{2} \sum_k \alpha_k \xi_k^2 + V_{vdw}^{r,s} + V^s \tag{2}$$

where E^0 is the vacuum energy of the quantum system (r) and the second and third terms denote interactions with permanent charges and induced dipoles in s, q_k, α_k, and ξ_k are the partial charge, electronic polarizability and electric field on the kth atom in s, respectively. Since the charge distributions of r and s are coupled through the second and third terms of eq. (1) one must solve the equation self-consistently. The fourth term of eq. (1) denotes the Van der Waal's interaction between r and s, while the last term represents the interaction energy of the surrounding system itself. These potential energy terms are most conveniently represented by standard classical force fields. eqs. (1) and (2) and the approach of ref. [12] as well as that of related works (e.g. see refs. [14, 15]) correspond to a description in which the solvent is explicitly represented and it does not rely on unknown parameters such as a cavity radius or an effective dielectric constant. This allows one to actually examine the validity of various hypotheses about enzyme catalysis (rather than to determine the effective parameters that reproduce the different hypotheses).

Calculation of the potential surface of the reaction according to eq. (2) using iterative SCF procedures in combination with energy minimization of the surrounding system does not pose any major difficulties, and this approach (which is now referred to as hybrid QM/MM approach) became quite popular recently as a strategy for studying enzymatic reactions and related processes (e.g. see refs. [13-17]). However, studies of processes in enzyme and solution should also involve an iterative evaluation of the solvent polarization in response to the solute (r) charge distribution (which requires additional computing time) and the resulting change in the solute charge distribution. Furthermore, in order to evaluate the actual free energy surface one must be able to perform statistical averaging over many configurations of the system. This is most conveniently achieved with the free energy perturbation (FEP) technique. A problem with the MO formulation in this context, is that it is difficult to rigorously devise a proper mapping potential which allows one to gradually force the system to move from reactants to products. That is, one must be able to make the total (mapping) Hamiltonian an explicit function of a control parameter, λ, without constraining the coordinates to any particular value, and still be able to monitor the actual ground-state surface. Other approaches for evaluating the free energy of hybrid MO/MM potential surfaces do not treat properly the solvent contribution to the activation barrier (see discussion in ref. [18]). An attempt to use an FEP approach that was reported in ref. [19] involved a mapping between the ground and transition state configurations of the solute in the gas phase. Such an approach cannot and should not be used for reactions that involve large solvent effects (e.g. heterolytic bond cleavage reactions). Effective ways of overcoming these problems were provided recently [18] but not yet adopted by other research groups. Finally, and perhaps most importantly, it should be mentioned that MO approaches do not readily lend themselves to a convenient calibration by experimental information about bond breaking and bond forming reactions. In this respect, one should keep in mind that even for a simple process such as homolytic bond cleavage, SCF calculations can give very large errors if double configuration interaction is not included, and it is difficult to obtain analytical derivatives for such corrections.

Many of the problems involved on MO studies of enzymatic reactions can be minimized by using a VB formulation and in particular by using the Empirical Valence Bond (EVB) approach. The EVB method is based on the realization that if we can reasonably describe and calibrate potential surfaces for reactions in solutions then the most important effects of enzymes can be evaluated by determining the *change* in these potential surfaces upon transfer from solution to the enzyme site. This change is largely due to the difference in "solvation" of the charges of the reacting atoms in the two environments. Therefore it is frequently sufficient to obtain reliable solvation energies of the reacting atoms, a significantly simpler task than evaluating the ab initio potential surfaces of the reacting enzyme. Furthermore, the asymptotic regions of the potential surfaces of reacting systems can be reproduced quite accurately even without ab initio calculations. For example, it is possible to use experimentally determined Morse potentials to describe the potential surface along the bonds of molecular fragments. It is also possible (and recommended) to use thermodynamic information to determine the energy difference between the reactant and product states. Other regions of the potential surfaces can also be estimated by using observed activation barriers or by quantum mechanical calculations in some cases. Since the physics of solvation can be taken into account correctly by using classical force fields [20], the *difference* in energetics upon change of environment from solution to protein can be reproduced even if some aspects of the particular regions are not completely accurate. The EVB method exploits the above facts and provides both a simple way for constructing and calibrating analytical potential surfaces for chemical reactions and a very effective way for *transferring* such potential surfaces between solution and enzyme. Since the details of the EVB method have been discussed and illustrated extensively [20-22] we will focus only on key points.

To demonstrate the main points of the EVB approach, we will consider a specific case. Here we illustrate the characteristics of the EVB potential surface by considering a recent simulation of a proton transfer (PT) reaction between OH^- and H_2O in water. We then illustrate the application of the EVB approach to enzymatic reaction by considering a study of PT between two water molecules in the active site of carbonic anhydrase (CA).

A PT between OH^- and H_2O can be described by the three valence bond (VB) resonance structures:

Ψ_1 Ψ_2 Ψ_3

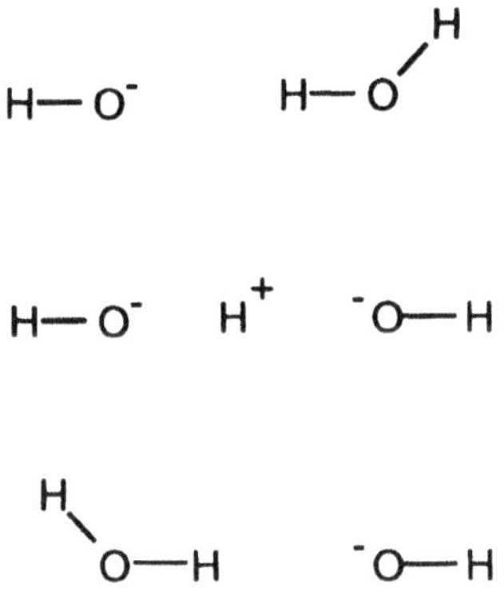

The basic idea in the EVB treatment is to obtain the relevant ground state potential, $\mathbf{E}_g$, by considering the mixing between the relevant resonance structures, where each resonance structure represents a specific bonding and charge conformation with well-defined interaction with the surrounding environment (protein and/or water). The ground state surface is obtained by solving the secular equation

$$\mathbf{HC}_g = \mathbf{E}_g\mathbf{C}_g \tag{3}$$

where $\mathbf{H}$ is the Hamiltonian of the system and $\mathbf{C}_g$ is the ground state eigenvector. The gas phase contribution to $\mathbf{H}$ can be calculated by ab initio VB approaches [23, 24]. We prefer, however, to represent this Hamiltonian by analytical functions which can be calibrated by empirical procedures [20, 21] and/or by fitting the calculated $\mathbf{E}_g$ and ground state charge distribution to ab initio results [18, 21, 25]. The solvent contributions to the Hamiltonians are then considered by evaluating the "classical" interaction of the solvent with the different resonance structures. Thus, we describe the diagonal elements of the Hamiltonian by the following force fields [20, 22]:

$$\varepsilon_i = H_{ii} = \sum_j \Delta M_j^{(i)}(b_j^{(i)}) + \frac{1}{2}\sum_l \gamma_l^{(i)} k_l^{(i)} (\theta_l^{(i)} - \theta_{0,l}^{(i)})^2 + V_{nb,rr}^{(i)} + \alpha^{(i)} + V_{nb,rs}^{(i)} + V_s \tag{4}$$

where $\Delta M_j^{(i)}$ denotes the Morse potential relative to its minimum value for the jth bond in the ith VB structure, and the second and third terms, respectively, denote the bond angle bending contribution and the nonbonded electrostatic and van der Waals interactions between the reacting groups, denoted by subscript r. The factor $\gamma_l^{(i)}$ in the second term is a coupling between bonds that are being broken or formed and those angles that depend on these bonds. The term $\alpha^{(i)}$ is the gas phase energy of the ith state when all the fragments are at infinite separation. The nonbonded interaction with the surrounding protein and solvent, denoted by subscript s, is given by $V_{nb,rs}^{(i)}$. The last term represents the internal potential energy of the protein/solvent system. The off-diagonal elements of the Hamiltonian are represented by a simple single exponential approximation

$$H_{ij} = A_{ij} \exp\{-\mu r\} \tag{5}$$

where r is the distance between atoms whose bonding is changed upon transfer from Ψ_i to Ψ_j. The parameters A_{ij} and μ can be adjusted to reproduce the observed barrier for the reaction in solution or to reproduce gas phase ab initio calculations.

In view of the simple form of the EVB Hamiltonian, one may question how it can reproduce reliable ab initio surfaces. The answer is simply that EVB offers what is perhaps the best way of fitting analytical surfaces to ab initio surfaces, rather than producing them from first principles. That is, the asymptotic limit of each resonance structure (where the corresponding fragments are at infinite separation) corresponds to a very well defined molecular fragment that can be reliably described by potential functions (as is now clear from force field descriptions of molecules). With correct physics at the asymptotes it is usually not difficult to obtain a reasonably accurate description of intermediate regions. In other words, the EVB method elegantly transforms what we *know* about a reacting system into a form that captures the main features in a natural way and allows "extrapolation" to the conditions in which one is actually *interested*. The best way to realize this point is to examine specific fitting cases. Several examples were given recently by our group [18] and others [25, 26]. We give an example in Fig. 1 that describes the similarity between the EVB and ab initio surface for our PT reaction. An even better fitting could have been obtained with more effort but this is not justified in most cases (see below).

Recently we started developing an automated fitting procedure where the EVB surfaces of the reactant and products are forced to reproduce the corresponding ab initio surfaces at each "corner" of the relevant potential surface (which include the reactant and product and intermediate states). It should also be pointed out that despite the progress in reliability of ab initio calculations we still consider verification and calibration of the EVB surface by solution thermodynamic information a major part of our procedure. At present, this is our main safeguard for reliability of the EVB method.

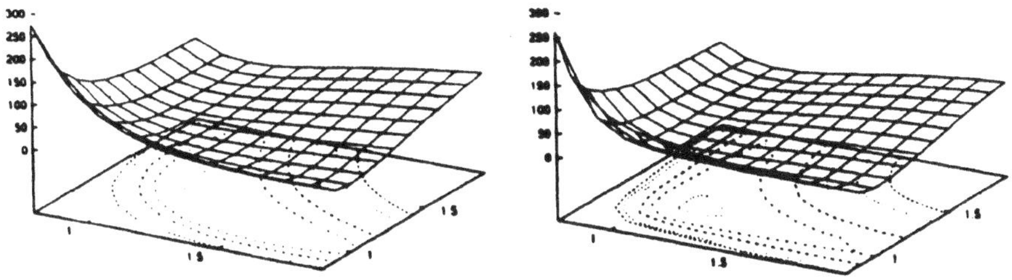

Figure 1. Comparison of ab initio (right) and EVB (left) gas phase potential energy surfaces for proton transfer in the $[HOHOH]^-$ ion. Results are in kcal/mol, relative to the R_{OO}=2.4Å minimum.

Calibrated EVB surfaces provide an extremely effective way of studying enzyme catalysis since they focus on the *difference* between reactions in enzyme and solution (this difference is what catalysis is all about). Furthermore, since the calibration process necessarily reproduces the correct experimental solution energetics, it also consequently prevents unrealistic results. In contrast, some of the current semiempirical studies of enzymatic reactions cannot reproduce the correct energetics of the reference reaction in water (e.g. reproducing the energy of a PT process) and most studies do not subject themselves to this crucial and useful test.

To see the effectiveness of our procedure in studies of enzymatic reactions, we describe the results of our study on the rate determining step in the catalytic reaction of CA. This step involves a PT between a zinc bound water to another water followed by a proton transfer to His 64. It is possible that this may be a concerted process, although it is more likely that the transfer follows a stepwise mechanism. Fig. 2 presents the results of our studies [27].

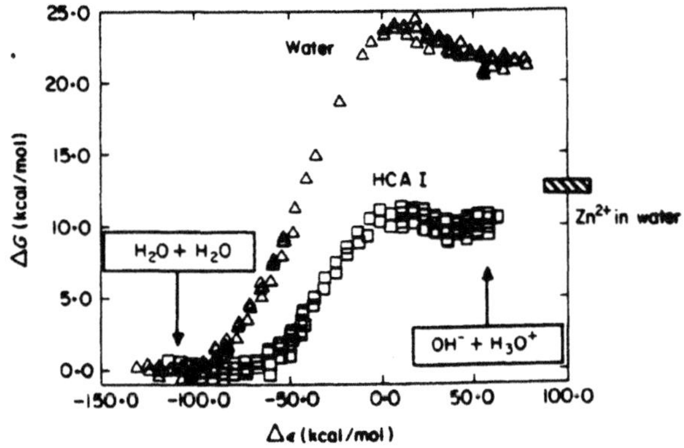

Figure 2. Calculated free energy profiles for the reference reaction ($2H_2O \rightleftharpoons OH^- + H_3O^+$) in water after calibration of the $\Delta\alpha$ and H_{12} parameters (upper curve, triangles) and for the proton transfer step in HCA I (lower curve, open squares). The calculated energetics of proton transfer from a zinc-bound water molecule in aqueous solution are also shown for comparison.

The calculated activation barrier for the reaction in the enzyme provides a very good approximation of the observed activation barrier (the calculated and observed values of Δg^{*} are 11.0 kcal/mol and 10.8 kcal/mol, respectively, with an error range of ~3 kcal/mol). While the good agreement might be fortuitous, it is important to note that our results were obtained without adjustment of any parameter for calculating the energy difference between the reaction in the protein and in water. It should also be noted that the observed activation barrier reflects some nuclear quantum mechanical effects which are not evaluated in classical FEP procedures. However, these effects are similar in the enzyme and solution reactions and are apparently absorbed into the effective H_{ij}. Finally, it is interesting to note that the effectiveness and success of this approach has led other research groups to adopt similar methodologies [25, 26, 28, 29].

Before completing this section, it is important to make some comments about the treatment of metal ions. In selecting the proper description of the metal, it is important to realize that the metal interaction with its ligands involves significant charge-transfer character. This means that a rigorous description should include the metal and its ligands in the quantum mechanical region.

A reasonable alternative to such a treatment (see discussion section) might be offered by calibrating force fields for the metal ligand interaction using experimental information about the metal ion in water. This approach has been taken in our studies where we required that the force field of the metal ion will reproduce its solvation and radial distribution function in water. The resulting model (which has some theoretical justification as will be argued below) corresponds to adding six fractional positive charges, $+\delta$, in octahedral geometry at a distance r_δ from the center of the ion and a corresponding charge of -6δ in the center, thus retaining a net charge of +2. The ion then interacts with One Lennard-Jones center (at the center of the ion) and seven electrostatic centers: the geometry of the six peripheral fractional charges is rigid, but overall rotation of the six-center frame about the nucleus is allowed, and no internal forces are associated with such rotations. The resulting model has been found to provide much better results than those obtained by using a single center model [30].

3. Simulation Studies

The EVB method has allowed for quantitative and semiquantitative progress in studies of a wide class of metalloenzymes [22, 27, 30, 31] as well as the effect of metal substitution in some of these systems. The action of Ca^{2+} ion in SNase has been explored in refs. [32] and [30]. This study reproduced the very large catalytic effect of the enzyme assuming the mechanism of Cotton and coworkers. In addition, the effects of several key mutations were also reproduced. It was concluded that the Ca^{2+} in its specific protein environment plays a major role in both stabilizing the generated OH^- ion and in helping the transfer of negative charge to the penta-coordinated transition state.

The effect of metal ion substitution was also examined with the help of the EVB free energy surfaces and the corresponding Marcus-like parabolas [32]. This instructive study explains why metals that stabilize the nucleophile too much reduce the catalytic effect of the enzyme.

The catalytic action of carbonic anhydrase was studied in a series of papers [27, 31, 33], demonstrating the robustness of the EVB approach and its usefulness in studying metalloenzymes. These studies reproduce quantitatively the catalytic effect of the enzyme. The role of the Zn^{2+} ion in stabilizing the generated OH^- ion and the carbonic acid was quantified. In this case, it is easy to verify our conclusion since the effect of the enzyme in the initial proton transfer step is directly related to the pK_a of the Zinc-bound water molecule which is known experimentally. One of the important conclusions of this study is that the exact orientation of the carbonic acid, which has been emphasized in studies that did not explore the actual activation free energy (e.g. refs. [34-36]) is much less important than previously thought. As expected, what counts most is the electrostatic effect of the zinc and the rest of the active site on the reacting fragments. In other words, the effect of the active site on the reactants in a slightly different reaction pathway has less importance than the overall stabilization effect exerted by the enzyme and its bound zinc ion. The EVB approach combined with centroid path integral calculation allowed us to explore quantum mechanical nuclear effect on catalysis in general and carbonic anhydrase in particular. The conclusions of this study have been summarized elsewhere [31].

The EVB approach has been applied in studies of the important class of G-proteins where Mg^{2+} ions play a major role [1-3, 37]. These studies were instrumental in elucidating the mechanism of $p21^{ras}$ and other related proteins such as transducin [38]. The calculations succeeded in correctly predicting the effect of the protein and its bound metal ion on the pK_a of the γ-phosphate [2, 37] and in calculating the effect of the metal ion on the binding of GTP and GDP ligands [39].

Another important system that was studied recently is the exo site of DNA polymerase [9]. This site is a prototype of enzymes that use two metal ions rather than one metal ion (the difference between these two catalytic motifs will be discussed below). All the above studies provided some general conclusions about the role of metal ions on enzyme catalysis and some of these conclusions will be discussed below.

4. General Conclusions

Computer simulations of the action of metalloenzymes appear to be quite useful in clarifying the role of metal ions in enzyme catalysis. This is not a trivial issue as might be assumed by those who consider electrostatic effects as a major catalytic factor. That is, traditional structural and biochemical studies have focused on qualitative description, suggesting that the metal "polarizes" bonds and "activates" nucleophiles without being able to define and address quantitative concepts which are crucial for understanding what factors are really important. For example, although our finding that divalent metals play a major catalytic role by electrostatic stabilization of the generated nucleophile is reasonable, it has not been generally appreciated. A case in point is the role of metals in phosphate hydrolysis. That is, some studies have assumed that RNA hydrolysis involves abstraction of a proton from a ribose hydroxyl (ROH) by an OH^- ion bound to Mg^{2+} in the active site [5, 6]. In this mechanism, $Mg^{2+}OH^-$ is viewed as a general base or proton acceptor [6]. A recent study has suggested that the metal ion helps to "facilitate" OH^- formation [7], but no specific details were given to show how this facilitation is accomplished, or what its magnitude is.

It was also suggested that enzymes work by "orienting" the nucleophile (e.g. ref. [5]) and a good example is the proposal that Glu 357 in DNA polymerase orients the OH^- ion (refs. [7] and [8]). Unfortunately, the negative charge of Glu 357 repels the negative charge of the OH^- ion and tries to keep it as far away as possible. This repulsion persists even when the hydrogen of the OH^- ion points toward the acid and provides the opportunity of forming an (O^- H–O^-) interaction. Perhaps the misunderstanding is associated with the confusion of the stable hydrogen bond of the (O^- H–O) system with the present (O^- H–O^-) system. Similarly, we do not find support for the idea [5] that the Mg^{2+} ion is to orient the attacking OH^- and supposedly help entropically by restricting the angular orientational entropy of this nucleophile. Our simulations do not produce a drastic angular constraint on the orientation of the OH^- ion. More importantly, if the OH fragment is fixed at the transition state, there is no way to use its orientational entropy in increasing k_{cat} / K_M (as explained in chapter 9 of ref. [20], this quantity reflects the difference between the free energy of the transition state in the protein and the reactants in water and it is not affected by intermediate states). Thus we feel that the seemingly simple fact that the metal helps in stabilizing the nucleophile has established and quantified the EVB simulation studies mentioned above. It is useful to mention in this respect that other theoretical studies have explored the role of metal ions, but most of them considered the effect of the metal gas phase (e.g. [40, 41]) where the enormous calculated effect is basically irrelevant to enzymatic reactions (see, for example, discussion in Chapter 9 of ref. [20]). Even calculations that include the enzyme (e.g. ref. [42]) drastically overestimated the effect of the metal since they did not treat properly the dielectric effect of the surrounding solvent and the protein (see discussion in ref. [43]). Apparently, reaching quantitative conclusion is crucial in assessing the effect of the metal since catalysis represents the difference between the energetics of the given reaction in protein and in water and non-quantitative studies can easily overestimate the stabilization in one of these cases and reverse the actual trend.

Another example of the difficulties in assessing the role of metal ions is provided by our study of the effect of the second metal in DNA polymerase [9]. Although the crucial role of this metal was pointed out in the original structural studies [5, 7, 8] the details did not involve recognition of the importance of the corresponding electrostatic stabilization effect. It has been proposed [7, 8] that the second metal ion acts both as a Lewis acid in facilitating the leaving of the 3' oxyanion and by inducing strain on the O–P–O bond angle. The first proposed effect is related to electrostatic stabilization of the product state and is quite reasonable if the formation of this state is the rate-limiting step. However, for cases where the formation of the pentacoordinated intermediate is the rate-limiting step, it is necessary to focus on the role of the second metal in stabilizing this state. The strain hypothesis can be analyzed by separating the electrostatic effect of the metal ion from its steric effect. A quantitative study of this issue [9] has established that the strain effect is anticatalytic.

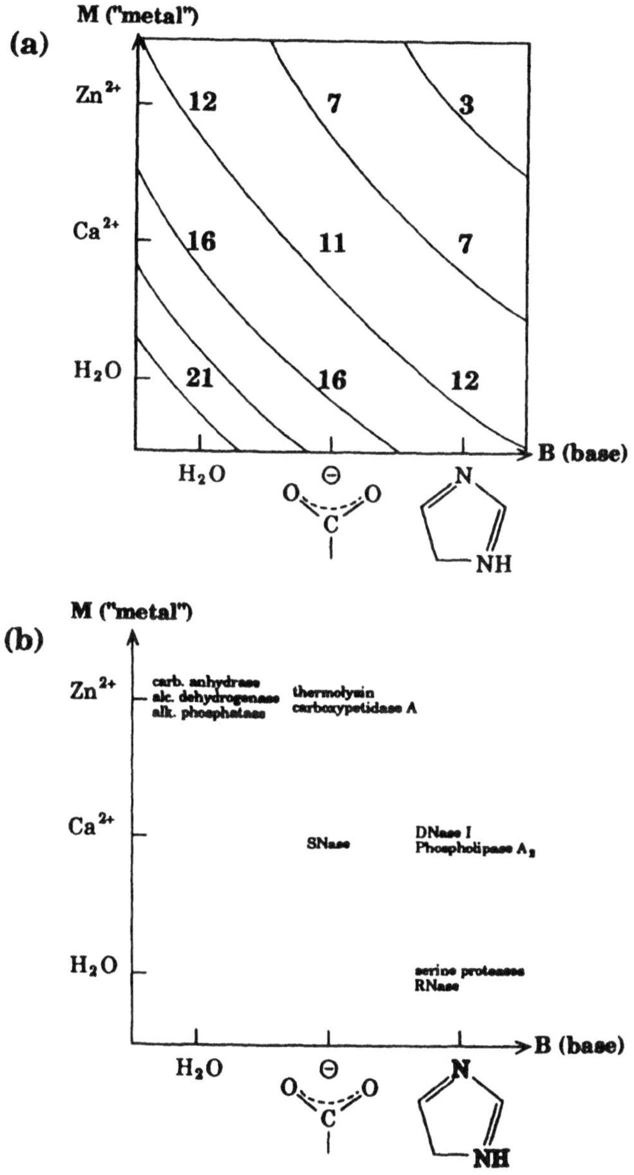

Figure 3. Classifying metalloenzymes according to their catalytic metal and the coupled general base. Part (*a*) of the figure shows the energetics (in kcal/mol) of transferring a proton from a metal-bound water to a general base in water. For example. a proton transfer from Ca^{2+}-bound water to a general base in water. Part (*b*) classifies different metalloenzymes according to the corresponding metal and general base. The figure illustrates that metalloenzymes are usually found in the low-energy part of the diagram.

EVB studies of the action of staphylococcal nuclease (SNase) have provided an interesting qualitative classification of metalloenzymes [30]. It was concluded that the nucleophile RO^- needed in many catalytic reactions can be prepared by either having a strong base that will reduce the energetics of the reaction

$$ROH + B \rightleftharpoons RO^- + BH^+$$

or by having a metal ion that stabilizes the generated nucleophile.

Considering the energetics of these two options led to the general roles shown in Fig. 3, where it is postulated that a given enzyme will try to be at the relevant low energy regions of figure 3a (see ref. [27] for more details). In fact, the enzyme considered appears to correspond nicely to this diagram.

Another interesting issue that can be explored by EVB simulations is the relationship between two-metal ions and a single metal-ion mechanisms. Such a study has been reported in ref. [9], which compares the catalytic mechanism of SNase and DNA polymerase. It was concluded that sites which hold to metal in a very rigid environment may reduce its ability to participate in a single-metal catalytic mechanism since the metal cannot move effectively toward the phosphate center during the nucleophilic attack step. However, such sites would be quite useful in a two-metal mechanism. Another possible advantage of a two-metal site can be provided in cases where the rate limiting step involves the dissociation of the leaving group, or when this step occurs in concert with the nucleophile attack step. This effect may be examined by substrates with better leaving groups where the energy of the product state is lowered (see ref. [9] for more details).

The role of the metal in G-proteins has also been analyzed using the EVB free energy functions [1, 2] and led to interesting Linear Free Energy relationships [4] which indicated that a substitution of Mg^{2+} by Mn^{2+} will fix oncogenic mutants of $p21^{ras}$. This prediction has been confirmed experimentally [37].

All the above points are just examples of the insight that can be provided by simulations of the action of metalloenzymes. Before completing this paper, it might be useful to comment about the future direction of simulations of metalloenzymes. As stated in the method section, our current strategy involves classical treatment of metal ions and their ligands. The obvious next step is to treat the metal and its ligand quantum mechanically. This can be done conveniently by the EVB formulation and this will require, of course, reparameterization of the reference state in solution which will now include a significant degree of explicit charge transfer. Another option involves a DFT description of the metal ion and its ligand. To this end, we developed the powerful Frozen DFT (FDFT) and Constraint DFT (CDFT) approaches that allow us to evaluate ab-initio DFT free energies of large systems [44, 45] by treating some regions of the system with constraint densities. These approaches will be further examined in our lab.

Acknowledgement

This work was supported by NIH grant GM24492.

References

1. Langen, R.; Schweins, T. and Warshel, A. (1992) On the Mechanism of Guanosine Triphosphate Hydrolysis in ras p21 Proteins. *Biochemistry* **31**, 8691.
2. Schweins, T.; Langen, R. and Warshel, A. (1994) Why Have Mutagenesis Studies Not Located the General Base in ras p21. *Nat. Struct. Biol.* **1**, 476.
3. Schweins, T. and Warshel, A. (1996) Mechanistic Analysis of the Observed Linear Free Energy Relationships in p21ras and Related Systems. *Biochemistry* **35**, 14232.
4. Schweins, T.; Geyer, M.; Kalbitzer, H. R.; Wittinghofer, A. and Warshel, A. (1996) Linear Free Energy Relationships in the Intrinsic and GTPase Activating Protein-Stimulated Guanosine 5'-Triphosphate Hydrolysis of p21ras. *Biochemistry* **35**, 14225.
5. Yarus, M. (1993) How many catalytic RNAs? Ions and the Cheshire cat conjecture. *FASEB J.* **7**, 31.
6. Haydock, K. and Allen, L. C. (1985) In *Progress in Clinical and Biological Research,* R. Rein and A. R. Liss, Ed.; Vol. 172A, pp 87-98.
7. Steitz, T. A. and Steitz, J. A. (1993) A General Two-Metal-Ion Mechanism for Catalytic RNA. *Proc. Natl. Acad. Sci.* **90**, 6498.
8. Beese, L. S. and Steitz, T. A. (1991) Structural Basis for the 3'-5' Exonuclease Activity of *Escherichia coli* DNA Polymerase I: A Two Metal Ion Mechanism. *EMBO J.* **10**, 25.
9. Fothergill, M.; Goodman, M. F.; Petruska, J. and Warshel, A. (1995) Structure-Energy Analysis of the Role of Metal Ions in Phosphodiester Bond Hydrolysis by DNA Polymerase I. *J. Am. Chem. Soc.* **117**, 11619.
10. Lipscomb, W. N. and Sträter, N. (1996) Recent Advances in Zinc Enzymology. *Chem. Rev.* **96**, 2375-2433.
11. Page, M. I. (1977) Entropy, Binding Energy, and Enzymic Catalysis. *Angew. Chem. Int. Ed. Engl.* **16**, 449-459.

12. Warshel, A. and Levitt, M. (1976) Theoretical Studies of Enzymatic Reactions. I. Dielectric Electrostatic and Steric Stabilization of the Carbonium Ion in the Reaction of Lysozyme. *J. Mol. Biol.* **103**, 227.

13. Luzhkov, V. and Warshel, A. (1991) Microscopic Calculations of Solvent Effects on Absorption Spectra of Conjugated Molecules. *J. Am. Chem. Soc.* **113**, 4491.

14. Gao, J. (1995)*Reviews in Computational Chemistry*; VCH: New York; Vol. 7, pp 119-185.

15. Mulholland, A.J.; Grant, G.H. and Richards, W.G. (1993) Computer Modelling of Enzyme Catalysed Reaction Mechanisms. *Prot. Eng.* **6**, 133-147.

16. Hartsough, D.S. and K.M. Merz, Jr. (1995) Dynamic Force Field Models: Molecular Dynamics Simulations of Human Carbonic Anhydrase II Using a Quantum Mechanical/Molecular Mechanical Coupled Potential. *J. Phys. Chem.* **99**, 11266-11275.

17. Thompson, M. A. and Schenter, G. K. (1995) Excited States of the Bacteriochlorophyll *b* Dimer of *Rhodopsuedomonas viridis*: A QM/MM Study of the Photosynthetic Reaction Center That Includes MM Polarization. *J. Phys. Chem.* **99**, 6374.

18. Muller, R. P. and Warshel, A. (1995) Ab initio Calculations of Free Energy Barriers for Chemical Reactions in Solution. *J. Phys. Chem.* **99**, 17516.

19. Bash, P. A.; Field, M. J. and Karplus, M. (1987) Free Energy Perturbation Method For Chemical Reactions in the Condensed Phase: A Dynamical Approach Based on a Combined Quantum and Molecular Mechanics Potential. *J. Am. Chem. Soc.* **109**, 8092.

20. Warshel, A. (1991)*Computer Modeling of Chemical Reactions in Enzymes and Solutions*; John Wiley & Sons: New York.

21. Hwang, J.-K.; King, G.; Creighton, S. and Warshel, A. (1988) Simulation of Free Energy Relationships and Dynamics of S_N2 Reactions in Aqueous Solution. *J. Am. Chem. Soc.* **110**, 5297.

22. Åqvist, J. and Warshel, A. (1993) Simulation of Enzyme Reactions Using Valence Bond Force Fields and Other Hybrid Quantum/Classical Approaches. *Chem. Rev.* **93**, 2523.

23. Cooper, D. L.; Gerratt, J. and Raimondi, M. (1986) The Electronic Structure of the Benzene Molecule. *Nature* **323**, 699.

24. Bernardi, F.; Olivucci, M.; McDouall, J. J. W. and Robb, M. A. (1988) Parametrization of a Heitler-London Valence Bond Hamiltonian from Complete-Active-Space Self-Consistent-Field Computations: An Application to Chemical Reactivity. *J. Chem. Phys.* **89**, 6365.

25. Chang, Y.-T. and Miller, W. H. (1990) An Empirical Valence Bond Model for Constructing Global Potential Energy Surfaces for Chemical Reactions of Polyatomic Molecular Systems. *J. Phys. Chem.* **94**, 5884.

26. Bala, P.; Grochowski, P.; Lesyng, B. and McCammon, J.A. (1996) Quantum-Classical Molecular Dynamics Simulations of Proton-Transfer Processes in Molecular-Complexes and in Enzymes. *J. Phys. Chem.* **100**, 2535.

27. Åqvist, J.; Warshel, A. (1992) Computer Simulation of the Initial Proton Transfer Step in Human Carbonic Anhydrase I. *J. Mol. Biol.* **224**, 7.

28. Lobaugh, J. and Voth, G.A. (1996) The Quantum Dynamics of an Excess Proton in Water. *J. Chem. Phys.* **104**, 2056-2069.

29. Bursulaya, B.D.; Zichi, D.A. and Kim, H.J. (1996) Molecular Dynamics Simulation Study of Polarizable Solute Solvation in Water. 1. Equilibrium Solvent Structure and Solute Rotational Dynamics. *J. Phys. Chem* **100**, 1392-1405.

30. Åqvist, J. and Warshel, A. (1990) Free Energy Relationships in Metalloenzyme-Catalyzed Reactions. Calculations of the Effects of Metal Ion Substitutions in Staphylococcal Nuclease. *J. Am. Chem. Soc.* **112**, 2860.

31. Hwang, J-K. and Warshel, A. (1996) How Important are Quantum Mechnical Nuclear Motions in Enzyme Catalysis? *J. Am. Chem. Soc.* **118**, 11745.

32. Åqvist, J. and Warshel, A. (1989) Calculation of Free Energy Profiles for the Staphylococcal Nuclease Catalyzed Reaction. *Biochemistry* **28**, 4680.

33. Åqvist, J.; Fothergill, M. and Warshel, A. (1993) Computer Simulation of the CO_2/HCO^-_3 Interconversion Step in Human Carbonic Anhydrase I. *J. Am. Chem. Soc.* **115**, 631.

34. Liang, J.-Y. and Lipscomb, W. N. (1990) *Proc. Natl. Acad. Sci. USA* **87**, 3675.

35. Solá, M.; Lledòs, A.; Duran, M. and Bertràn, J. (1991) *J. Am. Chem. Soc.* **114**, 869.

36. Krauss, M. and Garmer, D. R. (1991) Active Site Ionicity and the Mechanism of Carbonic Anhydrase. *J. Am. Chem. Soc.* **113**, 6426.

37. Schweins, T.; Scheffzek, K.; Abheuer, R. and Wittinghofer, A. (1997) The role of the Metal Ion in the $p21^{ras}$ Catalyzed GTP-Hydrolysis: Mn^{2+} versus Mg^{2+}. *J. Mol. Biol.* In press.

38. Sondek, J.; Lambright, D. G.; Noel, J. P.; Hamm, H. E. and Sigler, P. B. (1994) GTPase Mechanism of G proteins from the 1.7 Å Crystal Structure of Transducin α-GDP-AlF^-_4. *Nature* **372**, 276-279.

39. Muegge, I.; Schweins, T.; Langen, R. and Warshel, A. (1996) Electrostatic Control of GTP and GDP Binding in the Oncoprotein p21 ras. *Structure* **4**, 475.

40. Nakagawa, S.; Umeyama, H.; Kitaura and Morokuma, K. (1981) *Chem. Pharm. Bull.* **29**, 1.

41. Osman, R. and Basch, H. (1984) *J. Am. Chem. Soc.* **106**, 5710.

42. Waszkowycz, B.; Hillier, I. H.; Gensmantel, N. and Payling, D. W. (1991) *J. Chem. Soc. Perkin Trans.* **2**, 1259.

43. Naray-Szabo, G.; Fuxreiter, M. and Warshel, A. (in press) In *Approaches to Biochemical Reactivity,* A. Warshel and G. Naray-Szabo, Ed.; Kluwer Academic Publishers: .

44. Wesolowski, T. and Warshel, A. (1994) *Ab Initio* Free Energy Perturbation Calculations of Solvation Free Energy Using the Frozen Density Functional Approach. *J. Phys. Chem.* **98**, 5183.

45. Wesolowski, T.; Muller, R. P. and Warshel, A. (1996) *Ab Initio* Frozen Density Functional Calculations of Proton Transfer Reactions in Solution. *J. Phys. Chem.* **100**, 15444.

THE ROLE OF THE PROTEIN IN MODULATING COFACTOR ELECTROCHEMISTRY IN PROTEINS: THE CALCULATION OF ELECTROSTATIC FORCES

MARILYN R. GUNNER AND EMIL ALEXOV
Department of Physics
City College of New York
138th and Convent Avenue
New York, NY 10031 U.S.A.

1. Introduction

Living organisms store information in a compact and flexible manner in nucleic acids and use proteins to create both structural and catalytic molecules. A wealth of information about the atomic 3-dimensional structure of many proteins and some nucleic acids is available and stored in a readily accessible form in the Brookhaven Data Bank [1]. The challenge is to understand how the structure of these biomolecules enables them to carry out their function. Each molecule of protein is a linear polymer with identical backbone units. The same is true for the nucleic acids DNA and RNA. It is the specific sequence of amino acid side chains in a protein or bases in a nucleic acid that makes each unique. Thus, it is non-bonded interactions between the side chains that influence the backbone geometry, defining regions of α-helix or β-sheet as well as the overall fold in proteins, the propensity for A, B, Z DNA, or the structure of an RNA molecule. In addition, it is the non-bonded interactions that enable proteins and nucleic acids to bind reaction partners and participate in the appropriate reactions.

The non-bonded interactions include the repulsion caused by the overlapping electron orbitals of closed shell atoms and the dispersion (induced-dipole:induced-dipole) interactions [2]. The energy resulting from these forces can be described by a Lennard-Jones potential where the repulsion falls off with distance between atoms at $1/r^{12}$ while

L. Banci and P. Comba (eds.), Molecular Modeling and Dynamics of Bioinorganic Systems, 361–390.

the attraction is attenuated at $1/r^6$. Thus, these interactions are very important between neighboring atoms, but have a negligible impact at distances greater than 3Å. The hydrophobic effect, which opposes the disruption of intra-water hydrogen bonds by solutes also influences the structure of biological polymers [3]. These forces yield molecules that are close-packed without van der Waals overlaps and with non-polar side chains primarily in the interior [4,5]. Electrostatic attractions and repulsions also contribute to the non-bonded forces [6,7]. At distances of more than 2Å the electrostatic forces between charged or dipolar groups are generally larger than the overlap and dispersion interactions. The electrostatic potential has a 1/r dependence on the separation of the charges. Thus, this falls off slowly with distance and so provides a mechanism for portions of a structure to have impact beyond its nearest neighbors. Macromolecule structures represent a balancing of all of these forces to achieve an equilibrium structure.

Proteins are made up of backbone units, each of which has a dipole that is larger than water, with non-polar, dipolar, and charged side chains. Nucleic acids have charged phosphate and dipolar sugar groups in their backbone and dipolar bases as side chains. Thus, the electrostatic interactions among groups of charges and dipoles are important in these molecules [8]. Important roles for electrostatic interactions are found for the ionization of acidic and basic residues [9], oxidation or reduction of redox cofactors [10], binding of polar or charged reaction partners [11,12], association and folding of proteins [13,14], and reaction catalysis since the reactants, transition states, and products will not have the same charge distribution [15]. When reactions occur, the changing charge distribution disturbs the original balance of forces so that the macromolecule must now come to equilibrium with the product's charges. These structural changes make the calculation of experimental results so challenging. While the focus of this review will be on proteins, electrostatic forces also play important roles in any reaction involving the highly charged nucleic acid molecules RNA or DNA.

2. Methods of calculation

Calculation of electrostatic interactions is simple for a group of charges immersed in a medium with a uniform propensity to be polarized by charges. Under these conditions the potential Ψ (in volts) at distance r (in Å) from a charge q (in multiples of the charge on an electron) is simply $14q/\varepsilon r$ where ε, the dielectric constant, accounts for the attenuation of the potential due to the polarization of the medium. However, the calculation of the electrostatic potential within proteins is complicated by the different response of the protein and the water to charges. Added complexity is found if the solvent contains mobile ions which can move in a field [16]. In addition, all amino acids are not equally polarizable and all regions of the protein are not uniformly flexible [17]. Each of these terms contribute to a system with a spatially heterogeneous response to charges [18]. In addition, each motion has a different rate of response, so ε and Ψ may vary with the time after the charge distribution is perturbed. The calculation of the interactions between charges in different regions of a protein requires a model that is capable of including as many of these terms as possible.

The most complete treatments of electrostatic interactions within a protein should incorporate electronic states of amino acids that can be polarized by charges, atoms that are constrained in the close packed protein interior, surface residues with greater flexibility, and a surrounding solvent with a very large polarizability. Molecular dynamics simulations put each of the types of nuclear motions explicitly into the calculations [19]. Varying flexibility of the protein to introduced charges is captured by the dynamics in different regions of the protein. Waters and solution ions are explicitly added. Methods such as QM/MM [20] which couple quantum mechanics with molecular mechanics allow the differences in electronic polarizability to play a role. Unfortunately, an enormous amount of computer time is needed to simulate sufficient real time to follow events that are interesting. For example, even the creation of an appropriate initial structure must deal with the many picoseconds that it takes waters and ions to come to equilibrium around a protein in simulations that need to recalculate forces at intervals of $\approx$1 fs. In addition, since electrostatic interactions are important between distant sites, treating them correctly means that the entire protein and many solvent molecules need to be included.

Two approaches, the Protein-Dipole Langevien-Dipole (PDLD) method [21] and methods dependent on the solution of the Poisson-Boltzmann equation [8,22,23] have been developed to provide electrostatic interaction energies for any molecular structure. The PDLD method [21] considers the protein and water as a system of point charges and dipoles immersed in vacuum [21,24]. Protein atoms are considered polarizable point charges. Surface waters are treated as Langevin dipoles, mapped onto a grid with appropriate polarizability. No atomic level information is included about more distant solvent molecules which are represented by a region with the dielectric constant of water. The potential is obtained by solving the self-consistent problem of the electrostatic potential and induced potential due to polarizability of atoms and waters.

The Poisson equation provides the electrostatic potential throughout space if the charge and polarizability distribution is defined [25]. The Poisson-Boltzmann (PB) equation allows the potential to be determined in a system where mobile ions come to equilibrium in the field of the fixed charges. The PB equation can be solved analytically only for very simple regions of differing polarizability such as spheres or cylinders immersed in a solvent. However there are many numerical techniques that can solve the Poisson-Boltzmann equation for systems with any distribution of dielectric constants and charges. These include finite difference [22,26,27], finite element [28], multi griding [29], conjugate gradient [30], and fast multiple [31] methods. The finite difference Poisson-Boltzman (FDPB) method maps the protein atoms and buried waters onto a 3-dimensional grid. The charges assigned to atoms are mapped to the grid points. The mid-grids are assigned one of two dielectric constants. Within the protein the dielectric constant (ε_{prot}) is small (from 2 to 20). Regions outside of the protein are assigned the dielectric constant of the solvent (ε_{solv}). A value of ε=80 is used when the solvent is water. Space can be further divided to model a protein in a membrane [32]. A concentration of movable ions can be assigned to the solvent [33]. Then the Poisson-Boltzmann equation is solved yielding the potential at each grid point. 'Focusing' techniques provide final grid spacing of less than 1Å, providing information about how the potential varies at this scale [34].

Despite their differences, both FDPB and PDLD methods represent appropriate treatments of the equations of classical electrostatics and should provide consistent results [35]. In addition, their implementation can be quite similar. Both start with the atomic structure of the protein with partial charges assigned to each atom. Thus, they use the same level of detail for the protein. The main difference is the treatment of polarization. In the PDLD method atoms are polarizable point charges, while the FDPB method introduces polarizability through a protein local dielectric constant. The rest of this review will focus on calculations that use the FDPB method implemented in the DelPhi program [27,31,34].

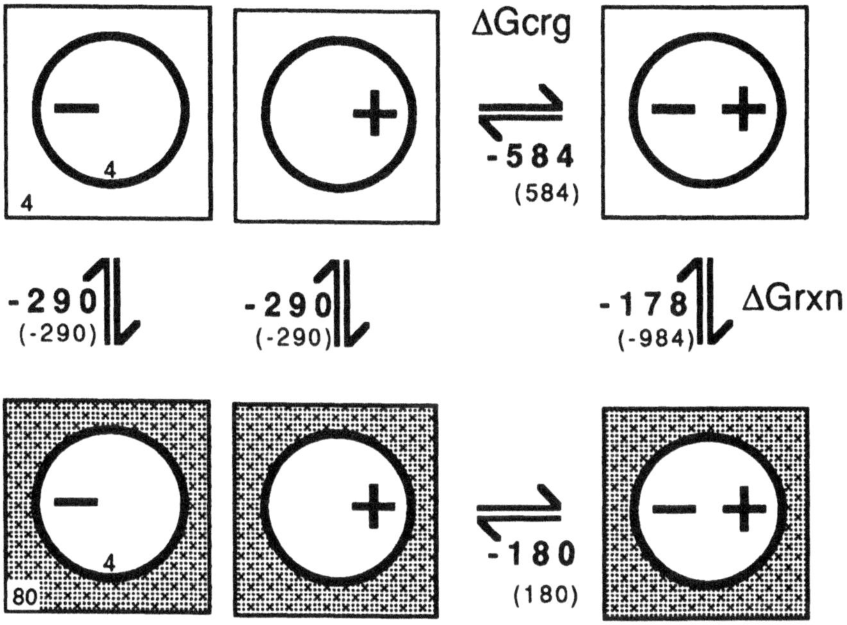

Figure 1. Two charges infinitely far apart ($\varepsilon_{in} = \varepsilon_{out} = 4$) are assembled into a 7Å sphere with ε_{in} of 4 and ε_{out} of 80. The charges are 6Å apart and 3Å from the center. The energy is in meV. The total energy of forming a dipole (or ion pair) from the two isolated charges is -760 meV. The 0.26% difference in the two assembly paths represents the numerical accuracy of the DelPhi program. The grid spacing for the calculation was 0.5 Å/grid. The two directions around the thermodynamic box yield the same total energy, but different values for ΔG_{rxn} and ΔG_{crg}. Most methods using FDPB take the counter-clockwise path.

The numbers in parentheses are the energies for bringing together 2 like charges (either ++ or --). The energy for assembling the system with net charge +2 or -2 is -400 meV. The free energy, while smaller than for the dipole is still favorable because of the very favorable reaction field stabilization of the doubly charged system in water.

The free energy of assembling the two charges is the sum of the energy of placing each charge into the system using the counter-clockwise direction around the thermodynamic box. The energy of a system with only one charge is -290 meV (ΔG_{rxn}). The addition of the second charge to form the dipole adds -290 meV (ΔG_{rxn}) + -180 meV ($\Delta\Delta G_{crg}$) = -470 meV. A third charge would add its self- energy (ΔG_{rxn}) plus its pair-wise interaction with each of the two original charges (ΔG_{crg}).

One advantage of methods based on the PB equation is speed of calculation. Interactions that are fully at equilibrium with the surrounding water and ions can be determined in less than a minute on a modern workstation. This is accomplished by treating only average properties of the solvent water and dissolved ions. A second advantage of the PB methodology is that the interaction of the whole protein with a site is a linear sum of the individual interactions (see Fig. 1). Thus, it is easy to describe the importance of individual residues to the potential at a site of interest. The two fundamental limitations to methods such as PDLD or FDPB to solve the electrostatic equations are that they are classical descriptions of matter, and in the calculations discussed here the response of matter to a changing field is assumed to be linear [33].

The use of a classical electrostatic model assumes that charged and polar groups surrounding a site change the relative energy of different electronic states without significantly modifying the states themselves. Thus, the electron distribution in each state is not changed, states do not mix together, and new electronic states do not become accessible due to the potential imposed by the protein. However, this is not always the case. In chemical reaction pathways or interactions of ligands with a metal center the changing charge distributions are not dominated by classical electrostatic forces, although these can influence the outcome. One simple example that has been described by electrostatic forces influencing charge distributions is that dipole moments are larger in solvent than in vacuum. This can be viewed as the solvent reaction field stabilizing a charge separation within the solute. There are a number of methods that include the polarization of the solute electrons by the solvent inputting information from the Poisson equation into the quantum mechanical analysis [39,84,85]. However, at an atomic/molecular level approximations are made anytime a process is treated classically. Thus, in some sense similar simplifications are used in molecular electrostatic analyses and in molecular mechanics calculations which rely on Newton's laws and the classical harmonic oscillator approximations.

The Poisson equation assumes that material has a linear response to a changing field. The non-linear Poisson-Boltzmann equation allows for non-linear effects of large potential (>kT) only on the screening of the potential by mobile ions [33]. Electrostatic interactions in a medium with a linear response may describe certain reactions such as site protonation or oxidation/reduction [36]. However, large, local motions of the protein or motions that are not a linear response to charge often occur. This is probably the most significant source of error in applications of a simple electrostatic analysis

[37]. Methods that are being developed to incorporate atomic information to determine the response of the protein to charges will be described below.

In addition to problems that are inherent in the use of a classical electrostatics formalism, there are approximations in current implementations that can be modified while still using a numerical solution of the Poisson-Boltzmann equation to obtain electrostatic energies. For example: (1) Current methods use simple atom centered charges which are the same for all residues of a given type. Charges sets are developed to reproduce experimental values such as solvation energies and dipole moments [38,39,40]. Atom centered charges are likely to be good for interactions between residues that are several Ångstroms apart. However, more complex charge distributions that better represent the multipolar nature of electronic distributions can be used when available. The charge density, derived from quantum mechanical calculations, could be directly mapped onto the grid used for solving the FDPB equation. (2) The polarizability of the protein is treated with a uniform value. This is not required in the Poisson-Boltzmann equation which can allow ε_{prot} to vary over space in any manner. Thus, it would be possible to acknowledge small differences in polarizability of different residues by varying ε_{prot}. As in all semi-empirical methods the atomic charge and atomic size are approximations that are tuned to fit experimental data. Thus, residues with different polarizability are likely to be implicitly included in a model using one value of ε_{prot} by differences in atom radius or partial charge. (3) The boundary between ε_{in} and ε_{out} for the protein and solvents is a step function. More sophisticated boundaries can be used.

2.1. WHAT IS THE PROPER DIELECTRIC CONSTANT OF THE SOLVENT AND PROTEIN?

In order to implement the Poisson-Boltzmann equation a dielectric constant must be assigned to all regions in and around the molecule of interest. The general use of the term 'dielectric constant' seems to imply that a macroscopic quantities are being used on a molecular level [18]. However, the Poisson equation results directly from the macroscopic Maxwell's equations and so is applicable at any scale where these equations are appropriate [41]. The macroscopic Maxwell's equations are derived from the vacuum Maxwell's equations by averaging induced dipoles over a vanishingly small region of space. The region used for the average must be large enough to hold the whole induced dipole. The resultant 'dielectric constant' ε should describe the average polarization produced in some region by an applied field. The dielectric constant can be formally connected to the more obviously microscopic susceptibility (χ) through the relationship $\varepsilon = (1 + 4\pi\chi)$. The polarization (**P**) is given by $\mathbf{P} = \chi\mathbf{E}$ where **E** is the total field. It should therefore be possible to define a dielectric constant for a small molecule or for a protein side chain.

For equilibrium properties at room temperature water has a dielectric constant of 78.5. However, the appropriate value of ε_{wat} may depend on the time scale under consideration [42] as well on the temperature. Experimentally the dielectric response of water is time dependent [43]. Thus, while ε_{wat} is $\approx$80 in a field that is constant or oscillating at a frequency less that 1 GHz, it drops to 4 at frequencies above 200 GHz and to 1.93 above 1000 GHz [44]. Thus, the motions of water that provide the largest contribution to its high dielectric constant require 0.010-1 ns to come to equilibrium with the field. Fluctuations of relatively fixed water dipoles yield a dielectric constant of only 4. This value is often used for protein, with its constrained dipoles. The small value of ε_{wat} seen at very high frequencies provides the contribution of electronic polarizability alone. A value of $\varepsilon \approx 2$ is the high frequency value for most material and is the both the low and high frequency dielectric constant of many non-polar molecules [44].

Bulk water at long time scales has a dielectric constant of $\approx$80. However, there are two other regions of waters that may require special treatment [45]. Specific bound waters are often found in proteins. Their position may even be evolutionarily conserved [46]. These molecules are appropriately treated in the same atomic detail as any side chain of the protein [37]. In addition, proteins are generally surrounded by a hydration sphere of waters. Bound surface waters are seen in crystal structures. They will have dielectric properties that are different from bulk water [45]. However, calculations often provide better results when these surface waters are replaced by the high dielectric constant solvent rather than treated as specific bound dipoles with a low dielectric constant [47]. Thus, the errors involved in assuming that these waters are rigid is greater than assuming that they have an ε of 80 [37]. One reason may be that there are two changes in water structure that occur near a charge. The waters become highly oriented so that additional charge cannot increase the orientation. This dielectric saturation reduces ε. However, oriented waters can move closer to a charge increasing the charge-dipole interaction energy, effectively increasing ε. Thus, electrostriction allows even fully oriented waters to react to an increase in the field [48].

The choice of an internal dielectric constant depends on what information is to be averaged in the calculations. For example, an active site in a protein can be treated with a quantum mechanical analysis while the surrounding protein and solvent is treated as a uniformly polarizable, classical structure obeying the Poisson-Boltzmann equation. Here, the interior region should have $\varepsilon=1$ in a protein with $\varepsilon=\varepsilon_{prot}$, embedded in a solvent with $\varepsilon=\varepsilon_{solv}$. Electronic polarizability yields the high frequency dielectric constant of 2 to 2.5 for most organic molecules [44]. This is the correct value for ε_{prot} when only electronic polarizability is incorporated within the PB equation, an appropriate choice for a region allowed to move in a molecular mechanics simulation. Use of higher values for ε_{prot} includes an average atomic rearrangement in response to changes in field. The dielectric constant of dry protein powders is 3-4 [49]. Estimations of ε_{prot} from calculations on proteins in vacuum also provide low values [50,51]. Calculations of the dielectric constant of the protein shows that motions of charged side chains make the largest contribution to the protein's response to changing field [17]. Thus, if the flexibility of charged side chains were also to be averaged in the dielectric response, then ε_{prot} should be larger (10-20) and should increase approaching the surface of the protein [17]. The value of ε_{prot} can be tested by comparing calculated and experimental results. For a 4 heme cytochrome subunit in the reaction center protein of *Rps. viridis* an ε_{prot} of 4 provided better results than 2 or 6 [52]. In contrast, the calculation of pK_a's in a number of small soluble proteins provided better correspondence with the measured results when $\varepsilon_{prot}=20$ [53,54]. The higher values are an indication that there are more substantial motions that occur on residue ionization in these proteins. The use of $\varepsilon_{prot}=20$ may provide good estimates of equilibrium pK_a's. However, all information about what regions of the protein respond to ionization is lost. Methods to allow for specific conformational changes in response to changes in residue or cofactor charges are being developed [55,37,56].

The pair-wise interactions between two charges can be measured by the perturbation of an equilibrium property of one site due to a change in charge at the other site. Thus, a pK_a or an electrochemical midpoint may change when a distant site is changed. The distal site can be modified by site-directed mutation or by changes in protonation states as the pH is changed. The dielectric constant for the protein that is derived from this measurement is ε_{eff} which is very different from the value of ε_{prot} used in the calculations. The dielectric constant of the protein is the averaged value of the polarization of the electrons and nuclei in the protein to change in field. In contrast, ε_{eff} is a value that would be measured for a specific site and contains contributions from both protein and solvent. The interactions between charges in a protein are always partially screened by the surrounding solvent. Therefore, even though calculations use low, uniform values of ε_{prot} the pair-wise interactions between groups are characterized by ε_{eff}'s that are very non-uniform and almost always significantly larger than the value used for ε_{prot} [37]. The distance between the charges and their position relative to the surrounding solvent has a large effect on their resultant interaction.

3. The contribution of electrostatic forces to the free energy of site ionization

The electrostatic free energy of an assembly of charges can be obtained as described in Fig. 1. The contributions to the observed free energy of a heme is shown in Fig. 2. The analysis in Fig. 2 would also be used for calculating the pK_a of an acidic or basic group in a protein.

(A) The simplest reference state is the cofactor alone in vacuum. The free energy of ionization in vacuum can be estimated from a variety of *ab initio* and semi-empirical quantum mechanical calculations [57].

(B) The addition of a ligand modifies the free energy of the oxidized and reduced heme differently. The difference in the effects of the ligand in the oxidized and reduced state is $\Delta\Delta G_{lig}$. Estimation of this term by electrostatic analysis is generally inappropriate. The ensuing analysis assumes that the free energy of the electronic states of the cofactor and ligands will be modified by their surroundings but not the nature of the electronic states themselves. This will not always be the case (For example, see [58]).

(A*) An alternate approach for generating a standard state is to use experimentally measured reaction ΔG's in model systems (Fig. 3) [59]. The difference between this experimental ΔG and the ΔG for the desired reference state of the cofactor with ligand in vacuum (B) is just the $\Delta\Delta G_{rxn}$. If measurements in water can be found with the same ligands as in the protein calculation of $\Delta\Delta G_{rxn}$ is relatively straightforward. Amino acid pK_a's in capped tetrapeptides have been measured in aqueous solution [60]. Heme midpoints have been measured in heme octapeptides that preserve the ligation found in cytochromes [61,62]. However, not all experimental values provide easily used reference states. Electrochemical measurements are often carried out in aprotic solvents rather than water. Computational analysis may be limited since the parameters used for

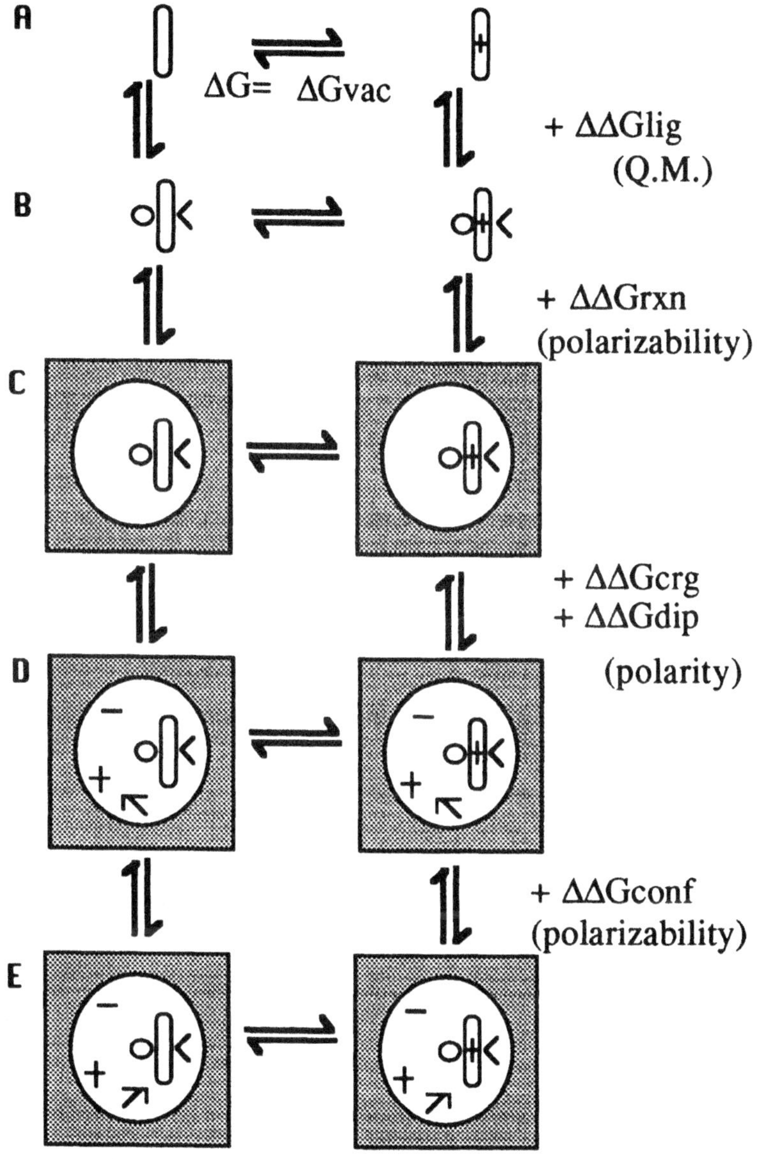

$$\Delta G = \Delta G_{vac} + \Delta\Delta G_{lig} + \Delta\Delta G_{rxn} + \Delta\Delta G_{crg} + \Delta\Delta G_{dip} + \Delta\Delta G_{conf}$$

Figure 2. Factors that contribute to the free energy of site ionization.

calculations in water may not be suitable in other solvents [63]. In addition, the relevant reactions may not be measurable in water. For example, the photosynthetic reaction center forms the semiquinone from a bound quinone in a one electron transfer that does not involve proton transfer [64]. However, in neutral aqueous solution the semiquinone is unstable and quinone reduction forms the dihydroquinone with the addition of two electrons and two protons [65,66]. Thus, results of experiments in an aprotic solvent such as dimethyl formamide must be used to estimate the ΔG for the quinone to semiquinone reduction in water [67]. Another problem is that there may be no model system which mimics ligand and cofactor orientation and electron coupling found in the protein. For example, the primary electron donor in photosynthetic electron transfer is a dimer of bacteriochlorophylls (in bacteria) or chlorophylls (in green plants). No model systems contain this unique structure. Estimations of the dimer reference electrochemical midpoints rely on extrapolations from the experimentally determined dependence of the midpoint on the solvent and degree of aggregation [68].

(C) ΔG_{rxn} is the effect of reorganizing the atoms and electrons in the protein and its surrounding solvent and ions to lower the energy of each charge distribution (see Fig. 1). The energy (in eV) to transfer a sphere of radius r with a charge q from a solvent with a dielectric constant (ε_{in}) to a medium with dielectric constant (ε_{out}) is $\Delta G_{rxn} = -14\ q^2/(2r)*(1/\varepsilon_{in} - 1/\varepsilon_{out})$ [21,69,70]. This response of the surroundings to the introduction of charge or dipole is known as the Born, reaction field, solvation, or self energy. Moving a charge or dipole into an environment with a smaller effective dielectric constant is unfavorable. The larger the charge the greater the penalty. Thus, removing any ionizable group from water ($\varepsilon_{eff} = 80$) into a protein ($\varepsilon_{prot} = 4$) $\Delta\Delta G_{rxn}$ stabilizes the neutral form (see Fig. 4) [71,72]. The reaction field energy stabilizes positive and negative charges to the same extent. If a residue or cofactor in water is used as the reference state then moving the group into the protein can be seen as increasing its radius, diminishing the reaction field energy.

(D) ΔG_{crg} is the change in the free energy of a charge state because of any static, non-zero potential from fixed charges within the protein. The contribution of a particular group (j) to the redox potential of the heme (i) is: $\Delta G_{crg}(i,j) = q_{hem} * \Psi_{crg}(i,j)$ where $\Psi_{crg}(i,j)$ is the electrostatic potential at atom i from the charges on site j at the atoms in the heme; and q_{hem} are the charges on the heme atoms [52,68]. $\Psi_{crg}(i,j)$ is calculated in a manner that accounts for the screening of the charge by the polarizability of the protein and the solvent. The total ΔG_{crg} is the sum of the contributions from individual charged groups. Corrections to the ionization state of acids and bases as a function of pH or heme ionization state can be included [36,37,53,73]. ΔG_{dip} is calculated in a similar manner given the potential from the backbone dipoles and polar side chains. $\Delta\Delta G_{dip}$ and $\Delta\Delta G_{crg}$ are the differences between the pair-wise interactions of charges and dipoles in the protein with the charged and neutral state of the heme. The differences between the two oxidation states are contained in differences in q_{hem}.

(E) The reorganization of nuclei contribute to the polarizability of the system. Thus, as a site changes ionization state there are small and large amplitude motions that occur on different time scales. Motions that are appropriately averaged over the whole protein are included in ΔG_{rxn} by using a value of ε_{prot} greater than 2. However, motions can be very different in different regions of the protein. The protein backbone dipoles are likely to be fairly rigid [19]. In contrast, surface groups and some buried atoms may move significantly [55,17]. Polar groups including the hydroxyls of Ser, Tyr, and Thr, water, and the amide groups of Asn and Gln can reorient their dipole moments to change from a hydrogen bond donor into a hydrogen bond acceptor [37]. Reorganization can also involve changes in the ionization state of surrounding residues [74,75,76] or changes in the occupancy of ion binding sites.

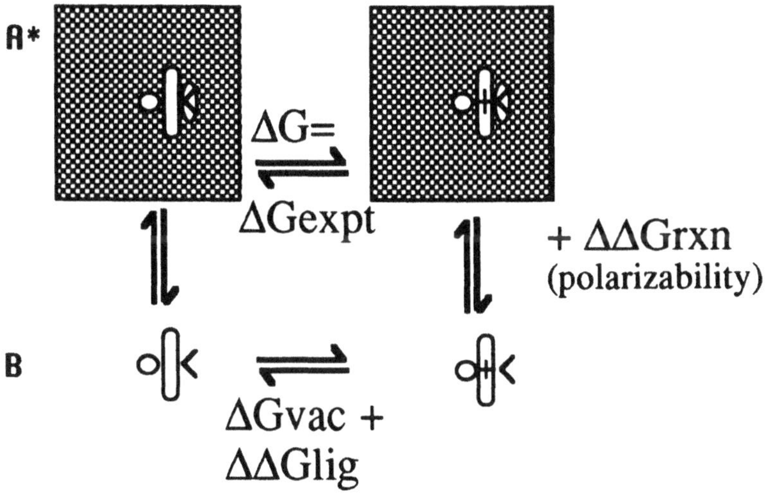

Figure 3: Creation of a reference state from experimental data. $\Delta\Delta G_{rxn}$ is most easily estimated if the solvent for the experiments is vacuum or water.

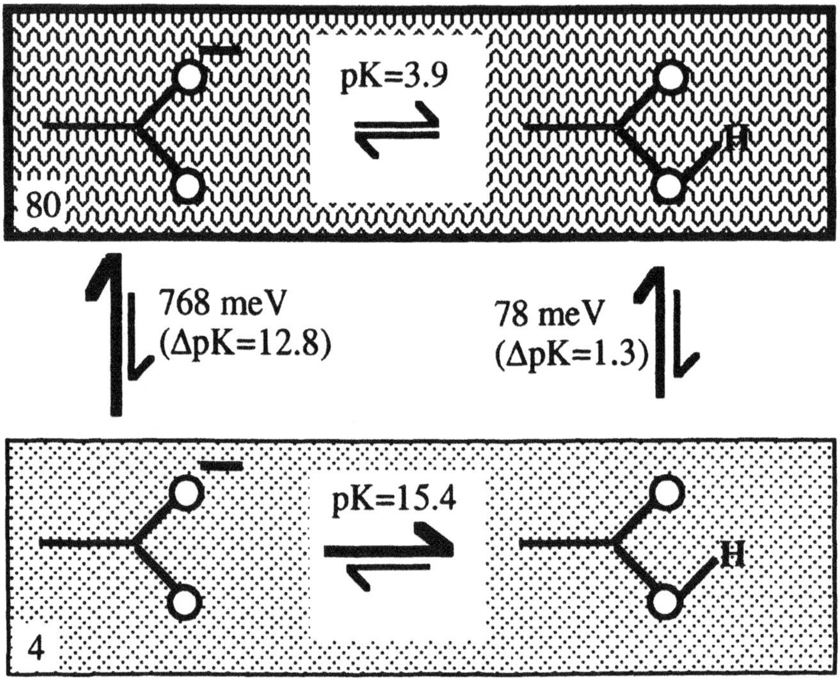

Figure 4. Change in the pK_a of an Asp residue (ε_{in}=4) as it is moved from water (ε=80) to a solvent with ε=4. (See reference [37] for atomic partial charges and radii).

3.1. INCORPORATING PROTEIN CONFORMATION CHANGES INTO ELECTROSTATICS CALCULATIONS

A number of methods have been developed to find the Boltzmann distribution of ionization states of a protein with many acidic and basic groups. Calculation of equilibrium properties requires either enumeration of all states or Monte Carlo sampling of many states. For N sites which can be ionized or neutral there are 2^N states. Current calculations of the pK_a's of residues generate energies of these states rapidly, relying on the principle that the electrostatic energies are a sum of the self and pair-wise energies. No higher terms are required. Thus, the PB equation need not be solved for each state. Rather one calculation is required for the ionized and neutral state of each site. Then all state energies can be assembled from knowledge of 2N self-energies (reaction field energies of the residue in its ionized and neutral form). The same 2N calculations can provide the interactions with polar groups in the protein as well as the 2Nx2N matrix that will give the pair-wise interaction between each residue in its ionized or neutral state (see [37] for a more complete description). The energy of any state of the protein, defined by the charge state of each residue, can be calculated from their self and pair-wise energies. Then, the Boltzmann distribution of ionization states is obtained from the statistical mechanical average or by Monte Carlo sampling [74,75,76]. This method seems to provide calculated values that match experimental results for many residues where data is available. However, it is becoming clear that problems arise when the protein is treated as a mostly rigid structure. When a uniform ε_{prot} is used, the best value is 20 [53,54] which implies that there can be significant rearrangements of the protein during a titration [8].

Methods are being developed that incorporate multiple conformations into the analysis of site pK_a's and electrochemical midpoints. Addition of conformational flexibility increases the number of possible states of the protein dramatically. For example, if each ionizable residue has M different conformations in its ionized and neutral form and K non-ionizable groups have L conformations, the number of possible states is $(2M)^N(L)^K$. The increase in the number of states is a minor problem if state energies can still be assumed to be calculated from self and pair-wise energies. In this case the difficulty of the problem scales linearly with the number of conformers. This is manageable given the speed of the individual calculations. However, conformational changes can breakdown the assumption of the independence of self and pair-wise interactions. For example, if side chain A buries site B, the reaction field, and pair-wise interaction energies with other residues for the n_b conformers of B will be different for each of A's n_a conformers. Thus, $n_b^{n_a}$ calculations would be required to obtain B's contribution to the state energies. The rotation of hydrogens in hydroxyls and waters change the surface of the protein a small amount so that the state free energies remain a sum of self and pair-wise terms. Conformer flexibility has be implemented for these groups [37]. However, larger motion of side chains often occur on site ionization, and ignoring these motions leads to significant errors in the calculations. Two side chain rotamers have been considered by Beroza and Case, retaining the assumption that state energies contain no higher terms [55]. Scheraga and co-workers [77] have taken a different approach and generated several thousand different conformers of a 17-residue polypeptide and calculated ensemble pK_a's from this small group of all possible states. Thus, while it is certain that the protein moves as sites are ionized it is not yet clear how to best incorporate these motions into the calculations.

Including specific nuclear motions in response to changes in the field accounts for a heterogeneous polarizability of the protein [37]. In addition, atomic level information about the process of relaxation can be derived from the analysis. For example, specific hydroxyl groups can be identified that change from accepting a hydrogen bond from a neutral acid to donating a hydrogen bond to the ionized acid. In all cases, conformational freedom allows the protein to stabilize both ionized and neutral forms of a residue to some extent. The ΔG of site ionization is therefore smaller than in a rigid protein. In addition, the change in potential at adjacent residues when a site is ionized now includes the added change in the potential due to this reorientation. The two contributions to $\Delta\Psi$ are generally of opposite sign. Thus, the total impact of site ionization on the surroundings is smaller. This effect can be viewed as a larger 'dielectric constant'.

4. The importance of charges in proteins.

The dipoles in a protein backbone and side chains that are dipoles or formally charged allow proteins to create a unique environment that can be viewed as a spatially heterogeneous solvent. Dipoles and charged groups can be arranged to stabilize a particular charge state. Charged residues have a large, spatially varying impact within the protein. For example, two hemes less than 7 Å apart in a 4 heme cytochrome have potentials due to charged groups in the protein that differ by ≈250 mV [52]. The potential from the backbone dipoles can also provide large potentials that are different in positions that are fairly close together [73].

Charges such as those on ionized residues, cofactors, and bound metals can have a long-range impact on the electrostatic potential in a protein. Since the electrostatic potential from a charge decreases slowly with distance the impact of charges can be very important but challenging to calculate. Using the distance dependence of the potential in a medium with a uniform dielectric constant, the potential will be 100 mV 35Å from a unit charge in $\varepsilon=4$, a distance which is as large as many proteins. In contrast, in water the same potential is found 1.75 Å from the charge. This is now a distance which is close to the van der Waals limit for the separation of atoms. The pair-wise interactions between charges in proteins will fall between these two extremes [37,68,78].

The importance of the long-range potential from a charge is connected to the mechanism used to stabilize the charge within the protein. As seen in Fig. 4, in the absence of stabilizing interactions the concentration of ionized residue or affinity of bound charged ligands would be vanishingly small in a medium of $\varepsilon=4$ [79,80]. In the example shown, the Asp is half ionized when the proton concentration is $10^{-3.9}$ (pH 3.9) in water. In 'hexane' where ε=4, the concentration would be $10^{-15.4}$ (pH 15.4) before the free energy of neutral and ionized Asp is equal. Thus, if the 'dielectric constant' of the protein is 4, it must create an environment that can replace the ΔG_{rxn} stabilizing the charge in the highly polarizable solvent water. There are four mechanism of stabilization, each of which has different implications for the long-range impact of a buried charge.

(1) The charge can be kept close to the surface of the protein so that the reaction field energy from the polarizable surrounding water is retained. This mechanism stabilizes positive and negative charges equally well. The water will be arranged differently in the presence or absence of the charge, always in its lowest energy conformation. However, the potential from the charge will not be felt at long distances. The difference in the importance of charges on the surface or in the interior of yeast cytochrome c is shown in Fig. 5. None of the charges on the surface of this protein influence the potential at the heme by more than 10 mV. Buried charges have a much greater impact at the heme. The largest contributor to the potential is 165 meV from the propionic acid on the A ring of this heme. The potential from a charge at a given distance is proportional to the reaction field energy of that charge [81]. The greater the reaction field energy the more the solvent screens the charge and the smaller its impact. An example is seen in Fig. 6 which compares the potential from an exposed and a buried propionic acid in cytochrome c' from *Chromatium vinosium*.

While surface charges have little effect on reactions within a protein they can be important in the association of reaction partners. The gradient of the potential from these charges can increase the rate of substrate binding to the active site [82]. In addition, the complementarity of charges at the surface of proteins is important for the correct docking of protein reaction partners [83].

(2) A charge can be stabilized by rigid dipoles properly arranged around the charge as in a chelator. Here ΔG_{dip} replaces ΔG_{rxn} to stabilize the charge. The protein must be designed to stabilize either a positive or a negative charge; the opposite charged group will be destabilized. The dipoles will not be in their lowest electrostatic free energy state in the absence of the charge. The dipoles are held in place by other forces contributed by the rest of the protein. Stabilization by this mechanism requires a system which is polar, with a large number of dipoles but not very polarizable so the dipoles can be rigidly held. Proteins fit this description. The long range impact of a charge stabilized by fixed dipoles will be significant. The potential from the dipoles

will roughly fall off as $\approx 1/r^2$ while the stabilized monopole charge will have its potential fall off as 1/r. An example is shown in Fig. 7, where a functionally important charge in photosynthetic reaction centers is stabilized by an α-helix dipole.

(3) A charge can be stabilized by a nearby oppositely charged site [79]. Here ΔG_{crg} replaces ΔG_{rxn} to stabilize the charge. However, while ion pairs can stabilize the protein structure, they may not contribute to the reaction mechanism for two reasons. The charges in the pair cannot undergo a change in ionization state without loosing most of their large pair-wise interaction energy. In addition, an ion pair is a dipole so the potential of the pair will be small at long distances.

(4) This first three contributors to the stabilization of a buried charge consider the protein relatively rigid (ε_{prot} = 4 and ΔG_{conf}=0). Large changes in protein conformation can also occur when the protein ionization state changes. This represents an increase in the polarizability of the environment. These movements create changes in potential that offset the potential from the new charge distribution and screen its long-range impact in a manner similar to the effect of water.

5. Conclusion

The interactions among charges and dipoles are important in proteins. Electrostatic forces provide a mechanism for interactions between distant regions of the protein. However, moving any charged group out of water into a protein destabilizes the charge in the absence of additional interactions. The mechanisms of charge stabilization within proteins determine the long-range impact of the charge.

Acknowledgments

We would like to thank Lucia Banci for organizing the meeting that stimulated this contribution. The financial support of N.I.H. GM-48726 is gratefully acknowledged.

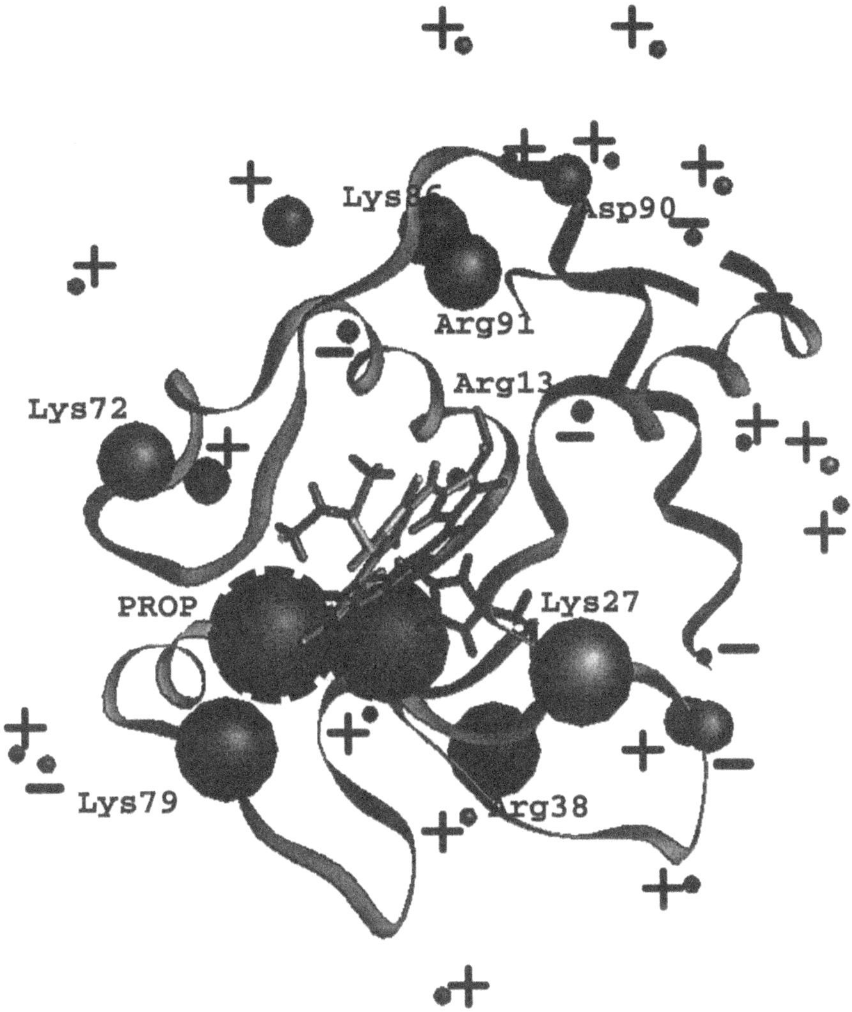

Figure 5. Yeast cytochrome c (1ycc) [86]. One ball for each charged residue. The relative size of the balls indicates the contribution of that residue to the potential at the heme. The smallest balls change the potential by less than 10 mV. The + and – indicate the sign of the potential from the site. The propionic acids, which change the potenital the most are surrounded by dashed lines. (Figure drawn with the program GRASP [87])

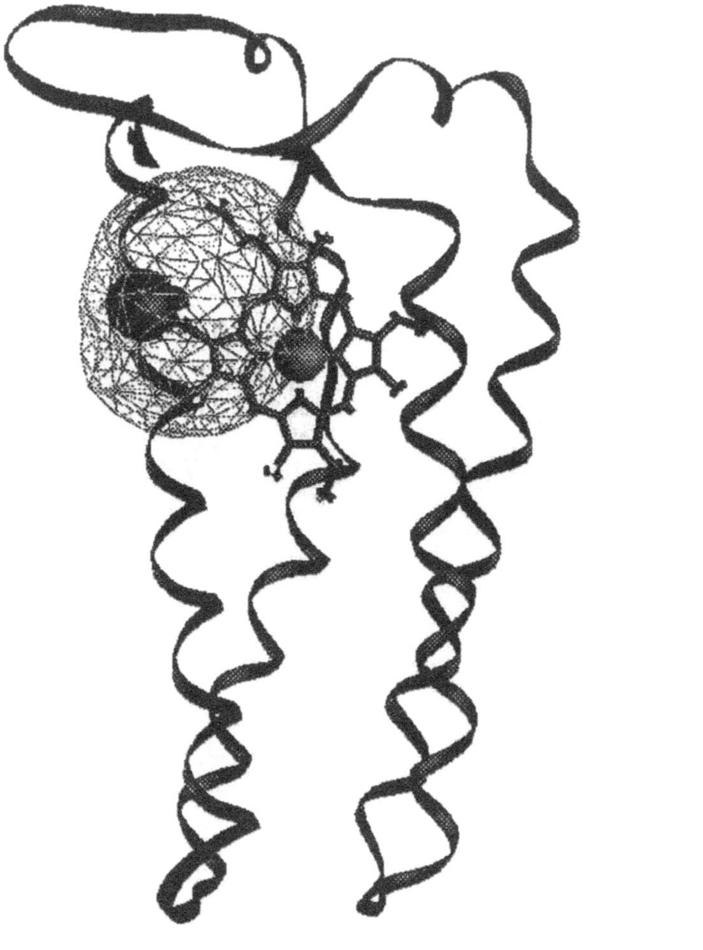

Figure 6a. Cytochrome c' from Chromatium vinosum (1bbh) [88]. A −1 charge is placed on the CG of the heme's ring A propionic acid, and the potential is calculated with ε_{in}=4, ε_{out}=80. The contour is drawn at the −50 mV level. The potential at the heme Fe is −69 mV. The reaction field energy of the acid is −605 meV. (Figure was drawn with the program GRASP [87].)

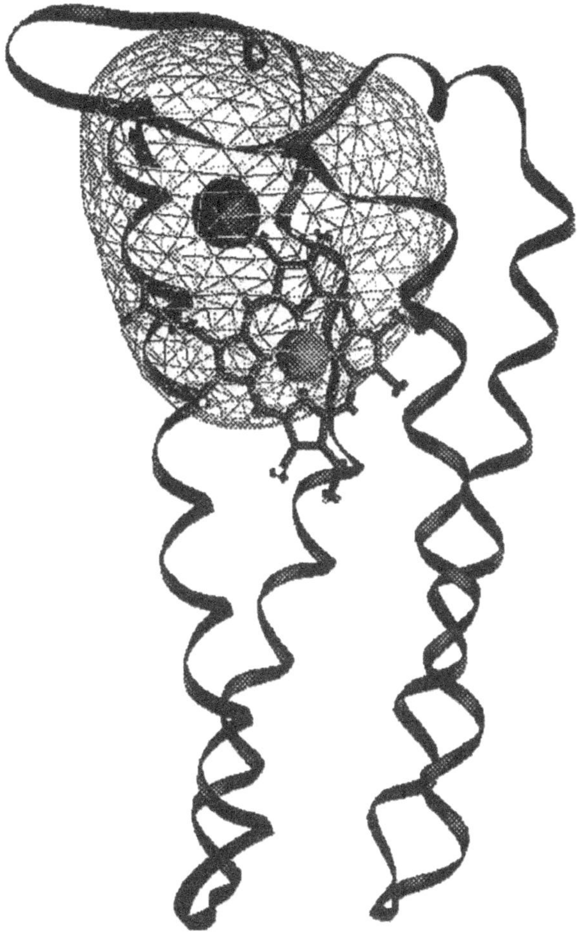

Figure 6b. Calculation as in Fig. 6a except that the charge is on the CG of the heme's ring D propionic acid. The potential at the heme Fe is −135 mV. The reaction field energy of the propionic acid is −217 meV.

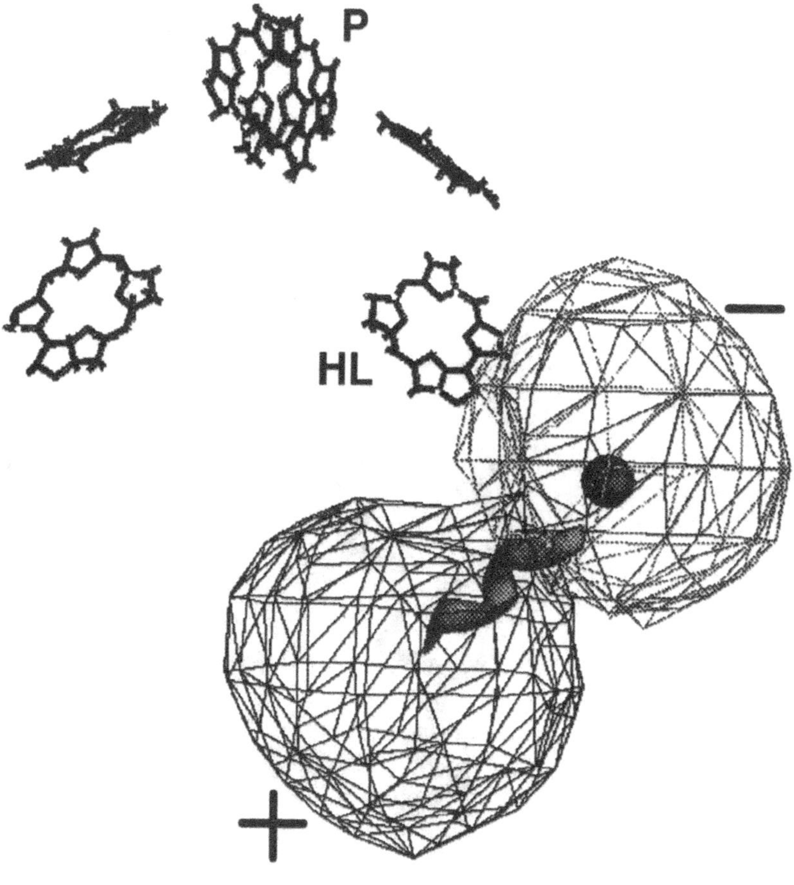

Figure 7a. The cofactors of the photosynthetic reaction centers (1PRC) [89]. Electron transfer occurs from P to H_L leaving a positive charge on P. Thus, a positive potential at HL facilitates the reaction. There are partial charges on the backbone dipoles of residues M 246 to 253 (helical worm shown). The electrostatic contours are drawn at the + and −100 mV level. The ball at the center of the region of negative potential is Arg L103 (CZ). H_L is effected by the helix dipole by less than 100 mV. The dielectric boundary in the calculation includes the protein and a membrane (see [68]).

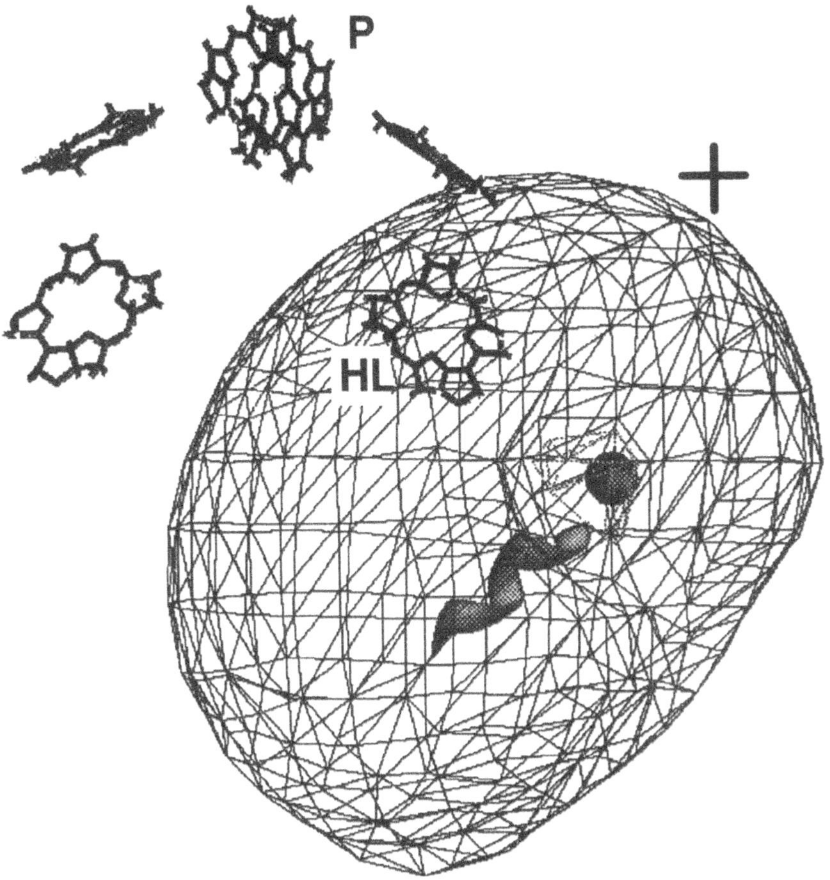

Figure 7b. There is a +1 charge on CZ of Arg L103 and partial charges on the backbone dipoles of residues M 246 to 253 (helical worm shown). The electrostatic contour is drawn at the +100 mV level. The contour at −100 mV cannot be seen. The ball is Arg L103 (CZ). The potential from the Arg L103 monopole extends over long distance. The potential from the helix dipole does not (see Fig. 7a).

References

1. Bernstein, F. C., Koetzle, T. F., Williams, G. J. B., Meyer, E. F., Brice, M. D., Rodgers, J. R., Kennard, O., Shimanouchi, T. F. and Tasumi, M. (1977). The protein data bank: A computer based archival file for macromolecular structures. *J. Mol. Biol.* **112**, 535-542.

2. Karplus, M. and Porter, R. N. (1970). *Atoms and Molecules*,.Benjamin/Cummings, Menlo Park.

3. Spassov, V., Karshikoff, A. and Ladenstein, R. (1995). The optimization of protein-solvent interactions: Thermostability and the role of hydrophobic and electrostatic interactions. *Protein Science* **4**, 1516-1527.

4. Richards, F. M. (1977). Areas, volumes, packing, and protein structure. *Ann. Rev. Biophys. Bioeng.* **6**, 151-176.

5. Ponder, J. W. and Richards, F. M. (1987). Tertiary templates for proteins Use of packing criteria in the enumeration of allowed sequences for structural classes. *J. Mol. Biol.* **193**, 775-791.

6. Spassov, V., Karshokoff, A. and Ladenstein, R. (1994). Optimization of the electrostatic interactions in proteins of different functional and folding type. *Protein Science* **3**, 1556-1569.

7. Spassov, V. Z. and Atanasov, B. P. (1994). Spatial optimization of electrostatic Interactions between the ionized groups in globular proteins. *Proteins : Struct. Func. and Gen.* **19**, 222-229.

8. Honig, B. and Nicholls, A. (1995). Classical electrostatics in biology and chemistry. *Science* **268**, 1144-1149.

9. Nakamura, H. (1996). Roles of electrostatic interactions in proteins. *Quart. Rev. Biophys.* **29**, 1-90.

10. Gunner, M. R. and Honig, B. (1996). Electrostatic simulation of cytochrome c. Cytochrome c Sourcebook. Mill Valley, University Science Books. 347-372.

11. Gilson, M. K. and Honig, B. H. (1987). Calculation of electrostatic potentials in an enzyme active site. *Nature* **330**, 84-86.

12. Bagdassararian, C. K., Schramm, V. L. and Schwartz, S. D. (1996). Molecular electrostatic potential analysis for enzymatic substrates, competitive inhibitors, and transition-state inhibitors. *J. Am. Chem. Soc.* **118**, 8825-8835.

13. Oliveberg, M. and Fersht, A. R. (1996). Formation of electrostatic interactions on the protein-folding pathway. *Biochem.* **35**, 2726-2737.

14. Rich, M. R. and Evans, J. S. (1996). Molecular dynamics simulations of adipocyte lipid-binding protein: Effect of electrostatics and acyl chain unsaturation. *Biochem.* **35**, 1506-1515.

15. Karshikoff, A., Reinemer, P., Huber, R. and Ladenstein, R. (1993). Electrostatic evidence for the activation of the glutathione thiol by Tyr7 in π-class transferases. *Eur. J. Biochem.* **215**, 663-670.

16. Ullner, M., Woodward, C. E. and Jonsson, B. (1996). A Debye-Huckel theory for electrostatics interactions in proteins. *J. Chem. Phys.* **105**, 2056-2065.

17. Simonson, T. and Brooks, C. L. (1996). Charge sScreening and the dielectric constant of proteins: Insights from molecular dynamics. *J. Am. Chem. Soc.* **118**, 8452-8458.

18. Simonson, T., Perahia, D. and Brunger, A. T. (1991). Microscopic theory of the dielectric properties of proteins. *Biophys. J.* **59**, 670-690.

19. McCammon, J. A. and Harvey, S. C. (1987). *Dynamics of Proteins and Nucleic Acids*,.Cambridge University Press, Cambridge.

20. Bakowies, D. and Thiel, W. (1996). Semiempirical treatment of electrostatic potentials and partial charges in combined quantum mechanical and molecular mechanical approaches. *J. Com. Chem.* **17**, 87-108.

21. Warshel, A. and Russell, S. T. (1984). Calculations of electrostatic interactions in biological systems and in solutions. *Q. Rev. Biophys.* **17**, 283-422.

22. Warwicker, J. and Watson, H. C. (1982). Calculation of the electric potential in the active site cleft due to a α-helix dipoles. *J. Mol. Biol.* **157**, 671-679.

23. Sharp, K. A. and Honig, B. (1990). Electrostatic interactions in macromolecules: Theory and applications. *Annu. Rev. Biophys. Biophys. Chem.* **19**, 301-332.

24. Lee, F. S., Chu, Z. T. and Warshel, A. (1993). Microscopic and semimicroscopic calculations of electrostatic energies in proteins by the POLARIS and ENZYMIX programs. *J. Comp. Chem.* 2-51.

25. Sharp, K., Gilson, M., Fine, R. and Honig, B. (1987). Electrostatic interactions in proteins. *Protein structure and Design* **2**, 235-244.

26. Gilson, M. K. and Honig, B. (1988). Calculation of the total electrostatic energy of a macromolecular system: Solvation energies, binding energies and conformation analysis. *Proteins* **4**, 7-18.

27. Nicholls, A. and Honig, B. (1991). A rapid finite difference algorithm utilizing successive over-relaxation to solve the Poisson-Boltzmann equation. *J. Comp. Chem.* **12**, 435-445.

28. Rashin (1990). *J. Phys. Chem.* **14**, 484.

29. Holst, M., Kozack, R., Saied, F. and Subramaniam, S. (1995). Treatment of electrostatic effects in proteins: Multigrid-based Newton iterative methods for solution of the full nonlinear Poisson-Boltzmann equation. *Proteins* **18**, 231-245.

30. Davis, M. E. and McCammon, J. A. (1990). Solving the finite difference linearized Poisson-Boltzmann equation: A comparison of relaxation and conjugate gradient methods. *J. Comp. Chem.* **10**, 386.

31. Bharadwaj, R., Windemuth, A., Sridharan, S., Honig, B. and Nicholls, A. (1995). The fast multipole boundary element method for molecular electrostatics: An optimal approach for large systems. *J. Comp Chem.* **16**, 898-913.

32. Karshikoff, A., Spassov, V., Cowan, S. W., Ladenstein, R. and Schirmer, T. (1994). Electrostatic properties of two porin channels from *Escherichia coli. J. Mil. Biol.* **240**, 372-384.

33. Sharp, K. A. and Honig, B. (1990). Calculating total electrostatic energies with the nonlinear Poisson-Boltzmann equation. *J. Phys. Chem.* **94**, 7684-7692.

34. Gilson, M. K., Sharp, K. A. and Honig, B. H. (1987). Calculating the electrostatic potential of molecules in solution: method and error assessment. *J. Comp. Chem.* **9**, 327-335.

35. Jean-Charles, A., Nicholls, A., Sharp, K., Honig, B., Tempczyk, A., Hendrickson, T. F. and Still, W. C. (1991). Electrostatic contributions to solvation energies: Comparison of free energy perturbation and continuum calculations. *J. Am. Chem. Soc.* **113**, 1454-1455.

36. Beroza, P., Fredkin, D. R., Okamura, M. Y. and Feher, R. (1995). Electrostatic calculations of amino acid titration electron transfer, $Q^{-}_{A}Q_{B}$---> $Q_{A}Q_{B}^{-}$, in the reaction center. *Biophys. J.* **68**, 2233-2250.

37. Alexov, E. G. and Gunner, M. R. (1997). Combining conformation changes and calculation of pH-titration. *Biophys. J.* **74**,

38. Jorgensen, W. L. and Tirado-Rives, J. (1988). The OPLS potential functions for proteins. Energy minimizations for crystals of cyclic peptides and crambin. *J. Am. Chem. Soc.* **110**, 1657-1666.

39. Tannor, D. J., Marten, B., Murphy, R., Freisner, R. A., Sitkoff, D., Nichols, A., Ringnalda, M., Goddard, I., W. A. and Honig, B. (1994). Accurate first principles calculation of molecular charge distributions and solvation energies from *ab Initio* quantum mechanics and continuum dielectric theory. *J. Am. Chem. Soc.* **116**, 11875-11882.

40. Sitkoff, D., Sharp, K. A. and Honig, B. (1994). Accurate calculation of hydration free energies using macroscopic solvent models. *J. Phys. Chem* **98**, 1978-1988.

41. Jackson, J. D. (1975). *Classical Electrodynamics*,.John Wiley & Sons, New York.

42. Barbara, P. F., Walker, G. C. and Smith, T. P. (1992). Vibrational modes and the dynamic solvent effect in electron and proton transfer. *Science* **256**, 975-981.

43. Pethig, R. (1979). *Dielectric and electronic properties of biological materials*,.John Wiley, New York.

44. Lide, D. R., Ed. (1990). *CRC Handbook of Chemistry and Physics*. Boca Ratom, CRC Press.

45. Pethig, R. (1990). Protein-water interactions determined by dielectric methods. *Ann. Rev. Phys. Chem.* **43**, 177-205.

46. Luntz, T. L., Schejter, A., Garber, E. A. E. and Margoliash, E. (1989). Structural significance of an internal water molecule studied by site-directed mutagenesis of tyrosine-67 in rat cytochrome c. *Proc. Natl. Acad. Sci., USA* **86**, 3524-3528.

47. Gibas, C. and Subramaniam, S. (1996). Explicit solvent models in protein pKa calculations. *Biophys. J.* **71**, 138-147.

48. Jayaram, B., Fine, R., Sharp, K. and Honig, B. (1989). Free energy calculations of ion hydration: An analysis of the Bborn model in terms of microscopic simulations. *J. Phys. Chem.* **93**, 4320-4327.

49. Harvey, S. (1989). Treatment of electrostatic effects in macromolecular modeling. *Proteins* **5**, 78-92.

50. Gilson, M. K. and Honig, B. H. (1986). The dielectric constant of a folded protein. *Biopolymers* **25**, 2097-2119.

51. Smith, P. E., Brunne, R. M., Mark, A. E. and VanGunsteren, W. F. (1993). Dielectric properties of trypsin inhibitor and lysozymes calculated from molecular dynamics. *J. Phys. Chem.* **97**, 2009-2014.

52. Gunner, M. R. and Honig, B. (1991). Electrostatic control of midpoint potentials in the cytochrome subunit of the *Rhodopseudomonas viridis* reaction center. *Proc. Natl. Acad. Sci. USA* **88**, 9151-9155.

53. Antosiewicz, J., McCammon, J. A. and Gilson, M. K. (1994). Prediction of pH-dependent properties in proteins. *J. Mol. Biol.* **238**, 415-436.

54. Antosiewicz, J., McCammon, J. A. and Gilson, M. K. (1996). The determinants of pKa's in proteins. *Biochemistry* **35**, 7819.

55. Beroza, P. and Case, D. (1996). Including side chain flexibility in continuum electrostatic calculations of protein titration. *J. Phys. Chem.* **100**, 20156-20163.

56. You, T. J. and Bashford, D. (1995). Conformation and hydrogen ion titration of proteins: A continuum electrostatic model with conformational flexibility. *Biophys. J.* **69**, 1721-1733.

57. Thompson, M. A. and Zerner, M. C. (1991). A theoretical examination of the electronic structure and spectroscopy of the photosynthetic reaction center from *Rhodopseudomonas viridis*. *J. Am. Chem. Soc.* **113**, 8210-8215.

58. Naray-Szabo, G. (1997). *J.B.I.C.* **2**, 135-138.

59. Parson, W. W., Chu, Z.-T. and Warshel, A. (1990). Electrostatic control of charge separation in bacterial photosynthesis. *Biochim. Biophys. Acta* **1017**, 251-272.

60. Richarz, R. and Wüthrich, K. (1975). Carbon-13 NMR chemical shifts of the common amino acid residues measured in aqueous solutions of the linear tetrapeptides H-Gly-Gly-X-L-Ala-OH. *Biopolymers* **17**, 2133-2141.

61. Harbury, H. A. and Loach, P. A. (1960). Oxidation-linked proton functions in heme octa- and undecapeptides from mammalian cytochrome c. *J. Biol. Chem.* **255**, 3640-3645.

62. Wilson, G. S. (1974). Electrochemical studies of porphyrin redox reactions as cytochromes models. *Bioelectrochem. and Bioenerg.* **1**, 172-179.

63. Sitkoff, D., Ben-tal, N. and Honig, B. (1996). Calculation of alkane to water solvation free energies using continuum solvent models. *J. Phys. Chem.* **100**, 2744-2752.

64. Gunner, M. R. (1991). The reaction center protein from purple bacteria: Structure and function. *Current Topics in Bioenergetics* **16**, 319-367.

65. Rich, P. R. and Bendall, D. S. (1979). A mechanism for the reduction of cytochromes by quinols in solution and its relevance to biological electron transfer reactions. *FEBS Lett.* **105**, 189-194.

66. Prince, R. C., Lloyd-Williams, P., Bruce, J. M. and Dutton, P. L. (1986). Voltammetric measurements of quinones. *Methods Enzymol.* **125**, 109-119.

67. Warncke, K. and Dutton, P. L. (1993). Influence of QA site cofactor structure on equilibrium binding, In situ electrochemistry, and electron-transfer performance in the photosynthetic reaction center protein. *Biochemistry* **32**, 4769-4779.

68. Gunner, M. R., Nicholls, A. and Honig, B. (1996). Electrostatic potentials in *Rhopseudomonas viridis* reaction center: Implications for the driving force and directionality of electron transfer. *J. Phys. Chem.* **100**, 4277-4291.

69. Bockris, J. O. and Reddy, A. K. N. (1973). *Modern Electrochemistry*,.Plenum, New York.

70. Rashin, A. A. and Honig, B. (1985). Reevaluation of the Born model of ion hydration. *J. Phys. Chem.* **89**, 5588-5593.

71. Kassner, R. J. (1972). Effects of nonpolar environments on the redox potentials of heme complexes. *Proc. Natl. Acad. Sci. USA* **69**, 2263-2267.

72. Churg, A. K. and Warshel, A. (1986). Control of the redox potential of cytochrome c and microscopic dielectric effects in proteins. *Biochemistry* **25**, 1675-1681.

73. Lancaster, C. R. D., Michel, H., Honig, B. and Gunner, M. R. (1996). The coupling of light-induced electron transfer and proton uptake: Electrostatic calculations on the photosynthetic reaction center of *Rhodopseudomonas viridis*. *Biophys. J.* **70**, 2469-2492.

74. Yang, A.-S., Gunner, M. R., Sampogna, R., Sharp, K. and Honig, B. (1993). On the calculation of pK_a's in proteins. *Proteins* **15**, 252-265.

75. Beroza, P., Fredkin, D. R., Okamura, M. Y. and Feher, G. (1991). Protonation of interacting residues in a protein by a Monte Carlo method: Application to Lysozyme and the photosynthetic reaction center of *Rhodobacter sphaeroides*. *Proc. Natl. Acad. Sci. USA* **88**, 5804-5808.

76. Bashford, D. and Karplus, M. (1990). The pK_a's of ionizable groups in proteins: Atomic detail from a continuum electrostatic model. *Biochemistry* **29**, 10219-10225.

77. Ripoll, D. R., Vorobjev, Y. N., Liwo, A., Vila, J. A. and Scheraga, H. A. (1996). Coupling between folding and ionization equilibria: Effects of pH on the conformational preferences of polypeptides. *J. Mol. Biol.* **264**, 770-783.

78. Honig, B. (1993). Theory and simulation. *Curr. Opin. Struc. Biol.* **3**, 223-224.

79. Honig, B. H. and Hubble, W. L. (1984). Stablilty of "salt bridges' in membrane proteins. *Proc. Natl. Acad. Sci. USA* **81**, 5412-5416.

80. Wolfenden, R. (1983). Waterlogged molecules. *Science* **222**, 1087-1093.

81. Gunner, M. R., Alexov, E., Torres, E. and Lipovaca, S. (1997). The importance of the protein in controlling the electrochemistry of heme metalloproteins: methods of calculation and analysis. *JBIC* **2**, 126-134.

82. Klapper, I., Hagstrom, R., Fine, R., Sharp, K. and Honig, B. (1986). Focusing of electric fields in the active site of Cu-Zn superoxide dismutase: Effects of ionic strength and amino-acid modification. *Proteins* **1**, 47-59.

83. Tiede, D. M., Vashishta, A.-C. and Gunner, M. R. (1993). Electron transfer kinetics and electrostatic properties of the *Rhodobacter sphaeroides* reaction center and soluble c-cytochromes. *Biochemistry* **32**, 4515-4531.

84. Liu, Y. and Newton, M.D. (1995). Solvent Reorganization and Donor/Acceptor Coupling in Electron-Transfer Processes: Self-Consistent Reaction Field Theory and Ab Inition Applications. *J. Phys. Chem.* **99**, 12382-12386

85. Sharp, K., Jean-Charles, A. and Honig, B. (1992). A local dielectric constant model for solvation free energies which accounts for solute polarizability. *J. Phys. Chem.* **96**, 3822-3828.

86. Berghuis, A. M. and Brayer, G.D. (1992). Oxidation state-dependent conformational changes in cytochrome c. *J. Mol. Biol.* **223**, 959-976.

87. Nicholls, A., Sharp, K. and Honig, B. (1991). Protein folding and association: Insights from the interfacial and thermodynamics properties of hydrocarbons. *Proteins* **11**, 281.

88. Ren, Z. and Meyer, T. (1993). Atomic structure of a cytochrome c' with an unusual ligand-controlled dimer dissociation at 1. 8 A resolution. *J. Mol. Biol.* **234**, 433-445.

89. Deisenhofer, J., Epp, O., Miki, R. and Michel, H. (1985). Structure of the protein subunits in the photosynthetic reaction center of *Rhodopseudomonas viridis* at 3 Å resolution. *Nature* **318**, 618-624.

MOLECULAR DYNAMICS STUDY OF H93G *Sperm Whale* DEOXYMYOGLOBIN MUTANTS WITH EXOGENOUS PROXIMAL LIGANDS

WIESŁAW NOWAK
Institute of Physics
N.Copernicus University
ul. Grudziądzka 5
87100 Toruń
Poland

1. Introduction

Life is based on organic molecules. However, many crucial biological functions are regulated by bioinorganic compounds. Transition metal ions catalyze biochemical reactions, e.g. as cofactors in enzyme reactions, and they are important in transport proteins as reaction centers for the reversible binding of small ligands. Since the discovery of heme proteins the iron coordinated by the porphyrin ring and embedded in the specific protein environment was the subject of extensive studies. Of particular interest is the very precise regulation of the reactivity of the five-coordinate metalcenter of hemoglobins (Hb) towards diatomic ligands [1]. The ability of Hbs to discriminate between a "good" oxygen and a "bad" (toxic) carbon monoxide has puzzled researchers for many years [2,3] and many interesting theories were tested experimentally [4] and computer modeled theoretically [5].

Hundreds of X-ray structures of Hbs, myoglobins (Mb) and corresponding model compounds are known. Time resolved X-ray analysis provides unique information on photodissociated molecules [6]. The vast amount of structural information, together with advanced spectroscopic techniques, available for studies of heme proteins [7], are the basis for precise molecular models of the functioning of these transport proteins and related biological systems.

The complex of four interacting subunits in Hb is the classical example of a protein cooperativity, triggered by a subtle change in the coordination state of the heme ferrous iron [2]. After dissociation of the oxygen molecule from one of the prosthetic groups, within this subunit, the iron shifts from the average heme plane by 0.4-0.5 Å towards the fifth, proximal ligand. This small translational motion induces a reorganization of the protein structure which facilitates consecutive reactions in the remaining subunits. The physical origin of this structural change is still not quite clear: both, a change of the iron spin state from singlet to quintet [8] and steric interactions

L. Banci and P. Comba (eds.), Molecular Modeling and Dynamics of Bioinorganic Systems, 391–417.

of the proximal histidine with the porphyrin moiety [9] coupled with the F-helix pull [2,10,11] are considered. It is convenient to study the processes of heme doming and structural changes in the orientation of the proximal ligand which accompany the binding and dissociation of oxygen in simpler, monomeric molecules, such as - myoglobins or leghemoglobins (Lb).

Once the substrate molecule is trapped by the deoxyMb, the reaction of O_2 (or CO, NO, etc.) with Mb is divided into two major steps: (i) the diffusion of the substrate through a complex path in the protein to the so called distal cavity; (ii) the reaction of the substrate with Fe^{+2}. Both steps were investigated by theoretical methods. While the motion of the substrate within the protein matrix may be studied by a classical molecular dynamics (MD) approach, the reaction step has to be described by a quantum mechanical (QM) model.

One of the first theoretical descriptions of Hb functionality was proposed by Warshel [12]. In an important series of papers by Karplus and co-workers [13,14] the mechanism for the tertiary structural change in Hb was studied with empirical potential energy functions. Other early simulations revealed the fact that fluctuations are crucial for the ligand escape [15], and that there are numerous pathways for ligand diffusion in Mb [16,17]. A different picture has been observed for the related protein Lb, where only one path dominates [18]. Advances in reaction path calculations [19,20] make it possible to study conformational changes along such a path in detail [21], to calculate the free energy profile [22] and to estimate changes of the free energy changes induced by point mutations [23].

In recent years, MD studies of Mbs concentrated on simulations of the dynamics of photodissociated systems. The first study was published by Henry *et al.* [24]. The same straightforward approach of a sudden switching of force field parameters was used by Petrich *et al.* [25] and Kuczera *et al.* [26] in their picosecond studies of Mb relaxation which involved a possible link of the slow protein relaxation with nonexponential kinetics of a geminate recombination. The reactions of NO were main point of interest since it reacts with the heme iron on the picosecond time scale. Schaad *et al.* [27] developed a special set of classical potentials for NO modeling, but the first semi-quantitative estimates of reaction rates have been obtained by Elber *et al.* [28-30], using the Landau-Zener theory for the crossing of the electronic states curve. Extensive simulations performed by these authors concentrated on the distal side of the heme mutants since these are more important in affecting the bimolecular part of the reaction [30].

However, the accessibility of the iron center and its reactivity toward ligands is determined by the coordination geometry, including the pyrrole nitrogens (heme) and the nitrogen of the proximal histidine. Restraints that maintain the iron out of the plane of the pyrrole nitrogens are perhaps responsible for the lower reactivity. The role of the proximal side of the heme, especially the orientation of the proximal histidine, has been stressed by many authors (see for example [2,14,31]), and models of the Hb cooperativity were developed [10,11]. Unfortunately, rigorous MD simulations, which may take into account electronic factors related to small rearrangements of the proximal side, require quantum mechanical calculations for the heme-ligand subsystem [32]. These have yet to be done. There are numerous QM studies of simple heme models ([32] and references within), including a recent high level SAC-SI study of oxyheme by Nakatsuji *et al.* [33], a potential energy surface evaluation by Gosh *et al.* [34] and estimates of the effect of the proximal side conformation on the geometry of the distal ligand [35]. However, combined QM-MD modeling of transition metal systems with an adequate level of quantum theory is still to be developed.

A systematic assessment of all the factors governing the heme protein functionality is facilitated by enormous progress in protein engineering. Tailored mutants [3] help to test molecular models. Recently, DePillis *et al.* [36] presented a simple and promising method for obtaining Mb mutants with exogenous proximal ligands. The X-ray structure of H93G(Im) was obtained with 2.17 Å resolution [37]. It has been shown [36] that a large proximal space in the H93G mutant of SW Mb allows the binding of various histidine analogues. Spectroscopic and kinetic features of these proteins were characterized [8,38-41], and substantial differences in the rebinding kinetics of CO were observed [36]. The absence of the covalent bond to the protein backbone gives a unique possibility for studies of net effects of bulky substituents on the reactivity of the iron center and the structure of the heme complex [42]. In particular, the stabilizing role of proximal Im - Leu89, Im - Ser92 hydrogen bonds in Mbs and steric interactions with pyrrole nitrogens may be addressed [43-45]. New H93G mutants were also used recently in an interesting study of factors govering affinity of apomyoglobins to the heme [46-48].

In this paper we report on our first attempt to develop computer models for a series of SW Mb H93G mutants with the proximal ligands shown in Fig. 1. The goal was to check, in an approximate but systematic way, the effect of the cleavage of the prosthetic group-apomyoglobin bond on the structure, stability and dynamics of SW Mb in the deoxy form. In particular, the orientation of various proximal ligands was studied in a detail, and the results of simulations were compared with estimates of structures of related systems based on NMR experiments [49]. The computational scheme is the same as that used in our recent study of human Mb H93G mutants [50].

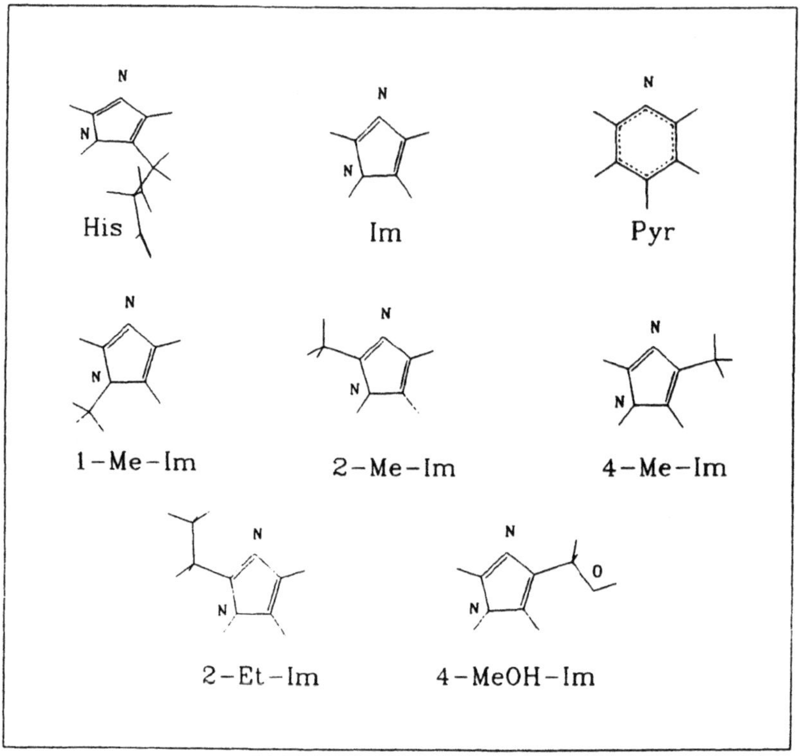

Fig. 1. Ligands used in SW Mb H93G simulations as the fifth ligand of iron instead of the proximal His93.

2. Methods

The structure of SW metaquoMb H93G(Im) [37] has been obtained from the Protein Data Bank (pdb) [51]. The structure is not complete, coordinates of 10 side chains of sufrace amino acids (mainly Lys, Glu and Arg) were absent. The positions of the missing atoms were generated with the „Biopolymer" module of the InsightII v95.0 molecular modeling package [52]. The missing last Gly residue at the C-terminus of the polypeptide chain has also been added. All the X-ray waters, including the water molecule coordinated to the ferric iron, were deleted from the pdb structure and no solvent molecules were used in the simulations at this stage.

In order to generate the coordinates of all polar hydrogen atoms in consistence with pH = 7, the modified pdb structure file was used as the input for the *puth* module of the MOIL (version 6.2) suite of programs [53]. The MOIL potential energy function is a combination of the AMBER, OPLS and CHARMM force fields:

$$E_{total} = \sum_{bonds} K_r(r-r_{eq})^2 + \sum_{angles} K_\Theta(\Theta-\Theta_{eq})^2 + \sum_{torsions} \frac{V_n}{2}[1+\cos(n\phi-\gamma)] + \quad \leftarrow \text{AMBER}$$
$$+\sum_i\sum_j\left(\frac{q_i q_j e^2}{r_{ij}} + \frac{A_{ij}}{r_{ij}^{12}} - \frac{C_{ij}}{r_{ij}^{6}}\right) + \quad \leftarrow \text{OPLS} \qquad (1)$$
$$+ \sum_{improper} K_\omega(\omega-\omega_0)^2 \quad \leftarrow \text{CHARMM}$$

Details and references for the parameterization may be found in [53]. It is worthwhile to note that the MOIL program has been succesfully used in many simulations of heme proteins [28-30, 53].

The parameters for exogenous ligands were developed by Czarnecka [54], and were the same as used in our previous study of human Mb H93G mutants [50]. They were based on the intrinsic MOIL parameters for the histidine imidazole. No special torsional potential was imposed on the proximal ligand rotation along the N_ε-Fe bond. Methyl groups were approximated by united atoms. Since we mainly address the question of „soft" degrees of freedom in this study, in the integration of the equation of motion the rattle (shake) algorithm [55] was used for all bonds.

The time step in the dynamics simulation was 0.5 fs, except for ImMb, where it was 1 fs. A constant temperature of 300 K (unless otherwise indicated) was maintained by velocity scaling each 10 or 5 steps. Electrostatic and Van der Waals interactions were limited to the close lying atoms with cutoffs of 9.5 Å and 10 Å, respectively. The list of nonbonded interactions was updated each 100 steps. A dielectric constant of 1 was used in all calculations.

The protocol applied in the present study was very similar to that used in our recent human Mb (HMb) simulations [50]. For all molecules studied with the exogenous ligands of Fig. 1, the pdb structure IRC1 served as a template for the construction of the molecules. Initial coordinates were generated by the overlap [56] of the heme pyrrole nitrogens of this structure with the minimized structures of the appropriate mutants of HMb obtained previously [50]. The coordinates of the exogenous ligands (Hb) were merged with those of the IRC atoms and then subjected to 5000 Powell minimization steps. The minimized structures were used to initiate dynamics simulations. All systems were subjected to 10 ps of heating (0-300K) and 20-60 ps of equilibration runs. MD trajectory data were collected each 0.1 ps for 50 ps except for PyrMb where only a 25 ps trajectory was computed.

Calculations were performed on SGI Indy and O2 workstations and a Sun Enterprise-6000 computer. Trajectories were analyzed using the MOIL-view v8.2 graphical tool [56] and home made software.

Below SW H93G(X) mutants will be named simply by the symbol of the proximal ligand. Note, that the imidazole ring atoms numbering adopted here is the same as in [36] but different from that used in [38].

3. Results

3.1 MINIMIZATION OF H93G(X) MUTANTS

Root-mean-square deviations (rms) between X-ray and minimized structures, defined by eq. (2) :

$$rms = \frac{1}{n}\sqrt{\sum_{i=1}^{n}(r_i - r_{av})^2} \ , \tag{2}$$

for n atoms with coordinates r_i are shown in Table 1.

TABLE 1. RMS differences of minimized structures with respect to the X-ray metaqoMb [37] structure.

protein	rms (Å)	
	C_α	all atoms [(a)]
Im	1.45	1.78
2-MeIm	1.28	1.63
4-MeIm	1.41	1.81
N-MeIm	1.25	1.70
2-EtIm	1.39	1.77
4-MeOHIm	1.30	1.64
Pyr	1.30	1.64

(a) except for the heme-proximal ligand complex

All minimized structures show the expected systematic deviations from the X-ray structure. The large values of the rms of C_α indicate that the differences are not related to the proposed *de novo* conformations of the side chains of the few residues but to the overall relaxation of the protein backbone, induced by the MOIL force field and the absence of the solvent in our systems. The direct comparison of computer models of Mbs with different ligands, presented in Table 2, leads to interesting observations.

The torsional orientation of the proximal ligand is measured by the azimuthal angle ϕ (defined in Fig. 2) which is related to the projection of the Im plane into the plane of the four pyrrole nitrogens. The angle is defined as positive if CD2 rotates clockwise from NC to NA. Note, that in the literature there are various definitions of torsional angles [37,49,50].

TABLE 2. RMS differences (in Å) between the minimized structures of H93G(X) mutants. „CA" denotes C_α atoms and „all", that all atoms except those of the heme complex are included.

	Im		2-MeIm	
	CA	all	CA	all
2-MeIm	0.955	1.218		
4-MeIm	0.478	0.563	1.029	1.316
N-MeIm	1.148	1.452	1.067	1.315
2-EtIm	0.789	1.031	1.232	0.971
4-MeOHIm	0.940	1.164	0.851	1.030
Pyr	0.801	1.046	1.205	1.501

The bending angle θ is also described in Fig. 2. Note that θ is not identical to the tilt angle discussed by other authors [11]. The values of φ and θ angles of the minimized structures, together with relevant H-bond distances, are presented in Table 3. The original metaquoMb(Im) pdb data are also included.

TABLE 3. Orientation of the proximal ligand plane and distances of the HD1 Im hydrogen (or an atom in the position 1, see Fig.2) to the carbonyl oxygen of Leu89 and the hydroxyl oxygen of Ser92.

protein	φ (deg)	θ (deg)	H-O(Leu89) (Å)	H-OG(Ser92) (Å)
Im(pdb) [37]	41.0	104.1	3.56	1.89
Im	26.4	124.9	2.54	1.86
2-MeIm	-105.7	118.4	3.13	4.81
4-MeIm	42.7	130.9	2.81	1.71
N-MeIm	12.4	126.9	4.17	3.31
2-Et-Im	-102.3	117.4	2.14	3.86
4-MeOHIm	28.4	135.3	2.78	1.77
Pyr	12.4	121.1	3.25	3.31

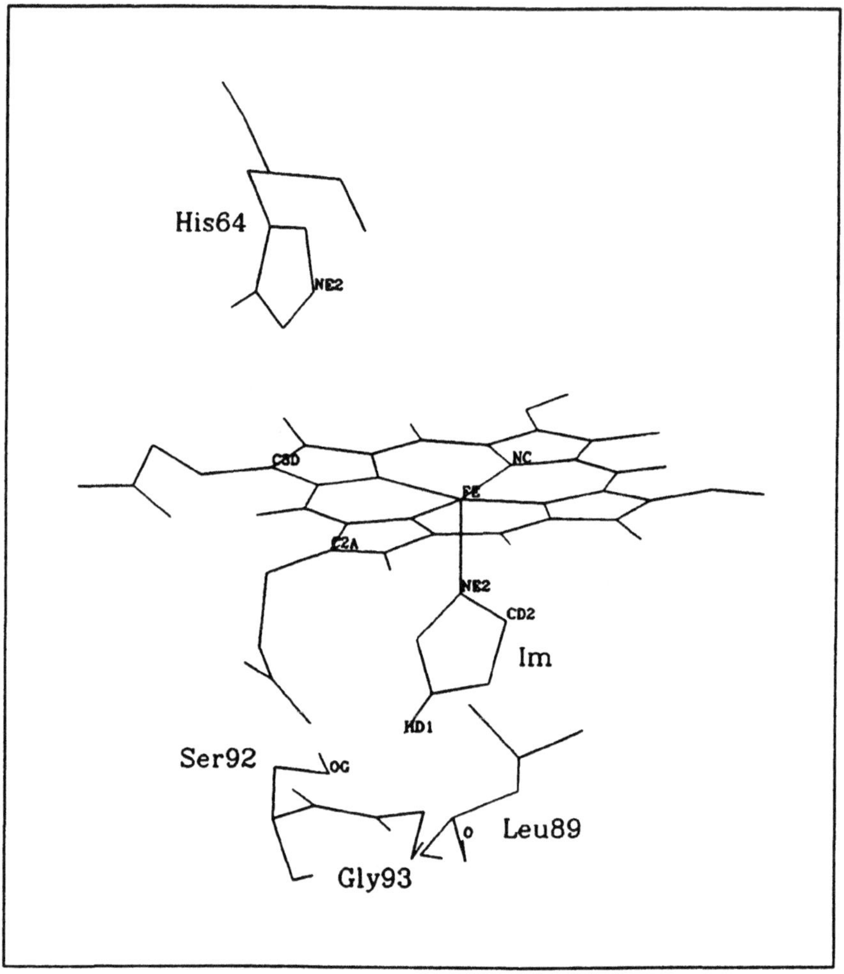

Fig. 2. Definitions of coordinates used in the analysis of the orientation of the proximal ligands: torsional angle ϕ = CD2-NE2-FE-NC, θ = CD2-NE2-FE.

Despite the fact that there are substantial differences in the conformations of the ligands of the proximal side and in the heme doming, the alpha-helical structure of Mb is well preserved after the minimization.

A comparison of the structure of the heme in the minimized MbIm model and the X-ray coordinates is shown in Fig. 3.

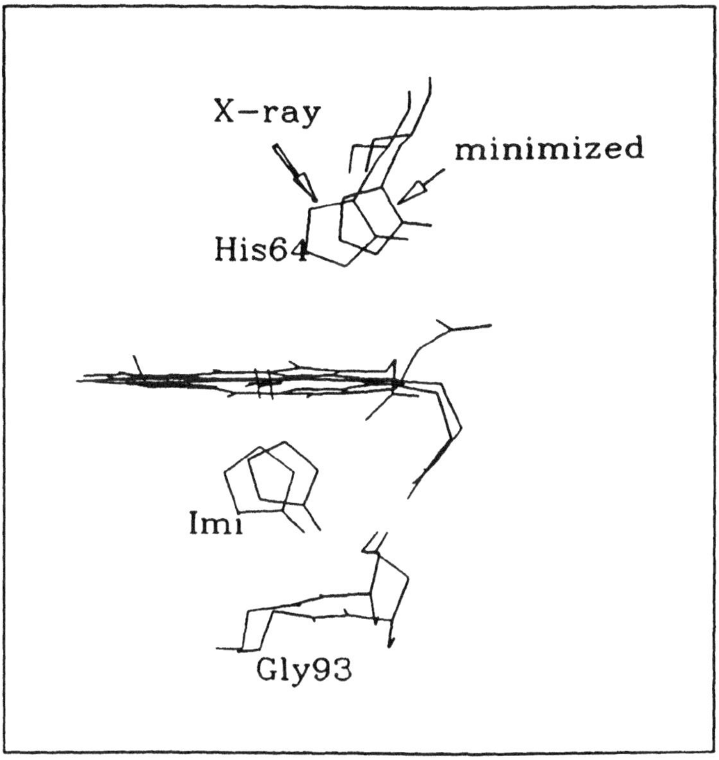

Fig. 3. A comparison of the MOIL minimized and the X-ray structure of Im-Mb.

3.2 DYNAMIC ORIENTATION OF PROXIMAL LIGANDS

The averaged values of ϕ and θ, and their standard deviations were calculated for these trajectories in which fluctuations around pronounced „equilibrium" positions were observed. These are presented in Table 4.

TABLE 4. Averaged values of ϕ and θ angles and their standard deviations (in deg.).

protein	$<\phi>$	σ_ϕ	$<\theta>$	σ_θ
Im (T=100K)	11.6	12.6	123.5	2.27
Im (T=200K)	-151.5 [b)]	54.8	121.3	3.66
	33.9	14.12		
Im (T=300K)	31.3	12.7	122.4	4.06
2-MeIm	-164.2	18.6	118.3	4.10
4-MeIm	-59.1 [b)]	20.9	128.9	3.7
	101.3	26.9		
N-MeIm	150.0	46.7	122.6	4.16
2-Et-Im(a)	-94.5 [c)]	12.4	118.1 [c)]	3.6
4-MeOHIm	-60.0	29.9	104.1	4.5
Pyr	---	----	121.8	5.7

b) multiple torsional conformations
c) data shown for the second trajectory denoted (a)

The distributions of the torsional angles observed with different ligands are more informative. We have therefore computed the number of structures in each trajectory that have a particular orientation of the proximal ligand plane. All values of the torsional ϕ angles ($-180° < \phi < +180°$) were divided into 32 bins, about 11.25 degrees wide, and the number of structures in each bin was used for a histogram plot of ϕ. The following bins are related to approximately staggered (S) or eclipsed (E) orientations of the ligands (the letters in parentheses refer to the names of pyrrole nitrogens used in the pdb nomenclature) S: 5, 13, 21, 29; E: 32-1 (AC), 8-9 (BD), 16-17 (AC) and 24-25 (BD). For example, bin#1 corresponds to $\phi \in (-180°, -168.75°)$, and all structures from this bin have the Im plane oriented along the NA-NC axis. The range of ϕ may be calculated from the formula:

$$\phi(n) \in (-180°+(n-1)*11.25°, -180°+n*11.25°) \quad (3)$$

Histogram plots of ϕ for each compound are presented in Figs. 4-13.

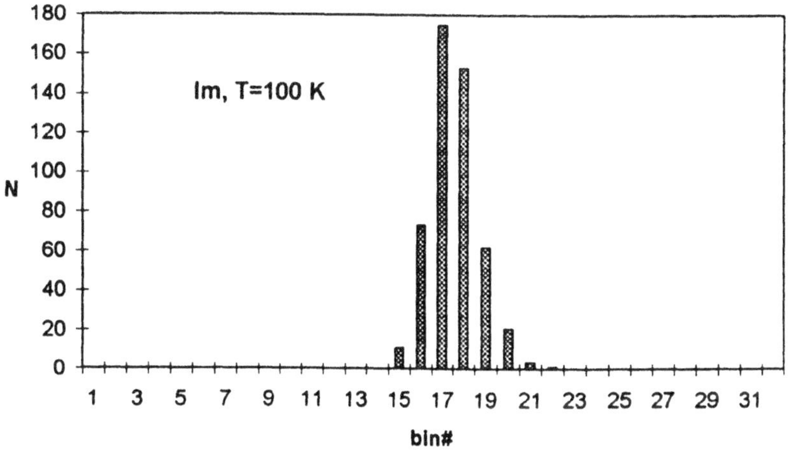

Fig. 4. The distribution of torsional angles ϕ in 50 ps, T= 100 K trajectory of H93G(Im)

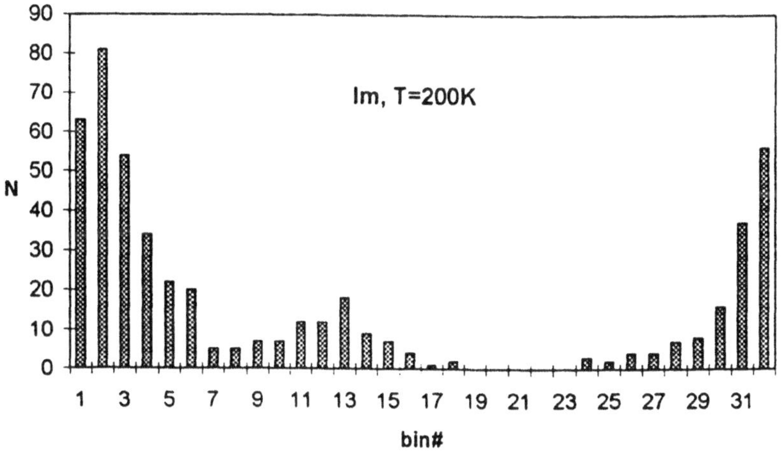

Fig. 5. The distribution of torsional angles ϕ in the 50 ps, T=200 K trajectory of H93G(Im).

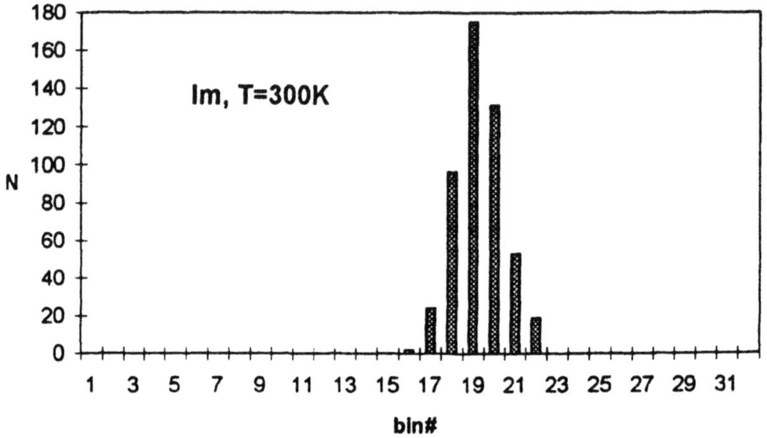

Fig. 6. The distribution of torsional angles ϕ in the 50 ps, T=300 K trajectory of H93G(Im).

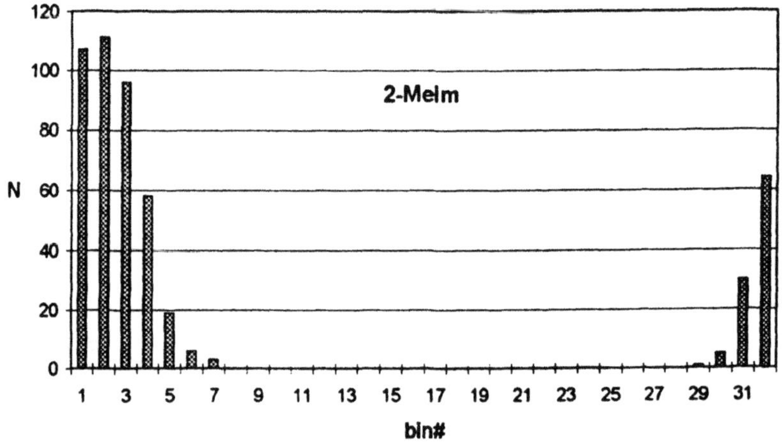

Fig. 7. The distribution of torsional angles ϕ in the 50 ps trajectory of H93G(2-MeIm).

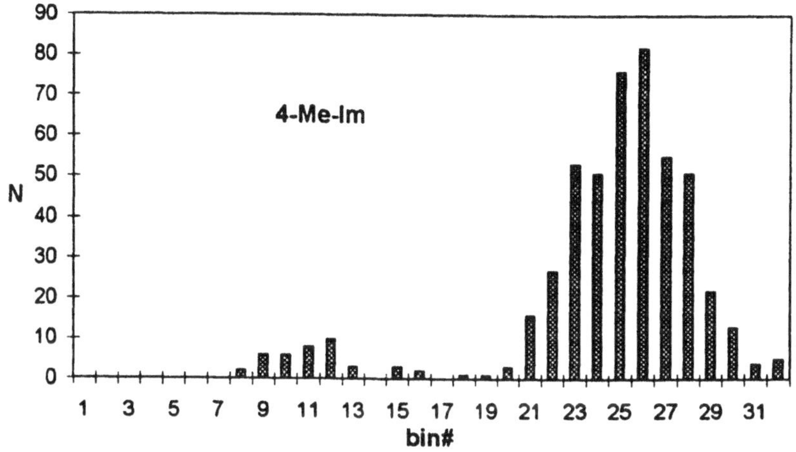

Fig. 8. The distribution of torsional angles φ in the 50 ps trajectory of H93G(4-MeIm).

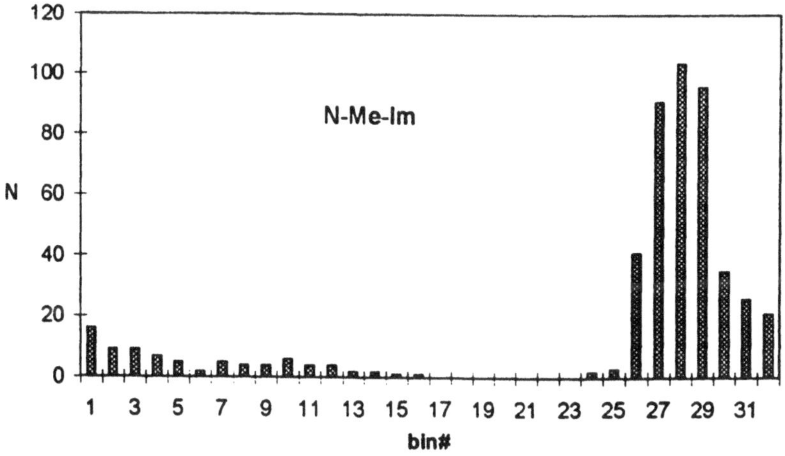

Fig. 9. The distribution of torsional angles φ in the 50 ps trajectory of H93G(N-MeIm).

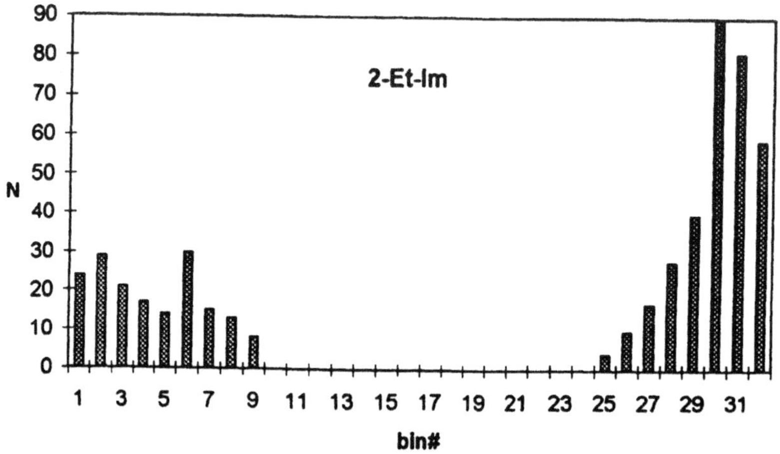

Fig. 10. The distribution of torsional angles ϕ in the initial 50 ps trajectory of H93G(2-EtIm).

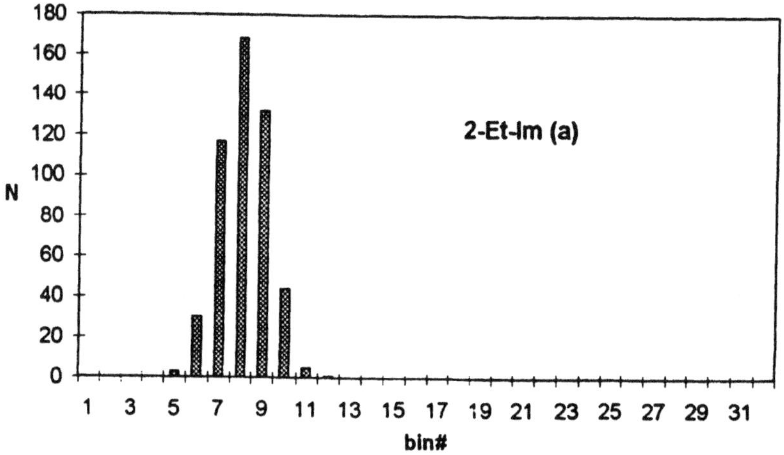

Fig. 11. The distribution of torsional angles ϕ in the second 50 ps trajectory (a) of H93G(2-EtIm).

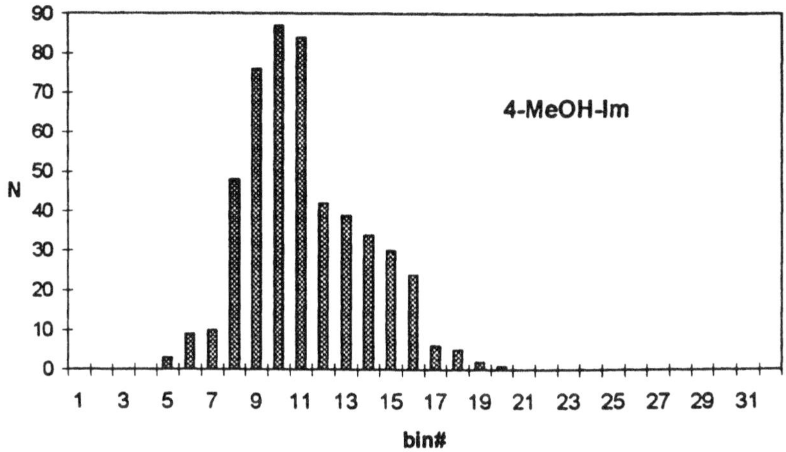

Fig. 12. The distribution of torsional angles ϕ in 50 ps trajectory of H93G(4-MeOHIm).

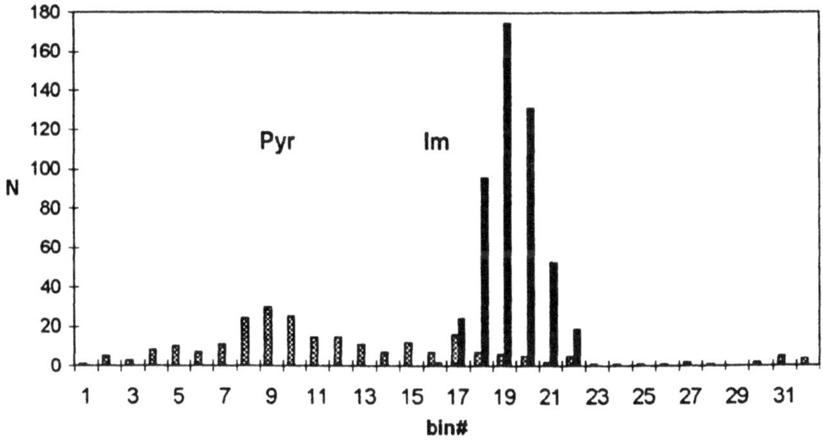

Fig. 13. A comparison of distribution of torsional angles ϕ in 50 ps trajectory of H93G(Im) and 25 ps trajectory of H93G(Pyr).

4. Discussion

4.1 STRUCTURES OF MUTANTS.

The protocol used in the minimization cannot give global minima for the structures studied. However, an analysis of low energy conformations gives useful hints on the effects related to structural variations of the proximal ligand. Although the structures of all mutants were prepared in the same way, with the H93G(Im) X-ray coordinates as a template, the rms data shown in Table 2 indicate that side chains in positions 1, 2 and 4 induce some changes in the backbone with distinct local minima for each system. Typical rms between these minima, calculated for all polypetide chain atoms, is about 1 Å. A particularly large rms distance of 1.45 Å is observed between Im and N-MeIm structures. This may be due to the absence of a hydrogen bond between the N_{δ} atom of the proximal ligand and the Leu89 carbonyl oxygen which is typical for Mbs. The distance between the methyl group of N-MeIm and Leu89 oxygen is 4.17 Å (see Table 3). In all the other mutants it does not exceed 3.2 Å. This distance is 3.6 Å in the X-ray structure [37]. The OG atom of the Ser92 residue in this mutant does also not form a hydrogen bond with N-Im, but instead the OH Ser92 group donates the hydrogen to the O2A atom of the heme-7-propionate. The distance from the OG atom of Ser92 to the methyl „united atom" in N-MeIm in the minimized structure is 3.3 Å. The corresponding distance in the other structures with bifurcated HD1 hydrogen bonds varies between 1.7 Å and 1.9 Å. In the 2-EtIm system this H-bond to the propionate is also observed. The largest HD1-OG(Ser92) distance is present in 2-MeIm (see Table 3), where the imidazole plane rotates away from Ser92 and makes a H-bond with the OH group of Tyr146.

As expected, the rms difference to the Barrick's X-ray structure is rather large, i.e. 1.6 - 1.7 Å for all atoms (1.3-1.5 Å for CA). This is related to at least 3 factors: (i) the absence of water in computer model (ii) the adopted force field (iii) specific interactions induced by the bulky proximal ligand side chains. We believe that factors (i) and (ii) affect all systems studied uniformly, so MD simulations may reveal differences in the interactions of various ligands with the Mb protein matrix.

These differences may be observed already in ϕ and θ angles of the minimized structures (see Table 3). In the deoxyMb steric interactions of the ligand with the heme are perhaps weaker then in the metaquo form, and long range forces are important for the orientation of the proximal ligand. In contrast to the X-ray structure, in the domed heme systems we do not observe a large tilting of the proximal ligand plane from the orientation perpendicular to the pyrrole nitrogens plane. The Ω angle of 75 °, defined in [37], is here nearly 90°.

The effect of the energy minimization on the heme group is presented in Fig. 3. Clearly, the distal histidine is considerably shifted towards the surface of the protein which may be related to the absence of water.

The torsional (azimuthal) angle ϕ is different for each ligand (see Table 3), but all values belong to two groups: (i) ca. -100° (2-MeIm, 2-EtIm); this corresponds to the „flipped" orientation of the imidazole, (ii) ca. 30° (all other ligands); this is close to the value of 41° found in the X-ray structure, and it allows for hydrogen bonding to Leu89 and Ser92. One should be careful with the interpretation of the values of these angles ϕ since they correspond only to local minima of the mutants. The analysis of MD trajectories gives better estimates of real torsional preferences of the proximal ligands.

The θ angles in the minimized systems indicate that the main deviation from an axial orientation of the ligands (which are not covalently linked to the backbone) results from the steric interactions with the porphyrin ring. All θ values for derivatives with bulky side chains are slightly larger (4-substituted molecules) or smaller (2-substituted molecules) than those for the unsubstituted Im and Pyr. This effect is even more pronounced in the time averaged θ values presented in Table 4.

4.2. DYNAMICS

Barrick [37,42] and Ghia *et al.* [36] obtained a series of stable SW Mb structures with exogenous heterocyclic ligands. Systems with nonbonded prosthetic groups were characterized experimentally in the past [57], see also references in [42]. However, besides our preliminary study of human Mb [50], no MD simulations were performed for protein-modified-heme complexes. Polarization spectroscopic studies by Ansari *et al.* [58], combined with MD simulations by Henry [59], indicate that the cavity in apomyglobin is large enough to accomodate rather large amplitude motions of the heme.

Thus, the first interesting question is, whether the SW mutants constructed here are stable on a close to 100 ps time scale. We have found that, in contrast to HMb, some systems show very large structural changes during the dynamics simulation, and that the first stage of the heme dissociation could be observed. Below, we discuss the characteristics of the MD trajectories for each system. The second problem addressed here is the flexibility and conformation of proximal ligands with substituents of various size.

4.2.1 H93G(Im)

For this molecule MD trajectories at 100 K, 200 K and 300 K were calculated (T100-T300). At all temperatures the molecule is stable. The heme group remains in the interior of Mb but its orientation is only similar to that of the native protein at 100K and 300K, where the classical HD1-Leu89 hydrogen bond stabilizes the system.

In the T200 trajectory (30-40 ps region) this bond is suddenly broken. Simultanously Im makes a 180° rotation and His97 moves away from the proximal ligand pocket to the outside of the protein. Due to periodic collisions an „activated" torsional mode of Im is observed which results in a major reorganization of the prosthetic group orientation. In His97 the χ_1 transition occurs, and a new H-bond with Glu109 is created. Simultanously, the C-terminal end of the polypeptide chain approaches the proximal ligand, and the HD1 hydrogen atom of Im „fishes" acceptors from that part of the protein. Then, most of the time HD1 remains bound to Tyr151. The overlap of C_α atoms of the last structure of the T200 trajectory with the pdb structure indicates that the iron position shifts outside by nearly 3.5 Å, and the heme unit rotates by approximately 40° around the C_α-C_γ meso axis, in such a way that Im points to the FG corner direction.

These rearrangements result in a different distribution of the angles ϕ of Im in T200 than in T100 and in T300 (see Figs. 4-6). The widths of the distributions of ϕ in T100 and in T300 are similar. The maximum of the distribution at ϕ=12° in T100 approximately corresponds to the E orientation of Im, while in T300 the maximum is at ϕ=31°, which is close to the S orientation, found by Barrick in the 6-coordinated complex [37]. The stabilization for this orientation results from the H-bond to Leu89, and the interactions of the Im ring with His97 (oriented parallel to heme) and Ile99 residues. Similar to our recent findings for HMb [45], the HD1-OG Ser92 H-bond is not preserved in the modeled structures. In T300 the distance for a H-bond is too large, and oscillates around 4 Å (min. 2.29 Å, max. 5.26 Å). Fluctuations of ϕ and θ at T=300K are presented in Figs. 14 and 15, respectively.

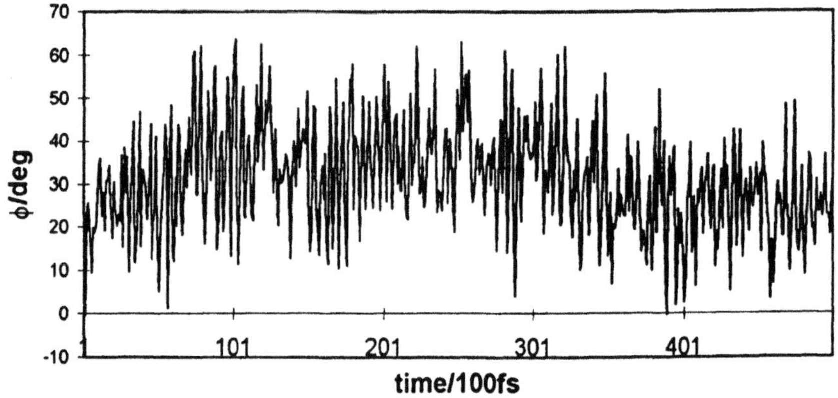

Fig. 14. Time-dependence of the torsional angle ϕ in H93G(Im), T=300K.

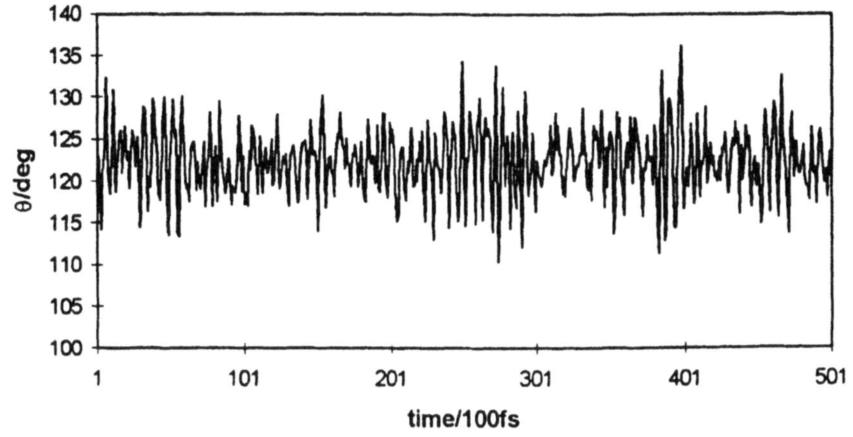

Fig. 15. Time-dependence of the bending angle θ in H93G(Im), T=300K.

4.2.2 H93G(2-MeIm)

The trajectory calculated for 2-MeIm is less stable than that of Im. Larger changes in the protein backbone are observed, especially in the regions Ala16-Ala23, Thr96 and Lys103. The C_α RMSD with respect to the minimized structure raises in the equilibration phase by 4 Å and during „production" stage further increases by 0.8 Å . As one can see, the distribution of the angle ϕ presented in Fig. 7 has a maximum at -165° (bin#2), which corresponds to the E orientation of Im. Fluctuations of ϕ are slightly larger than those in the unsubstituted imidazole, but this ligand is clearly fixed in space by a stabilizing factor, namely a HD1 - OH Tyr 151 hydrogen bond. The Tyr151 residue moves more than 10 Å to make this bond possible. The heme moiety undergoes a rigid body rotation around the C_α- C_γ meso carbon axis (approx. 20°) in the course of the trajectory.

4.2.3 H93G(4-MeIm)

This mutant trajectory has only in the equilibration phase and during the first 10 ps of dynamics a stable Mb structure. Perhaps, strong interactions of charged propionate oxygens induce the in-plane rotation of the heme group and its partial motion to the surface. In the major part of the trajectory A and B pyrrole rings of heme are exposed to the solvent. Two locations of the Im are observed in the trajectory; in the first part ϕ ~ -56° (close to S) and in the second part ϕ ~ 100° (close to E), see Fig. 8. Due to this flip the bulky Me group is exposed to the solvent. The second part of the trajectory does not correspond to a realistic orientation of the heme group. This result may be useful, however, in the analysis of the mechanism of heme dissociation from haloproteins [46-48].

4.2.4 H93G(N-MeIm)

In this derivative no hydrogen bonds to the protein atoms are observed. Inspection of Figs. 9 and 16 indicate that there is a large torsional flexibility of the N-MeIm ligand. It may adopt virtually any conformation, but the most probable are those in the vicinity of 160° (S).

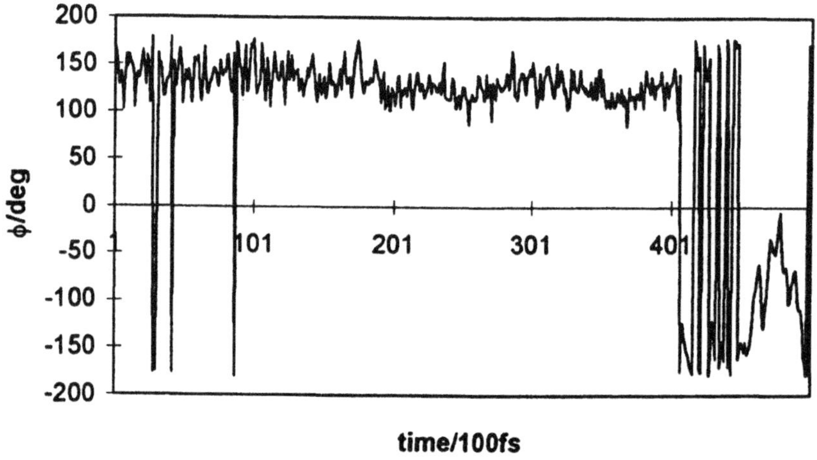

Fig. 16. Time-dependence of the torsional angle ϕ in H93G(N-MeIm), T=300K.

This particularly large rotational freedom is perhaps related to the fact that the heme system shifts towards the distal side in comparison to the X-ray structure of H93G(Im). A rotation of the heme group along the fourfold axis, and short distance (2.2 Å) interactions of Arg45 and O2D of the heme propionate are also observed. His97 is located far from the heme and does not stabilize the proximal ligand ligand N-MeIm , which is located inside the protein. Occasional collisions of the Me group produce large (~30°) tilts of the Im plane.

4.2.5 H93G(2-EtIm)

For this derivative two trajectories were analyzed, that produced for all systems (Fig. 10) and an additional 50 ps trajectory (2-EtIm(a)), Fig. 11, where EtIm is trapped in a local minimum. Large steric interactions of the ethyl group are expected in this derivative. Indeed, the first trajectory shows a significant rearrangement of the orientation and position of the heme group, corresponding to a nearly 90° clockwise rotation (view from the distal side of the heme). Major conformational changes are observed in the FG corner region and in the D helix. The heme group remains inside the Mb but Et points outwards. The changes of the angle ϕ in the first trajectory are large (Fig. 10), but in the second one the proximal ligand is trapped in a cavity created by Leu86, Leu89 and Gln91 (Fig. 11). The position of Gln91 is stabilized by a hydrogen bond with the heme-7-propionate. HD1 points towards the carbonyl oxygen of Leu89 but the large distance of ~ 3 Å excludes the possibility of a H-bond in this pair of residues.

4.2.6 H93G(4-MeOHIm)

In this trajectory the bulky MeOH side chain induces a very fast transition of the heme complex to the exterior of the protein. From Fig. 17 it emerges that, due to electrostatic forces of propionate heme side groups, the heme remains connected to the apoMb in the timescale of our simulation. In explict solvent simulations a full dissociation may possibly be observed. The detailed mechanism of such a major conformational transition in this system has not yet been scrutinized. Thus, data shown in Fig. 12 refer to the almost free rotation of the proximal ligand that only occasionally is hydrogen bonded to the protein.

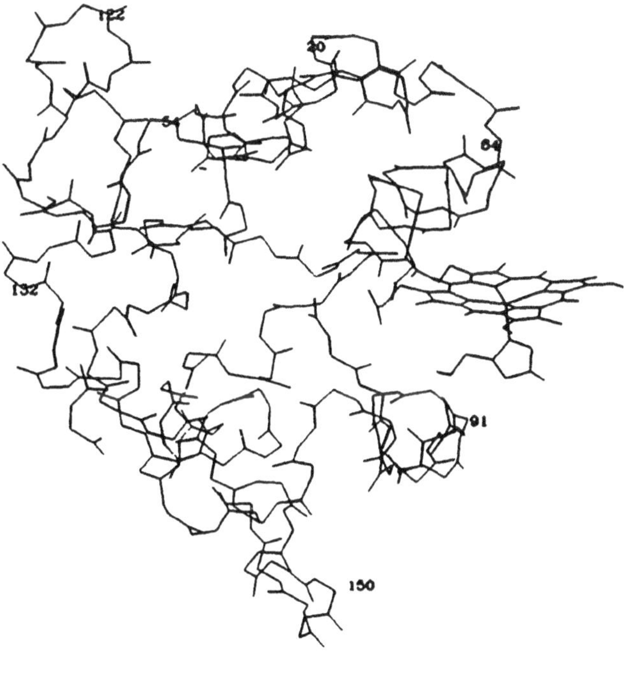

Fig. 17. A randomly selected conformation from the last part of the 4-MeOH-Im trajectory, showing the position of the heme complex outside the protein matrix.

4.2.7 H93G(Pyr)

Pyridine as a proximal ligand of H93G has already been reported in the literature [39]. It has been included in the present study becauce it may help to estimate the effect of the elongated shape of the ligand that is not stabilized by hydrogen bonds. As expected, and in constrast to Im, Pyr may adopt almost any torsional conformation (see Fig. 13). In SW Mb the tilt of the Pyr plane was much smaller than that observed in HMb [50], however for small periods (2-3 ps) the tilt may be as large as 50°. A large ring induces fast rearrangements of the system and the first phase of the dissociation of heme may be observed: the propionates penetrate more deeply the protein interior and frequent collisions with the F helix residues Leu86, Gln91, Ala90 and Ser92 are observed.

4.3 GENERAL REMARKS

Kinetic experiments show substantial differences in CO rebinding to H93G mutants in dependence of the proximate ligands [37,39]. These differences were attributed to size, shape and pK_a values of the ligands. Proximal ligand azimuthal angles were estimated for the met H93G(Im)H_2O [37] and the H93G(X)CO [39] forms of Mb. The very small differences in ^1H NMR spectra of the WT MbCO and the H93G(Im)CO mutant (0.15 ppm for the α-meso proton and 0.2 ppm for the γ-meso proton) were correlated with changes in the conformation of the imidazole ligand [39]. The change of conformation and H-bonding of the proximal imidazole are also a possible explanation for the differences in the ν_{CO} IR spectra. In particular, the lower intensity of the A3 band of the mutants is correlated with the increase of the angle ϕ with respect to that of the WT [11, 39].

In a publication by Yamamoto *et al.* [49], the observed hyperfine $^1H/^{13}C$-COSY spectra of metcyano Mb from various species were empirically correlated with the angle ϕ found in crystals. Results of our preliminary MD simulations and analysis the angle ϕ cannot be directly linked to experimental data for ligated species, where the ligand is closer to the porphyrin ring. However, they clearly show how large the torsional freedom of proximal ligands in H93G mutants is. Therefore, it seems that these fluctuations should be included in the interpretation of experiments [11,39,49] and in theoretical studies on structural determinants of IR spectra of heme proteins [60,61].

One may expect that at lower temperatures some exogenous proximal ligands may be „trapped" in local conformations. Thus, a more pronounced structure of IR marker bands is expected when the temperature is decreased. Multiple components in the $\nu_{Fe\text{-}His}$ profile are usually linked to different proximal side conformers [62,63]. To our knowledge such low temperature experiments for these systems have not been done yet, and they are worth pursuing.

It would also be interesting to learn more about the electron donating properties of the ligands studied here. To this end, redox potentials of the mutants should be measured, and correlated with the HD1 imidazole NMR shifts, along the lines proposed by Banci *et al.* [64].

In our previous study of HMb [50], the torsional flexibilty of many ligands was much smaller than that calculated for SW (σ_ϕ=3-8° and σ_ϕ=13-47°, respectively) and well defined preferential orientations of proximal ligands were observed. Thus, within our model, one may estimate that the human Mb protein structure is more rigid and keeps the heme prosthetic group more tightly than the SW Mb. This observation may be related to problems with early attempts to create exogenous ligands in H93A mutants of human Mb, reported in footnote in [38]. According to Boxer [38], the failure of the formation of stable complexes with Im may be associated with the inability of the human apomyoglobin to fold properly.

Obviously, results from classical MD modeling of bioinorganic systems are determined by the approximations inherent to the force fields, the trajectory leghths, etc.. It is therefore desirable to make similar calculations with a different set of parameters. In our problem, the electrostatics play a very important role [65], so this part of the model should be particularly carefully treated.

Another bias, inherent to classical simulations, comes from a neglect of possible interactions between π electrons of Im and p electrons of the iron, postulated by Scheidt and Chipman [66] on the basis of a semi-empirical Hückel model. Such an interaction was not very important in our *ab inito* calculation of the NO rebinding [32], but a systematic quantum study of the Im rotation may lead to additional arguments for the E conformation observed in many proteins and better force field parameters for the torsional motion. Quantum effects may restrict large fluctuations of the Im. Therefore, one should note that the results presented here must be considered as an „upper" estimate for the proximal ligand torsional flexibility.

5. Conclusions

MD simulations show that the structure of the proximal side of the heme in H93G mutants may substantially change with time. On a 50 ps simulation time scale only Im and 2-Im ligands show stable heme-protein complex conformations. In N-Me-Im, Pyr-Im and Et-Im mutants a rotation of the heme group is observed. The first phase of heme dissociation is seen in systems with bulky side chain substituents (OH-Im, Et-Im, 4-Me-Im). This result is based on a simple computer model and probably is due to the neglect of explicit water molecules. It indicates that hydrophobic interactions are major stabilizing force in H93G(X) systems. In future, the explicit water molecules and long-range electrostatic interactions should be included in simulations of protein systems with non-covalently bound heme. Such simulations are in progress in our laboratory.

The range of torsional angles adopted by the proximal ligands is large and depends on the presence of imidazole nitrogen H-bonds to Leu89 or other residues. In general, we do not observe stabilizing interactions with Ser92. The bending angles may vary substantially and modulate the imidazole ring-porhyrine coupling strongly. This conformational flexibility should be taken into account in the future quantum-mechanical modeling of heme proteins and also in direct comparisons of NMR hyperfine shifts with the internal proximal side coordinates.

Acknowledgments

Support from the Polish State Committee for Scientific Research (grant no. 6 P04A 032 11), project BiMol (FNP) and UMK grant 389-F is acknowledged.

References

1. Antonini, E. and Brunori, M. (1971) Hemoglobin and myoglobin in their reaction with ligands, North-Holland Publishers, London.
2. Perutz, M.F. (1979) Regulation of oxygen activity of hemoglobin: influence of structure of the globin on the heme iron, Ann. Rev. Biochem. **48**, 327-386.
3. Springer, B.A., Sligar, S.G., Olson, J.S., and Phillips, Jr., G.N. (1994) Mechanism of ligand recognition in myoglobin, Chem. Rev. **94**, 699-714.
4. Lim, M., Jackson, T.A., and Anfinrud, P.A. (1997) Ultrafast rotation and trapping of carbon monoxide dissociated from myoglobin, Nature Struct. Biol. **4**, 209-214.
5. Vitkup, D., Petsko, G.A., and Karplus, M. (1997) A comparison between molecular dynamics and X-ray results for dissociated CO in myoglobin, Nature Struct. Biol. **4**, 202-208.
6. Schlichting, I., Berendzen, J., Phillips, Jr. G.N., and Sweet, R.M. (1994) Crystal structure of photolyzed carbonmonoxymyoglobin, Nature **371**, 808-812.
7. Martin, J.-L. and Vos, M.H. (1992) Femtosecond biology, Annu. Rev. Biophys. Biomol. Struct. **21**, 199-222.
8. Franzen, S., Bohn, B., Poyart, C., DePillis, G.D., Boxer, S.G., and Martin, J.-L. (1995) Functional aspects of ultra-rapid heme doming in hemoglobin, myoglobin and the myoglobin mutant H93G, J. Biol. Chem. **270**, 1718-1720.
9. Olafson, B.D. and Goddard, W.A, Molecular description of dioxygen bonding in hemoglobin, Proc. Natl. Acad. Sci. USA **74**, 1315-1319.
10. Friedman, J. (1985) Structure, dynamics and reactivity in hemoglobin, Science **228**, 1273-1280.
11. Friedman, J.M., Campbell, B.F., and Noble, R.W. (1990) A possible new control mechanism suggested by resonance Raman spectra from a deep ocean fish hemoglobin, Biophys. Chem. **37**, 43 - 59.
12. Warshel, A. and Weiss, R.M. (1981) Energetics of heme-protein interactions in hemoglobin, J. Am. Chem. Soc. **103**, 446-451.
13. Gelin, B.R. and Karplus, M. (1977) Mechanism of tertiary structural change in hemoglobin, Proc. Natl. Acad. Sci. (USA) **74**, 801-805.
14. Gelin, B.R., Lee, A.W.M., and Karplus, M. (1983) Hemoglobin tertiary structural change on ligand binding. Its role in cooperative mechanism, J. Mol. Biol. **171**, 489-559.
15. Case, D.E. and Karplus, M. (1979) Dynamics of ligand binding to heme proteins, J. Mol. Biol. **132**, 343-368.
16. Tilton, R.F.,Jr., Singh, U.C., Weiner, S.J., Connolly, M.L., Kuntz, Jr. I.D., Kollman, P.A., Max, N., and Case, D.A. (1986) Computational studies of the interaction of myoglobin and xenon, J. Mol. Biol. **192**, 443-456.
17. Elber, R. and Karplus, M. (1990) Enhanced sampling in molecular dynamics: Use of Time-Dependent Hartree approximation for the simulation of carbon monoxide diffusion through myoglobin, J. Am. Chem. Soc. **112**, 9161-9175.
18. Czerminski, R., Elber, R. (1991) Computational studies of ligand diffusion in globins: I. Leghemoglobin, Proteins **10**, 70-80.

19. Elber, R., Czerminski R., Nowak, W., Choi, C., and Ulitsky, A. (1990) Dynamics of conformational transitions in peptides and ligand diffusion in proteins, Chem. Des. Autom. News **5**, 17-21.
20. Elber, R. (1996) Reaction Path Studies of Biological Molecules , in R. Elber (ed.), *Recent Developments in Theoretical Studies of Proteins*, World Scientific, Singapore, in press.
21. Nowak, W., Czerminski, R., and Elber, R. (1991) Reaction path study of ligand diffusion in proteins: Application of the Self Penalty Walk (SPW) method to calculate reaction coordinates for the motion of CO through leghemoglobin, J. Am. Chem. Soc. **113**, 5627-5637.
22. Verkhivker, G., Elber, R., and Gibson, Q.H., (1992) Microscopic modeling of ligand diffusion through the protein leghemoglobin: Computer simulations and experiments, J. Am. Chem. Soc. **114**, 7866-7878.
23. Verkhivker, G., Elber, R., and Nowak W. (1992) Locally enhanced sampling in free energy calculations: Application of mean field approximation to accurate calculation of free energy differences, J. Chem. Phys. **97**, 7838-7841.
24. Henry, E.R, Eaton W.A. and Hochstrasser, R. (1986) Molecular dynamics simulations of cooling in laser-excited heme proteins, Proc. Natl. Acad. Sci. (USA) **83**, 8982-8986.
25. Petrich, J.W., Lambry, J.-C., Kuczera, K., Karplus, M., Poyart, C., and Martin, J.-L. (1991) Ligand binding and protein relaxation in heme proteins: A room temperature analysis of NO geminate recombination., Biochemistry **30**, 3975 - 3987.
26. Kuczera, K., Lambry, J.-C., Martin, J.-L., and Karplus, M. (1993) Nonexponential relaxation after ligand dissociation from myoglobin. A molecular dynamics simulations, Proc. Natl. Acad. Sci. (USA) **90**, 5805-5807.
27. Schaad, O., Zhou, H.-X., Szabo, A., Eaton, W.A. and Henry, E.R. (1993) Simulation of the kinetics of ligand binding to a protein by molecular dynamics: geminate rebinding of nitric oxide to myoglobin, Proc. Natl. Acad. Sci. (USA) **90**, 9547-9551.
28. Li, H., Elber, R. and Straub, J. (1993) Molecular dynamics simulation of NO recombination to myoglobin mutants, J. Biol. Chem. **268**, 17908-17916.
29. Carlson, M.J., Regan, R., Elber, R., Li, H., Phillips, Jr., G.N., Olson, J.S., and Gibson, Q.H. (1994) Nitric oxide recombination to double mutants of myoglobin: role of ligand diffusion in fluctuating heme pocket, Biochemistry **33**, 10597-10606.
30. Quillin, M.L., Li, T., Olson, J.S., Phillips, Jr., G.N., Dou, Y., Ikeda-Saito, M., Regan, R., Carlson, M., Gibson, Q.H., Li, H., and Elber, R. (1995) Structural and functional effects of apolar mutations of the distal valine in myoglobin, J. Mol. Biol. **245**, 416-436.
31. Duprat, A.F., Traylor, T.G., Wu, G.Z., Coletta, M., Sharma, V.S., Walda, K.N., and Magde, D. (1995) Myoglobin-NO at low pH: Free four-coordinated heme in the protein pocket, Biochemistry **34**, 2634-2644.
32. Nowak, W. and Martin, J.-L. (1996) Towards an understanding of quantum factors in small ligand geminate recombination to heme proteins, in D. Bicout and M. Field (eds.), *Quantum Mechanical Simulation Methods for Studying Biological Systems*, Springer, Les Editions de Physique, Berlin , pp. 255-270.
33. Nakatsuji, H., Hasegava, J., Ueda, H., and Hada, M. (1996) Ground and excited states of oxyheme: SAC/SAC-CI study, Chem. Phys. Letters **250**, 379-386.
34. Gosh, A. and Bocian, D.F. (1996) Carbonyl tilting and bending potential energy surface of carbon monoxyhemes, J. Phys. Chem. **100**, 6363-6367.
35. Jewsbury, P., Yamamoto, S., Minato, T., Saito, M., and Kitagawa, T. (1995) The proximal residue largely determines the CO distortion in carbonmonoxy globin proteins. An *ab initio* study of a heme prosthetic unit, J. Phys. Chem. **99**, 12677-12685.

36. DePillis, G.D., Decatur, S., Barrick, D., and Boxer, S.G. (1994) Functional cavities in proteins: a general method for proximal ligand substitution in myoglobin, J. Am. Chem. Soc. **116**, 6981-6982.
37. Barrick, D. (1994) Replacement of the proximal ligand of sperm whale myoglobin with free imidazole in the mutant His-93->Gly, Biochemistry **33**, 6546-6554.
38. Decatur, S.M. and Boxer, S.G (1995) ^{1}H NMR characterization of myoglobins where exogenous ligands replace the proximal histidine, Biochemistry **34**, 2122-2129.
39. Decatur, S.M., DePillis, G.D., and Boxer, S.G. (1996) Modulation of protein function by exogenous ligands in protein cavities: CO binding to a myoglobin cavity mutant containing unnatural proximal ligands, Biochemistry **35**, 3925-3932.
40. Decatur, S.M., Franzen, S., DePillis, D.G., Dyer, R.B., Woodruff, W.H. and Boxer, S.G. (1996) *Trans* effects in nitric oxide binding to myoglobin cavity mutant H93G, Biochemistry **35**, 4939-4944.
41. Franzen, S. and Boxer, S.G. (1997) On the origin of heme absorption band shifts and associated protein structural relaxation in myoglobin following flash photolysis, J. Biol. Chem. **272**, 9655-9660.
42. Barrick, D. (1995) Depletion and replacement of protein metal ligands, Curr. Opin. Biotech. **6**, 411-418.
43. Smerdon, S.J., Krzywda, Sz., Wilkinson, A.J., Brantley, Jr., R.E., Carver, T.E., Hargrove, M.S. and Olson, J.S. (1993) Serine^{92} (F7) contributes to the control of heme reactivity and stability in myoglobin, Biochemistry **32**, 5132-5138.
44. Shiro, Y., Iizuka, T., Marubayashi, K., Ogura, T., Kitagawa, T., Balasubrama- nian, S., and Boxer, S.G. (1994) Spectroscopic study of Ser92 mutants of human myoglobin: hydrogen bonding effect of Ser92 to proximal His93 on structure and property of myoglobin, Biochemistry **33**, 14986-14992.
45. Nowak, W. (1997) Computer modelling of dynamics of Ser92X deoxymyoglobin mutants, Computers & Chemistry, accepted.
46. Hargrove, M.S, Barrick, D., and Olson, J.S (1996) The association rate constant for heme binding to globin is independent of protein structure, Biochemistry **35**, 11293-11299.
47. Hargrove, M.S, Wilkinson, A.S., and Olson, J.S (1996) Structural factors govering hemin dissociation from metmyoglobin, Biochemistry **35**, 11300-11309.
48. Hargrove, M.S. and Olson, J.S (1996) The stability of holomyoglobin is determined by heme affinity, Biochemistry **35**, 11310-11318.
49. Yamamoto, Y., Nanai, N., Chujo, R., and Suzuki T. (1990) Heme methyl hyperfine shift pattern as a probe for determining the orientation of the functionally relevant proximal histydyl imidazole with respect to the heme in hemoproteins, FEBS Lett. **264**, 113 - 116.
50. Nowak, W. (1997) Molecular dynamics simulations of human deoxymyoglobin with a modified proximal side of the heme group, J. Mol. Struct. (Theochem), in press.
51. Barrick, D. (1996) Protein Data Bank File PDBIRC1.
52. Insight II, v.95.0, Molecular Simulations, San Diego, USA, 1995.
53. Elber, R., Roitberg, A., Simmerling, C., Goldstein, R., Li, H., Verkhivker, G., Keasar, C., Zhang, J., and Ulitsky, A. (1995) MOIL: a program for simulation of macromolecules, Comp. Phys. Comm. **91**, 159-189.
54. Czarnecka, J., MSc thesis, UMK, Poland, 1996, unpublished.
55. Ryckaert, J.P., Ciccotti, G., and Berendsen, H.J.C. (1977) Numerical integration of the cartesian equation of motion of a system with constraints: molecular dynamics of n-alkanes, Comput. Phys. **23**, 327-341.
56. Simmerling, C., Elber, R., and Zhang, J. (1995) MOIL-View - A program for visualisation of structure and dynamics of biomolecules and STO - a program for

computing stochastic paths, A.Pullman, J.Jortner and B.Pullman (eds.), *Modelling of Biomolecular Structure and Mechanisms*, Kluwer, Netherlands, pp. 241-265.

57. Neya, S., Funasaki, N., Shiro Y., Iizuka, T., and Imai, K. (1994) Consequence of rapid heme rotation to the oxygen binding of myoglobin, Biochim. Biophys. Acta **1208**, 31-37.
58. Ansari A., Jones, C.M., Henry, E.R., Hofrichter, J., and Eaton, W.A. (1993) Photoselection in polarized photolysis experiments on heme proteins, Biophys. J. **64**, 852-868.
59. Henry, E.R. (1993) Molecular dynamics simulations of heme reorientational motions in myoglobin, Biophys. J. **64**, 869-885.
60. Kushkuley, B. and Stavrov, S.S. (1996) Theoretical study of the distal-side steric and electrostatic effects on the vibrational characteristics of the FeCO unit of the carbonylheme proteins and their models, Biophys. J. **70**, 1214-1229.
61. Kushkuley, B. and Stavrov, S.S. (1997) Theoretical study of the electrostatic and steric effects on the spectroscopic characteristics of the metal-ligand unit of heme proteins.2. C-O vibrational frequencies, ^{17}O isotropic chemical shifts, and nuclear quadrupole coupling constants, Biophys. J. **72** , 899-912.
62. Parak, F. and Knapp, E.W. (1984) A consistent picture of protein dynamics, Proc. Natl. Acad. Sci. (USA) **81**, 7088-7092.
63. Glich, H., Dreyboldt, W., and Schweitzer-Stenner, R. (1995) Thermal fluctuations between conformational substates of the Fe^{+2}-His^{F8} linkage in deoxymyoglobin probed by the Raman active Fe-Nε(HisF8) stretching vibration, Biophys. J., **69**, 214-227.
64. Banci, L., Bertini, I., Turano, P., Tien, M., and Kirk, T.K. (1991) Proton NMR investigation into basis for the relatively high redox potential of lignin peroxidase, Proc. Natl. Acad. Sci. (USA) **88**, 6956-6960.
65. Warshel, A., Papazyan, A., and Muegge, I. (1997) Microscopic and semimacroscopic redox calculations: what can and cannot be learned from continuum models, JBIC **2**, 143-152.
66. Scheidt, W.R. and Chipman, D.M. (1986) Prefered orientation of imidazole ligands in metallophorphyrins, J. Am. Chem. Soc. **108**, 1163-1167.

THE ROLE OF ELECTROSTATICS AT THE CATALYTIC METAL BINDING SITE IN XYLOSE ISOMERASE ACTION

B. ASBÓTH
Institute for Biochemistry and Protein Research, Agricultural Research Center, P.O. Box 170, H-2101 Gödöllõ, Hungary

Z. BÖCSKEI, M. FUXREITER and G. NÁRAY-SZABÓ
Department of Theoretical Chemistry, Eötvös Loránd University, H-1117 Budapest, Pázmány Péter st. 2, Hungary

1. Introduction

D-xylose isomerase (EC 5.3.1.5) catalyses the reversible conversion of D-xylose to D-xylulose and is also capable of converting other sugars from aldose to ketose [1]. It is the latter activity on glucose and fructose, that accounts for the fact that it is one of the most widely used industrial enzymes. It is active in oligomeric forms and requires Mg^{2+}, Co^{2+} or Mn^{2+} for activation while other divalent cations, e.g Zn^{2+}, Ba^{2+}, Cu^{2+} and Ca^{2+} inhibit catalysis [2]. The effect of the latter cation is of practical importance since a preceding enzyme in the starch bioconversion process, α-amylase, is only active in the presence of calcium ions, the removal of which requires an ion exchange step.

Crystal structures of several D-xylose isomerases have been determined (*Actinoplanes missouriensis* [3,4], *Streptomyces rubiginosus* [5,6], *Arthrobacter strain B3728* [7,8], *Streptomyces olivochromogenes* [9,10]) and were found to be very similar. Each subunit folds into a TIM motif and the well conserved active site contains two metal cations per subunit. The "structural" metal (at Site 1) is required for substrate

L. Banci and P. Comba (eds.), Molecular Modeling and Dynamics of Bioinorganic Systems, 419–439.

binding, whereas the "catalytic" metal (at Site 2) plays a role in isomerization.

On the basis of the tertiary structures of the *Arthrobacter* enzyme with various metal ions and inhibitors, a detailed stereochemical mechanism was proposed [8] that implies binding of the α-D-pyranose form of the substrate, followed by ring opening, extension and isomerization steps, the last one taking place via an anionic transition state. The remaining steps are the formal reversal of the above outlined process and a proton shuttle step [6], in which a structural water molecule coordinating the metal at Site 2, (W-690), abstracts the OH_2 proton of the substrate as a general base prior to the hydride shift. Recently it has been argued that this mediator is a bound hydroxyl ion rather than a water molecule [11].

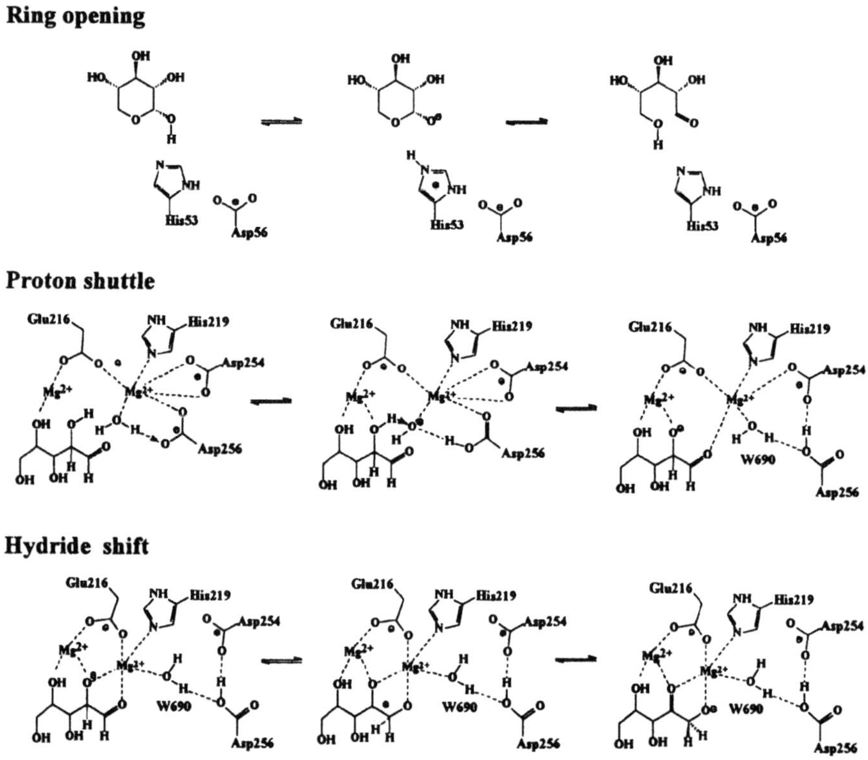

Fig. 1. A proposed reaction mechanism for xylose-xylulose conversion by D-xylose isomerase.

The mechanism of the xylose-xylulose conversion is shown in Fig. 1. Details were confirmed by kinetic [12], site-directed mutagenesis [4,13,14] and molecular modeling studies [15-18]. On the basis of deuterium kinetic isotope effects [19] and kinetic measurements with various metal ions [11] isomerization of the open chain substrate by hydride shift was proposed as the rate-limiting step of catalysis.

In this work we present an overview of our molecular orbital and X-ray diffraction studies on *Arthrobacter* D-xylose isomerase and its D254E/D256E double mutant that were done in the last few years. First we focus on questions related to the active-site structure, then details of the mechanism of action are presented with special emphasis on electrostatic aspects.

2. Methods

Three-dimensional models for molecular orbital calculations were constructed from the atomic coordinates deposited in the Protein Data Bank [20-22]. The geometry of various models were first optimized with a TRIPOS force field [23] using Kollman charges [24], a dielectric constant of 1.0 and an 8 Å cut-off radius. Constraints were imposed in order to reproduce essential interatomic distances obtained for enzyme-substrate complexes by X-ray diffraction [6]. Residues from the minimized structures that directly affect the mechanism were selected and transition-state geometries were determined both for the ring opening and hydride shift steps by the semiempirical NDDO molecular orbital method [25] with the PM3 parametrisation [26]. The MOPAC package [27] was used for calculations of the models in Fig. 2. Protein environment and bulk solvent were not considered at this stage of study, entropy effects were also neglected.

Charge transfer was estimated via ab initio molecular orbital calculations with the STO-3G minimal basis set using the MONSTERGAUSS program [28]. It is known that this yields, in general, atomic charges and dipole moments smaller than those given by the higher quality 6-31G** basis set, however, reasonably reproduces experimental trends [29]. The protein electrostatic potentials were calculated with the DelPhi program [30,31] using a geometric model of

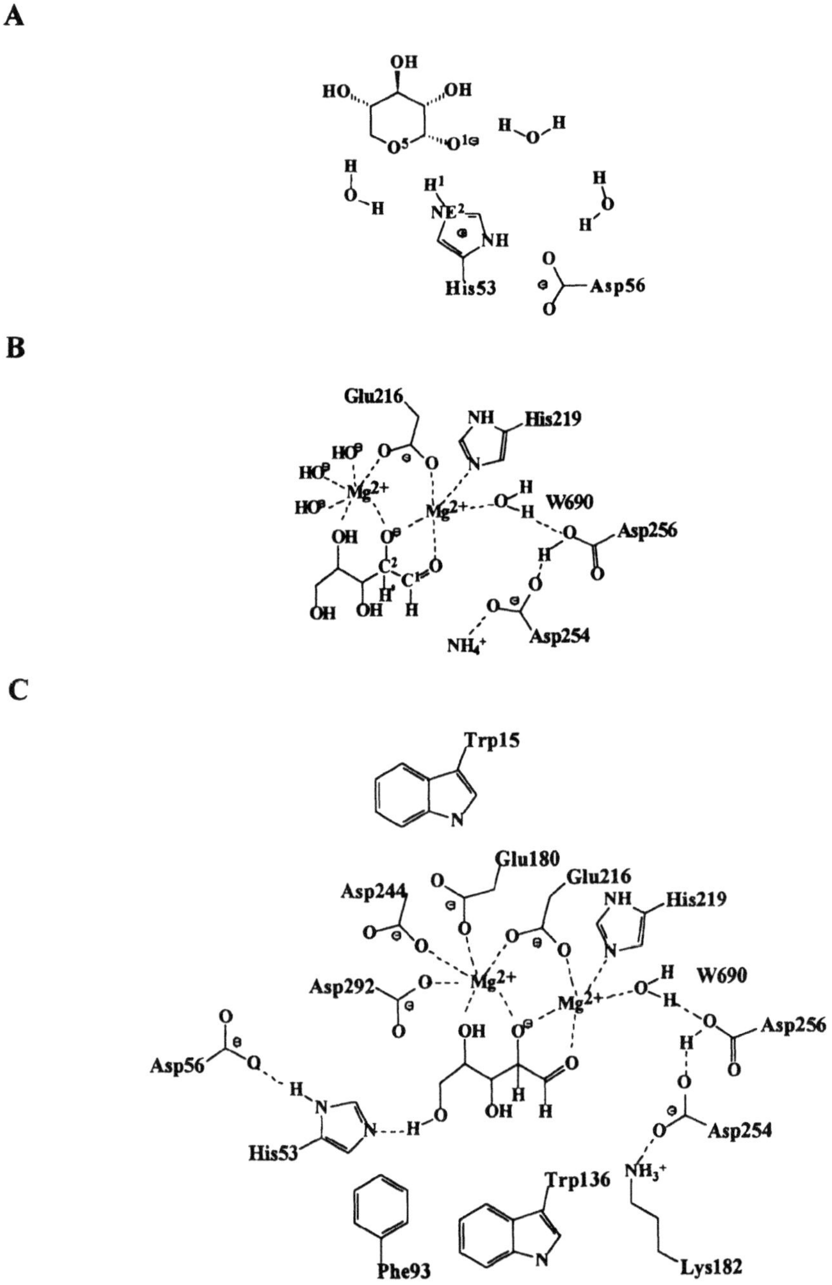

Fig. 2. Active site models of D-xylose isomerase. A: ring opening, B: hydride shift small model, C. hydride shift large model.

the wild-type enzyme from the Protein Data Bank [20] and from our own X-ray diffraction studies [32].

We performed X-ray crystallographic studies for the D254E/D256E double mutant (for preparation, enzyme purification, assay and kinetic parameters see [32]). Crystals of the mutant were grown as described elsewhere [8] with slight modifications. Crystals of the binary complex were grown at room temperature in 2-4 days with the hanging drop vapor diffusion method from a solution containing 1.5 M ammonium sulfate, 3 mM thymol and 9.5 mM magnesium chloride. Ternary complex crystals were obtained by co-crystallization similarly as for the binary complex.

Plate shaped crystals of 0.7×0.7×0.2 mm size were mounted in a quartz capillary and data sets were collected at room temperature on a RIGAKU R-AXIS II imaging plate system mounted on a RIGAKU RU200 X-ray generator using a copper anode. Crystals of the binary complex with magnesium are isostructural with the wild-type counterpart (space group $P3_121$). The ternary complexes co-crystallized with xylitol and the metal cofactors belong to the space group C222 with cell parameters a = 139.66 Å, b = 147.77 Å, c = 83.60 Å, $\alpha = \beta = \gamma = 90°$. In all cases the crystallographic asymmetric unit contained a dimer. Crystal structures were determined via the molecular replacement technique, the model structures contained no inhibitor or co-factor in any of the cases studied. We also removed the mutated side chains from the models and detected no major movements of the main chain in any of the structures.

All crystallographic calculations were done by the CCP4 program package [33], using the PROLSQ program [34] for refinement. Strong stereochemical restraints were applied to keep the geometry close to the ideal one. Model building and display were done by the crystallographic modeling package O [35]. The model derived from co-ordinates of the Protein Data Bank file PDB1XLA [36] was refined against data collected on the binary complex at 2.5 Å resolution, while for ternary complexes we used the file PDB1XLD [37] excluding the mutated residues, inhibitor and metals (for data at 2.2-2.5 Å resolution). Water structures were totally rebuilt in all cases using difference density maps. The R factor decreased from the initial value of 30 % to 16-19 % in 50-100 cycles of refinement. Basic crystallographic data are summarized in the original publication [32].

3. Structure

The catalytic metal ion is known to occupy different subsites in xylose isomerase during isomerization [4,6,10]. In the enzyme-substrate complex the catalytic metal is hexa-coordinated by Glu-216, His-219, Asp-256, W-690 and a bidentate bond to Asp-254 (see Fig. 3; denoted as 2_1). As the reaction proceeds the metal moves to a site (denoted as 2_2, located 1.8 Å away from the previous position), where it has a substantially different coordination. Here, the metal binds to O1 and O2

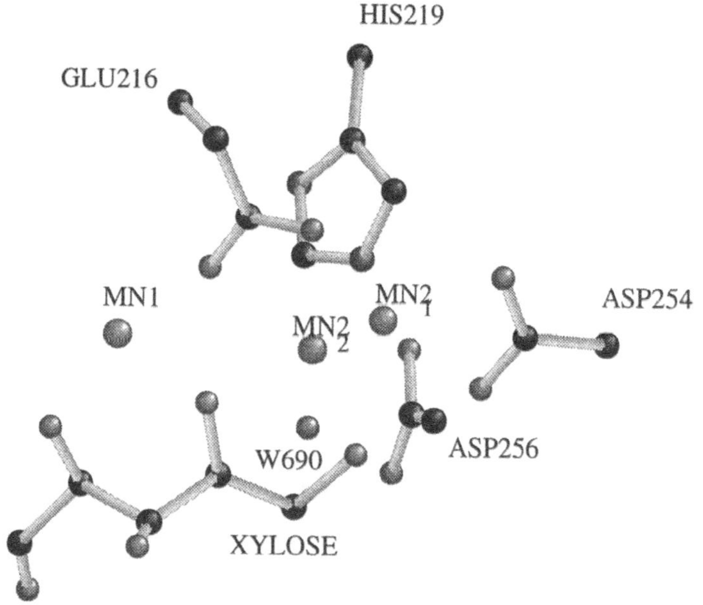

Fig. 3. Catalytic metal binding site of D-xylose isomerase with the bound substrate model and the metal center (coordinates from Whitlow et al. [6]).

of the substrate with distances of 2.4 and 2.6 Å, respectively. The coordination by Glu-216, His-219 and W-690 is maintained, whereas the bond with Asp-254 and Asp-256 breaks.

When Asp-254 and Asp-256 were replaced by glutamates via site-directed mutagenesis [32], our expectation was that the size reduction of metal site 2 would exclude the larger calcium ion without seriously perturbing the active site structure. X-ray studies revealed that, indeed, the active site of the binary complex of the D254E/D256E mutant with Mg^{2+} does not considerably differ from the wild-type structure (see Fig. 4). The root mean square deviation was 0.16 Å for backbone atoms and 0.49 Å when side chains were also included (excluding the mutated residues and metals). The backbone of the engineered glutamates deviates with an rms deviation of 0.13 Å from the wild-type structure. Superimposing the active sites of the mutant and the wild-type enzyme, we found that the distance between the oxygen atoms of Glu-254...Asp-254 and Glu-256...Asp-256 pairs are 2.33 Å (OE1-OD1), 2.52 Å (OE2-OD2), and 1.04 Å (OE1-OD1), 2.49 Å (OE2-OD2), respectively. The intrusion of longer side-chains upon the Asp-Glu double mutation apparently reduced the space available for binding at the catalytic metal site.

In the binary complex of the double mutant with Mg^{2+}, cofactors could be observed at both the structural and the catalytic metal sites, in contrast to the wild type enzyme from *Arthrobacter* [8]. The coordination of the structural metal ion is similar in the two structures, with a 0.61 Å difference in their positions. The side chain of Glu-216 is displaced to the same extent. As this residue binds to both metal ions, the difference between the wild-type and mutant structures can be due to the shift in the position of the catalytic metal ion. Using the binary complex with Mn^{2+} for comparison [8], the rms. deviation of the active site backbone is 0.18 Å, the distance between the structural metal ions is 0.75 Å. In both structures the metal ion at the catalytic site coordinates to residues 216, 219, 254, 256 and to the water molecule that mediates the proton shuttle. Slight differences appear in the coordination, i.e. the metal ion forms a monodentate bond with Glu-254 in the mutant and a bidentate one with Asp-254 in the wild type enzyme, and a few distances (e.g. to His-219) are also larger in the mutant than in the native enzyme. The distance between the position of the catalytic metal ions in the superimposed active sites of the native and mutant enzymes is 2.02 Å. This agrees fairly well with the movement of the carboxylate of residue 254, as the distance between Asp-254 OD1 and Glu-254 OE1 is 1.98 Å. Thus, apparently, the longer side-chain pushes the catalytic metal ion away from its position in the wild type protein. Due to this effect the distance between the structural and the catalytic metal ions is also somewhat reduced, from 4.8 Å to 4.1 Å.

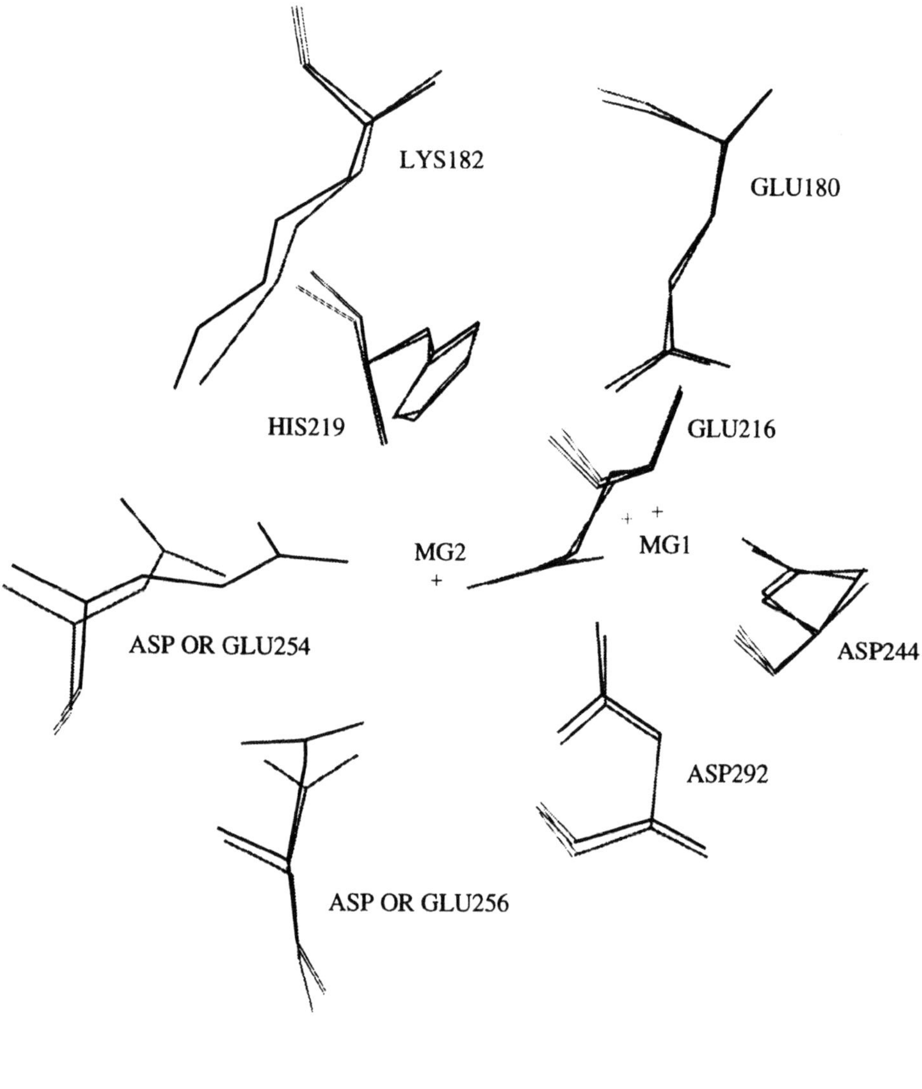

Fig. 4. Geometries of the catalytic metal-binding sites in binary xylose isomerase complexes with magnesium. Wild-type structures (light lines, from the PDB1XLD file [37]) superimposed on the double mutant (heavy lines, from [32]). Graphics by the SYBYL software [23].

The binding of the inhibitor in the double mutant caused further changes in the position of the metal ions and the glutamates. In the ternary complex containing Mg^{2+} and xylitol, the structural metal ion moved by 0.39 Å (similarly to Glu-216), whereas the catalytic ion is shifted by a further 0.82 Å compared to the binary complex with Mg^{2+}. The carboxylate oxygens of Glu-254 are shifted by 0.63 Å from their position in the binary complex, and those of Glu-256 by 0.67 Å (OE1) and 0.86 Å (OE2). The distance between the metal ions at the structural and catalytic sites is reduced by another 0.7 Å from 4.1 in the binary complex to 3.4 Å in the ternary one. The structural metal ion in the ternary complex of the mutant is coordinated octahedrally by the O2 and O4 atoms of xylitol, and by the Asp-244, Asp-292, Glu-180 and Glu-216 side chains, similar to the wild-type enzyme. The metal ion at the catalytic site is also octahedrally coordinated by the protein residues and the inhibitor. The position occupied by Mn^{2+} at the catalytic metal Site 2 is close to Site 2_2 described for the native enzyme. However, some significant differences in coordination were also observed. As in the wild type enzyme, the catalytic metal is coordinated by O1 and O2 of xylitol, NE2 of His-219 and OE2 of Glu-216. In contrast to the wild-type enzyme the bonds to residues Glu-254 and Glu-256 do not break when the catalytic metal binds to the inhibitor, thus the latter is coordinated by both engineered glutamates.

A striking difference from all structures determined to date is the absence of the electron density pattern corresponding to water molecule W-690 in the ternary complexes. Inspection of the space-filling model of the mutant based on the coordinates obtained by X-ray diffraction revealed that in the redesigned active site there is not sufficient space near the catalytic metal to accommodate a water molecule. Since in the binary complex of the mutant with Mg^{2+} this water molecule could still be present at a distance of 2.8 Å from the catalytic metal, it is probably expelled when the inhibitor (or the substrate) binds. This changes the mechanism of the proton shuttle step preceding hydride shift (see below).

Despite the considerably reduced space at the catalytic metal binding site, difference Fourier maps clearly indicated the presence of two metal ions in all double mutant structures determined. Accordingly, the redesign of the catalytic metal binding site resulted in a pocket of smaller size that is large enough for Ca^{2+}, but not sufficiently large to accommodate both a water molecule and a metal ion when the substrate (or inhibitor) is also present. A further important observation is that in the ternary complexes coordination of the activating metals Mg^{2+} and Mn^{2+} and the inhibiting Ca^{2+} by the protein residues and xylitol are identical.

4. Mechanism

The main features of the enzymatic process had already been elucidated by X-ray diffraction, kinetic, molecular modeling and other studies [3-19]. In the following sections details of ring opening, proton shuttle and hydride shift steps will be analyzed in terms of our molecular orbital and electrostatic calculations. Consequences of the D254E/D256E double mutation are discussed in separate subsections in an attempt to shed light on the catalytic competence of metals.

4.1. RING OPENING

We studied the conditions required for the opening of the α-pyranose ring. Details of the reaction path as obtained on the basis of semiempirical molecular orbital calculations for the minimum-size model are depicted in Fig. 5. It was found that the sugar ring opens only after the proton starts moving from NE2 of His-53 towards O5 of the xylose substrate [17], the transition state structure being closer to a ring than to

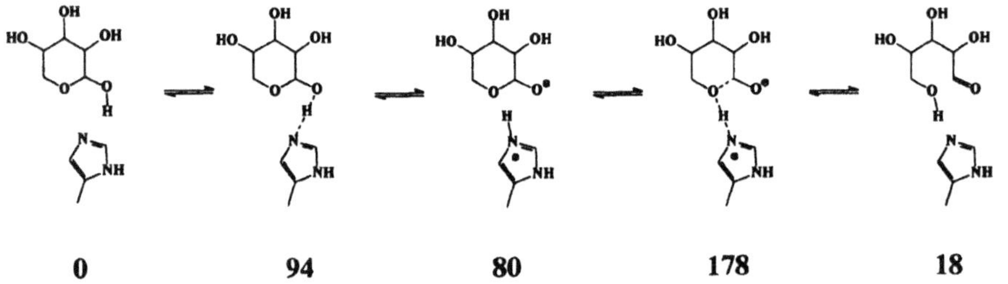

Fig. 5. Local equilibrium and transition state structures for the ring opening step of xylose isomerase reaction. Relative energies for the bare model (not including water and the Asp-56 as in Fig. 2A) in kJ/mol.

an open form [18]. This is at variance with earlier proposals [8,12,16]. The route of proton transfer is summarized as follows. There are three local minima on the potential energy surface connected by two transition states. In the first transition state a proton is located between the O1 hydroxyl oxygen of the substrate and NE2 of His-53 (barrier height 94 kJ/mol), in the other one the proton is shared by O5 of the sugar ring and NE2 (barrier height 98 kJ/mol). This latter proton transfer leads to ring opening.

Since ring opening takes place only in the final stage of the reaction a (- + -) charge distribution forms with participation of the substrate, protonated His-53 and deprotonated Asp-56 side chains, respectively. A similar charge arrangement is also present in catalytic intermediates of serine proteases [38,39], lipases [40,41], and lysozyme [42]. It has been stressed that the anionic-cationic-anionic charge distribution is embedded in an electrostatically complementary protein environment that displays similar features in the aforementioned enzymes despite their wide variety [43,44]. The complementarity for D-xylose isomerase is illustrated in Fig. 6. Beside the protein dipoles, the structural Mg^{2+} ion also plays an important role in determining the electrostatic potential pattern and stabilizing the transition-state structure of the ring opening step, thus strongly increasing the rate of the reaction.

The movement of the proton along a different trajectory (i.e. transfer of the His-53 ND1 proton to Asp-56, as suggested by the original charge relay mechanism in serine proteases) was excluded on the basis of a higher value for the computed activation energy. Such a proton transfer may be hindered by the electrostatic effect of the environment favouring the formation of the ion pair [17]. Our calculations revealed a motion of Asp-56 towards His-53 until transition state was reached, whereas afterwards we observed movement in the opposite direction. Thus, we considered that ring opening is initiated by the approach of Asp-56 towards His-53, which enhances the basicity of His-53. The removal of the proton from O1H1 of this ring mediates it towards O5, and this leads to ring opening. The simulation also accounts for the anomeric specificity of the enzyme, since only the α-anomer is properly positioned to transfer a proton from O1H1 to NE2. This supports the proposal by Lambeir et al. [13] who conclude that it is the His-53 residue which is responsible for the conformational specificity of the enzyme.

A

B

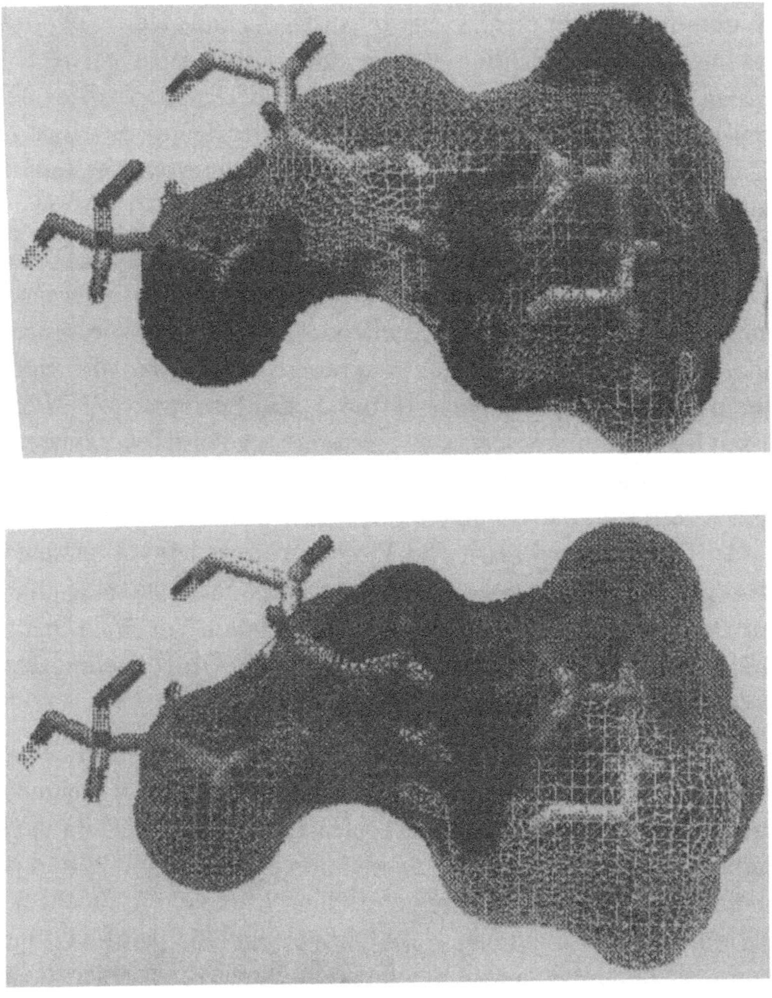

Fig. 6. Complementarity of the electrostatic potential patterns of the Asp-57...His-54...substrate active site of D-xylose isomerase (A) and its environment (B). Positive and negative potential regions are indicated by light and dark grey shading, respectively. Figure made by Ms. Gérczei (Budapest) using the DelPhi program [30,31].

On the basis of semiempirical MNDO molecular orbital calculations on larger models we compared activation energies of ring opening and hydride shift steps (cf. Table 1). These data are in agreement with the conclusions of Rangarajan and Hartley [12] who considered the hydride

TABLE 1. Activation energies (kJ/mol) for the ring opening and hydride shift reaction steps with activating (Mg) and inactivating (Zn) divalent metals [18]. Models from Fig.2.

	Mg		Zn	
	Model A or B	Model C or C'	Model A or B	Model C or C'
Ring opening	50[a]	42[c]	50[a]	63[c]
Hydride shift	105[b]	71[d]	109[b]	138[d]

[a] model A

[b] model B

[c] model C, Protein Data Bank File PDB9XIA [22]

[d] model C', Protein Data Bank File PDB5XIA [21]

shift as rate determining. This conclusion is indirectly supported by the results of Jenkins *et al.* [4] with the D56N mutant of D-xylose isomerase. They found that replacement of aspartate by asparagine has virtually no effect on the catalytic rate. This means that, if the hydride shift is rate determining, the mutation does not alter the activation barrier of this reaction step. On the other hand, in serine proteases, where an anionic-cationic-anionic charge distribution, similar to that during ring opening of D-xylose isomerase forms, the Asp-Asn mutation decreases the rate by approximately four orders of magnitude, i.e. it increases the barrier by 25 kJ/mol [45,46]. Extended calculations on a sophisticated model indicated that this rate decrease is exclusively of electrostatic origin [47]. Since substantial changes in the kinetic parameters of the D56N mutant of D-xylose isomerase were not detected, the activation energy difference between the two steps must be larger than the increase in energy barrier due to the mutation. This is in agreement with the data in Table 1, the activation energy for hydride shift being 29 kJ/mol larger than that for ring opening. Our calculations also explain why zinc inhibits the catalytic reaction. The activation energy of both reaction steps is much larger for this metal than for the activating magnesium.

4.2. PROTON SHUTTLE

Semiempirical molecular orbital calculations have been carried out on Model B of Fig. 2 with the initial geometry parameters of Whitlow *et al.* [6] for Site 1 in manganese-xylose complexes. With a complete geometry optimization carried out we could propose a detailed mechanism outlined in Fig. 1. First W-690 approaches O2 of Asp-256 and a hydrogen bond forms with almost ideal geometry (H_{water}...O2 distance, 1.78 Å, O2...H_{water}- O_{water} angle 180°). After the water proton has been abstracted by O2 of Asp-256, the hydroxyl ion is hydrogen-bonded to O2H with a H...O_{water} distance of 1.68 Å. Coordinating to the catalytic magnesium, W-690 breaks the Mg^{2+}...Asp-256 bond and forms a hydrogen bond with the aspartate side chain. Thus, the proton of O2 is abstracted by the catalytic water molecule and transferred to Asp-256, thereby providing a hydrogen-bonded network from W-690 to Asp-254. As the proton shuttle involves displacement of W-690, it provides a possibility for magnesium to move to Site 2.

The role of Lys-182 was studied in Model C of Fig. 2. This residue does not belong to the coordination sphere of the metal, but crucially affects catalysis [4] since none of the enzymes mutated at Lys-182 was found to be active. A variety of reasons may account for the importance of this side chain, e.g. stabilization of the open substrate conformation, participation in the proton shuttle and polarization of the sugar carbonyl bond. The Lys side chain with its conserved hydrogen-bonded network (including Asp-254 and protonated Asp-256) has a tendency to move away from the active center. However, if we excluded it from the model our calculations indicated that proton transfer, instead of hydride shift, takes place [18]. From our studies Lys-182 appears to play a role in fixing the proton abstracted by Asp-256 with the mediation of W-690 and in the electrostatic stabilization of the transition state.

Our X-ray diffraction studies of the ternary complexes of the D254E/D256E mutant indicated that there is no water molecule present in the coordination sphere of the catalytic metal ion [32]. Inspection of a space-filling computer model of the active site also showed no room for a water molecule. Since W-690 is attributed a crucial role in the proton shuttle step, its absence should change the catalytic choreography. In order to assess the catalytic consequences of the mutation, the enzyme was assayed with D-fructose as substrate. The double mutant displays maximum activity only at rather high magnesium concentrations, and even then it is by almost two orders of magnitude less effective than the wild-type enzyme (see Table 2). The activity reduction results from both an increase of K_M and a decrease of k_{cat}. Apparently, the double mutant is not only less effective in isomerization but also binds the substrate less tightly than the wild-type enzyme.

The decrease of k_{cat}/K_M by two orders of magnitude (cf. Table 2) might well have been caused by the rate decrease of the proton shuttle step, although alteration of electrostatics may also increase the energy barrier of the hydride shift (see below). In the light of such substantial changes it is rather surprising that the mutant is still active.

TABLE 2. Catalytic parameters of wild-type and D254E/D256E mutant xylose isomerase, at pH 8.0 in 0.05 M Tris.HCl, at 37 °C, in the presence of $MgCl_2$. Initial D-fructose concentration was in the range of 0.0-2.5 M.

	K_M (M)	k_{cat} (s^{-1})	k_{cat}/K_M ($M^{-}s^{-1}$)
wild type[a]	0.30	1.57	5.1
D254E/D256E[b]	2.1	0.17	0.08

[a] 25 mM $MgCl_2$

[b] 200 mM $MgCl_2$

Catalysis requires that either a protein side chain acts as a general base, or catalytic fluctuations of the active site allow a water molecule from the bulk to move in and take over the role of W-690.

4.3. HYDRIDE SHIFT

X-ray structures of the double mutant show that the replacement of aspartates by glutamates causes only minor changes in the backbone conformation, while it results in an approach of the negatively charged carboxylates to the substrate and the metal. This obviously changes the electrostatic environment of the substrate. In order to investigate, whether this has an effect on the enzymatic reaction we calculated charge distributions for ground and transition state models of wild-type and mutant enzymes containing magnesium (Table 3). There is a charge flow from atom O2 to O1 during the hydride shift which is reflected in the changes of charges from ground to transition states of both the wild-type and the mutant enzymes. This charge flow is smaller in the mutant than in the wild-type enzyme. Approach of the negative carboxylates of Glu-254 and Glu-256 to the substrate results in an increase of the negative

TABLE 3. Net atomic charges (millielectrons) calculated for wild-type xylose isomerase and its D254E/D256E double mutant.

Enzyme	state	O1	O2	H'
wild type	ground	-456	-523	-125
	transition	-590	-426	-241
D254E/D256E	ground	-327	-487	42
	transition	-475	-426	-103

electrostatic potential at these, mainly around O1 [32]. Due to this effect the double mutant stabilizes the transition-state complex of the hydride shift less effectively than does the wild type enzyme, thus reducing the rate of the rate-limiting reaction.

4.4. METAL COMPETENCE

There is unanimity in the literature that metal ion binding at two sites is required for enzymatic activity of xylose isomerase. Kinetic studies demonstrated that Mg^{2+}, Co^{2+}, Mn^{2+}, and Fe^{2+} activate catalysis, while Ni^{2+}, Ca^{2+}, Zn^{2+}, Cu^{2+}, and Hg^{2+} are inhibitors [2]. Although a number of 3D structures of ternary xylose isomerase complexes with different metals were determined, and extensive studies on binding of metal ions to several species were carried out [48] the reason why a particular metal can or cannot activate the enzyme is not yet known. Without structural information on binding of Ca^{2+}, Collyer *et al.*[8] suggested that this cation may be unable to adopt an octahedral coordination like Mg^{2+} or Mn^{2+}, and that may account for the catalytic non-competence.

We solved the first structure with calcium bound at both metal sites of a xylose isomerase-xylitol ternary complex [32]. This structure enabled us to compare the stereochemistry with inhibitory Ca^{2+} to that with activating metals. As mentioned above, in the double mutant D254E/D256E, in contrast to the hypothesis of Collyer *et al.* [8], calcium ion was found to coordinate to the enzyme residues and to the xylitol molecule octahedrally, and essentially in the same site and geometry as Mg^{2+} or Mn^{2+}. Moreover, this position corresponds to Site 2_2 in the wild type enzyme, suggested to be catalytically relevant in the course of the enzyme action [11,49]. Despite the similar stereochemistry, Mg^{2+} and Mn^{2+} are active with the mutant enzyme but Ca^{2+} is not, indicating that steric factors are not a decisive factor in catalytic competence.

In earlier theoretical studies on the wild type enzyme we found that charge transfer to the metal ion plays a crucial role in the catalytic rate acceleration [17,18], and a correlation exists between the charge transfer towards the catalytic metal and activity. We performed ab initio molecular orbital calculations with a minimal STO-3G basis set on models derived from the X-ray structures of ternary complexes containing various metals (see Fig. 7). Charge distributions are summarized in Table 4. Kinetic data for these complexes show a correlation with charge transfer, i.e. an increasing transfer of the negative charge to the catalytic metal reduces the efficiency of the catalysis. It should be mentioned that only those charges can be compared that have

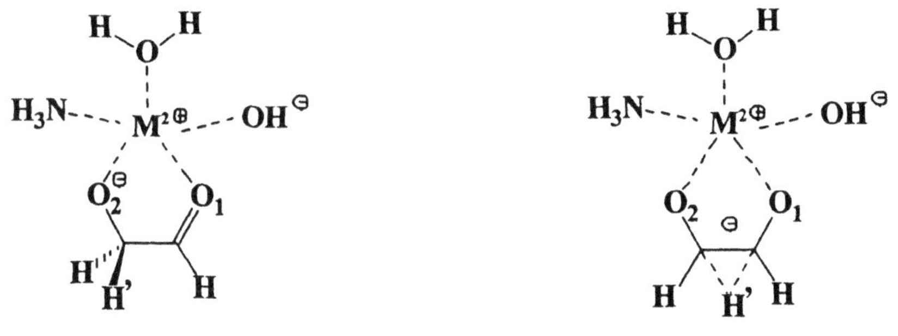

Fig. 7. Models for ab initio molecular orbital calculations of charge transfer in the initial (left) and transition (right) state of the hydride shift [18].

TABLE 4. Charge transfer to metal (millielectrons) as obtained for models of Fig. 7 by ab initio molecular orbital calculations [18].

metal	initial state	transition state
restricted Hartree-Fock calculations		
Al^{3+} (inactive)	1,862	1,769
Zn^{2+} (inactive)	1,240	1,200
Mg^{2+} (active)	971	919
unrestricted Hartree-Fock calculations		
Cu^{2+} (inactive)	1,722	1,461
Co^{2+} (active)	1,563	1,393
Mn^{2+} (active)	1,348	1,281

been calculated with the same (restricted or unrestricted Hartree-Fock) method. A comparison of the amount of charge transfer to the metal with the calculated activation barrier could support this hypothesis. These time consuming ab initio calculations are in progress.

5. Conclusions

Molecular orbital and X-ray diffraction studies on wild-type and D254E/D256E double mutant xylose isomerase allowed to draw the following conclusions on atomic details of the enzymatic catalysis.

1. His-53 acts as the general base mediating a proton from O1 to O5 of the substrate and this is the driving force for ring opening. There is no proton transfer (charge relay) to Asp-56; however, this side chain moves towards N1 of His-53 and enhances its basicity.
2. In the course of the ring opening step, a characteristic (- + -) charge pattern is formed of deprotonated Asp-56, protonated His-53 and the open chain structure of the substrate. This pattern, stable on its own, is further stabilized by the electrostatic field of protein dipoles, similarly to serine proteases, lipases, and lysozyme.
3. In the D254E/D256E mutant the inhibiting Ca^{2+} occupies the same position as activating Mg^{2+} and Mn^{2+} cations. Furthermore, this position is the catalytically relevant one (Site 2_2), the same as found

with the cofactors in the wild-type enzyme. The octahedral coordination of Ca^{2+} by enzyme and inhibitor groups is also similar to that of the cofactors, thus it is not the stereochemical difference that account for catalytic competence.

4. *Ab initio* molecular orbital calculations indicate that charge transfer to the catalytic metal ion from the coordinating substrate oxygens may account for the catalytic rate enhancement.
5. An important factor for the catalytic rate decrease observed in the mutant is the altered electrostatic environment that is less capable of stabilising the transition-state complex. However, the absence of W690 at the active site, influencing the proton shuttle step of the wild-type enzyme, also may reduce the catalytic efficiency.

Acknowledgment

This work was supported by a grant from the Hungarian Ministry of Culture and Education MKM/38.

References

1. Yamanaka, K. (1968) Purification, crystallisation and properties of the D-xylose isomerase from *Lactobacillus* brevis, *Biochim. Biophys. Acta* **151,** 670-680.
2. Callens, M., Tomme, P., Kerstens-Hilderson, H., Cornelis, W., Vangrysperre, W., and De Bruyne, C.K. (1988) Metal ion binding to D-xylose isomerase from *Streptomyces violacoeruber*, *Biochem. J.* **250,** 285-291.
3. Rey, F., Jenkins, J. Janin, J., Lasters, I. Alard, P., Claesens, M., Matthyessens, G., and Wodak, S. (1988) Structural analysis of the 2.8 Å model of xylose isomerase from *Actinoplanes missouriensis*, *Proteins: Struct. Funct. Genet.* **4,** 165-173.
4. Jenkins, J., Janin, J., Rey, F., Chiadmi, M., van Tilbeurgh, H., Lasters, I., Maeyer, M.D., Belle, D.V., Wodak, S., Lauwereys, M., Stanssens, P., Mrabet, N.T., Snauwaert, J., Lambeir, A.M. (1992) Protein engineering of xylose (glucose) isomerase from *Actinoplanes missouriensis*. 1. Crystallography and site-directed mutagenesis of metal binding sites, *Biochemistry* **31,** 5449-5458.
5. Carrell, H.L., Glusker, J.P., Burger, V., Manfre, F., Tritsch, D., Biellmann, J.-F. (1989) x-ray analysis of D-xylose isomerase at 1.9 Å: Native enzyme in complex with substrate and with a mechanism-designed inactivator, *Proc. Natl. Acad. Sci. U.S.A.* **86,** 4440-4444.
6. Whitlow, M., Howard, A.J., Finzel, B.C., Poulos, T.L., Winborne, E., Gilliland, G.L. (1991) A metal-mediated hydride shift mechanism for xylose isomerase based on the 1.6 Å *Streptomyces rubiginosus* structures with xylitol and D-xylose, *Proteins: Struct. Funct. Genet.* **9,** 153-173.
7. Henrick, K., Collyer, C.A., Blow, D.M. (1989) Structures of D-xylose isomerase from *Arthrobacter* strain B3728 containing the inhibitors xylitol and D-sorbitol at 2.5 Å and 2.3 Å resolution, respectively, *J. Mol. Biol.* 208, 129-157.
8. Collyer, C.A., Henrick, K., Blow, D.M. (1990) Mechanism for aldose-ketose interconversion by D-xylose isomerase involving ring opening followed by a 1,2-hydride shift, *J. Mol. Biol.* **212,** 211-235.
9. Farber, G.K., Machin, P., Almo, S.C., Petsko, G.A., Hajdu, J. (1988) X-ray Laue diffraction from crystals of xylose isomerase, *Proc. Natl. Acad. Sci. U.S.A.* **85,** 112-118.

10. Allen, K.N., Lavie, A., Farber, G.K., Glasfeld, A., Petsko, G.A., Ringe, D. (1994) Isotopic exchange plus substrate and inhibition kinetics of D-xylose isomerase do not support a proton-transfer mechanism, *Biochemistry* **33**, 1481-1487.
11. Allen, K.N., Lavie, A., Petsko, G.A., Ringe, D. (1995) Design, synthesis, and characterisation of a potent xylose isomerase inhibitor, D-threonohydroxamic acid, and high-resolution X-ray crystallographic structure of the enzyme-inhibitor complex, *Biochemistry* **34**, 3742-3749.
12. Rangarajan, M., Hartley, B.S. (1992) Mechanism of D-fructose isomerization by *Arthrobacter* D-xylose isomerase, *Biochem. J.* **283**, 223-233.
13. Lambeir, A.M., Lauwereys, M., Stanssens, P., Mrabet, N.T., Snauwaert, J., van Tilbeurgh, H., Matthyssens, G., Lasters, I., Maeyer, M.D., Wodak, S., Jenkins, J., Chiadmi, M., Janin, J. (1992) Protein engineering of xylose (glucose) isomerase from *Actinoplanes missouriensis*. 2. Site-directed mutagenesis of the xylose binding site, *Biochemistry* **31**, 5459-5466.
14. van Tilbeurgh, H., Jenkins, J., Chiadmi, M., Janin, J., Wodak, S., Mrabet, N.T., Lambeir, A.M. (1992) Protein engineering of xylose (glucose) isomerase from *Actinoplanes missouriensis*. 3. Changing metal specificity and the pH profile by site-directed mutagenesis, *Biochemistry* **31**, 5467-5471.
15. Smart, O.S., Akins, J., and Blow, D.M. (1992) Molecular mechanics simulations of a conformational rearrangement of D-xylose isomerase, *Proteins: Struct. Funct. Gen.* **13**, 100-111.
16. Zheng, Y.J., Merz, K.M., Jr., Farber, G.K. Theoretical examination of the mechanism of aldose-ketose isomerization, *Protein Engn.* **6**, 479-486.
17. Fábián, P., Asbóth, B., and Náray-Szabó, G. (1994) The role of electrostatics in the ring opening step of xylose isomerase catalysis, *J. Mol. Struct. THEOCHEM* **307**, 171-178.
18. Fuxreiter, M., Farkas, Ö., and Náray-Szabó, G. (1995) Molecular modelling of xylose isomerase catalysis: the role of electrostatics and charge transfer to metals, *Protein Engn,* **8**, 925-933.
19. Lee, C., Bagdasarian, M., Meng., and Zeikus, J.G. (1990) Catalytic mechanism of xylose (glucose) isomerase from *Clostridium thermosulfogenes*. Characterisation of a structural gene and function of active site histidine, *J. Biol. Chem.* **265**, 19082-19090.
20. Abola, E.E., Bernstein, F.C., Bryant, S.H., Koetzle, T.F., and Weng, J. (1987) in Protein Data Bank, Allen, F.H., Bergerhoff, G., and Sievers, R., (eds.) *Crystallographic Databases - Information Content, Software Systems, Scientific Applications*, Data Commission of the Union of Crystallography, Bonn, p. 107.
21. Henrick, K., Collyer, C.A., Blow, D.M. (1989) Protein Data Bank File PDB5XIA.
22. Carrell, H.L., Glusker, J.P. (1989) Protein Data Bank File PDB9XIA.
23. SYBYL Program Version 6.0a (1993) TRIPOS Associates, St. Louis, MO.
24. Weiner, S.J., Kollman, P.A., Nguyen, D.T., and Case D.A. (1986) An all-atom force field for simulations of proteins and nucleic acids, *J. Comput. Chem.* **7**, 230-252.
25. Dewar, M.J.S., Zoebisch, E.G., Healy, E.F., and Stewart, J.J.P. (1985) Development and use of quantum mechanical molecular models. 76. AM1: a new general purpose quantum mechanical molecular model. *J. Am. Chem. Soc.* **107**, 3902-3909.
26. Stewart, J.J.P. (1989) Optimisation of parameters from semiempirical methods, *J. Comput. Chem.* **10**, 209-217.
27. MOPAC Program Version 5.0 (1989) *Quantum Chemistry Program Exchange*, No. 455, Indiana University, Bloomington, IN, USA.
28. Peterson, M.I. and Poirier, R.A. (1985) Program MONSTERGAUSS, University of Toronto, Toronto, Canada.
29. Hehre, W.J., Radom, L., Schleyer, P.v.R., and Pople, J.A. (1986) *Ab Initio Molecular Orbital Theory*, Wiley, New York.
30. Klapper, I., Hagstrom, R., Fine, R., Sharp, K. and Honig, B. (1986) Focusing of electric fields in the active site of Cu-Zn superoxide dismutase: effects of ionic strength and amino-acid modification. *Proteins: Struct. Funct. Gen.* **1**, 47-59
31. Sharp, K.A. and Nicholls, A. (1989) Program DelPhi, Department of Biochemistry and Molecular Biophysics, Columbia University, Bloomington, Indiana, USA.
32. Fuxreiter, M., Böcskei, Z., Szeibert, A., Szabó, E., Dallmann, G., Náray-Szabó, G., and Asbóth, B. (1997) The role of electrostatics at the catalytic metal binding site in xylose isomerase action. Ca2+ inhibition and metal competence in the double mutant D254E/D256E, *Proteins: Struct. Funct. Gen.*, in press.
33. The CCP4 Suite: programs for protein crystallography (1994) *Acta Cryst.* **A50**, 157-163.

34. Hendrickson, W.A. and Konnert, J.H. (1980) PROLSQ program, in R. Diamond, S. Ramesseshan and K. Venkateshan (eds.), Indian Academy of Sciences, Bangalore, 1301-1323.
35. Jones, T.A., Zhou, J.Y., Cowan, S.W., and Kjeldgaard, M. (1991) Improved methods for building protein models in electron density maps and the location of errors in these models, *Acta Cryst.* **A47,** 110-119.
36. Collyer, C.A., Henrick, K., and Blow, D.M. (1989) Protein Data Bank File PDB1XLA.
37. Collyer, C.A., Henrick, K., and Blow, D.M. (1989) Protein Data Bank File PDB1XLD.
38. Kraut, J. (1977) Serine proteases: structure and mechanism of catalysis, *Annu. Rev. Biochem.* **46,** 331-358.
39. Sussman, J.L., Harel, M., Frolow, F., Oefner, C., Goldman, A., Toker, L. and Silman, I. (1991) Atomic structure of acetylcholinesterease from *Torpedo californica*: a prototype acetylcholine-binding protein, *Science*, **253,** 872-879.
40. Brady, L., Brzozowski, M., Derewenda, S., Dodson, E., Dodson, G., Tolley, S., Turkenburg, J., Christiansen, L., Huge-Jensen, B., Norskov, L., Thim, L., and Menge, U. (1990) A serine protease triad forms the catalytic centre of a tryglycerol lipase, *Nature*, **343,** 767-770.
41. Winkler, F.K., D'Arcy, A., and Hunziker, W. (1990) Structure of human pancreatic lipase, *Nature*, **343,** 771-774.
42. Blake, C.C.F., Johnson, L.N., Mair, G.A., North, A.C.T., Phillips, D.C., and Sarma, V.R. (1967) Crystallographic studies of the activity of hen egg-white lysozyme, *Proc. Roy. Soc. Ser. B* **167,** 378-388.
43. Ángyán, J. and Náray-Szabó, G. (1983) Comparison of protein electrostatic potential along the catalytic triad of serine proteases, *J. Theor. Biol.* **103,** 349-356.
44. Náray-Szabó, G., and Gérczei, T. (1996) The (- + -) charge distribution: A common pattern in the transition state of some enzymes, *Croat. Chem. Acta*, **69,** 955-965.
45. Craik, C.S., Roczniak, S., Largeman, C., and Rutter, W.J. (1987) The catalytic role of the active site aspartic acid in serine proteases, *Science* **237,** 909-912.
46. Carter, P. and Wells, J.A. (1988) Dissecting the catalytic triad of a serine protease, *Nature* **332,** 564-567.
47. Warshel, A., Náray-Szabó, G., Sussman, F., and Hwang, J.K. (1989) How serine proteases really work? *Biochemistry* **28,** 3629-3637.
48. Van Bastelaere, P.B.M., Callens, M., Vangrysperre, W.A.E., and Kersters-Hilderson, H.L.M. (1986) Catalytic properties of D-xylose isomerase from *Streptomyces violaceoruber*, *Enzyme Microb. Technol.* **8,** 696-700.
49. Whitaker, R.D., Cho, Y., Cha, J., Carrell, H.L., Glusker, J.P., Karplus, P.A., and Batt, C.A. (1995) Probing the roles of active-site residues in D-xylose isomerase, *J. Biol. Chem.* **270,** 22895-22906.

COPPER(II) AND ZINC(II) COMPLEXES OF PEPTIDES AS MODELS FOR COLLAGENASE INHIBITORS

KATALIN VÁRNAGY[a] and H. SÜLI-VARGHA[b]
[a] *Department of Inorganic and Analytical Chemistry*
Lajos Kossuth University
H-4010 Debrecen, P.O.B. 21, Hungary
[b] *Research Group of Peptide Chemistry*
Hungarian Academy of Sciences
H-1518 Budapest 112, P.O.B. 32, Hungary

1. Introduction

Imidazole nitrogen donor atoms of histidyl residues are common binding sites in various metalloenzymes. For example, in carbonic anhydrase zinc(II) is coordinated to three imidazoles [1], in carboxypeptidase to two imidazoles and a carboxylate [2]. In blue copper proteins the copper(II) ion is coordinated to two histidine imidazole nitrogen and two sulfur atoms from cysteine and methionine, respectively [3]. In metalloproteins the three-dimensional structure of the macromolecular ligand facilitates the coordination of metal ions by independent side-chain residues. Therefore, ligands containing two or more imidazole donors linked by aliphatic carbon chains may be able to mimic the binding sites and catalytic activities of the enzymes. Due to the remarkable stability of metal complexes of polyimidazole ligands these may serve as inhibitors in zinc enzymes. Specific enzyme inhibitors may be obtained by attaching bis(imidazolyl) ligands to the peptide site which is responsible for enzyme cleavage.

2. Ligands

We have studied copper(II) and zinc(II) complexes of peptide molecules containing bis(imidazolyl) ligands, aimed at inhibiting collagenase activity [4]. These peptides

L. Banci and P. Comba (eds.), Molecular Modeling and Dynamics of Bioinorganic Systems, 441–463.

were fragments (and their derivatives) of the specific sequence of collagen (Pro-Leu-Gly–Ile-Ala-Gly), cleaved by vertebrate collagenases at the glycyl-isoleucine bond.

The copper(II)- and zinc(II) complexes of some simple bis(imidazolyl) containing compounds were studied to allow a meaningful comparison of the complexation capabilities of the protein derivates (Fig. 1.)

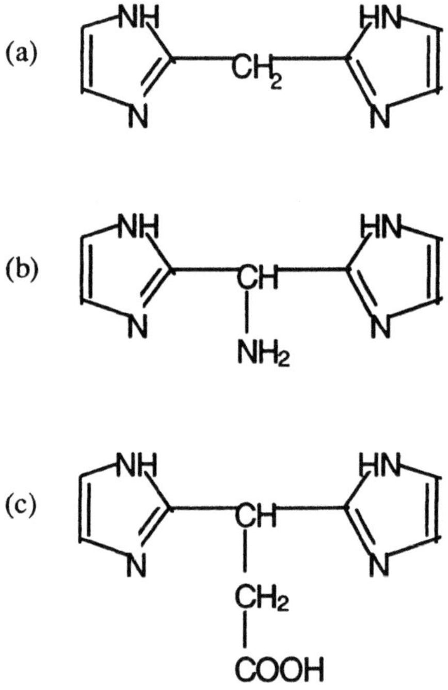

Fig. 1. Structural formulae of the ligands: (a) bis(imidazol-2-yl)methane (BIM) (b) bis(imidazol-2-yl)methylamine (BIMA) (c) 3-[bis(imidazol-2-yl]propionic acid (BIP)

With bis(imidazol-2-yl)methylamine and 3-[bis(imidazol-2-yl)propionic acid it was possible to prepare derivatives of peptides containing the bis(imidazol-2-yl) residues either at their C- or their N-termini [5]. Within these peptide derivatives there are two types of ligands, with or without a side-chain donor group. The bis(imidazol-2-yl) group is connected to an aliphatic peptide chain in the ligands Ac-Pro-Leu-Gly-BIMA and BIP-Ile-Ala-Gly-OEt (Fig. 2). The other type of ligand contains a histidine in the first, second or third position of the peptide chain, respectively (Fig. 3).

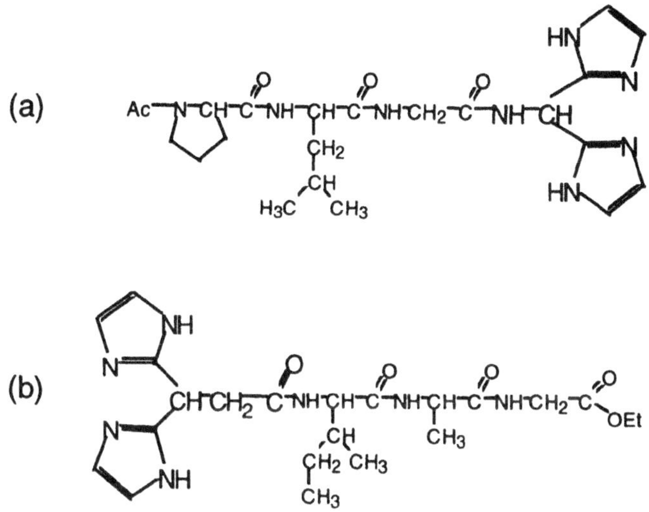

Fig. 2. Structural formulae of the ligands (a) Ac-Pro-Leu-Gly-BIMA (b) BIP-Ile-Ala-Gly-OEt

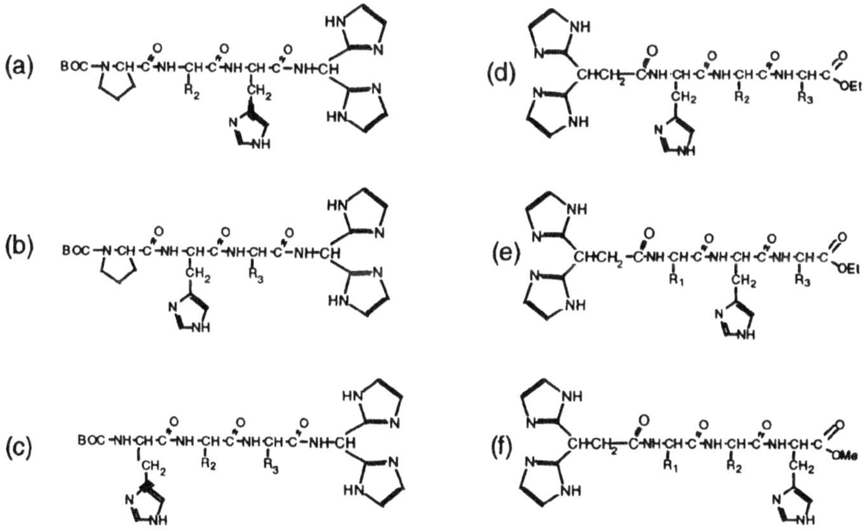

Fig.3. Structural formulae of the ligands (a) Boc-Pro-Leu-His-BIMA (b) Boc-Pro-His-Gly-BIMA (c) Boc-His-Leu-Gly-BIMA (Boc=tert.butyloxycarabonyl group) (d) BIP-His-Ala-Gly-OEt (e) BIP-Ile-His-Gly-OEt (f) BIP-Ile-Ala-His-OMe

3. Equilibrium studies

The stability constants of the proton, copper(II) and zinc(II) complexes and the coordination mode of the different ligands were determined by potentiometry [6-8] and spectroscopy.

3.1. PROTONATION CONSTANTS

The protonation constants indicate that the deprotonation of the imidazole groups occur at a lower pH than in imidazole (Table 1).

TABLE 1. Protonation constants of the ligands

Ligand	pK(COOH)	pK(Im)$_1$	pK(Im)$_2$	pK(His)	pK(NH_3^+)
BIM		4.74(1)	6.93(1)		
BIMA		-	4.07(1)		6.49(1)
Ac-ProLeuGly-BIMA		3.31(1)	5.67(1)		
Boc-ProLeu**His**-BIMA		2.85(2)	5.25(1)	6.64(1)	
Boc-Pro**His**Gly-BIMA		3.11(1)	5.42(1)	6.38(1)	
Boc-**His**LeuGly-BIMA		2.78(2)	5.24(2)	6.65(1)	
BIP	2.79(1)	4.69(1)	6.90(1)		
BIP-IleAlaGlyOEt		3.89(1)	5.99(1)		
BIP-**His**AlaGlyOEt		3.73(1)	5.77(1)	6.77(1)	
BIP-Ile**His**GlyOEt		4.01(1)	5.67(1)	6.65(1)	
BIP-IleAla**His**OMe		3.73(1)	5.80(1)	6.79(1)	
Imidazole		6.95			
GlyHisGly*				6.35	
GlyGlyHis*				6.64	

* Data from ref. [10]

The interaction with aromatic rings decreases the basicity of nitrogen donors. This effect depends on the length of carbon chain between the aromatic rings and on substituents on the α-carbon atom. This also emerges from an investigation of a series of bis(pyridyl)alkanes with various alkyl spacer groups [9]. It was found that longer aliphatic chains had less effect on the protonation processes (Table 2).

TABLE 2. Protonation constants of α,ω-bis(2-pyridyl)alkanes

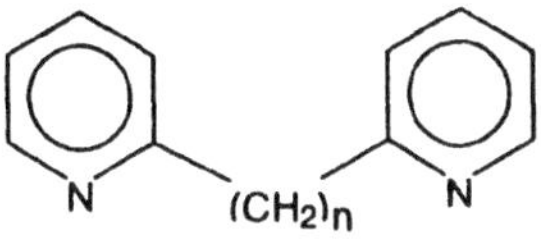

n	0	1	2	3	4	5	6	pyridine
pK_1^*	-	2.69	3.99	4.80	5.27	5.45	5.61	5.31
pK_2^*	4.49	5.18	5.80	6.15	6.25	6.33	6.38	

* Data from ref. [9].

From Table 1 it follows that the pK values of the bis(imidazol-2-yl)methane derivates are also influenced by the substituents of the methine group. The free amino group in bis(imidazol-2-yl)methylamine has the most pronounced effect in this respect. The charge of the protonated ammonium group further decreases the basicity of the aromatic nitrogen donors, and the first pK value cannot be determined potentiometrically.

When two imidazole rings are attached to propionic acid, the pK values for the imidazole nitrogen donors are barely affected. However, if a peptide chain is connected to the bis(imidazol-2-yl) group, either to the amine of bis(imidazol-2-yl)methylamine or to the carboxylate of 3-[bis(imidazol-2-yl)]propionic acid, a decrease in the basicity of the imidazole nitrogen donors is observed. However, the decrease of the pK values is smaller than in the case of bis(imidazol-2-yl)methylamine.

There is an additional imidazole in the BIMA and BIP derivatives with a histidine in the peptide substituent. When histidine is the second amino acid in the peptide chain, the histidyl imidazole acts as an independent binding site, and the pK values are very similar to those of the Gly-His-Gly tripeptide, where histidine is an internal histidyl residue [10].

In the case of Boc-Pro-Leu-His-BIMA, Boc-His-Leu-Gly-BIMA, BIP-His-Ala-Gly-OEt and BIP-Ile-Ala-His-OMe the pK values of the histidyl imidazole nitrogen corresponds to either that of Gly-His-Gly or that of Gly-Gly-His [9]. A comparison with the pK values with those of His-Gly (pK(His)=6.01) and His-Gly-Gly (pK(His)=5.52) [11] is not possible, because of the overlap in the deprotonation of the terminal amino group and the histidyl imidazole nitrogen. However, the first pK value of the bis(imidazol-2-yl) group is lower than that of internal histidyl derivatives. In those cases there is a possibility to form a hydrogen bond between the bis(imidazol-2-yl) fragment and the histidine imidazole nitrogen. When the bis(imidazol-2-yl) group is directly bound to His the two residues are close enough to form this interaction. When the histidine is at the other end of the ligand, it may interact with the bis(imidazol-2-yl) group by forming a "loop". These interactions result in a further decrease of the basicity of the bis(imidazol-2-yl) group.

3.2. COPPER(II) AND ZINC(II) COMPLEXES

The formation of various mono- and bis(ligand) complexes with copper(II) or zinc(II) ions is characteristic for the ligands discussed here.

3.2.1. Complexes of simple bis(imidazol-2-yl) ligands

The imidazole donors of bis(imidazol-2-yl)methane provide binding sites for metal ion coordination, which lead to stable chelate rings. The formation of these complexes occurs in a pH-range below 7, and this leads to rather high values for the stability constants. The coordination of the second ligand is sterically slightly hindered and this is reflected in a high ratio of the stepwise stability constants.

In $[ML]^{2+}$ and $[ML_2]^{2+}$ two or four imidazole nitrogen atoms, respectively, are coordinated equatorially. This is supported by the visible and EPR spectroscopic data of the corresponding copper(II) complexes (Table 4). Theoretical studies predicted the absorption maximum for this chromophore at 581nm [12]. Thus, absorption maxima for the bis(ligand) complexes indicate to the equatorial coordination of four imidazole nitrogen donors around copper(II).

TABLE 3. Stability constants of copper(II) and zinc(II) complexes of bis(imidazol-2-yl)methane, bis(imidazol-2-yl)methylamine and 3-[bis(imidazol-2-yl)]propionoic acid

	BIM		BIMA		BIP	
	Cu(II)	Zn(II)	Cu(II)	Zn(II)	Cu(II)	Zn(II)
logβ(ML)	9.64(1)	5.53(1)	9.89(1)	5.38(2)	10.13(1)	5.63(1)
logβ(ML_2)	17.03(1)	10.22(1)	16.89(2)	9.92(5)	17.29(3)	10.10(1)
logβ(MLH)	-	-	14.0(2)	9.66(4)	13.36(2)	-
logβ(ML_2H)	-	-	21.50(2)	14.25(17)	21.20(3)	-
logβ(ML_2H_2)	-	-	25.21(1)	18.85(15)	-	-
logβ($M_2L_2H_{-1}$)	-	-	18.43(5)	-	-	-
logβ($M_2L_2H_{-2}$)			12.78(4)	6.27(6)	-	-
lg(K_1/K_2)	2.25	0.84	2.89	0.84	2.97	1.16
$pK^{ML_2H_2}_{ML_2H}$	-	-	3.71	4.60	-	-
$pK^{ML_2H_2}_{ML_2H}$	-	-	4.61	4.33	3.94	-
pK^{MLH}_{ML}	-	-	4.11	4.28	3.23	-

TABLE 4. Spectroscopic parameters of the copper(II) complexes

	CuL/CuLH		CuL_2		
	$g_\parallel$	$A_\parallel$	λ_{max}	$g_\parallel$	$A_\parallel$
BIM	2.307	181	578	2.237	199
BIP	2.298	179	580	2.235	202
BIMA	2.303	176	597	2.227	200

The EPR data of $[CuL]^{2+}$ and $[CuL_2]^{2+}$ with bis(imidazol-2-yl)methane are in good agreement with the coordination of two and four nitrogen donors, respectively, and the appearance of the well resolved ^{14}N superhyperfine splitting in the spectrum supports this coordination mode (Fig. 4).

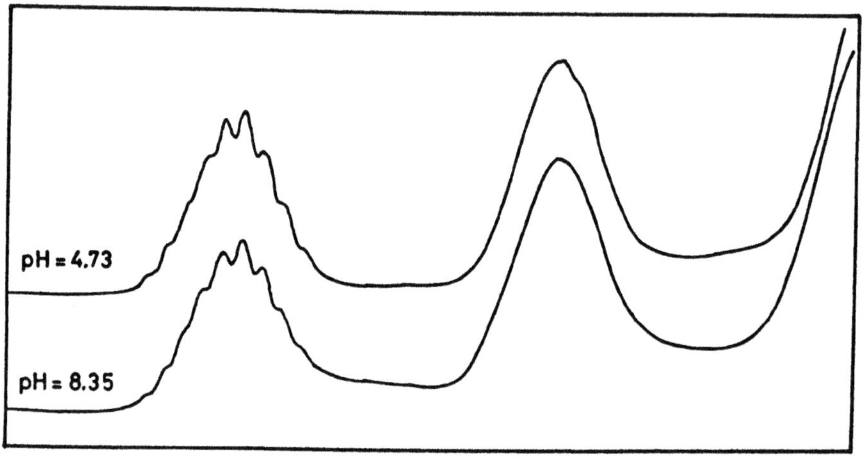

Fig. 4. Frozen solution EPR spectra ($g_{||}$ region) of $[CuL_2]^{2+}$ (L = bis(imidazol-2-yl)methane) at pH= 4.73 and pH=8.35

The bis(imidazol-2-yl) groups are the main bindings sites in the Cu(II) and Zn(II) complexes of bis(imidazol-2-yl)methylamine and 3-[bis(imidazol-2-yl)]propionic acid. However, the substituents on the α-carbon atom have some influence on the complex formation.

Protonated and deprotonated complexes are formed with ligands containing additional donor groups, i.e. carboxylate in 3-[bis(imidazol-2-yl)]-propionic acid and the amine in bis(imidazol-2-yl)methylamine. In the case of 3-[bis(imidazol-2-yl)]propionic acid the imidazole donors are the main binding sites in the entire pH-range studied and they are coordinated equatorially. The species $[CuLH]^{2+}$ and $[CuL_2H]^+$ are complexes with a protonated carboxylate, while $[CuL_2H_2]$ and the protonated zinc(II) complexes are not present in measurable concentration, due to the relatively low protonation constant of the carboxylate group. Upon deprotonation of these species there is a weak axial interaction of the carboxylate group ($[ML]^+$ and $[ML_2]$). This emerges from an increase of the ratio of the stepwise stability constants and the higher values obtained for the stability constant of $[ML]^+$.

The visible and EPR spectroscopic parameters of the copper(II) complexes with 3-[bis(imidazol-2-yl)]propionic acid are very similar to those with bis(imidazol-2-yl)methane (Table 4), suggesting that the imidazole nitrogens are the main binding sites, and that they are coordinated equatorially. The superhyperfine splitting in the EPR spectrum of the $[CuL_2]$ species (L = 3-[bis(imidazol-2-yl)]propionic acid) supports the equivalence of nitrogen donor atoms around the copper(II) ion in the equatorial plane (Fig. 5).

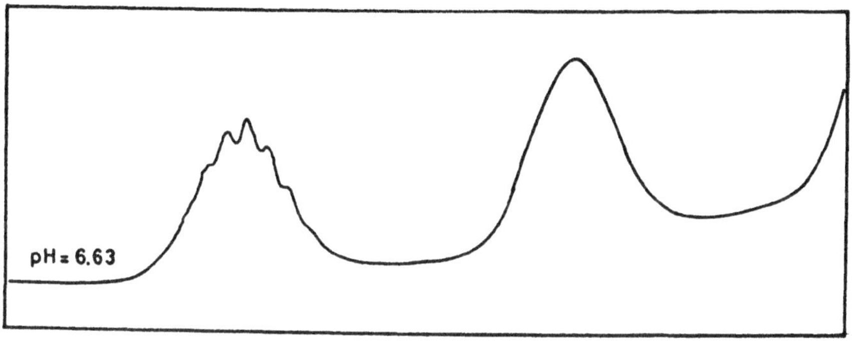

Fig. 5. Frozen solution EPR spectrum ($g_{||}$ region) of $[CuL_2]^{2+}$ (L = 3-[bis(imidazol-2-yl)]propionic acid) at pH= 6.63

The complex formation processes with bis(imidazol-2-yl)methylamine are more complex. The amino group coordinates very efficiently to the metal ion, and it can form a five-membered chelate ring with the metal ion and the imidazole nitrogen donors.

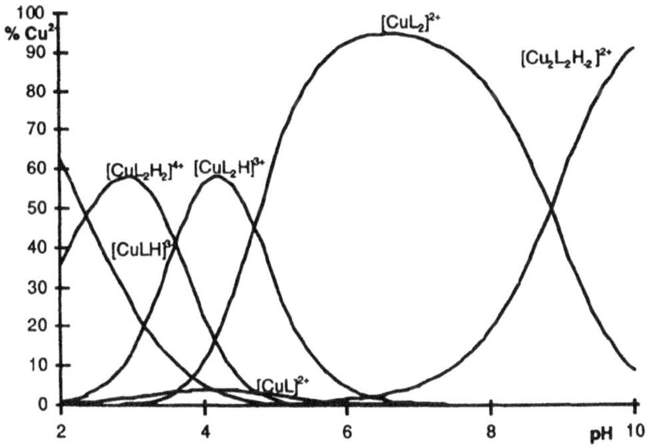

Fig. 6. Species distribution of complexes formed in the copper(II)-BIMA system as a function of pH (c_L=4·10^{-3} mol/dm^3, c_M=2·10^{-3} mol/dm^3)

The formation of protonated mono- and bis complexes with coordination of two or four imidazole nitrogens, respectively, occurs below pH 2. Deprotonation of the non-coordinated ammonium groups is accompanied by characteristic changes in both the visible and EPR spectra, and these changes depend on metal ion to ligand ratios. In the presence of excess ligand the absorption spectra indicate the axial coordination of at least one nitrogen donor in the species $[CuL_2]^{2+}$ (small red shift). The EPR measurements suggest that one of the imidazole nitrogens is coordinated axially, and it is replaced by the amino group in the equatorial position. The loss of detectable superhyperfine coupling supports the non-equivalence of nitrogen donors around copper(II) (Fig. 7). The small decrease in $g_{\parallel}$ also indicates the different coordination modes of BIMA and bis(imidazol-2-yl)methane (see Table 4). A similar coordination mode has been observed in $[CuL_2]^{2+}$ (L = bis(pyridin-2-yl)methylamine) [13].

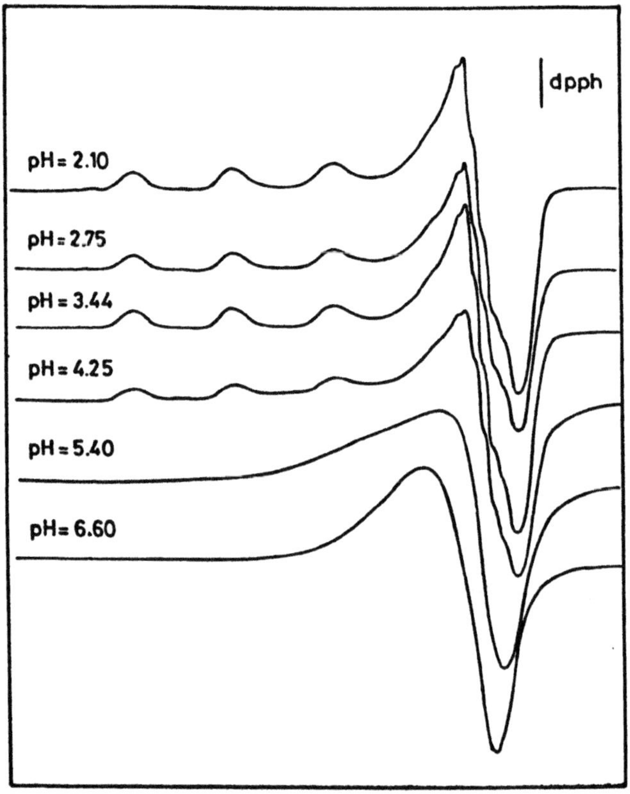

Fig. 7. Frozen solution EPR spectra of frozen solutions of copper(II) and bis(imidazol-2-yl)methylamineat an 1:2 ratio as a function of pH

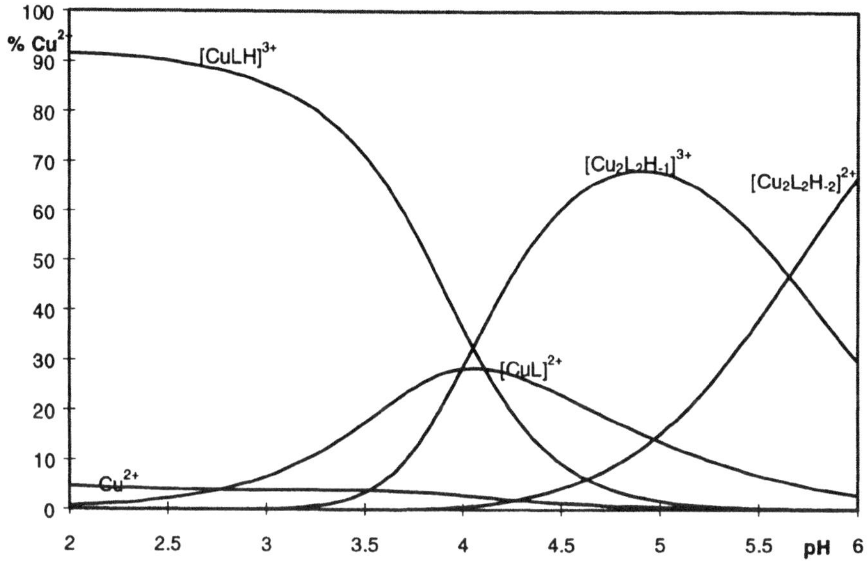

Fig. 8. Species distribution of complexes formed in the copper(II)–bis(imidazol-2-yl)-methylamine system as a function of pH (c_L=4·10^{-3} mol/dm^3, c_M=4·10^{-3} mol/dm^3)

The potentiometric measurements suggest that there may be the formation of hydroxo complexes. However, the titration curves can also be fitted with equilibria including either monomeric $[MLH_{-1}]^+$ or dimeric $[M_2L_2H_{-2}]^{2+}$ species. The visible and EPR spectra suggested the formation of dinuclear complexes. Dinuclear hydroxo compounds of copper(II) are generally hydroxo bridged dimers but the absorption maximum around 595 nm suggests equatorial coordination of at least three nitrogen donors.

For steric reasons, all three nitrogen donors of BIMA cannot coordinate equatorially to the same metal ion. We suggest a structure, where the three coordination sites of each copper(II) ion are occupied by an imidazole nitrogens, an amino nitrogen and a hydroxo group, while the second imidazole is bridging to the second copper(II) ion (Fig. 9).

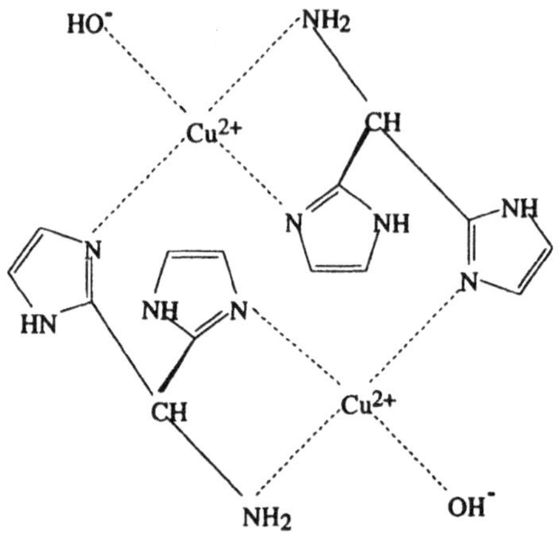

Fig. 9. The suggested structure of the $[Cu_2L_2H_{-2}]$ species (L = bis(imidazol-2-yl)methylamine)

3.2.2. Complexes of bis(imidazol-2-yl) substitute peptide ligands

In Ac-Pro-Leu-Gly-BIMA and BIP-Ile-Ala-Gly-OEt an aliphatic peptide chain is connected to the bis(imidazol-2-yl) group. When peptides are coordinated to copper(II) the amide nitrogens are generally involved in complex formation. However, this usually requires the presence of an anchoring donor group, which should be in a chelating position with respect to the amide groups [14, 15].

The coordination of the bis(imidazol-2-yl) group can prevent the metal ion induced deprotonation and coordination in both ligands discussed here. The bis(imidazol-2-yl) group is then the only binding site for metal ions, resulting in an equatorial coordination of two or four imidazole nitrogens in the species $[ML]^{2+}$ and $[ML_2]^{2+}$, respectively. The absorption maximum at 583 nm for $[CuL_2]$ (L = Ac-Pro-Leu-Gly-BIMA) is similar to that with L = bis(imidazol-2-yl)methane.

TABLE 5. Stability constants of copper(II) and zinc(II) complexes of Ac-Pro-Leu-Gly-BIMA and BIP-Ile-Ala-Gly-OEt

	Ac-ProLeuGly-BIMA		BIP-IleAlaGlyOEt	
	Cu(II)	Zn(II)	Cu(II)	Zn(II)
$\log\beta(ML)$	8.65(3)	4.85(8)	8.92(3)	4.92(3)
$\log\beta(ML_2)$	15.24(3)	8.58(4)	15.52(3)	8.90(5)
$\log(K_1/K_2)$	2.06	1.12	2.32	0.94

There is an additional imidazole residue in the derivatives of BIMA and BIP containing histidine. This is, however, too far away from the bis(imidazol-2-yl) moiety to form a stable chelate ring. On the other hand, the imidazole nitrogen donor atom of the histidyl residue could also be an anchor group for copper(II)-induced amide deprotonation and coordination of peptide molecules, especially in the case of Boc-Pro-Leu-His-BIMA and BIP-His-Ala-Gly-OEt. However, both potentiometric and spectroscopic measurements indicate that in these cases the amide coordination of the peptide residues cannot compete with the coordination of bis(imidazol-2-yl) residues. The stability constants indicate that protonated and deprotonated mono- and bis complexes are formed in all six cases (Table 6).

TABLE 6. Stability constants (logβ) of the copper(II) complexes of histidyl containing peptide derivatives

	logβ(CuLH)	logβ(CuL)	logβ(CuL_2H_2)	logβ(CuL_2H)	logβ(CuL_2)
Boc-ProLeuHis-BIMA	14.37(5)	9.17(15)	27.34(4)	21.85(14)	15.89(10)
Boc-ProHisGly-BIMA	14.70(2)	9.51(2)	27.44(2)	21.88(3)	15.69(2)
Boc-HisLeuGly-BIMA	14.66(3)	11.4112)	27.44(3)	23.06(4)	-
BIP-HisAlaGlyOEt	15.08(1)	10.63(2)	27.97(3)	22.44(7)	16.03(7)
BIP-IleHisGlyOEt	14.64(2)	10.05(4)	26.68(6)	21.08(12)	14.74(10)
BIP-IleAlaHisOMe	17.05(28)	13.09(7)	29.74(11)	-	-

In the protonated complexes the ligands are coordinated via bis(imidazol-2-yl) nitrogens. The deprotonation of $[MLH]^{3+}$ and $[ML_2H_2]^{4+}$ corresponds to the deprotonation of the non-coordinated and protonated histidine imidazole nitrogen. The pK values characteristic for these deprotonation processes (Table 7) are lower than the pK values of the histidine imidazole nitrogens of metal-free ligands. This indicates some interaction between the third imidazole donor and the metal center. For Boc-Pro-Leu-His-BIMA and Boc-Pro-His-Gly-BIMA the absorption maximum of the species $[CuL_2]^{2+}$ is slightly shifted to higher wavelengths, also suggesting axial coordination of the side-chain imidazole residue.

TABLE 7. Equilibrium data of the copper(II) complexes of histidyl containing peptide derivatives

	pK^{MLH}_{ML}	$pK^{ML_2H_2}_{ML_2H}$	$pK^{ML_2H}_{ML_2}$	$\log(K_1/K_2)^*$	$\log(K_1/K_2)$
Boc-ProLeu**His**-BIMA	5.20	5.49	5.96	1.40	2.45
Boc-Pro**His**Gly-BIMA	5.19	5.56	6.19	1.96	3.33
Boc-**His**LeuGly-BIMA	3.25	4.39	-	1.88	-
BIP-**His**IleAlaGlyOEt	4.45	5.53	6.41	2.19	5.23
BIP-Ile**His**GlyOEt	4.62	5.60	6.34	2.66	5.36
BIP-IleAla**His**OMe	3.96	-	-	4.36	-

$\log(K_1/K_2)^* = \log K(MLH) - \log K(ML_2H_2) = 2\log\beta(MLH) - \log\beta(ML_2H_2)$

The ratio of the stepwise stability constants for protonated species is similar to that of bis(imidazol-2-yl)methane or other 4N-coordinated species. The ratio of $\log(K_1/K_2)$ is, however, increased for the species $[CuL_2]^{2+}$, which suggests some change in the coordination geometry. This supports the tridentate coordination of at least one of the ligands. When the histidine residue is the third amino acid in the peptide chain, the imidazole nitrogen is too far away for coordination. The coordination of the bis-imidazolyl group is then too weak to prevent the formation and precipitation at copper hydroxide.

The ratio of the stepwise stability constants is particularly high for the copper(II)–BIP-His-Ala-Gly-OEt and copper(II)–BIP-Ile-His-Gly-OMe systems. The large difference between the stepwise stability constants suggests a change of geometry, probably resulting from the coordination of the histidyl residue. Visible and EPR spectroscopic measurements support this conclusion and reveal that the coordination mode of the side-chain imidazole residue depends on the metal-ion to ligand ratios.

The absorption maxima in equimolar solution are shifted progressively to lower wavelenghts and the maximum at about 640 nm supports the equatorial coordination of all three imidazole nitrogens in the species $[CuL]^{2+}$. The existence of this species can be explained by the formation of a macrochelate or a "loop".

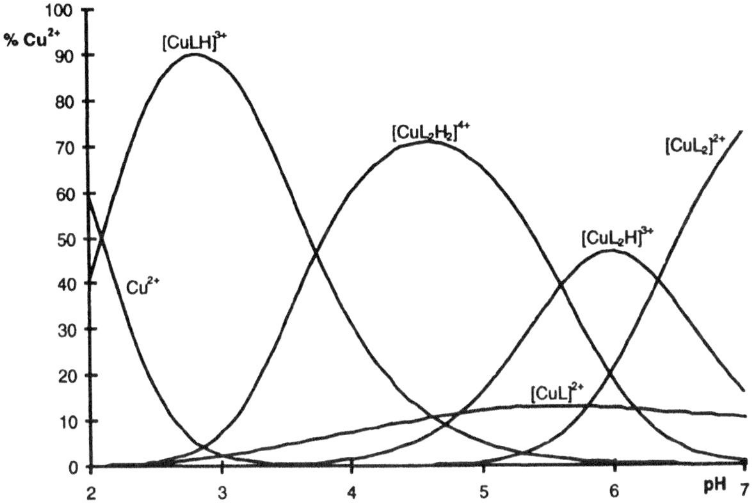

Fig. 10. Species distribution of complexes formed in the copper(II)-BIP-Ile-His-Gly-OEt system as a function of pH (c_L=4·10^{-3} mol/dm^3, c_M=2·10^{-3} mol/dm^3)

For a metal ion to ligand ratio of 1:2 the absorption maxima are shifted to higher energy in moderately acidic solution. This corresponds to the coordination of the bis(imidazol-2-yl) nitrogen donors. Upon deprotonation of the histidyl side-chain there is a significant red shift in the absoption spectra, suggesting a decreasing tetragonality at the copper(II) center and axial coordination of one (or two) additional imidazole donors.

TABLE 8. Visible spectroscopic parameters of the copper(II)–BIP-Ile-His-Gly-OEt system as a function of pH at a copper(II)–ligand ratio of 1:2

pH	2.71	3.03	4.06	4.52	5.03	6.01	6.55
λ_{max}	680	677	630	620	625	652	683
ε (dm^3 mol^{-1}cm^{-1})	38.0	40.0	45.5	48.0	50.0	54.0	59.0

TABLE 9. Binding modes of copper(II) complexes

Ligand	**Species**	**Coordination**
Ac-ProLeuGly-BIMA	$[ML]^{2+}$	2 N(Im)
	$[ML_2]^{2+}$	4 N(Im)
BIP-IleAlaGlyOEt	$[ML]^{2+}$	2 N(Im)
	$[ML_2]^{2+}$	4 N(Im)
Boc-ProLeu**His**-BIMA	$[MLH]^{3+}$	2 N(Im)
	$[ML_2H_2]^{4+}$	4 N(Im)
	$[ML]^{2+}$	2 N(Im) + His(ax)
	$[ML_2]^{2+}$	4 N(Im) + His(ax)
Boc-Pro**His**Gly-BIMA	$[MLH]^{3+}$	2 N(Im)
	$[ML_2H_2]^{4+}$	4 N(Im)
	$[ML]^{2+}$	2 N(Im) + His(ax)
	$[ML_2]^{2+}$	4 N(Im) + His(ax)
Boc-**His**LeuGly-BIMA	$[MLH]^{3+}$	2 N(Im)
	$[ML_2H_2]^{4+}$	4 N(Im)
BIP-**His**AlaGlyOEt	$[MLH]^{3+}$	2 N(Im)
	$[ML_2H_2]^{4+}$	4 N(Im)
	$[ML]^{2+}$	3N(Im) (His (eq))
	$[ML_2]^{2+}$	4 N(Im) + His(ax)
BIP-Ile**His**GlyOEt	$[MLH]^{3+}$	2 N(Im)
	$[ML_2H_2]^{4+}$	4 N(Im)
	$[ML]^{2+}$	3N(Im) (His (eq))
	$[ML_2]^{2+}$	4 N(Im) + His(ax)
BIP-IleAla**His**OMe	$[MLH]^{3+}$	2 N(Im)
	$[ML_2H_2]^{4+}$	4 N(Im)

From the results of this series of bis(imidazol-2-yl) containing peptides it can be concluded, that the bis(imidazol-2-yl) groups are the main binding site in the copper(II) and zinc(II) complexes and that the peptide residues do not have a significant effect on complex formation, regardless of the location of the bis(imidazol-2-yl) moiety in the molecules. In the bis(imidazol-2-yl) containing peptides the histidine imidazole nitrogen cannot behave as an anchor group. The deprotonation and coordination of the peptide nitrogen donors does not occur in any of the peptide ligands.

3.3. COMPLEXES OF A BIS(IMIDAZOL-2-YL) DERIVATIVE CONTAINING AN AMINO ACID

In the case of common peptides the terminal amino group is the most frequent anchoring group [14,15]. The ligand, in which an amino acid with a free amino group is connected to the bis(imidazol-2-yl) group, has a good anchoring group with respect to the amide and imidazole nitrogens. The simplest member of this type of ligand is H-Gly-BIMA (Fig. 11).

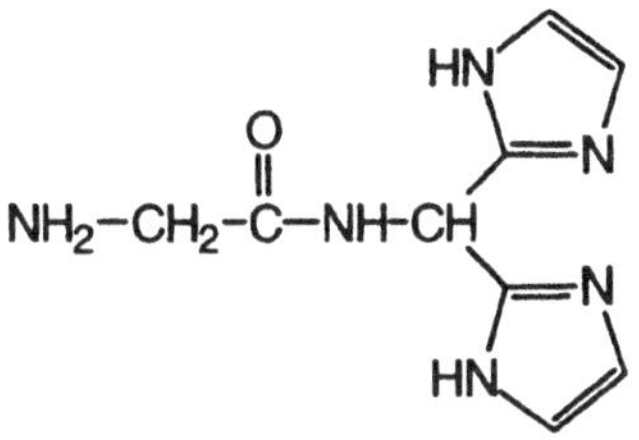

Fig. 11. Structural formula of H-Gly-BIMA

The pK values (Table 10) of the ligand are in good agreement with those of peptides containing the bis-imidazol-2-yl group at the C-terminus and pK value of Gly-Gly (pK(NH_3)=8.15) [16].

TABLE 10. Stability constants of the proton complexes of H-Gly-BIMA

logβ(HL)	logβ(H_2L)	logβ(H_3L)
7.95(1)	13.46(1)	16.68(1)
pK(NH_3)=7.95	pK$(Im)_2$= 5.51	pK$(Im)_1$= 3.22

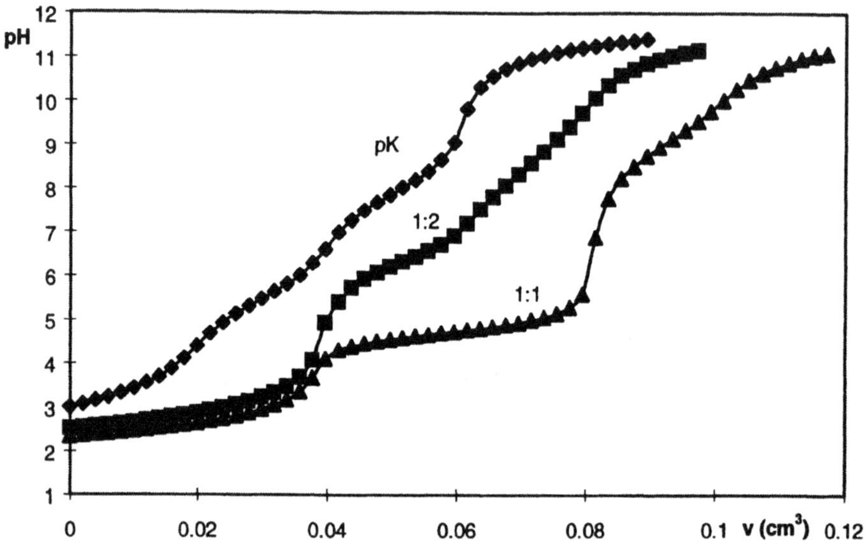

Fig. 12. Titration curves of the copper(II)-H-Gly-BIMA system

From the titration curves of the free ligand and the copper-ligand system at a 1:2 and a 1:1 ratio (Fig. 12) it follows that in the copper(II) containing system there is an extra base consuming process. This may correspond to the formation of hydroxo complexes or the deprotonation of the amid nitrogen.

TABLE 11. The stability constants of copper(II) and zinc(II) complexes of H-Gly-BIMA

	Cu(II)	Zn(II)
$\log\beta(MLH)$	17.11(9)	12.40(1)
$\log\beta(ML_2H_2)$	31.64(9)	24.04(4)
$\log\beta(ML_2H)$	-	17.17(6)
$\log\beta(ML_2)$	18.96(5)	10.28(5)
$\log\beta(ML_2H_{-1})$	11.30(8)	-
$\log\beta(M_2L_2H_{-2})$	18.44(3)	0.40(2)
$\lg(K_1/K_2)^*$	2.58	0.76
$pK^{ML_2H_2}_{ML_2H}$	–	6.87
$pK^{ML_2}_{ML_2H_{-1}}$	7.66	–

$\log(K_1/K_2)^* = \log K(MLH) - \log K(ML_2H_2) = 2\log\beta(MLH) - \log\beta(ML_2H_2)$

Protonated mono- and bis-ligand complexes are formed in the pH range below 5. The ratio of the stepwise stability constants for protonated species is similar to that of bis(imidazol-2-yl)methane, which suggests the coordination of two or four imidazole nitrogen donors in $[MLH]^{2+}$ and $[ML_2H_2]^{2+}$, respectively, while the terminal amino group is protonated. This coordination mode is supported by visible spectroscopic data.

The deprotonation of the amino group is accompanied by an extra base consuming process, which indicates the deprotonation and coordination of the amide nitrogen. The metal(II) ion is probably coordinated to the amine, the deprotonated amide and two imidazole nitrogen donor atoms. The second imidazole donor can behave as a bridging ligand, forming a dimeric $[M_2L_2H_{-2}]$ species (Fig. 13).

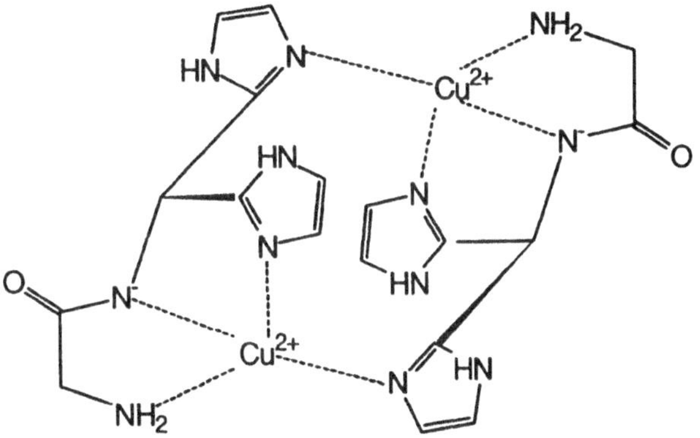

Fig. 13. Suggested structure of the $[M_2L_2H_{-2}]$ species

The titration curves can be fitted to a model involving either the monomer $[MLH_{-1}]^+$ or the dimer $[M_2L_2H_{-2}]$. In equimolar solution the species with a deprotonated amide nitrogen predominates in the pH range above 5 (Fig. 14). The visible spectra suggest the formation of dimeric species. Conclusive evidence for the dimer could be obtained from EPR spectra.

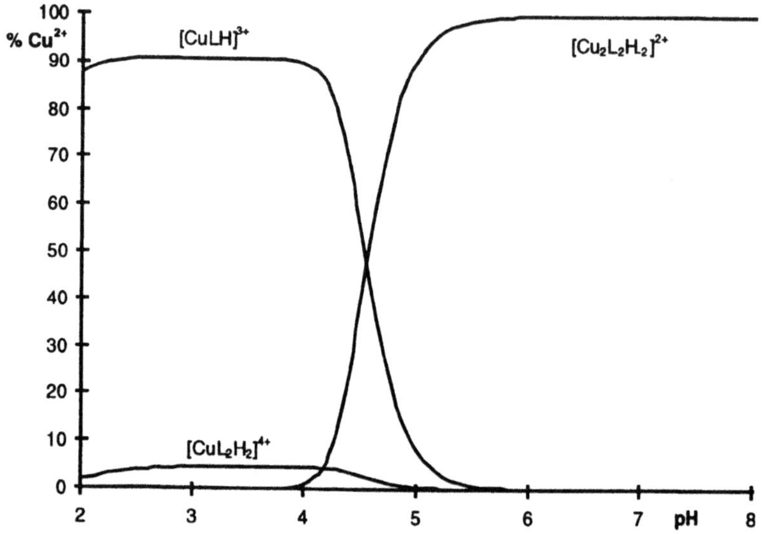

Fig. 14. Species distribution of complexes formed in the copper(II)-H-Gly-BIMA system as a function of pH (c_L=4·10^{-3} mol/dm^3, c_M=4·10^{-3} mol/dm^3)

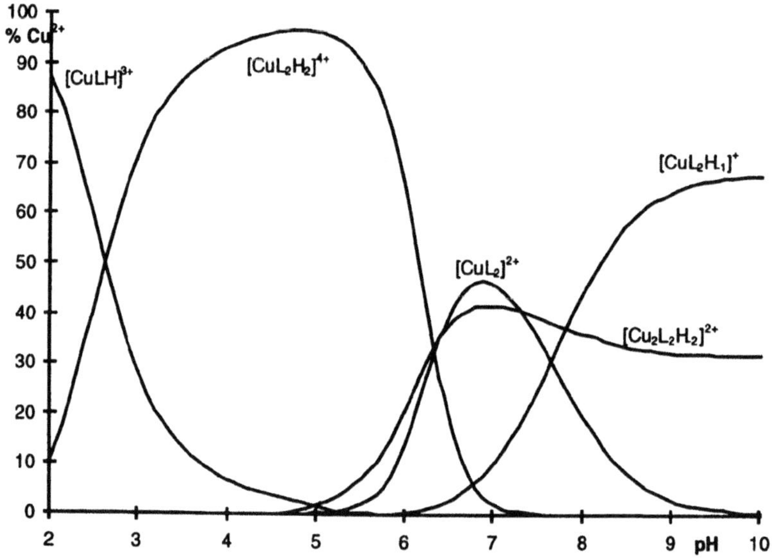

Fig. 15. Species distribution of complexes formed in the copper(II)-H-Gly-BIMA system as a function of pH (c_L=4·10^{-3} mol/dm^3, c_M=2·10^{-3} mol/dm^3)

From the calculated distribution curve of metal-ligand system at a ratio of 1:2 (Fig. 15) it emerges that also in the presence of excess ligand the deprotonation and coordination of the amid nitrogen can occur. However, the excess of ligand leads to the formation of $[MLH_{-1}LH]^{2+}$ and $[MLH_{-1}L]^{+}$, where the second ligand is probably coordinated by an imidazole nitrogen donor (Fig. 16). This results in the formation of $[ML_2]^{2+}$ and $[ML_2H_{-1}]^{+}$.

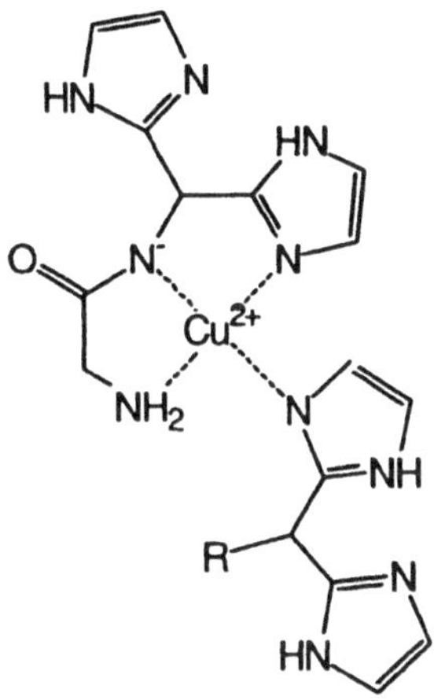

Fig. 16. Suggested structure of the bis(ligand) complexes

4. Summary

We have compared the complex formation abilities of peptides and amino acids containing a bis(imidazol-2-yl) group. The main binding site in both cases is the bis(imidazol-2-yl) group. The peptide residues do not have a significant effect on complex formation, they can only slightly modify the complex stability and coordination mode of the ligand. Even the imidazole nitrogen of a histidyl residue cannot compete with the bis(imidazol-2-yl) fragment, independent of its position in the tripeptide sequence. This coordination mode prevents the metal(II) ion induced deprotonation and coordination of peptide amide nitrogen donors.

In the case of the amino acid derivative the amino group is a good anchor group for metal ion induced deprotonation and coordination of the amide nitrogen. The coordination of the bis(imidazol-2-yl) group cannot prevent the coordination of the amine and the amide nitrogens in the higher pH range. This suggests that the bis(imidazol-2-yl) group is not the exclusive binding site in this ligand, in contrast to the peptide derivatives.

Acknowledgement

This work was supported by the Hungarian National Research Fund OTKA-T015709, OTKA-1646 and OTKA-007512.

References

1. Bertini, I., Lucinat, C. and Scozzafava, A. (1982) An insight into the zinc binding site and active cavity through metal substitution, in M.J. Clarke, P. Hemmerich, C.K. Jorgenson, D. Reinen and R.J.P Williams (eds.), *Struct. Bonding* **48,** Springer-Verlag, New York, pp. 45-92
2. Quichio, F.A. and Lipscomb, W.N. (1971) Carboxypeptidase A, *Adv. Protein Chem.* **25,** 1-78
3. Collman, P.M., Freeman, H.C., Guss, J.M., Murata, M., Norris, V.A., Ramshwa, J.A.M. and Venkatapp, M.P. (1978) X-ray crystal structure analysis of plastocyanin at 2.7 A resolution, *Nature* **272,** 319-324
4. Várnagy, K., Sóvágó, I., Ágoston, K., Likó, Z., Süli-Vargha, H., Sanna, D. and Micera, G. (1994) Potentiometric anc spectroscopic studies on copper(II) and zinc(II) complexes of peptides containing bis(imidazol-2-yl) ligands, *J. Chem. Soc., Dalton Trans.* 2939-2945
5. Likó, Z. and H. Süli-Vargha (1993) Synthesis of peptides containing bis-imidazole ligands, *Tetrahedron Letter* **34,** 1673-1676
6. Gergely, A. and Nagypál, I. (1977) Studies on transition-metal-peptides complexes. Part 1. Equilibrium and thermochemical study of the cooper(II) complexes of glycylglycine, glycyl-DL-α-alanine, DL-α-alanylglycine and DL-α-alanyl-DL-α-alanine, *J. Chem. Soc., Dalton Trans.* 1104-1108
7. Irving, H., Miles, G. and Pettit, L.D. (1967) A study of some problems in determining the stoichiometric proton dissociation constants of complexes by potentiometric titrations using a glass electrode, *Anal. Chim. Acta* **38,** 475-488
8. Zékány, L. and Nagypál, I. (1985) in D.J. Leggett (ed), PSEQUAD, A comprehensive program for the evaluation of potentiometric and/or spectrophotometric equilibrium data using analytical derivatives, *Computational Methods for the Determination of Stability Constants*, Plenum, New York, pp. 291-299
9. Bühler, H. and Anderegg, G. (1970) Metallkomplexe mit α,ω-bis(2-pyridyl)-alkanen, *Chimia* **24,** 433-436
10. Farkas, E., Sóvágó, I. Kiss, T. and Gergely, A. (1984) Studies on transition metal complexes Part 9., Copper(II) complexes of tripeptides containing histidine, *J. Chem. Soc., Dalton Trans.* 611-614
11. Aiba, H., Yokoyama, A. and Tanaka, H. (1974) Acid dissociation constants of some histidine-containing peptides and formation constants of their metal complexes, *Bull. Chem. Soc. Jap* **47**, 112-117
12. Billo, E.J. (1974) Copper(II) chromophores and the rule of average environment, *Inorg. Nucl. Chem. Lett.* **10,** 613-717
13. Bernhardt, P.V., Comba, P., Mahu-Rickenbach, A., Stebler, S., Steiner, S., Várnagy, K. and Zehnder, M. (1992) Transition metal complexes of the novel tridentate di-2-pyridylmethanamine (dipa) *Inorg. Chem.* **31,** 4194-4200

14. Sigel, H. and Martin, R.B. (1982) Coordination properties of the amide bond. Stability and structure of metal ion complexes of peptides and relataed ligands, *Chem. Rev.* **82,** 385-426

15. Sóvágó, I. (1990) Metal complexes of peptides and their derivatives, in K. Burger (ed), *Coordination chemistry: Coordination equilibria in biologically active systems,* Ellis Horwood, New York, pp. 135-184.

16. Sigel, H. (1975) Ternary complexes in solution. XXIII. Influence of alkyl side chains on the stability of binary and ternary copper(II)-dipeptide complexes, *Inorg. Chem.* **14,** 1535

SUBJECT INDEX

GPSR Compliance

The European Union's (EU) General Product Safety Regulation (GPSR) is a set of rules that requires consumer products to be safe and our obligations to ensure this.

If you have any concerns about our products, you can contact us on ProductSafety@springernature.com

In case Publisher is established outside the EU, the EU authorized representative is:

Springer Nature Customer Service Center GmbH
Europaplatz 3
69115 Heidelberg, Germany

Batch number: 10816188

Printed by Printforce, the Netherlands